Dynamics of Solutions and Fluid Mixtures by NMR

Dynamics of Solutions and Fluid Mixtures by NMR

Edited by

Jean-Jacques Delpuech
Université Henri Poincaré, Nancy I,
Vandoeuvre-les-Nancy, France

JOHN WILEY & SONS
Chichester · New York · Brisbane · Toronto · Singapore

Copyright (©) 1995 by John Wiley & Sons Ltd,
Baffins Lane, Chichester,
West Sussex, PO19 1UD, England
Telephone: National   Chichester (01243) 779777
International   +44 1243 779777

Other Wiley Editorial Offices

John Wiley & Sons, Inc. 605 Third Avenue,
New York, NY 10158-0012, USA

Jacaranda Wiley Ltd, 33 Park Road, Milton,
Queensland 4064, Australia

John Wiley & Sons (Canada) Ltd, 22 Worcester Road,
Rexdale, Ontario M9W ILI, Canada

John Wiley & Sons (SEA) Pte Ltd, 37 Jalan Pemimpin #05-04,
Block B, Union Industrial Building, Singapore 2057

Library of Congress Cataloging-in-Publication Data

Dynamics of solutions and fluid mixtures by NMR / edited by J.-J.
Delpuech.
p.   cm.
Includes bibliographical references and index.
ISBN 0-471-95411-X
1. Solution (Chemistry)   2. Fluids.   3. Nuclear magnetic resonance
spectroscopy.   I. Delpuech, Jean-Jacques.
QD541.D96   1994
541.3'4—dc20

British Library Cataloguing in Publication Data

A catalogue record for this book is available from the British Library

ISBN 0 471 95411 X

Typeset in 10/12pt Times by Mathematical Composition Setters Ltd, Salisbury, Wiltshire
Printed and bound in Great Britain by Biddles Ltd, Guildford, Surrey

CONTENTS

PREFACE

Besides structural information brought about through spectrum analysis, an outstanding contribution of nuclear magnetic resonance chemistry concerns dynamical information at the molecular level, as the consequence of the time-dependence of NMR spectra. Progress in resolution and sensitivity now allows NMR to study systems of higher complexity such as synthetic polymers or biological molecules. Large molecules are intrinsically mobile, and the knowledge of internal motions is necessary for an accurate description of three-dimensional structures and an understanding of their chemical reactivity. Dynamic aspects of NMR span a number of areas most of which have been covered in specialized articles scattered throughout the literature. This book aims at providing a comprehensive coverage of time-dependent phenomena in NMR and of the information on molecular dynamics brought in this way: rotational and translational diffusion, segmental motions, restricted and self-diffusion, conformational equilibria, inter- and intramolecular rate processes. Many institutions now have the facilities for using NMR to study dynamic processes. A timely review of what can be done and what use it is seemed to be of importance for workers in the fields of chemistry, physico-chemistry and biology.

This is indeed a multiauthored book. However, we have tried to take up the challenge of writing an authentic textbook, all the chapters being coherently related to each other. The text is divided into ten chapters. A short introduction to NMR spectroscopy is given in Chapter 1, with a special emphasis on dynamical aspects. Chapters 2 and 3 give an overview on two fundamental aspects of molecular dynamics, namely NMR relaxation and its relationship with molecular reorientation on the one hand, and magnetization transfer phenomena induced by molecular rate processes (dynamic NMR) on the other. Specific mechanisms of relaxation encountered in paramagnetic systems or with quadrupolar nuclei are dealt with in the next two chapters, 4 and 5. These chapters have, of necessity, a significant component of mathematical or physical theory. An effort has been made, however, to show the use and the limitations of the theoretical framework on concrete examples. Among the possible fields of application to solution chemistry, ionic solvation is dealt with in depth in Chapter 6, using for this purpose a somewhat unusual approach, namely DNMR under high pressure.

A further step is taken in Chapter 7, which introduces the necessity for field gradients when studying translational diffusion in solution. A related topic is NMR imaging where molecular mobility is a prime necessity for signal detection. Medical imaging is, however, discarded from this presentation, which will rather focus on aspects of interest for chemists and biochemists: localized spectroscopy and microimaging of fluid inclusions in solid materials. Rotational and translational diffusion processes examined in the above chapter allow the reader to be introduced to the dynamics of complex systems where extensions of existing theories are required to account for collective internal motions and for restricted diffusion: microheterogeneous solutions, micelles and emulsions in Chapter 8; large molecules: synthetic polymers in solution and in melts, and biopolymers, in Chapter 9; liquid-like molecules contained in rigid matrices and in soft matter, adsorption and fluid inclusions in hetero-geneous materials, with emphasis on a few illustrative examples: swollen polymers and gels, zeolites and catalysts, microimaging and food science, in the last Chapter 10.

We hope that this book will contribute to the development of dynamical aspects of NMR, especially for chemical and biochemical applications, by collecting all the existing basic information necessary for newcomers to these fields. And all those who already have some practice in these areas should hopefully discover, chapter after chapter, ignored facets of this highly versatile spectroscopy.

J.-J.D.

LIST OF CONTRIBUTORS

Barsukov, I. L.
Biological NMR Centre, University of Leicester, Leicester, LE1 9HN, UK

Canet, D.
Laboratoire de Méthodologie RMN (URA CNRS 406-LESOC) Université de Nancy I, 54506 Vandoeuvre-les-Nancy Cedex, France

Cohen-Addad, J. P.
Laboratoire de Spectrométrie Physique, Université de Grenoble I, BP87, 38402 Saint-Martin d'Hères, Cedex, France

Decorps, M.
Laboratoire de Neurobiophysique (INSERM U 318), Hopital A. Michallon, BP 217X, 38043 Grenoble Cedex, France

Delpuech, J.-J.
Laboratoire d'Etude des Systèmes Organiques et Colloïdaux (URA CNRS 406-LESOC), Université de Nancy I, 54506 Vandoeuvre-les-Nancy Cedex, France

Fraissard, J. P.
Laboratoire de Chimie des Surfaces, Université Pierre et Marie Curie, 4 Place Jussieu, 75252 Paris Cedex 05, France

Frey, U.
Institute of Inorganic and Analytical Chemistry, University of Lausanne, 3 Place du Château, CH-1005 Lausanne, Switzerland

Grandjean, J.
Université de Liège, Sart-Tilman par 4000 Liège, Belgium

Hills, B. P.
Institute of Food Research, Norwich Research Park, Colney, Norwich, NR4 7UA, UK

Krajewski-Bertrand, M-A.
Laboratoire de Physico-Chimie Structurale et Macromoléculaire associé au CNRS, ESPCI, 10 rue Vauquelin, 75231 Paris Cedex 05, France

Laszlo, P.
Ecole Polytechnique, Laboratoire de Chimie, F-91128 Palaiseau Cedex, France

Lauprêtre, F.
Laboratoire de Physico-Chimie Structurale et Macromoléculaire associé au CNRS, ESPCI, 10 rue Vauquelin, 75231 Paris Cedex 05, France

Lian, L. Y.
Biological NMR Centre, University of Leicester, Leicester, LEI 9HN, UK

Lindman, B.
Division of Physical Chemistry 1, Chemical Center, University of Lund, S-221 00 Lund, Sweden

Merbach, A. E.
Institute of Inorganic and Analytical Chemistry, University of Lausanne, 3 Place du Château, CH-1005 Lausanne, Switzerland

Monnerie, L.
Laboratoire de Physico-Chimie Structurale et Macromoléculaire associé au CNRS, ESPCI, 10 rue Vauquelin, 75231 Paris Cedex 05, France

Olsson, U.
Division of Physical Chemistry 1, Chemical Center, University of Lund, S-221 00 Lund, Sweden

Powell, D. H.
Institute of Inorganic and Analytical Chemistry, University of Lausanne, 3 Place du Château, CH-1005 Lausanne, Switzerland

Robert, J. B.
CRTBT/CNRS, Avenue des Martyrs, 38042 Grenoble Cedex, France

Söderman, O.
Division of Physical Chemistry 1, Chemical Center, University of Lund, S-221 00 Lund, Sweden

Westlund, P.-O.
Department of Physical Chemistry, University of Umea, S-901 87 Umea, Sweden

1 INTRODUCTION: DYNAMIC PHENOMENA IN NMR

J.-J. Delpuech

Université Henri Poincaré, Nancy I, France

1   Time as the Fourth Dimension in NMR Structure Determinations

NMR has been used in a variety of ways to provide structural information. NMR parameters measured for this purpose are classically: frequency shifts (chemical shifts), line intensities and coupling constants. In a further step of development, time-dependence of NMR spectra—through lineshape analysis of the frequency-domain signal or measurement of the decay of the time-domain signal—was used to obtain dynamic information at the molecular level. Early NMR spectrometers allowed the experimentalists to study small-size molecules only, typically encountered in organic and inorganic chemistry. It was soon realized that a static description of non-rigid molecules was

Dynamics of Solutions and Fluid Mixtures by NMR
Edited by J.-J. Delpuech   © 1995 John Wiley & Sons Ltd

unsufficient to fully account for their physical or chemical properties, due to the presence of a variety of dynamic processes. In this respect, time is often considered as a fourth dimension in NMR structure determination. The term 'stereodynamics' has been coined to emphasize this necessary overlapping of structural and dynamical information to describe mobile systems. Stereodynamic processes became accessible to measurement with the advent of advanced spectroscopic techniques in the 1960s. In this respect, NMR proved to be particularly well fitted to study *in situ* processes in labile systems and to make subtle distinctions between different types of motion, and also some bond-breaking processes with a low barrier of activation, through the observation of various nuclei within molecular structures.

Progress in resolution and sensitivity now allows NMR to study systems of higher complexity, such as synthetic polymers or biological molecules. Large molecules are intrinsically mobile and the knowledge of internal motions is still more necessary for an accurate description of three-dimensional structures [1]. In addition, physico-chemical properties of their solutions are often monitored by motions at the molecular level. This is particularly true for biological molecules in which internal motions often control biological function, such as channel opening for ion transport, fibre contraction, surface activity of proteins.

For all these systems, a variety of NMR methods can be used to study dynamic processes occurring in a time ranging from about one picosecond to a few seconds, as it will be explained in the next two chapters.

2   General View of the Fundamentals of NMR [2-5]

As any spectroscopy, NMR consists in performing transitions between the energy levels of particles by submitting them to the action of an electromagnetic wave at the Planck frequency ν_0 (in Hz) or ω_0 (in rad s^{-1}). The energy levels are the eigenvalues of a time-independent Hamiltonian $\mathcal{H}_0$. Transitions are induced by a time-dependent Hamiltonian $\mathcal{H}_1(t)$ representing a sinusoidal perturbation of frequency ν_0, which is off-diagonal in the representation of the eigenstates of $\mathcal{H}_0$.

2.1   THE ZEEMAN INTERACTION

In NMR, a system of nuclear spins $\mathbf{I}$ is placed in an intense static magnetic field $\mathbf{B}_0$. Nuclei of angular momentum $\hbar\mathbf{I}$ are endowed with a magnetic moment $\boldsymbol{\mu}$

$$\boldsymbol{\mu} = \gamma\hbar\mathbf{I} \tag{1.1}$$

where γ is the gyromagnetic ratio and $\hbar$ is the Planck constant in J s rad^{-1}. Each nuclear dipole $\boldsymbol{\mu}$ acquires a magnetic energy $E = -\mu_z B_0$ if the z-axis of the laboratory frame is taken along the direction of $\mathbf{B}_0$, the *longitudinal* direction. This energy may assume a series of discrete values, in number $2I+1$, where I is the spin number of the observed nucleus. The energy levels E_m

$$E_m = -\gamma\hbar B_0 m, \qquad \text{with } m = -I, -I+1, ..., I-1, I \qquad (1.2)$$

correspond to different z projections (I_z) of the spin vector $\mathbf{I}$, and, consequently, to different orientations of the magnetic moment $\boldsymbol{\mu}$ with respect to the z-axis. They are the eigenvalues of the *Zeeman Hamiltonian* $\mathcal{H}_z^0$

$$\mathcal{H}_z^0 = -\gamma\hbar_0 I_z \qquad (1.3)$$

A case of fundamental importance in practice, which will be the only one considered in this chapter, is that of spin $1/2$ nuclei, where $I=1/2$ and $m=\pm 1/2$, with two energy levels $E_{\alpha,\beta} = \pm\gamma\hbar_0/2$ for the two quantum states α $(m=1/2)$ and $\beta(m=-1/2)$. Transitions between α and β states are performed at *the resonance frequency*

$$\omega_0 = \gamma B_0 \quad \text{or} \quad \nu_0 = \gamma B_0/2\pi \qquad (1.4)$$

with a net absorption of energy from the RF wave as the result of slightly unequal Boltzmann populations N_α, N_β, of α and β states (Figure 1.1).

2.2   CHEMICAL SHIFT

The Zeeman Hamiltonian represents the major contribution to $\mathcal{H}_0$. In frequency units, $\mathcal{H}_z^0$ is expressed as

$$\mathcal{H}_z^0 = \nu_0 I_z (\text{Hz}) \quad \text{or} \quad \omega_0 I_z (\text{rad s}^{-1}) \qquad (1.5)$$

and the corresponding energy levels $(\pm\nu_0/2)$ amount to a few hundred megahertz, depending on the field strength B_0 (a few Tesla) and the type of nucleus (through γ). There are two additional contributions which, although of much smaller size (zero to a few hundred hertz), are of fundamental importance to NMR spectroscopists. First, the electrical charges surrounding the observed nucleus in its molecular environment slightly modify the applied magnetic field B_0 into $B_0(1-\sigma)$, and consequently $\mathcal{H}_z^0$ into $\mathcal{H}_z = \nu_0(1-\sigma)I_z$, where the screening constant σ is a small number, conventionally expressed in ppm for this reason. This produces in turn a frequency shift from the resonance value ν_0, by a factor $-\sigma$, the so called *chemical shift δ*

$$\delta(\text{ppm}) = -10^6\sigma, \quad \text{or else} \quad \delta = 10^6(\sigma_{\text{ref}} - \sigma) \qquad (1.6)$$

if the shift is referred to the resonance $\nu_0(1-\sigma_{\text{ref}})$ of a standard compound. The Hamiltonian $\mathcal{H}_z$ may be alternatively considered to be the sum of the above Zeeman Hamiltonian $\mathcal{H}_z^0$ and of a time-independent perturbation due to chemical shift (CS) effects, $\mathcal{H}_{\text{cs}}$

$$\mathcal{H}_{\text{cs}} = \nu_0 \delta I_z \qquad (1.7)$$

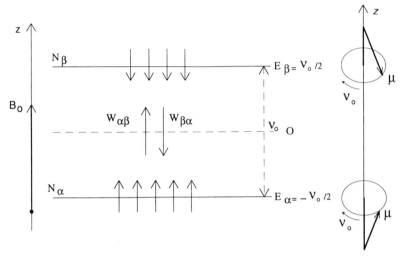

Figure 1.1 Energy levels, Larmor free precession, Boltzmann populations of an assembly of spin $1/2$ nuclei. Straight or inverted arrows are conventionally used to represent z magnetization in α and β states

Chemical shifts are used in common practice to identify the molecular species containing the observed nucleus and the site of this nucleus within this molecule.

2.3 SCALAR COUPLING

A second contribution to $\mathcal{H}_0$ is due to scalar interactions between two neighbouring spins $\mathbf{I}$ and $\mathbf{S}$. The magnitude of the spin–spin coupling Hamiltonian $\mathcal{H}_s$

$$\mathcal{H}_s = J\mathbf{I}.\mathbf{S} \tag{1.8}$$

is governed by the value of the coupling constant J which may range from zero to plus or minus a few hundred hertz. This contribution brings about a splitting of the chemical shift resonance into a multiplet of lines. J values deduced from the fine structure of NMR resonances are dependent on the number and the spatial proximity of interacting spins, and are thus of a great help in line assignment and in stereochemistry.

2.4 THE LARMOR PRECESSION AND THE ROTATING FRAME

Coming back to the Zeeman experiment, the magnetic field produces a couple of strength $\boldsymbol{\mu} \wedge \mathbf{B}_0$ on each magnetic dipoles $\boldsymbol{\mu}$. According to the fundamental law of dynamics, the rate of change of each individual spin angular momentum is:

$$d(\hbar\mathbf{I})/dt = \boldsymbol{\mu} \wedge \mathbf{B}_0 \qquad (1.9)$$

Introducing equation (1.1) into equation (1.9) results in the classical equation of motion of nuclear dipoles

$$d(\boldsymbol{\mu})/dt = \gamma\boldsymbol{\mu} \wedge \mathbf{B}_0 \qquad (1.10)$$

This motion, which requires no consumption of energy, takes place even in the absence of the RF wave. This is the *Larmor precession* in which the nuclear dipoles move round a cone about $\mathbf{B}_0$ at a uniform angular velocity

$$\omega_0 = -\gamma B_0 \qquad (1.11)$$

which is precisely equal to the above resonance frequency.

This point may be understood if it is recalled that NMR transitions are effected by a transverse magnetic field $\mathbf{B}_1$, of small amplitude $B_1 \ll B_0$, rotating at the frequency ω of the transmitter—in fact, $\mathbf{B}_1$ results from a linearly oscillating magnetic field $B_x = 2B_1 \cos \omega t$ along the X-axis of the laboratory frame OXYz, which is decomposed into two circularly polarized half-waves. This perturbation corresponds to an off-diagonal time-dependent Hamiltonian $\mathcal{H}_1(t)$

$$\mathcal{H}_1(t) = \gamma B_1 e^{i\omega t} I_x \qquad (1.12)$$

From a classical point of view, each nucleus is submitted to the combined action of two magnetic fields $\mathbf{B}_0$ and $\mathbf{B}_1$. The rotating field $\mathbf{B}_1$ may be considered as a static field along the x-axis of a reference frame $Oxyz$ rotating about the z-axis at frequency ω, *in phase* with the transmitter (Figure 1.2). However, the fundamental equation (1.9) is valid only in the laboratory frame, and should be modified if we wish to express the motion of magnetic dipoles in the rotating frame. This may be simply performed [2–8] by adding a fictitious magnetic field $\boldsymbol{\Omega}/\gamma$, where $\boldsymbol{\Omega}$ is a vector along the z-axis of algebraic value ω. In the rotating frame, nuclear dipoles thus experience a *static* magnetic field $\mathbf{B}_{\text{eff}}$

$$\mathbf{B}_{\text{eff}} = \mathbf{B}_0 + \mathbf{B}_1 + \boldsymbol{\Omega}/\gamma \qquad (1.13)$$

Magnetic dipoles $\boldsymbol{\mu}$, and their resultant $\mathbf{M}$, are therefore precessing about the direction of $\mathbf{B}_{\text{eff}}$ at angular frequency $-\gamma B_{\text{eff}}$.

If the frequency of the transmitter is adjusted to the resonance frequency ω_0, the term $(\mathbf{B}_0 + \boldsymbol{\Omega}/\gamma)$ vanishes in equation (1.13). This shows that, at resonance, the magnetic dipoles $\boldsymbol{\mu}$ are submitted to a second precession, called *nutation*, about the x-axis, with an associated Larmor frequency ω_r

$$\omega_r = -\gamma B_1 \qquad (1.14)$$

which is of an order of magnitude ranging from about 0.01 Hz (continuous

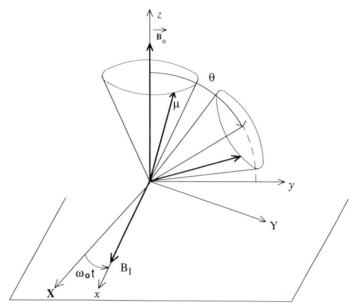

Figure 1.2 Nutation of individual nuclear moments $\boldsymbol{\mu}$ by a flip angle θ about the x-axis in the rotating frame carrying the RF magnetic field $\mathbf{B}_1$ at the resonance frequency ω_0

wave NMR) to a few tens of kHz (pulsed NMR). This nutation alternatively brings the z magnetization of each nucleus from its equilibrium value, say μ_z, parallel to $\mathbf{B}_0$, to the opposite position $-\mu_z$, *antiparallel to* $\mathbf{B}_0$, and vice versa. This actually amounts to performing transitions of individual nuclei between quantum states α and β. Such a motion changes the energy of each nucleus alternatively from $\pm\mu_z B_0$ to $\pm\mu_z B_0$, and thus gives rise to an NMR signal detected in a transverse coil tuned at the resonance frequency ω_0.

2.5   THERMAL EQUILIBRIUM AND LONGITUDINAL RELAXATION

From the above treatment, it is clear that the RF wave is inducing transitions between states α and β with equal probabilities W in both senses. This would rapidly result into an equalization of Boltzmann populations N_α, N_β, and thus to an extinction of the NMR signal, this is the so called *saturation* phenomenon. In fact, *relaxation* processes ensure a fast return to equilibrium of the longitudinal overall magnetisation $M_z(t)$

$$M_z(t) = \Sigma \ \mu_z$$

where the summation is over all the nuclei in the sample, either in α or β state. From a macroscopic point of view, this means that the overall nuclear magnetization M_z, after being rotated by an angle θ (the flip angle) from its equilibrium position M_0 along $\mathbf{B}_0$, is spontaneously driven back to this position after the transmitter has been switched off, according to the phenomenological equation:

$$\frac{dM_z}{dt} = \frac{M_0 - M_z}{T_1} \qquad (1.15)$$

T_1 is the *longitudinal relaxation time*, typically a few seconds.

From a microscopic point of view, this means that each nucleus spends an average residence time $2T_1$ in either quantum state, α or β, before it suffers a non-radiative transition to the other state, β or α. This corresponds to average transition probabilities per second $W_{\beta\alpha}$ and $W_{\alpha\beta}$

$$W_{\beta\alpha} \approx W_{\alpha\beta} = 1/2T_1 \qquad (1.16)$$

In fact, $W_{\alpha\beta}$ and $W_{\beta\alpha}$ are slightly different, being in the ratio $\exp[-(E_\alpha - E_\beta)/k_b T]$, so as to ensure slightly unequal Boltzmann populations N_α, N_β. Saturation phenomena are negligible as long as W (the probability of radiative transition) is maintained clearly smaller than $W_{\alpha\beta}$.

Under these conditions, the equilibrium magnetisation at temperature T is given by the Langevin equation:

$$M_0 = \gamma^2 \hbar^2 I(I+1) N_0 / 3 k_B T \qquad (1.17)$$

where k_B represents the Boltzmann constant and $N_0 = N_\alpha + N_\beta$, the total number of resonant nuclei.

2.6　FREE INDUCTION DECAY AND TRANSVERSE RELAXATION

It should be recalled at this point that the observation of the NMR signal involves the transverse magnetization only. This means that a *transverse relaxation* time T_2 should be introduced to account for the return of transverse magnetization to equilibrium. In thermal equilibrium, that is in the sole presence of the static field $\mathbf{B}_0$, all the nuclear dipoles have transverse components μ_{xy} freely precessing at the Larmor frequency ω_0, with a random phase angle ϕ. Their macroscopic resultant M_{xy} is thus averaged to zero over an ensemble of nuclei. This means that nuclear magnetization $\mathbf{M}$ is purely longitudinal at equilibrium with a component $M_z = M_0$.

To observe a signal by pulsed NMR experiments, the RF field is switched on during a short time (a few tens of microseconds), thus driving the equilibrium magnetization into the transverse plane (Figure 1.3). The transverse magnetisation obtained at the beginning of the receiving period, $M_{xy}(0) = M_0$, is then freely precessing about $\mathbf{B}_0$ at the Larmor frequency, thus inducing an electrical signal of frequency ω_0 in a transverse coil. This signal, the so-called free induction decay (FID), indeed decreases to zero in the course of time as the result of *transverse relaxation*.

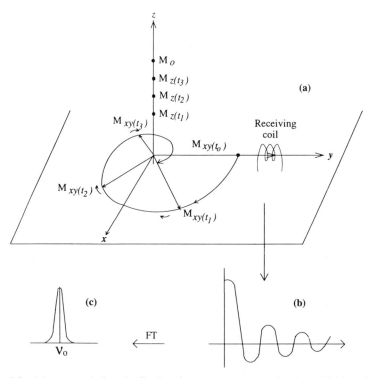

Figure 1.3 Macroscopic longitudinal and transverse magnetizations $M_z(t)$ and $M_{xy}(t)$ during the receiving period at successive times 0, t_1, t_2, t_3 ... (a), the time-domain signal (b), and the absorption spectrum (c)

From a macroscopic point of view, this decay is characterised by a transverse relaxation time T_2 and a rate law of the same type as that written above for longitudinal relaxation

$$\frac{\mathrm{d}M_{xy}}{\mathrm{d}t} = \frac{-M_{xy}}{T_2} \tag{1.18}$$

The time-domain signal (FID) is then Fourier transformed to retrieve the resonance frequency ω_0. The frequency-domain signal is represented by a Lorentzian curve centred at the frequency ω_0, with a linewidth at half height

$$\Delta\nu_{1/2}\,(\text{or } \Delta\omega_{1/2}) = 1/\pi T_2\,(\text{or } 2/T_2) \qquad (\text{in Hz or rad s}^{-1}) \tag{1.19}$$

2.7 BLOCH EQUATIONS

From a classical point of view, the motion of the magnetic dipoles can be obtained by combining the effects of relaxation and of the magnetic fields in equations (1.15), (1.18), (1.10) and (1.13) into a new set of equations for the nuclear magnetisation **M**, described by its components u, v, and M_z along the axes xyz of the rotating frame. These are called the Bloch equations [9]

$$du/dt = (\omega - \omega_0)v - u/T_2 \tag{1.20}$$

$$dv/dt = \gamma B_1 M_z - (\omega - \omega_0)u - v/T_2 \tag{1.21}$$

$$dM_z/dt = -\gamma B_1 v + (M_0 - M_z)/T_1 \tag{1.22}$$

In continuous-wave NMR, the RF field is set on for a long time, a steady state is reached, where $du/dt = dv/dt = dM_z/dt = 0$, and:

$$u = \gamma M_0 B_1 T_2^{\,2}(\omega - \omega_0)/D \tag{1.23}$$

$$v = \gamma M_0 B_1 T_2/D \tag{1.24}$$

$$M_z = M_0(1 + T_2^2(\omega - \omega_0)^2)/D \tag{1.25}$$

with $D = 1 + T_2^2(\omega - \omega_0)^2 + \gamma^2 B_1^2 T_1 T_2$.

The NMR signal is proportional to v, the component of the rotating nuclear magnetisation *out of phase* with the RF field. The absorption signal $v(\omega)$ is thus centred at the resonance frequency ω_0. In the absence of *saturation* ($M_z \approx M_0$), i.e. when the power of the transmitter is adjusted to a value such that

$$\gamma^2 B_1^{\,2} T_1 T_2 \ll 1 \tag{1.26}$$

the absorption curve assumes a *Lorentzian* shape:

$$v(\omega) = \gamma M_0 B_1 T_2/(1 + T_2^{\,2}(\omega - \omega_0)^2) \tag{1.27}$$

In complex notation, the transverse magnetization M_{xy}, can be represented by the quantity [10]

$$G = u + iv \tag{1.28}$$

Equations (1.20) and (1.21) can then be condensed into a single equation

$$dG/dt + \alpha G = i\gamma B_1 M_0 \tag{1.29}$$

where α is a complex quantity

$$\alpha = 1/T_2 + i(\omega - \omega_0) \tag{1.30}$$

In the steady state described above ($dG/dt = 0$):

$$G = i\gamma B_1 M_0/[1/T_2 + i(\omega - \omega_0)] \tag{1.31}$$

The absorption signal v, taken as the imaginary part of G

$$v = imag.(G) \tag{1.32}$$

is found equal to the value given in equation (1.24).

Finally, it should be mentioned that in pulsed NMR, the pulse duration t_p, is so short that relaxation forces and saturation phenomena may be ignored during the transmitting period, this means that the effect of pulses can be described merely as rotations of the equilibrium magnetization about the RF field direction $\mathbf{B}_1$ by an angle θ

$$\theta = \omega_r t_p \tag{1.33}$$

During the receiving period, the RF field is switched off, and the nuclear spins perform a damped precession about $\mathbf{B}_0$ at their own Larmor frequency ω_0 (or $\omega_0 (1 - \sigma)$ if we include chemical shift effects), in which the rotating transverse component M_{xy} decays to zero with a characteristic time T_2, while M_z is restored to its equilibrium value M_0 with a characteristic time T_1.

3 Relaxation Times as a Source of Dynamical Information

Relaxation processes are related to the existence of parasitic time-dependent magnetic fields $\mathbf{B}(t)$ arising from neighbouring magnetic dipoles and quadrupoles and from electronic spins. Molecular reorientation and translational diffusion rapidly change the relative positions of interacting centres with respect to the observed nucleus. Thus the magnetic field $\mathbf{B}(t)$ and the corresponding perturbation Hamiltonian $\mathcal{H}_p(t)$ are randomly fluctuating, being averaged to zero over time [6–8, 11].

3.1 LONGITUDINAL RELAXATION

Transverse components $B_x(t)$, $B_y(t)$ of $B(t)$, which correspond to the off-diagonal part of $\mathcal{H}_p(t)$ may induce transitions between states α and β, and thus can participate to longitudinal relaxation. This is not the case for the longitudinal component $B_z(t)$ which is collinear to the z projection of nuclear moments μ. Transitions are possible because the Fourier spectrum of $\mathbf{B}(t)$ contains a component at the Larmor frequency ω_0. More precisely, as the fluctuating field $\mathbf{B}(t)$ does not decay to zero over an infinite period of time, the Fourier transform is applied not to the perturbation $\mathcal{H}_p(t)$ itself, but to the effect of these perturbations, which is very limited in time and described [9] by an *autocorrelation function* $g(\tau)$

$$g(\tau) = \overline{\mathcal{H}_p(t).\mathcal{H}_p(t+\tau)} \tag{1.34}$$

The bar denotes an ensemble average over all the nuclei, the result being independent of the time t when the observation begins (Chapter 2). A simplified model of perturbation is a magnetic field randomly jumping between two opposite values $\pm B_x$ (which may be imagined to represent the average maxima and minima of fluctuations). It can be shown in this case that

$$g(\tau) = B_x^2 \, e^{-\tau/\tau_c} \tag{1.35}$$

where τ_c is a *correlation time* which characterizes the mean time elapsed between two successive jumps, itself closely related to molecular motions.

Fourier transformation of the autocorrelation function will result in a distribution of frequencies, the *spectral density* $J(\omega)$, which is a Lorentzian function of ω

$$J(\omega) = B_x^2 \frac{\tau_c}{1 + \omega^2 \tau_c^2} \tag{1.36}$$

The contribution of the perturbation to longitudinal relaxation, i.e. the transition probability $W_{\alpha\beta}$, thus reaches a maximum when τ_c^{-1} approaches the resonance frequencies of $\mathcal{H}_0$, 10^{10}–10^{11} rad s^{-1} in most NMR spectrometers, and is small at long and short τ_c. This gives a first example of an NMR *timescale*

$$\omega_0 \tau_c \approx 1 \tag{1.37}$$

which is precisely that encountered in the study of molecular motions in fluids where correlation times τ_c are ranging from *ca.* 10^{-8} to 10^{-12} s. This is the clue to all the dynamical information obtained by NMR at the molecular level as explained in the next chapter.

3.2 TRANSVERSE RELAXATION

Relaxation times T_2 may be determined in principle either from the decay of the time-domain signal or from the linewidth of the resonance (see Chapter 2). Like T_1 values, relaxation times T_2 are closely related to molecular dynamics. After the RF pulse, the individual transverse vectors μ_{xy}, in phase at the beginning of the receiving time, are progressively dephasing under the action of fluctuating magnetic fields $\mathbf{B}(t)$, thus averaging their resultant M_{xy} to zero after a time of the order of T_2. Fluctuating fields will have two effects on the NMR signal. A first contribution to transverse relaxation is brought by the transverse components $B_x(t)$, $B_y(t)$ through the operation of longitudinal relaxation. The limitation of the residence time of nuclei in quantum states α and β to a duration of $2T_1$ results, according to the fourth uncertainty principle of Heisenberg, into a frequency dispersion of $1/2T_1$. This contribution to the relaxation rate $1/T_2$ is called the *lifetime* or *non-secular broadening* (with reference to the perturbation Hamiltonian $\mathcal{H}_p(t)$). A second contribution arises from the action of the longitudinal component $B_z(t)$. The fluctuations of $B_z(t)$ result into variations of the Larmor frequency, which produce in turn an *extra* phase angle $\delta\phi$ of the observed nucleus over its normal precession. We still assume that fluctuations $B_z(t)$ are fastly occurring between two extreme values $\pm B_z$ and that a progressive *extra* dephasing takes place through a random walk process by small steps, each of duration τ_c equal to the correlation time of the process. In each step, the elemental dephasing assumes any of two opposite values, at random with equal probabilities:

$$\delta\phi = \pm\gamma B_z \tau_c = \pm\Delta\omega\tau_c \qquad (1.38)$$

if we call $\Delta\omega$ half the amplitude of the Larmor frequency fluctuation around the central value ω_0. According to statistical theories, the mean square dephasing $\overline{\Delta\phi^2} = n\delta\phi^2$ increases linearly with the number n of steps during a total observation time $t = n\tau_c$. We take as relaxation time T_2' for this process the time for a spin assembly in phase at $t = 0$ to get about one radian out of step, that is when

$$1/T_2' = \Delta\omega^2\tau_c \qquad (1.39)$$

This is the second contribution to the overall transverse relaxation rate, the so called *secular broadening*:

$$1/T_2 = 1/T_2' + 1/2T_1 \qquad (1.40)$$

Secular broadening is *frequency-independent*. The shorter τ_c, that is the more rapid the molecular motion concerned, the narrower the resonance, this is the phenomenon of *motional narrowing*. In most cases, linewidths are of the order of one rad s^{-1}, this means that secular broadening will be measurably large as long as

$$\Delta\omega^2\tau_c > 1 \quad \text{or} \quad \tau_c > \Delta\omega^{-2} \qquad (1.41)$$

This is another timescale to consider in applications to molecular dynamics. Typical values for spin 1/2 nuclei in diamagnetic solutions of small molecules are $T_1 \sim 10$ s, $\Delta\omega = 10^5$ rad s^{-1}, $\tau_c = 10^{-11}$ s, hence $1/T_2 \sim 0.15$ s^{-1} and $\Delta\omega_{1/2} \sim 0.3$ rad s^{-1} or $\Delta\nu_{1/2} \sim 0.05$ Hz.

At another extreme, longer correlation times do not result into infinite relaxation rates, as shown by equation (1.39). A random walk process, indeed, requires a large number of small steps to be performed during the time T_2 of observation, so as to avoid the spins being fully dephased between two successive jumps. This condition amounts to saying that

$$\delta\phi \ll 1 \quad \text{or} \quad \Delta\omega\tau_c \ll 1 \qquad \text{or else } \tau_c \ll T_2 \qquad (1.42)$$

With the typical values of $\Delta\omega$ and τ_c mentioned above, this condition is usually met in liquid samples.

It can be concluded from this section that variable frequency measurements of T_1 and T_2 are complementary to each other as sources of dynamical information at the molecular level.

4   Molecular Rate Processes

Up to now, we have considered dynamic processes inducing non-radiative transitions through the intervention of random fields. Another possibility for a spin to leave a given quantum state with subsequent lifetime broadening effects is the occurrence of a chemical process which forces the observed nucleus either to change its position within its parent molecule or to be carried into another species. The process is assumed to be reversible, this means that a given nucleus may jump forth and back between non-equivalent positions within a set of chemical species standing at equilibrium. The process may be intra- or intermolecular, involving one or several species, respectively. Typical examples are isomerization or ligand exchange processes respectively (see Chapter 3). The non-equivalent positions successively occupied by a nucleus in the course of time are called nuclear sites A, B. C... and are characterized by their resonance frequencies ν_A, ν_B, ν_C ... (or ω_A, ω_B, ω_C ...). This denomination is also given to the NMR lines concerned themselves. The effects of the kinetic process, primarily a *chemical exchange* (of nuclei), is often pictured with some abstraction as jumps between lines, A, B, C, ..., hence the traditional denomination of (nuclear) *site exchange*. It thus appears that molecular rate processes can modify NMR lineshapes and apparent relaxation times, and that kinetic data can be obtained by studying these effects. This aspect of NMR spectroscopy has been given somewhat improperly the specific name of dynamic NMR or DNMR.

In this introductory section, we only wish to show the full analogy between relaxation and DNMR processes. Let us consider the very simple example of an intramolecular two-site exchange resulting from the chemical exchange of the methyl groups of *N,N*-dimethylformamide (DMF) between *cis* and *trans* positions, itself arising from a reversible hindered rotation (the chemical process) about the N–C partial double bond:

This example has been the object of many investigations in the past because of its relevance to the chemists, namely the determination of the energy barrier to rotation. It illustrates well the power of DNMR. There is indeed no net chemical reaction in this process—we are thus obliged to distinguish the two exchanging groups by using a typographical mark (*)—and the kinetics of such a reaction cannot be measured by any other method than NMR.

It can be shown on this example that the two-site exchange between *cis* (A) and *trans* (B) positions is equivalent to the simple model of a longitudinal fluctuating field randomly jumping between two extreme values $\pm B_z$, as envisaged in the above treatment of secular transverse relaxation. Let us call ω_0 the average frequency over the two sites, $\omega_0 = (\omega_A + \omega_B)/2$, and $B_0 = \omega_0/\gamma$ the associated value of the static field. In the absence of exchange, the total Hamiltonian $\mathcal{H}_0$ is obtained by adding to or subtracting from the Zeeman Hamiltonian $\mathcal{H}_z^0$ the chemical shift Hamiltonian $\mathcal{H}_{CS}^0 = \nu_0 \delta I_z$, where $\delta = (\omega_A - \omega_B)/2\omega_0$, for sites A and B, respectively. Two single lines of equal intensities are then observed at frequencies ω_A and ω_B (Figure 1.4). In the presence of chemical exchange, a given nucleus randomly jumps from site A to site B, and vice versa, with a probability per unit time equal to the pseudo first-order rate constant k_{ex} of the site exchange, and an average lifetime τ_{ex} of one spin in each site equal to the reciprocal of k_{ex}

$$\tau_{ex} = 1/k_{ex} \qquad (1.43)$$

The site-exchange process amounts to submitting the observed nucleus to a time-independent Hamiltonian $\mathcal{H}_z^0$ and to a time-dependent perturbation $\mathcal{H}_{CS}(t) = \pm\mathcal{H}_{CS}^0$, or else to a fluctuating magnetic field $B_z(t) = \pm B_z$, where:

$$B_z = \omega_0 \delta/\gamma = B_0 \delta \qquad (1.44)$$

In the case of fast fluctuations, the spectrum will thus consist of one single line centred at θ_0, with an extra line-broadening (due to exchange and not to relaxation) computed according to equation (1.39):

$$\delta(\Delta\omega_{1/2}) = \Delta\omega^2 \tau_{ex}/2 \qquad (1.45)$$

where $\Delta\omega = (\omega_A - \omega_B)/2$ is half the frequency shift between sites A and B. In this treatment, the exchange lifetime τ_{ex} is twice as large as the correlation time used in Section 3.2 which referred to the average time in a succession of steps with or without a change of the random variable with equal probabilities. The exchange is said to be fast in this case, this means by analogy with equations (1.42) that

$$\Delta\omega\tau_{ex} \ll 1 \quad \text{or} \quad |\omega_A - \omega_B|\,\tau_{ex} \ll 1 \qquad (1.46)$$

The NMR timescale for exchange is, however, of a quite different order of magnitude, since $\Delta\omega$ amounts in the present case to a few hundred hertz at most, instead of a few hundred megahertz for relaxation. The faster the exchange, the shorter the line-broadening. This is an *exchange narrowing*, completely analogous to *motional narrowing* in the case of relaxation. However, extra line-broadening can be seen only if it is of the same order of magnitude as the natural linewidth $\Delta\omega_{1/2}$ or $\Delta\nu_{1/2}$ (*ca.* 0.1 to 1 Hz) produced by relaxation processes. Beyond this limit, that is when

$$\Delta\omega^2\tau_{ex} < \Delta\omega_{1/2} \sim 1 \qquad (1.47)$$

the site exchange is immeasurably fast. Equation (1.47) is often quoted to give NMR limitations towards the fast exchange side (Figure 1.4).

At the other extreme, the rate process is said to be very slow when:

$$\Delta\omega\tau_{ex} \gg 1 \quad \text{or} \quad |\omega_A - \omega_B|\,\tau_{ex} \gg 1 \qquad (1.48)$$

In this case, the nuclear spin leaves all phase coherence between two successive jumps. The nuclei with the instantaneous Hamiltonian $\mathcal{H}_z^0 + \mathcal{H}_{CS}^0$ can now be distinguished from those with the instantaneous Hamiltonian $\mathcal{H}_z^0 - \mathcal{H}_{CS}^0$. The resonance line is thus a doublet, with two components at the frequencies of sites A and B. Both lines for each site are resolved, but they are broadened with an extra linewidth at half-height

$$\delta(\Delta\omega_{1/2}) = 2/\tau_{ex} \quad \text{or} \quad \delta(\Delta\nu_{1/2}) = 1/\pi\tau_{ex} \qquad (1.49)$$

This is a lifetime broadening since each site exchange, indeed, contributes to limit the mean lifetime of a given quantum state to the average time τ_{ex} before the next jump. Exchange rates are immeasurably slow ('no exchange') when the extra half line-broadening $1/\tau_{ex}$ is of the same order of magnitude as the natural linewidth.

Intermediate exchange rates require recourse to the calculation of the full spectral density $J(\omega)$ of the transition [12] at frequency ω_0, which is shown to display two maxima centred on ω_0 (Figure 1.4) as long as

$$\Delta\omega\tau_{ex} > \sqrt{2} \qquad (1.50)$$

By increasing progressively exchange rates (e.g. by raising the temperature), individual lines A and B broaden and shift to each other with a line separation $\Delta\omega(\tau_{ex})$

$$\Delta\omega(\tau_{ex}) = (\Delta\omega^2 - 2/\tau^2_{ex})^{1/2} \qquad (1.51)$$

until they merge into one single line when condition (1.50) ceases being

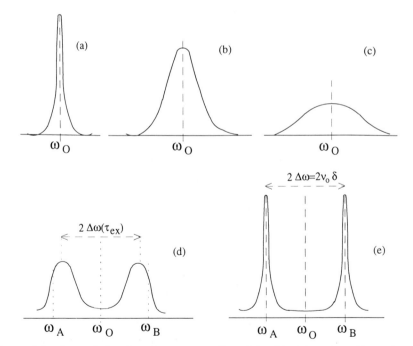

Figure 1.4 Spectral density of a system of nuclei subject to a two-site exchange, when the exchange rate τ^{-1}_{ex} is (a) $\gg\Delta\omega^2$, (b) $>\Delta\omega$, (c) $= 1/\sqrt{2}\Delta\omega$, (d) $<\Delta\omega$, (e) $\ll 1$

fulfilled. This is the *coalescence phenomenon*. The value of τ_{ex}^{-1} at coalescence determines *the NMR timescale* for the rate process under consideration. There is a range of measurable exchange rates τ_{ex}^{-1} around coalescence, loosely limited between $\Delta\omega_{1/2}$ and $\Delta\omega^2/\Delta\omega_{1/2}$, where neither of conditions (1.46) and (1.48) is satisfied. This range of values, *the NMR window*, is centred on the rate at coalescence, *the NMR timescale*, when

$$\Delta\omega\tau_{ex} \sim 1 \qquad (1.52)$$

It can be observed that NMR timescales are given by analogous expressions for both relaxation and exchange processes, as expected from this common presentation of both phenomena.

The existence of NMR timescales determines the applicability of NMR methods to molecular reorientation and exchange processes, as shown in Chapters 2 and 3. Two specific mechanisms of relaxation, using paramagnetic and quadrupolar probes, are examined in Chapters 4 and 5. Chapter 6 is devoted to high-pressure NMR, an informative methodology which is not yet much used to investigate the dynamics of solutions. The second type of molecular motion, translational diffusion, is presented in Chapter 7 together with the use of field gradients in NMR and applications to NMR imaging and localized spectroscopy. Further chapters are devoted to specific applications of recent interest (see below).

5   Domains of Application

The dynamics of small molecules has been intensively studied with the advent of the first NMR spectrometers. Typical examples of such investigations are given throughout the first seven chapters to illustrate theoretical and method-ological treatments. As stated at the beginning of this chapter, the most recent developments concern systems of higher complexity, such as synthetic or natural polymers (Chapter 9), where the major contribution to dynamics arise from internal motions. Other systems of high complexity are organized assemblies of amphiphilic molecules found in natural or artificial colloids and membranes. A striking property of fluid layers is their tendency to perform out-of-plane undulations which contribute to stabilize lamellar surfactant phases with respect to disordered dispersions [13–14]. NMR proved to be an invaluable tool to bring information on the dynamics of such microhetero-geneous systems (Chapter 8).

Solid-state materials and NMR are beyond the scope of this book. However, heterogeneous materials may contain fluid inclusions which can be studied by high-resolution NMR or NMR imaging. Three examples of such investigations have been selected and presented in the last chapter because of their novelty and practical relevance, (a) swollen polymers and gels, (b) fluids in and on inorganic materials, mainly the xenon gas used as a probe to study porous materials, and (c) water in the biological tissues found in food science.

References

1. R. J. P. Williams, *Eur. J. Biochem*, **183**, 479–497 (1989).
2. J. A. Pople, W. G. Schneider and H. J. Bernstein, *High Resolution Nuclear Magnetic Resonance*, McGraw-Hill, New York, 1959.
3. A. Carrington and A. D. McLachlan, *Introduction to Magnetic Resonance*, Harper & Row, London, 1967.
4. M. L. Martin, J.-J. Delpuech and G. J. Martin, *Practical NMR Spectroscopy*, Heyden–Wiley, London, 1980.
5. D. Canet, La RMN, *Concepts et Méthodes*, InterEditions, Paris, 1991.
6. A. Abragam, *The Principles of Nuclear Magnetism*, Clarendon Press, Oxford, 1961.
7. N. Bloembergen, E. M. Purcell and R. V. Pound, *Phys. Rev.*, **73**, 679 (1948).
8. C. P. Slichter, *Principles of Magnetic Resonance*, Harper & Row, London, 1963.
9. F. Bloch, *Phys. Rev.*, **70**, 460 (1946).
10. H. S. Gutowsky, D. W. McCall and C. P. Slichter, *J. Chem. Phys.*, **21**, 279 (1953).
11. A. G. Redfield, *Adv. Magn. Reson.*, **1**, 1 (1965).
12. J. Seliger, *J. Magn. Reson.*, **A 103**, 175–182 (1993).
13. W. Helfrich, *Z. Naturforsch*, **33A**, 305 (1978).
14. P. G. De Gennes and C. Taupin, *J. Phys. Chem.*, **86**, 2294 (1982).

2 TIMESCALES IN NMR: RELAXATION PHENOMENA IN RELATION WITH MOLECULAR REORIENTATION

D. Canet
Université Henri Poincaré, Nancy I, France

and J. B. Robert
Université Joseph Fourier, Grenoble I, France

Dynamics of Solutions and Fluid Mixtures by NMR
Edited by J.-J. Delpuech　© 1995 John Wiley & Sons Ltd

1 Introduction

The study of the dynamics of molecules in the liquid phase may be approached by a large variety of experimental techniques which correspond each to a different interaction of the molecules under study with an external perturbation. In this chapter, we shall examine the information which may be obtained on molecular dynamics from Nuclear Magnetic Resonance relaxation times, namely the spin-lattice relaxation time T_1, the spin–spin relaxation time T_2 and the spin–lock relaxation time $T_{1\rho}$ and other relaxation parameters defined in Section 4. We shall limit ourselves to the case of medium size molecules $(M \leqslant 300)$, with a special emphasis on the overall molecular motion which typically corresponds to correlation times in the range of 10^{-9} s to 10^{-11} s. Internal molecular motions such as ring inversions, rotations around single bonds or pseudo rotations which take place in ring compounds (e.g. saturated five-membered rings) or around highly co-ordinated atoms (e.g. penta-co-ordinated phosphorus) may be often studied by lineshape analysis and are examined elsewhere (Chapter 3). The present study will only consider molecules in pure liquids or highly diluted in a solvent. We shall exclude from this chapter the problems concerning (i) polymers and colloids; (ii) liquid crystals and molecules dissolved in liquid crystals; (iii) quantum liquids such as ^{4}He or simple atomic liquids such as molten metals; (iv) liquids confined in fine pores or capillaries (see Chapters 9 and 10).

Nowadays, the problem of organization and motion in molecular liquids remains extremely difficult to tackle, in spite of the considerable improvements which have been made in recent years in the available spectroscopic techniques and methods, including the theoretical approaches which use highly efficient computer programs and large memory size computers. The complexity of the problem originates from the large number of geometrical parameters needed to describe the relative position of the interacting molecules and from the diversity of the mathematical representations of the interaction which must be taken into account [1–5], and which are all of electric origin. Indeed, except in very special cases such as molecules of high anisotropic magnetic susceptibility placed in a high magnetic field [6–8], the magnetic interaction can be neglected.

Before analysing the physical description of the forces which take place in liquids, it must be emphasized that the most important features which characterize a liquid are the harsh repulsions that appear at short distances and the attractions which act much more smoothly at large distances. In the very special case of interacting atoms or molecules of spherical symmetry (e.g. C_{60} fullerene) or close to it (e.g. point group T_d, O_h) the potential of interaction $V(r)$ may be accounted for by the hard sphere model: $V(r) = \infty$ for $r < d$ and $V(r) = 0$ for $r > d$. A more elaborated description is given by the 12-6 potential of Lennard-Jones of the form

$$V(r) = 4\varepsilon \left[\left(\frac{\sigma}{r} \right)^{12} - \left(\frac{\sigma}{r} \right)^{6} \right]$$

ε is the potential depth at the minimum r_m ($r_m = \sigma\, 2^{1/6}$). The description of the interaction between non-spherical molecules poses greater problems, as the potential is a function of their distance and of their mutual orientation. Moreover, the situation may be even more complicated in some cases, owing to the non-splitting of the potential in pair interactions. Denoting by $\mathbf{r}_i$ the position of the ith molecule, the potential of interaction $\phi(\mathbf{r}_1, \mathbf{r}_2, ..., \mathbf{r}_N)$ cannot be written as the sum of pair potential $\phi(\mathbf{r}_1, \mathbf{r}_2, .., \mathbf{r}_N) = \Sigma_{i<j}\, U_2(\mathbf{r}_i, \mathbf{r}_j)$, but additional terms must be taken into account and $\phi(\mathbf{r}_1, \mathbf{r}_2, ..., \mathbf{r}_N) = \Sigma_{i<j}\, U_2(\mathbf{r}_i, \mathbf{r}_j) + \Sigma_{i<j<k}\, U_3(\mathbf{r}_i, \mathbf{r}_j, \mathbf{r}_k) + ... $. There exist cases where extra terms to U_2 are necessary to interpret experimental data.

In a somewhat formal way, the intermolecular interactions may be classified in long-range and short-range interactions. Denoting by r a characteristic intermolecular distance, the long-range distance forces decrease as r^{-m}, where m is a positive integer. The short-range forces which result from overlap of molecular electronic distribution vary as $\exp(-ar)$. By analysing the nature of the interactions, the following classification may be given:

(i) the coulombic interaction between the unperturbed charge distributions which characterizes the isolated ions or molecules;

(ii) the induction interaction which results from the distortion of the electric charge distribution in a molecule because of the presence of its neighbours;

(iii) the dispersion interaction which is of quantum nature, but can, however, be described from a classical viewpoint. Owing to the electron motion, at any instant, there exist in a molecule multipolar electric moments of different orders, which induce electric moments in the neighbouring molecules. The dispersion energy arises from the mutual interaction of these instantaneous electric charge distributions;

(iv) the short-range interactions existing when a significant overlap takes place between the electronic clouds of the interacting species which lose their identity and lead to a supermolecule.

To this list, which considers the main molecular interactions, one may add the resonance energy which occurs when the interacting molecules have degenerate energy levels which are split by the interaction thus producing some sort of stabilization.

Thus, there exists a large body of interactions which must be taken into account in order to describe the organization and motion in molecular liquids, making this problem quite difficult. Moreover, one must always bear in mind that, at room temperature, the usual temperature for studying and analysing molecular liquids, an energy difference as small as $2\,\text{kcal.mol}^{-1}$ ($8.36\,\text{kJ}$ mol^{-1}) corresponds, according to the Boltzmann law, to 96.5% of the lowest energy form. Thus, to predict molecular motions or organizations, the potential of interaction must be known with a high degree of accuracy.

Besides NMR spectroscopy, there exist several other experimental techniques which may be used to get information on molecular dynamics, among which we shall include the computer dynamic simulations. All these techniques use in their data analysis similar basic mathematical concepts such as correlation functions, spectral density functions and correlation times which will be briefly presented here first on the classical viewpoint, then transposed in a quantum mechanics view.

The time correlation function, which has been introduced in the theory of noise and stochastic problems [9], has therefore been largely used in different spectroscopies and also for the study of transport property measurements. Let $q(t)$ and $p(t)$ denote all the coordinate and conjugate momenta which are necessary to describe a system in its space phase coordinate system, and let $A(q(t), p(t)) = A(q, p, t) = A(t)$ some function of the space phase coordinate. A may be for example any scalar quantity or the components of a vector or of a tensor which describe a molecular property (velocity $\mathbf{v}$, position $\mathbf{r}$, electric dipole momenta $\boldsymbol{\mu}$, magnetization $\mathbf{M}$, electric polarization tensor ...). The classical time correlation $g(t)$ is defined as the ensemble average of the product of A at a given and arbitrary initial time by its value at a later time. The initial time may be considered as the origin, thus

$$g(t) = \iint A(q, p, 0) A(q, p, t) f(q, p)\, \mathrm{d}q\, \mathrm{d}p \qquad (2.1)$$

$f(q, p)$ is the probability that a molecule has q and p in the range $(q + \mathrm{d}q)$ and $(p + \mathrm{d}p)$ respectively. One may simply write $g(t) = \langle A(0)A(t)\rangle = \langle A(\tau)A(t + \tau)\rangle$, where $\langle\ \rangle$ denotes an ensemble average. One may notice that $g(0)$ which is used to define the correlation time in molecular motions is given by

$$g(0) = \iint A^2(q, p, 0) f(q, p)\, \mathrm{d}q\, \mathrm{d}p = \langle A^2(0)\rangle.$$

$g(0)$ is simply the mean square value of A which is in principle calculable from time independent statistical mechanisms.

According to the ergodic hypothesis, which states that for a stationary random process a large number of observations made on a single system at N arbitrary instants of time have the same statistical properties as observing at the same time N arbitrary systems chosen from an ensemble of similar systems, the autocorrelation function may also be defined, not by an ensemble-average quantity, but as

$$g(t) = \lim_{T \to \infty} \frac{1}{T} \int_0^T A(\tau)A(t + \tau) \, d\tau \qquad (2.2)$$

and the two definitions of $g(t)$ given by equations (2.1) and (2.2) are identical. $g(t)$ may be positive or negative. Thus, for the free rotation in a plane of a linear molecule of momentum of inertia I whose rotational velocity is governed by a Boltzmann distribution function, denoting by $\mathbf{u}$ a unit vector along the molecule, $\langle \mathbf{u}(t).\mathbf{u}(0) \rangle = \exp(-k_B T t^2 / I)$, k_B being the Boltzmann constant. For a three-dimensional free rotator, $\langle \mathbf{u}(t).\mathbf{u}(0) \rangle$ goes negative [10].

From its definition, the time correlation function has special mathematical properties [10]. (i) $g(t)$ is an even function of t, $g(t) = g(-t)$, thus $\exp(-t/\tau_C)$ is not a correlation function, but $\exp(-|t|/\tau_C)$ is; (ii) $g(t)$ satisfies the inequality $-1 \leqslant g(t)/g(0) \leqslant +1$.

On the physical point of view, in most cases, the autocorrelation function that one has to consider is a decreasing function of time with $g(t) \to 0$ as $t \to \infty$. Thus the zero-frequency Fourier transform of the dimensionless function $g(t)/g(0)$ gives an idea of the decreasing behaviour of $g(t)$, and it is natural to define the correlation time as

$$\tau_C = \int_0^\infty [g(t)/g(0)] \, dt$$

As an example, for $g(t) = A \exp(-\alpha |t|)$, one has $\tau_c = \alpha^{-1}$.

In molecular dynamic studies, depending upon the technique which is used, one may consider a single particle molecular autocorrelation function such as $\langle A_i(0)A_i(t) \rangle$, where A_i is a property of the ith molecule, or a many-particle correlation function $\Sigma_i \langle A_i(0)A_i(t) \rangle$. Resonance techniques such as infrared and Raman along with NMR and incoherent neutron scattering yield single particle correlation functions, whereas Rayleigh scattering and dielectric relaxation yield multiparticle correlation functions.

In defining single particle or multiparticle correlation functions, one considers the product of a given quantity A at time 0 and at time t, $A(0).A(t)$. In some studies, it may be necessary to consider (i) the product of two quantities which are of different physical nature, let say A and B, (ii) different components of a given tensor, $A_{rr}(0).A_{ss}(t)$, (iii) the same quantity on different atoms (e.g. electric dipole moment of different groups in a polymer). Thus one defines a cross-correlation function

$$C(t) = \iint A(q,p,0)B(q,p,t)f(q,p) \, dq \, dp.$$

As a generalization, we shall also denote cross-correlation function, when we have the same functions A and B but taken at different points of the phase space.

Another mathematical function which plays a very important role in the analysis of spectroscopic data, and which is often experimentally measured is the spectral density. The spectral density $I_A(\omega)$ of a time function $A(t)$ is the Fourier transform of its autocorrelation function

$$I_A(\omega) = \int_{-\infty}^{+\infty} \langle A(t)A(0) \rangle \exp(-i\omega t) \, dt \qquad (2.3)$$

The Fourier inverse theorem leads to

$$\langle A(t)A(0) \rangle = \frac{1}{2\pi} \int_{-\infty}^{+\infty} I_A(\omega) \exp(i\omega t) \, d\omega.$$

The time correlation function to consider is very often monoexponential (e.g. NMR), thus one obtains a spectral density of the form

$$\frac{2\tau_c}{1 + \omega^2 \tau_c^2} - \frac{2i\omega\tau_c^2}{1 + \omega^2 \tau_c}.$$

The real part of such expressions appears in the formulae which give T_1, T_2 and $T_{1\rho}$.

The definitions which have been previously defined from a classical point of view can be generalized in quantum mechanics. The function A is replaced by its quantum mechanics operator, and the function $f(q, p)$ which is the space phase distribution function becomes the equilibrium density matrix ρ, thus $\langle A \rangle = \text{Tr}\langle \rho A \rangle$.

2   A Brief Survey of Techniques other than NMR

Besides the NMR relaxation time measurements, there are several other experimental methods which may be used to get information on the difficult problem of the description of molecular motions in solution. There are numerous books or review articles which are fully or partly devoted to some particular aspects of this problem [11–15]. The methods which are the more often used for analysing molecular motions are Infrared Absorption; Raman and Rayleigh Scattering; Coherent and Incoherent Neutron Scattering; Dielectric and Kerr Relaxation and Fluorescence Depolarization. Each of these methods has its advantages and disadvantages and is related to a particular correlation function and spectral density, which characterizes the time evolution of a given parameter of the molecule. An accurate or realistic description of the motion of a molecule in a solution can only be reached in considering more than one experimental approach. We shall not go into a detailed presentation and analysis of the previously mentioned methods, but we shall show on a particular example, the infrared absorption, how the algebraic expression is established which exists between the line intensity shape, expressed as a function of the frequency, namely $I(\omega)$, and the molecular electric dipole moment correlation function [16].

2.1 INFRARED ABSORPTION

For the sake of simplicity, we consider the case of a purely rotational transition, in such a case the transition moment operator for an individual molecule is the permanent electric dipole operator **m**, which is denoted **M** for the assembly of the N molecules under study. According to the time dependent perturbation theory, one may calculate the probability P_{if} that a system which is in a stationary state $|i\rangle$ at time 0, would be in a stationary state $|f\rangle$ at time t under the influence of a time dependent perturbation $W(t)$. One obtains

$$\frac{1}{\hbar^2}\left|\int_0^t \langle i|W(t')|f\rangle \exp{(i\omega_{fi}t')}\,dt'\right|^2$$

with $\omega_{fi} = \omega_f - \omega_i$. By considering a sinusoidal electromagnetic wave **E** sin ωt, the probability P_{if} per unit time that a transition from state $|i\rangle$ to state $|f\rangle$ takes place is given by [17]

$$P_{if}(\omega) = \frac{\pi}{2\hbar^2}\left|\langle i|\mathbf{M}.\mathbf{E}|f\rangle\right|^2 [\delta(\omega_{fi} - \omega) + \delta(\omega_{fi} + \omega)]$$

If we specialize to the case of Boltzmann distribution ρ, for non-degenerate states, one has $\rho_f = \rho_i \exp(-\beta\hbar\omega_{fi})$ with $\beta = (k_B T)^{-1}$. Thus the rate of energy loss from the electromagnetic wave to the system is $-\dot{W} = \Sigma_{i,f}\,\rho_i\hbar\omega_{fi}\dot{P}_{if}(\omega)$ $\delta(\omega_{if} - \omega)$. After some algebraic manipulations, and introducing the factor $[1 - \exp(-\beta\hbar\omega)]^{-1}$ (where $\beta = 1/k_B T$) as a multiplying factor in the definition of the absorption shape $I(\omega)$, it becomes $I(\omega) \propto \Sigma_{i,f}\,\rho_i|\langle i|\mathbf{M}.\mathbf{E}|f\rangle|^2\delta(\omega_{fi} - \omega)$. Using the Heisenberg representation of an operator, which allows one to define the orientation of the resulting dipole moment at time t, as a function of its value at time 0, that is $\mathbf{M}(t)$ as a function of $\mathbf{M}(0)$, one ends up with

$$I(\omega) = \frac{1}{2\pi}\int_{-\infty}^{+\infty}\langle \mathbf{M}(0).\mathbf{M}(t)\rangle \exp{(-i\omega t)}\,dt.$$

Thus, the lineshape function is written as the Fourier transform of the time-correlation function of the dipole moment of the absorbing molecules in the absence of the perturbing electromagnetic wave. However, it must be noticed that **M** is the dipole moment of the entire assembly of molecules; hence, in the expression of $\mathbf{M}(0)\mathbf{M}(t)$, cross terms will appear. If the N molecules under study are dilutely dissolved in a non-polar solvent, $\langle \mathbf{M}(0)\mathbf{M}(t)\rangle = N\langle \mathbf{m}(0).\mathbf{m}(t)\rangle$, where **m** is the individual dipole moment of a molecule. Therefore the infrared lineshape analysis provides information on the rotational correlation function of a vector fixed in the molecular frame. Typical frequency range obtained from infrared spectroscopy is 3×10^{10} Hz–10^{10} Hz.

Recently the far infra-red transmittance of liquid acetonitrile has been measured at temperatures ranging from 238 K to 343 K, using a source of synchrotron radiation [18]. The analysis of the second derivative of the electric dipole moment vector correlation function with respect to time, $f(t) = (d^2/dt^2)\langle \mathbf{M}(t)\mathbf{M}(0)\rangle$ which is directly connected to the absorption and to the refractive index provides information on the molecular dynamics. The rotational motion of liquid acetonitrile would be libration in nature.

2.2 RAMAN AND RAYLEIGH LIGHT SCATTERING

Light scattering corresponds to the collision process of a photon with a molecule. Two cases may be considered. First, the incident photon is absorbed and the molecule is excited from state $|i\rangle$ to state $|f\rangle$, and the molecule returns to state $|i\rangle$ if the wavelengths of absorbed and emitted photons are identical; this elastic process corresponds to the Rayleigh scattering. Second, the emitted photon undergoes a frequency shift with respect to the incident frequency, this inelastic process corresponds to the Raman scattering (Stokes line, lower frequency shift; anti-Stokes line, higher frequency shift).

Treatments basically similar for the infra-red absorption and the light scattering can be developed. The oscillating electric field $\mathbf{E}_i$ of the incident wave induces an electric moment $\boldsymbol{\mu}$, and denoting by $\mathbf{a}$ the molecular polariz ability tensors $\boldsymbol{\mu} = \mathbf{a}\mathbf{E}_i$ considered in the laboratory frame, $\boldsymbol{\mu}$ is modulated by both the electromagnetic incident wave and by the molecular motion. $\mathbf{a}$ may be split in two parts $\mathbf{a} = \mathbf{a}_1 + \mathbf{a}_2$, with $\mathbf{a}_1 = \frac{1}{3}\mathrm{Tr}(\mathbf{a})\mathbf{I}$ and $\mathbf{a}_2 = \mathbf{a} - \frac{1}{3}\mathrm{Tr}(\mathbf{a})\mathbf{I}$, where $\mathbf{I}$ is the dyadic tensor, $\mathbf{a}_1$ induces an electric dipole moment which is constant, does not depend upon the molecular orientation and is parallel to $\mathbf{E}_i$; this does not hold for $\mathbf{a}_2$.

From simple geometry considerations concerning the polarization of the wave emitted by a radiant dipole, it turns out that the light scattered at right angle by $\mathbf{a}_1$ is polarized parallel to $\mathbf{E}_i$ and that the light scattered by $\mathbf{a}_2$ has a component perpendicular to $\mathbf{E}_i$. Thus one may study separately the purely vibrational and purely orientational effects.

In a way similar to the one briefly presented for the infra-red absorption line, the intensity distribution may be obtained for the light scattering. In the following, i and f refer to laboratory frame directions. For an incident light polarized along the i direction, and a scattered light polarized along the f direction, the scattering intensity depends upon the $\langle \alpha_{if}(0)\alpha_{if}(t)\rangle$ correlation function, which involves the molecular tensor polarizability element α_{if} in the laboratory fixed coordinate system. For the specific case of a symmetric top molecule, the autocorrelation function needed in the expression of the total intensity scattered reduces to $\langle P_2[\mathbf{u}(0).\mathbf{u}(t)]\rangle$, where $\mathbf{u}$ is the unit vector along the symmetry axis of the molecule, and $P_2(x)$ the Legendre polynomial $(3x^2 - 1)$.

As an example, we report some results concerning the Raman scattering investigation of the reorientational dynamics of methyl iodide as a function of temperature and dilution in cyclohexane [19]. The rotational correlation function $g_2(t)$ which represents the reorientation of the symmetry axis is analysed. At short and intermediate times, $g_2(t)$ does not decay exponentially as expected from Debye's rotational diffusion model. From this study, one may introduce two correlation times τ_c and τ_s. τ_c, as previously defined, is the zero frequency value of the spectral density

$$\tau_c = \int_0^\infty g_2(t)\, \mathrm{d}t.$$

τ_s corresponds to the experimentally observed decay of $g_2(t)$ at long times, $g_2(t) = A\, \exp{-(t/\tau_s)}$. Typical values given in picoseconds are: $T = 295$ K; $\tau_c = 1.14$; $\tau_s = 1.30$; $T = 215$ K, $\tau_c = 3.60$; $\tau_s = 4.65$.

2.3   INELASTIC NEUTRON SCATTERING

Neutrons are scattered by molecules in solution. In a diamagnetic species, the most important scattering arises from the interaction with the nucleus, thus the scattered neutrons provide information about molecular motion. The neutron–nucleus interaction is weak and may be described through a perturbation theory, in a way somewhat related to the molecule photon scattering. However, while light scattering is generally coherent, the neutron scattering may be coherent or incoherent. This comes from nuclear spin–neutron spin interaction which may randomize the phase of the scattered neutron. If all the nuclei in the sample interact in exactly the same way with a neutron beam, there will be interference between the scattered wave of each nucleus, this corresponds to the coherent neutron scattering. If the interaction of the neutron with the nucleus randomizes the phase, there is no correlation between the scattered amplitude in space of the different nuclei, this is the incoherent scattering. As a consequence, the incoherent scattering depends on the motion of an individual atom (self-correlation function) and the coherent scattering depends on the motion of all the atoms (pair correlation function).

The differential neutron cross-section $\mathrm{d}^2\sigma/\mathrm{d}\Omega\mathrm{d}E$, defined as the flux of neutrons scattered in the solid angle $\mathrm{d}\Omega$ in the energy range $\mathrm{d}E$, may be obtained through a mathematical treatment quite similar to the one used for infrared absorption and light scattering. For an incident neutron of momentum $\hbar\mathbf{k}_0$, scattered with a momentum $\hbar\mathbf{k}_1$, the differential cross-section is proportional to

$$\int_0^\infty \sum_{m,n} \langle \exp[-\mathrm{i}\mathbf{q}(\mathbf{r}_m(0) - \mathbf{r}_n(t))] \rangle b_m^* b_n \exp(-\mathrm{i}\omega t)\, \mathrm{d}t.$$

In this formula, the quantities $\mathbf{r}$ and b with a subscript are the position and the bond atom cross-section, respectively, m and n are the atom labels, $\mathbf{q} = (k_0 - k_1)$. Thus, the differential cross-section appears as a function of $\mathbf{q}$ and ω; the basic mathematical transformation is to consider this function as the double Fourier transform with respect to the variables $\mathbf{r}$ and t. At the end, in the classical limit $d^2\sigma/d\Omega\, dE$ appears as the sum of two terms: the coherent differential cross-section, $d^2\sigma_{coh}/d\Omega\, dt$, and the incoherent differential cross-section, $d^2\sigma_{incoh}/d\Omega\, dt$ [20].

$$\frac{d^2\sigma_{coh}}{d\Omega\, dE} = \frac{\langle a\rangle^2 N}{2\pi\hbar}\frac{k}{k_0}\int \exp\left[i(\mathbf{qr} - \omega t)\right]G(\mathbf{r}, t)\, d\mathbf{r}\, dt$$

$$\frac{d^2\sigma_{incoh}}{d\Omega\, dE} = \frac{(\langle a^2\rangle - \langle a\rangle^2)N}{2\pi\hbar}\frac{k}{k_0}\int \exp\left[i(\mathbf{qr} - \omega t)\right]G_s(\mathbf{r}, t)\, d\mathbf{r}\, dt$$

$$G_s(\mathbf{r}, t) = N^{-1}\left\langle\sum_{j=1}^{N}\delta[\mathbf{r} + \mathbf{r}_j(0) - \mathbf{r}_j(t)]\right\rangle;$$

$$G_d(\mathbf{r}, t) = N^{-1}\left\langle\sum_{j \neq 1=1}^{N}\delta[\mathbf{r} + \mathbf{r}_1(0) - \mathbf{r}_j(t)]\right\rangle$$

$G = G_r + G_d$, and a_j is an operator that depends upon the spin of the jth nucleus and $\langle a\rangle$ denotes an ensemble average over the spin states of the particle.

Thus, as mentioned in the introduction of this section, the incoherent scattering depends on the motion of an individual atom (self-correlation function) and the coherent scattering depends on the motion of all the atoms (pair correlation function). Neutron scattering may be inelastic, quasi-elastic (QENS) or elastic. By choosing appropriate momentum and energy transfer ranges of observation, one may obtain information on the whole molecule translation or rotation and possibly on the conformational dynamics of flexible molecules. This has been applied to analyse the molecular dynamics of 16-crown-6-ethers [21]. The following values are obtained for the rotational diffusion constant D_r: $T = 323$ K, $D_r = (1.5 \pm 0.8)10^9$ s^{-1}; $T = 353$ K, $D_r = (2.3 \pm 0.8)10^9$ s^{-1}. A mean jump frequency of $1.5\ 10^{22}$ s^{-1} is associated with the conformational changes.

2.4   FLUORESCENCE DEPOLARIZATION EXPERIMENTS

A pulse of light of frequency ω_i, polarized along a direction $\mathbf{e}_i$ impinges on a molecule whose absorption dipole moment for the frequency ω is $\boldsymbol{\mu}_a$. The light absorption probability is proportional to $|\boldsymbol{\mu}_a\mathbf{e}_i|^2$. Then the molecule emits a light of frequency ω_f, characterised by an emission dipole $\boldsymbol{\mu}_e$. The light intensity at frequency ω_f detected along a direction $\mathbf{e}_f$ by an analyser is proportional to $|\boldsymbol{\mu}_e\mathbf{e}_f|^2$. The light intensity I may be detected for different and independent configurations of polarizers and analysers. If e_i and e_f are parallel, $I_\parallel \propto k_1 + \langle P_2(\mathbf{\mu}_{a(0)}.\mathbf{\mu}_e(t)) \rangle$, and if e_i and e_f are perpendicular, $I_\perp \propto k_2 + \langle P_2(\mathbf{\mu}_a(0).\mathbf{\mu}_e(t)) \rangle$. k_1 and k_2 are two numerical constants. The information obtained from the fluorescence depolarization experiments is thus somewhat similar to that obtained from Raman scattering experiments, but generally for longer times.

2.5   MOLECULAR DYNAMICS

In addition to the experimental methods which have been briefly mentioned, one may consider the molecular dynamics (MD) where the spectrometer is a computer. Modern computers have reached such a degree of sophistication that Newton's or Lagrange's equations may be solved for a large assembly of particles. The basic ingredients in all the MD calculations are the potential of interaction $\varphi(r)$, which is always assumed to have a finite range of interaction, $\varphi(r) = 0$ for $r > r_0$, and to be pairwise and additive. The equations of motion are solved by finite difference technique, with steps which may be of the order of 10^{-14} to 10^{-15} s, a time which is three or four orders of magnitude smaller than the correlation times of the molecules which are considered in this chapter ($M \leqslant 300$), and which are in the range from 10^{-10} to 10^{-12} s (see Section 3.1). The initial conditions are defined according to the Maxwell distribution of velocity at a given temperature. MD simulations are often confined to considering the microcanonical ensemble which is defined by a constant number of molecules N, total energy E and volume V. The N molecules are located in a box of edge L and satisfy boundary conditions: if $\{x_i, y_i, z_i\}$ defines the position of particle i, there are 26 periodic images at $\{x_i \pm L, 0; y_i \pm L, 0; z_i \pm L, 0\}$. The interactions take place between the particles inside the box and with the periodic images.

Besides the MD which requires high computer resources, some approximate models have been proposed and, in the computer study of the behaviour of a many particle system, some degrees of freedom are treated explicitly, while some others are represented by their stochastic influence. As an example, one may assume that the solvent effect on the solute molecules is mimicked by a stochastic force plus a frictional force [22], and, as a result, obeys a Langevin equation.

In ending this brief review of some of the techniques which may be used to analyse molecular motions, we shall stress some particularities of the NMR parameters which are a source of information for dynamical studies. (i) NMR spectral transitions and its parameters are highly frequency dependent, and today, if necessary, one may record the spectra over almost a frequency decade (60 MHz up to 700 MHz for ^{1}H). Thus several regions of the spectral density $J(\omega)$ may be explored. However, it must be pointed out that the spectral density used in NMR often taken as $(1 + \omega^2 \tau_c^2)^{-1}$ is not very sensitive to ω in the range of correlation times found for liquids of low molecular weight

$(10^{-10} \, s < \tau_c < 10^{-12} \, s)$. A list of several values of correlation times of various molecules obtained from NMR spectroscopy along with their viscosity is given in reference [23]. For spherical or nearly spherical-shaped molecules, the correlation times range in two decades around 1 ps. Typical values in ps are given hereafter: CD_4 (0.20); C_6D_{14} (1.5); Sn I_4 (3.67); $TiCl_4$ (4.53); $PbCl_4$ (17.2); (ii) using spin labeling (of isotopes of different spins), one may generally obtain several local molecular probes for analysing molecular motions; (iii) the NMR relaxation times may have different physical origins, which will be presented in Section 5. Some of them probe the motion of a given vector in the molecule (dipole–dipole interaction), some others probe the overall molecular motion (chemical shift anisotropy). Thus NMR spectroscopy is an extremely valuable source of data for analysing molecular dynamics, however, depending on the model used to describe the molecular motion, the algebraic expressions of T_1, T_2 and $T_{1\rho}$ are not that easy to obtain.

3 Description of Molecular Motions

3.1 OVERALL MOTIONS OF RIGID SYSTEMS

The dynamical study of molecular liquids has been of interest for physicists and chemists for many years [24], and several models have been proposed to describe the translation and rotation of molecules by diffusion equations. As already stressed in the introduction to this chapter, the problem of describing molecular motions remains extremely difficult today, even for rigid molecules of low molecular weight. The complexity of the molecular movement is often masked by the averaging and smoothing processes which are recorded. Thus several simplified models may well fit the experimental data and it remains difficult to discriminate between them.

3.1.1 The sphere with one rotational diffusion coefficient

The random walk may be applied to describe the rotation of a spherical molecule of radius a, in a similar way it is applied to describe the translational motion. A given vector attached to the rotating molecule, and which passes through its centre starts with a given orientation and rotates by an angle θ_1 within a time τ_1. At this time, its history is forgotten and it starts to rotate by an angle θ_2 in a time τ_2. The duration of each step is short compared with the time required for a molecule to make a complete revolution. The probability density function $\psi(\Omega, t)$ that a given molecular axis which is in a direction Ω_0 at time 0, will be in direction Ω at time τ obeys the diffusion equation [25]

$$\frac{\partial \psi(\Omega, t)}{\partial t} = \frac{D}{a^2} \Delta_\Omega \psi(\Omega, t) \qquad (2.4)$$

Δ_Ω is the angular part of the Laplacian operator and D is the diffusion coefficient, with $D = k_B T / 8 \pi a \eta$, where η is the kinematic viscosity coefficient.

One may find a solution of the diffusion equation (2.4) as an expansion of spherical harmonics $Y_{l,m}$, $\psi(\Omega, t) = \Sigma_{l,m} c_{l,m}(t) Y_{l,m}(\Omega)$; the time dependent coefficients of the expansion are easily obtained: $c_{l,m}(t) = c_{l,m}(0) \exp(-t/\tau_1)$ with $\tau_1^{-1} = Dl(l+1)/a^2$. Using the expansion of the Dirac delta function $\psi(\Omega, 0) = \delta(\Omega - \Omega_0)$ as a sum of spherical harmonics, $\delta(\Omega - \Omega_0) = \Sigma_{l,m} Y^*_{l,m}(\Omega_0) Y_{l,m}(\Omega)$, the angular part of the function $\psi(\Omega, t)$ is obtained and the complete expression may be written $\psi(\Omega, t) = \Sigma_{l,m} Y_{l,m}(\Omega_0) Y_{l,m}(\Omega) \exp(-t/\tau_1)$. If one assumes that $\psi(\Omega, 0)$ is symmetric with respect to an arbitrary axis z, as it may be the case for an isotropic liquid, equation (2.4) may be solved in rather a different manner, since ψ depends only on θ and t, but not on φ. In terms of the variable $u = \cos \theta$, equation (2.4) may be rewritten, and $\psi(\theta)$ expanded in terms of Legendre polynomials [26].

The problem of the rotational Brownian motion of a sphere has also been solved by means of quaternions. In this method, the operation of turning through the angle α is described by a point within or on the surface of a unit sphere. The direction from the origin to the point gives the direction of the rotation axis, and its distance from the origin is equal to $\sin \alpha/2$.

3.1.2 The ellipsoid with three rotational diffusion coefficients

In two extremely well documented papers, Perrin established the rotational and translational Brownian diffusion equations for ellipsoids [27]. These papers are based on a very elegant way of using the basic equations of classical mechanics. The orientation of a solid from a position defined as the origin may be characterized by three parameters, two angles θ and φ which define the direction of the axes of rotation, and a third angle which gives the rotation amplitude. Thus the probability density ψ of finding the ellipsoid in a given orientation may be defined by a point on an hypersphere of dimension four. The differential equation which is satisfied by ψ is derived, it depends on generalized diffusion coefficients for reorientation around three perpendicular axes of the molecule. Perrin applied his results to analyse the motion of an ellipsoid in the presence of an electric field. For a rigid body of unspecified shape, a diffusional rotation equation may be established [28, 29], following an approach which presents some similarities to the one used for translation diffusion motion [30]: $\psi(\Omega, t)$ denotes the probability function of finding the molecule in direction Ω at time t; $p(\theta, \Delta t)$ is the probability that the rotation defined by θ, which rotates the molecule from orientation Ω_0 to orientation Ω (where $\Omega - \Omega_0 = \theta$) occurs in time Δt. Obviously, from the definition given above, one has

$$\psi(\Omega, t + \Delta t) d^3\Omega = \int p(\theta, \Delta t) d^3\theta \psi(\Omega_0, t) d^3\Omega_0 \qquad (2.5)$$

Using the rotation operator which is defined from the quantum angular momentum L, $\psi(\Omega_0, t)d^3\Omega_0 = \exp(i\theta.L)\psi(\Omega, t)d^3\Omega$, and considering small angular rotations, one may write $\exp(i\theta.L) = 1 - i\theta.L - \frac{1}{2}(\theta L)(\theta L) + \dots$. Expanding by a Taylor series the left-hand side and the right-hand side of equation (2.5), it comes:

$$\psi(\Omega, t) + \frac{\partial \psi}{\partial t}(\Omega, t)\Delta t + \dots =$$

$$\int p(\theta, \Delta t)(1 + i\theta.L - \frac{1}{2}(\theta.L)(\theta.L)\dots)d^3\theta\psi(\Omega, t)d^3\Omega$$

$$\frac{\partial \psi}{\partial t}(\Omega, t) = \frac{1}{\Delta t}\left\{ i\int p(\theta, \Delta t)\theta.Ld^3\theta\psi(\Omega, t) - \frac{1}{2}\int p(\theta, \Delta t)(\theta.L)(\theta.L)d^3\theta\psi(\Omega, t) \right.$$

which may be written as

$$\frac{\partial \psi}{\partial t}(\Omega, t) = (iL.A - LDL)\psi(\Omega, t)$$

where A is a vector operator and D a symmetric second rank tensor

$$A_r = \frac{1}{\Delta t}\int p(\theta, \Delta t)\theta_r d^3\theta; \quad D_{rs} = \frac{1}{2\Delta t}\int p(\theta, \Delta t)\theta_r\theta_s d^3\theta, \quad rs = x, y, z$$

In a completely random and isotropic motion, the components of the vectorial operator A are equal and each equal to zero, thus $\partial\psi(\Omega, t)/\partial t = -i(LDL)$. This equation, except for the missing factor $\hbar$, is completely equivalent to a time-dependent Schrödinger equation. The second rank tensor D is symmetrical and may be diagonalized. Thus $\partial\psi(\Omega, t)/\partial t = -i\Sigma_r D_{rr}L_r^2\psi(\Omega, t)$.

3.1.3   J and M extended diffusion models

In the Debye model, the assumption of small-angle diffusion implies that the molecule undergoes many changes in angular momentum before the orientation has been changed appreciably. The reorientational correlation time is long as compared to the angular momentum correlation time. Gordon proposed for linear molecules [16] an extension of this classical diffusion model, by removing the restriction of small angular diffusion steps. The byname of extended diffusion means that we are not dealing exclusively with rotations of small angle as assumed in the rotational diffusion model. In this extended diffusion model, either both magnitude and direction of the angular momentum are randomized at every collision (J diffusion, EDJ model), or only the direction of the angular momentum is randomized while its magnitude remains constant (M diffusion, EDM model). The molecules are assumed to rotate freely, as in the gas phase in the time intervals between collisions. This assumption is not very realistic for the liquid state, where the molecules are quite close one to the others. This model is limited to molecules or molecular fragments (e.g. CH_3) of low molecular weight which do not exhibit specific interactions in the liquid state (e.g. hydrogen bond). However, this model has been extensively used to describe and interpret some important results obtained in the liquid phase. The duration of each step is random and characterized by a mean duration time τ_j. The probability that a molecule is making its nth diffusion step at time t is given by a Poisson distribution

$$p(n, c) = \left(\frac{t}{\tau_j}\right)^{n-1} \frac{\exp(-\tau/\tau_j)}{(n-1)!}.$$

In a slight generalization of the extended diffusion model, the time τ_j which appears in the Poisson distribution $p(n, t)$ is taken as a function of the molecular rotation frequency.

The EDJ and EDM models which have been initially devoted to linear molecules have been extended to spherical [31], symmetric [32] and asymmetric [33] top molecules. Denoting by $\mathbf{m}$ the unit vector along the transition dipole moment and by m_z its component along the z molecular axis, the calculated correlation function $g(t) = \langle P_2(m_z(0).m_z(t))\rangle$ has been compared for the EDM and EDJ models [13, 34]. For short correlation times, both show identical behaviour, irrespective to the correlation time value τ_j. At longer times, $g_M(t)$ decreases slower than $g_J(t)$, all other conditions being the same.

3.1.4 Fokker–Planck–Langevin model

In the so-called Fokker–Planck–Langevin model (FPL), the molecular motion is governed by a classical Langevin equation and the orientation-angular velocity conditional probability density function is given by a Fokker–Planck equation. According to the Langevin equation [35], the force acting on a particle of mass m and velocity $\mathbf{v}$ immersed in a fluid is the sum of two terms: (i) a general dynamic friction $-\Gamma\mathbf{v}(t)$, (ii) a fluctuating force of the medium $\mathbf{K}(t)$. Thus the time evolution of the vector $\mathbf{r}$ which defines the particle position is $m\dot{\mathbf{v}}(t) = -\Gamma\mathbf{v}(t) + \mathbf{K}(t)$. The rotational motion of a molecule taken as an appropriately shaped object may be characterized by a set of three Eulerian angles Ω which characterize its orientation with respect to the laboratory frame, and by its angular velocity ω with respect to the molecular axes. The rotation of a molecule is modulated by a retarding torque proportional to its angular velocity and by a rapidly fluctuating Brownian torque. In the FPL model, there exists a time interval $(0, \tau)$ such that ω changes only infinitesimally on this time, but τ is long compared to the time between fluctuations in the Brownian torques experienced by the molecule. The conditional probability density $\psi(\omega, \Omega, t/\omega_0, \Omega_0)$ that a molecule will have angular velocity between ω and $\omega + d\omega$ and orientation between Ω and $\Omega + d\Omega$ at time t, given that it had angular velocity between ω_0 and $\omega_0 + d\omega_0$ and orientation between Ω_0 and

$\Omega_0 + d\Omega_0$ at time zero, is governed by a Fokker–Planck equation: $\partial\psi/\partial t = -Z\psi$, where Z is a Liouville operator $Z = i\omega J - \tau_J^{-1} (\nabla_\omega \cdot \omega + \nabla_\omega^2)$, where τ_J is the angular correlation time, J is the infinitesimal rotation operator referred to the molecular coordinate system.

In 1969, Fixman and Rider [36] proposed the FPL model for molecular reorientation in liquids. Exact [36] and approximate [37, 38] expressions for some of the reorientational correlation functions and spectral densities have been obtained for linear and spherical top molecules. The asymmetric top case has also been considered [39].

3.2 MOLECULES WITH FAST INTERNAL MOTION

In the case of molecules with internal degrees of freedom, the representation of the molecular geometry becomes more complex, and several sets of Euler angles may be necessary to give a full description of the motion of every atom of the molecule with respect to the laboratory frame. A simple, but very often encountered case of internal motion, corresponds to the case of the rotation of a methyl group attached to the rigid part of a molecular skeleton (e.g. toluene). Even in such a simple molecule, several situations may be considered depending on (i) the model chosen to describe the overall motion of the molecule, (ii) the description of the methyl hydrogen motion, (iii) the correlated or uncorrelated nature of the overall molecular motion and of the rotation around the methyl hydrogen axis. Examples of these different possibilities have been considered in the literature [40], and the problem of intramolecular rotation in nuclear magnetic resonance has been extensively analysed in ref. [41].

In a pioneering work, Woessner [42] examined the problem of a rotating methyl group attached to a rigid ellipsoid of revolution which undergoes a rotational isotropic Brownian motion. The overall molecular motion and the methyl rotation are assumed to be independent. The analysis was made in order to calculate the dipolar contribution to the spin–lattice relaxation time T_1 and to the spin–spin relaxation time T_2. Thus, denoting by l, m and n respectively the director cosines with respect to the laboratory frame of the vector $\mathbf{V}$ defined by the two nuclei coupled through the dipolar interaction, the spectral density we are interested in is

$$J_h(\omega) = \int_{-\infty}^{+\infty} \langle F_h(0)F_h^*(\tau) \rangle \exp(i\omega t)\, d\tau.$$

$F_h(t)$ are time dependent functions with $F_0(t) = 1 - 3n^2$; $F_1(t) = (l + im)n$; $F_2(t) = (l - im)n$ respectively. Let us denote by φ_0, θ_0 and ψ_0 the Euler angles which define the orientation of a stationary reference frame S_0 attached to the ellipsoid, with respect to the laboratory frame. The probability density $P(\varphi_0, \theta_0, \psi_0, \tau)$ to find S_0 at orientation φ_0, θ_0, ψ_0 at time τ obeys Perrin's equation $\partial P/\partial\tau = \Lambda P$; a complete expression of the operator Λ may be found in [42]. The rotation of the vector $\mathbf{V}$ around the methyl group symmetry axis depends upon an angle φ' and its motion is described by a stochastic diffusion process. Denoting by $\pi(\varphi'_0 + \varphi', \tau)$ the probability of finding the internuclear vector at angle φ' relative to φ'_0 at time $t + \tau$, $\pi(\varphi'_0 + \varphi', \tau)$ is given by a Gaussian distribution. Thus, all that is necessary for calculating the J_h spectral density function is obtained.

The previous treatment has been extended to the more general case where the internal rotation occurs about an axis at any constant angle to the symmetry axis of a symmetric ellipsoid [43]. If the flexible part of a molecule has two stable conformations, the internal motion has been accounted for by an oscillation between two sites [44]. This motion is completely defined by the conditional probability $p(\gamma_i, \gamma_f, \tau)$ of finding the mobile part of the molecule undergoing oscillation at time $t + \tau$ in the final orientation γ_f, given that, at time t, it was in the orientation γ_i. More complex analysis becomes necessary in biomolecules where the number of geometrical parameters increases (see Chapter 9B).

The overall motion of the molecule and the internal rotation may be considered as independent or not. If the two motions are uncorrelated, the correlation function for fully anisotropic motion with internal motion can be written as the product of correlation function for the overall motion and for the internal motion [45]. The demonstration is performed using some of the well-known mathematical expressions which allow one to transform a tensorial expression from one coordinate system to another [46]. For a molecule which has several degrees of freedom, it may be necessary to consider, besides the laboratory frame (I), several local coordinate systems (F) and the set of Euler angles $\{ \alpha_I \rightarrow \alpha_F; \beta_I \rightarrow \beta_F; \gamma_I \rightarrow \gamma_F \}$ which relates (I) to (F). All the magnetic interactions which contribute to the NMR relaxation times may be written as the scalar product of two second-rank spherical tensors $\Sigma_q (-1)^q F_q A_q$, where A_q contains only spin operators. A second rank tensor transforms from one frame to another through the Wigner rotation matrix $D_{ab}^{(2)}(\alpha, \beta, \gamma)$, all the algebraic calculations are greatly simplified using the well-known expression $D_{ab}^{(2)}(\alpha, \beta, \gamma) = \exp-(ia\alpha)d_{ab}^{(2)} \exp(-ib\gamma)$, where $d_{ab}^{(2)}$ is a reduced matrix element. A detailed analysis of the molecular motion in steroids, combining fully anisotropic overall tumbling and internal rotation of methyl hydrogens has been discussed in [47].

One may also consider that the overall and internal motions are correlated. If the motion of the molecule is assumed to be governed by extended J diffusion, it seems natural to assume that collisions also randomize internal angular momentum j [48]. Denoting by J the total angular momentum, and by i and k two directions of the molecular axes, the overall angular momentum correlation function is $\langle J_i^*(t)J_k(0)\rangle = \delta_{ik}\langle| J_k(0)|^2\rangle\exp-(t/\tau_c)$ and the internal angular momentum correlation function is $\langle j^*(t)j(0)\rangle = \langle| j(0)|^2\rangle\exp-(t/\tau_j)$. It is assumed that the overall and internal angular momentum correlation times are the same.

In a quasi 'model-free' approach, Lipari and Szabo [49] proposed describing the motion of a flexible molecule with only three parameters: (i) a generalized order parameter S which is the measure of the degree of spatial restriction of the local motion; (ii) an effective correlation time τ_f which is a measure of the rate of the local motion, generally fast; (iii) a correlation time τ_s, describing the overall motion, generally slow. The order parameter S and the correlation time τ_f are defined in a model independent way. This approach is widely used in dynamical studies of large molecules, as explained in Chapter 9. It will be further considered in Section 5.1 of this chapter.

Owing to the complexity of the complete description of the motion of each nucleus of a molecule with several internal degrees of freedom, a great number of models have been proposed which are reviewed in references [48] and [49]. Special situations such as the motion of a molecule which diffuse on the surface of a freely tumbling prolate or oblate spheroidal aggregate or macromolecule have also been discussed [50].

4   NMR Relaxation Parameters and their Experimental Determination

Before going into the details of spin dynamics under relaxation phenomena, it may be worth providing some idea about relaxation mechanisms and their order of magnitude. As explained before (see e.g. Chapter 1), a relaxation mechanism is meant as an interaction undergone by the nuclear spins which is time dependent. The most important mechanisms are considered in detail in Section 5. For the moment, we shall rely only on the ubiquitous dipolar mechanism (mutual interaction between the magnetic moments associated with two nuclear spins) and the so-called chemical shift anisotropy (CSA) mechanism which arises from different principal values of the shielding tensor. The order of magnitude of the former can be appreciated by the quantity $0.285 \, \tau_c / r^6$ which represents roughly the contribution to the relaxation rate (inverse of the relaxation time in s^{-1}) of a given proton due to the interaction with another proton at a distance r expressed in Å; in that expression, which applies to fast molecular reorientation, τ_c is the correlation time expressed in ps. Because for non-viscous liquids τ_c is in the range $1-10$ ps, the dipolar contribution to the relaxation time is seen to be of the order of unity (relaxation time of the order of a second). If the dipolar interaction concerns one proton and one nucleus X of gyromagnetic ratio γ_x, the quantity $0.285 \, \tau_c / r^6$ must be multiplied by $(\gamma_X / \gamma_H)^2$ thus reducing somewhat the contribution to the relaxation rate (e.g. by a factor of 16 in the case of a $^{13}C-^1H$ dipolar interaction). In that case, the CSA mechanism may be a competitor. Its scaling factor to the longitudinal relaxation rate can be expressed as $(2/15) \, \nu_0^2 (\Delta\sigma)^2 \tau_c$, where ν_0 represents the Larmor frequency in MHz and $\Delta\sigma$ the shielding anisotropy in ppm (τ_c should be expressed here in s). This mechanism may become non-negligible provided that $\Delta\sigma$ is of the order of several hundreds of ppm and the magnetic field high enough for rendering the product $(\nu_0 \Delta\sigma)^2$ comparable to the inverse of τ_c.

4.1 THEORETICAL BACKGROUND

The various molecular motions described in the previous sections of this chapter can be probed by dynamical (or relaxation) parameters as already mentioned in the introductory chapter of this book. It was explained, from the Bloch equations, that the time constants governing the evolution of either longitudinal magnetization or transverse magnetization enable one to obtain some pieces of dynamical information, possibly on different timescales depending on the frequency at which the measurement is performed. However, meaningful information can be derived provided some caution is exercised. First, it should be realized that Bloch equations are strictly valid in only a few situations which are intimately related (i) to the nature of the spin system (in principle a single spin 1/2 nucleus which can be considered as isolated and subjected to randomly varying magnetic fields due to other spins, nuclear or electronic) (ii) to the nature of the interactions (mechanisms) which are responsible for the relaxation phenomena (in this respect, quadrupolar nuclei, in an isotropic medium, do obey Bloch equations in the limit of extreme narrowing, i.e. $\nu\tau_c \ll 1$, where ν is the measurement frequency and τ_c the characteristic time of the considered rotational motion). This is because Bloch equations have been derived within the hypothesis of a spin system possessing two eigenstates, e.g. α and β for a single spin 1/2 nucleus (a quadrupolar nucleus in extreme narrowing conditions and in an isotropic medium can be shown to behave equivalently). As soon as the spin system involves more than two eigenstates, its behaviour or its evolution under relaxation may become much more complicated and hence may require more relaxation parameters to be properly described. In turn, the determination of these additional parameters is generally rewarding because they contain more information about molecular motions. To make this point more clear, let us consider a system of two spin 1/2 nuclei, A and X, for which we limit ourselves to their longitudinal magnetizations (as will be seen later, and as was already apparent in Bloch equations, longitudinal and transverse relaxations are totally decoupled). Let us recall that such a system possesses four eigenstates; the corresponding eigenvectors can be denoted as $\alpha_A\alpha_X$, $\alpha_A\beta_X$, $\beta_A\alpha_X$ and $\beta_A\beta_X$ provided that one is dealing with a weakly coupled system; this implies that, if there exists an indirect coupling J_{AX} between A and X, it is much weaker than the difference of resonance frequencies $(J_{AX} \ll |\nu_A - \nu_X|)$. The important point is the existence of these four eigenstates, related to energy level populations, or in other words to longitudinal magnetization. If we want to describe longitudinal magnetization by a set of operators, we must therefore find out four independent quantities, which, in addition, should be easily detectable. Three of them are obvious: the identity $\hat{E}$, which insures spin conservation and which is of little relevance since it does not evolve; the A longitudinal magnetization $\hat{I}_{A,z}$, and the X longitudinal magnetization $\hat{I}_{X,z}$. A further condition can be imposed for making symmetric the relaxation matrix to be defined below: these operators, which can be denoted by $\{\hat{V}_i\}$ in a general way, should be normal-ized and orthogonal as indicated by the following equation:

$$\mathrm{Tr}(\hat{V}_i \dagger \hat{V}_j) = \delta_{ij} \tag{2.6}$$

Tr means 'Trace of', while δ_{ij} is the Kronecker symbol (equal to 1 if $j = i$ and to 0 otherwise), and the symbol $\dagger$ stands for complex conjugate. The previous relation can be better understood if one refers to the matrices associated with the operators $\hat{V}_i$ and $\hat{V}_j$ and constructed for instance on the eigenvector basis given above. From this remark, it can be easily seen that equation (2.6) is effectively verified for $\hat{E}$, $\hat{I}_{A,z}$ and $\hat{I}_{X,z}$. It can be shown that the missing operator satisfying equation (2.6) with respect to those three latter operators is the product $2\hat{I}_{A,z}\hat{I}_{X,z}$, sometimes called longitudinal order because it can be viewed as A and X longitudinal magnetizations pointing in opposite directions. The next step is to trace out the evolution of longitudinal magnetization under relaxation phenomena, recognizing that longitudinal magnetization is completely described by these four operators. In fact, we are interested in the expectation values of these operators or better in the deviations of these expectation values with respect to their equilibrium values. We are naturally led to the definition of three quantities, which can be dubbed *magnetization modes* [51, 52] and which are listed below.

$$\begin{aligned}
{}^{(a)}v_1 &= \langle \hat{I}_{A,z} - I_{A,\mathrm{eq}} \rangle \\
{}^{(a)}v_2 &= \langle \hat{I}_{X,z} - I_{X,\mathrm{eq}} \rangle \\
{}^{(s)}v_3 &= \langle 2I_{A,z}\hat{I}_{X,z} \rangle
\end{aligned} \tag{2.7}$$

These definitions deserve some comments: (i) The equilibrium value of $2\hat{I}_{A,z}\hat{I}_{X,z}$ is zero. (ii) The superscript (a) or (s) refers to the antisymmetric or symmetric nature of these quantities with respect to the *total spin inversion* operation (which amounts to exchanging the spin functions α and β in the eigenstates; this feature will be of some relevance when considering possible factorization of the relaxation matrix). (iii) It is useless to include the identity which is trivially invariant. Now, as will be seen very shortly, these three modes do not evolve independently, i.e. they couple into each other so that we can write three simultaneous first-order kinetic equations of the form

$$\frac{\mathrm{d}v_i}{\mathrm{d}t} = \sum_j \Gamma_{ij} v_j \tag{2.8}$$

In the present situation (two spin 1/2 nuclei), we must therefore construct a (3×3) matrix (made of the elements Γ_{ij}), called the *relaxation matrix*, which is symmetric owing to the orthogonality and normalization properties of the operators $\{\hat{V}_i\}$ (this latter feature arises from a general property of quantum mechanics).

To this end, we can rely on a convenient tool which is the *density operator $\hat{\sigma}$* [53]. In the present context, it is out of the question to go into the details of this formalism; we shall rather attempt to provide a physical picture and to recognize how relaxation parameters show up. The density operator can be viewed more easily through its associated matrix [54], constructed on the eigenvectors basis $\{\varphi_k\}$ (for instance $\alpha_A \alpha_X$, $\alpha_A \beta_X$, $\beta_A \alpha_X$, $\beta_A \beta_X$). The element σ_{kl} is of the form

$$\sigma_{kl} = \langle C_k C_l^* \rangle \tag{2.9}$$

where C_k is the coefficient of φ_k in the expansion of the function ψ describing an arbitrary state of the system ($\psi = \Sigma_k C_k \varphi_k$). The star means complex conjugate and the brackets denote an ensemble average, that is, over all the spin systems, identical in nature, contained in the sample under investigation. As an example of this statistical concept already introduced in Section 2.1, let us assume that the two spins A and X defining the system belong to a given molecule; the ensemble average must be calculated over all the molecules within the sample. From (2.9), it is clear that σ_{kk} represents the population of the kth energy level, for the arbitrary state described by ψ. The meaning of an off-diagonal element is more subtle. At thermal equilibrium, the phases of C_k and C_l are randomly distributed, hence σ_{kl} ($k \neq l$) is zero. Conversely, if, for the state described by ψ, we find $\sigma_{kl} \neq 0$, this reveals a coherence between k and l. These coherences are classified according to the difference of their $\hat{F}_z$ eigenvalues, where $\hat{F}_z$ is associated with the z component of the total spin operator (here $\hat{F}_z = \hat{I}_{A,z} + \hat{I}_{X,z}$). Hence, with $|(F_z)_{kk} - (F_z)_{ll}| = 0, 1, 2, \ldots$, σ_{kl} will represent zero quantum, one quantum (normally detected in conventional experiments), two quanta, ... coherences, respectively. Apart from the fact that the density operator is representative of the statistical state of the system, its interest lies in the simplicity of its evolution equation (the Liouville–von Neumann equation)

$$\frac{d\hat{\sigma}}{dt} = i[\hat{\sigma}, \hat{\mathcal{H}}'(t)] \tag{2.10}$$

where the symbol [,] stands for the commutator of the two operators involved, i.e. $\hat{\sigma}\hat{\mathcal{H}}' - \hat{\mathcal{H}}'\hat{\sigma}$. The Hamiltonian $\mathcal{H}'(t)$, in the absence of any radio-frequency field, can be decomposed as

$$\hat{\mathcal{H}}'(t) = \hat{\mathcal{H}}(t) + \hat{\mathcal{H}}_0 \tag{2.11}$$

$\hat{\mathcal{H}}_0$ is the 'static' Hamiltonian which includes, for an isotropic medium, the usual Zeeman terms (involving the chemical shifts) and the J coupling terms. What matters here is the time dependent Hamiltonian $\hat{\mathcal{H}}(t)$ which includes the time dependent interactions responsible for relaxation phenomena (see Chapter 1). Another valuable property of the density operator comes from the ability to calculate the expectation value of any operator $\hat{G}$ through the relation

$$\langle \hat{G} \rangle = \mathrm{Tr}(\hat{\sigma}\hat{G}) \tag{2.12}$$

Now, as far as longitudinal magnetization is concerned, $\hat{\sigma}$, as any other operator, can be expanded on the basis constituted by the operators $\{\hat{V}_i\}$. Referring to equations (2.6) and (2.12), we can recognize that the coefficients in this expansion are simply the magnetization modes $v_i(t)$ such as these defined by equation (2.7). Hence

$$\hat{\sigma} - \hat{\sigma}_{eq} = \sum_i v_i(t)\hat{V}_i \qquad (2.13)$$

It is then conceivable that the kinetic equations providing dv_i/dt can be derived from equation (2.10) once it has been solved. For carrying out such a calculation, the form of the Hamiltonian $\hat{\mathcal{H}}(t)$ must be specified. Examination of the various interactions prone to affect the spin system (see next section) leads to a Hamiltonian of the form

$$\hat{\mathcal{H}}(t) = \sum_r \sum_{m=-2}^{2} \left(\hat{A}_r^m\right)^\dagger F_r^m(t) \qquad (2.14)$$

where $\hat{A}_r^m$ is made exclusively of spin operators and F_r^m is a 'space function' characteristic of the interaction (mechanism), the latter being denoted by the subscript r. Although we do not intend to go into details of the calculations, we can mention that both $\hat{A}_r^m$ and F_r^m belong to the class of irreducible tensors. Solving equation (2.10) by means of a second-order time-dependent perturbation treatment and inserting the result into equation (2.7) yields the element Γ_{ij} of the relaxation matrix

$$\Gamma_{ij} = \sum_{r,r'} \sum_m (-1)^m \mathcal{J}^{r,r'}(m\omega_0) T_r\left\{[\hat{A}_r^{-m}, \hat{V}_i][\hat{A}_{r'}^m, \hat{V}_j]\right\} \qquad (2.15)$$

Leaving aside the trace and commutator calculations which do not cause, at least in principle, any problem, we are left with the quantity $\mathcal{J}_{r,r'}(m\omega_0)$ which can be called a generalized *spectral density*, and which is just an extension of the ones introduced in Chapter 1 and Section 1 of this chapter. In an isotropic medium, it can be expressed as

$$\mathcal{J}^{r,r'}(m\omega_0) = \int_0^\infty \langle F_r^0(t) * F_{r'}^0(0)\rangle \exp(-im\omega_0 t)\, dt \qquad (2.16)$$

It can be observed that we are still dealing with the Fourier transform of a *correlation function*. The main difference with expressions given in Chapter 1 arises from the fact that we have formally specified the relaxation mechanisms involved in that spectral density (r and r'), instead of resorting only on randomly fluctuating magnetic fields. In fact, if r and r' are identical, we shall be dealing with an *autocorrelation* spectral density, whereas if $r \neq r'$ this will be a *cross-correlation* spectral density. As indicated above, we shall limit ourselves to two types of cross-correlation spectral densities: (i) those involving two different dipolar interactions sharing one spin (which evidently does not exist in a two-spin system), (ii) and those involving the chemical shift anisotropy of a given spin and a dipolar interaction which affects that spin (this type of cross-correlation can exist in a two-spin system). The last point concerning the expression (2.16) is its frequency dependence: $m\omega_0$ must be understood as ω_{kl}, the energy difference (expressed in rad s^{-1}) between two $\mathcal{H}_0$ eigenstates $|k\rangle$ and $|l\rangle$ such that $\langle k|\hat{A}_r^0|l\rangle \neq 0$; $m\omega_0$ is, in fact, a multiple of the Larmor frequency as already mentioned in Chapter 1. Finally, an especially important property deserves to be stated at this stage; it can be shown that *antisymmenic and symmetric modes can be coupled only via csa–dipolar cross-correlation.*

Let us therefore come back to the two-spin system and let us assume that chemical shift anisotropy of either nucleus is negligibly small. The two antisymmetric modes v_1 and v_2 evolve therefore independently of the symmetric mode v_3 and longitudinal relaxation is fully described by the two simultaneous differential equations (remember that the matrix Γ is symmetric)

$$\frac{dv_1}{dt} = \Gamma_{11}v_1 + \Gamma_{12}v_2$$

$$\frac{dv_2}{dt} = \Gamma_{12}v_1 + \Gamma_{22}v_2$$

(2.17)

which can be recast in a more familiar form, widely known as Solomon equations

$$\frac{d}{dt}(I_{A,z} - I_{A,eq}) = -\left(\frac{1}{T_1^A}\right)(I_{A,z} - I_{A,eq}) - \sigma(I_{X,z} - I_{X,eq})$$

$$\frac{d}{dt}(I_{X,z} - I_{X,eq}) = -\sigma(I_{A,z} - I_{A,eq}) - \left(\frac{1}{T_1^X}\right)(I_{X,z} - I_{X,eq})$$

(2.18)

The relaxation parameters involved in these equations, namely the two specific relaxation rates $R_1^A(=1/T_1^A)$, $R_1^X(=1/T_1^X)$ and the so-called cross-relaxation rate σ are linear combinations of spectral densities as defined by (2.16). It turns out that all mechanisms enter the two specific relaxation rates while σ depends only on interactions involving simultaneously both nuclei A and X, i.e. dipolar interaction or chemical exchange. This latter process will be examined in detail in Chapter 3 and disregarded here. We are consequently left with dipolar interaction, which makes the measurement of σ invaluable because it depends solely on the internuclear distance AX and is affected by a dynamical factor describing the reorientation of the AX direction. The information provided by cross-relaxation rates will be discussed in more detail in the next section. It can be noticed that in the absence of dipolar interaction between A and X, σ is zero and we are dealing simply with Bloch equations.

Now, if the chemical anisotropy either at one nucleus or at both nuclei becomes non-negligible (with the advent of high field spectrometers this is more and more likely), the cross-correlation csa–dipolar spectral density couples the antisymmetric modes into the symmetric mode and we must deal with the complete set of three simultaneous differential equations which can be written in the following form [55] and viewed as an extension of equations (2.18)

$$\frac{d}{dt}(I_{A,z} - I_{A,eq}) = -R_1^A(I_{A,z} - I_{A,eq}) - \sigma(I_{X,z} - I_{X,eq}) - \sigma_A'(2I_{A,z}I_{X,z})$$

$$\frac{d}{dt}(I_{X,z} - I_{X,eq}) = -\sigma(I_{A,z} - I_{A,eq}) - R_1^X(I_{X,z} - I_{X,eq}) - \sigma_X'(2I_{A,z}I_{X,z}) \quad (2.19)$$

$$\frac{d}{dt}(2I_{A,z}I_{X,z}) = -\sigma_A'(I_{A,z} - I_{A,eq}) - \sigma_X'(I_{X,z} - I_{X,eq}) - R_1^{AX}(2I_{A,z}I_{X,z})$$

σ_A' and σ_X' can be expressed as a function csa–dipolar cross-correlation spectral densities (also called *interference terms*) involving chemical shift anisotropy of nucleus A and X respectively; as will be discussed later, these terms yield information about shielding tensors and the way in which they reorient. R_1^{AX} can be considered as the specific rate of the longitudinal spin order, $2I_{A,z}I_{X,z}$, this latter quantity being recalled to be zero at thermal equilibrium.

An especially important case of practical interest occurs whenever transitions of one nucleus are irradiated for decoupling purposes. Let us consider, for instance, the often encountered situation where A is carbon-13, J-coupled to the proton X, and suppose that we investigate the evolution of $I_{A,z}$ under continuous proton decoupling. This implies that $I_{X,z}$ is identically zero, as well as $2I_{A,z}I_{X,z}$. Of course, the second equation in (2.18) and the second and third equations in (2.19) are no longer valid since the driving term associated with the decoupling field should be introduced. The first equation of (2.18) or of (2.19) (the same result will be obtained regardless the importance of interference terms) can be written as

$$\frac{d}{dt}(I_{A,z} - I_{A,eq}) = -R_1^A(I_{A,z} - I_{A,eq}) - \sigma I_{X,eq}$$

which can be recast in a more usable form

$$\frac{d}{dt}(I_{A,z} - I_{A,stat}) = -R_1^A(I_{A,z} - I_{A,stat})$$

with
$$I_{A,stat} = I_{A,eq}\left(1 + \frac{\gamma_X}{\gamma_A}\frac{\sigma}{R_1^A}\right) \quad (2.20)$$

where we have introduced a new 'equilibrium value', or rather a stationary state value, $I_{A,stat}$, for A magnetization under proton decoupling (γ_X and γ_A

stand for gyromagnetic ratios of X and A nuclei respectively). This yields the well-known *Nuclear Overhauser Effect* (NOE) [56], which represents an enhancement of the A magnetization under X decoupling conditions and which is generally quantified by the so-called NOE factor η:

$$\eta = \frac{I_{A,\,\text{stat}} - I_{A,\,\text{eq}}}{I_{A,\,\text{eq}}} \tag{2.21}$$

whose maximum value, equal to $\gamma_X / 2\gamma_A$, occurs if two conditions are fulfilled: (i) extreme narrowing prevails, (ii) R_1^A is dominated by the dipolar interaction between A and X. Another very important feature of equation (2.20) is the retrieval of an expression identical to the Bloch equation relative to the z component of nuclear magnetization. This means that, in such a situation, the specific relaxation rate R_1^A can be determined unambiguously without worrying about possible multi-exponentiality.

Whenever one is dealing with a system involving more than two spins, complexity increases, although, in many instances, extension of Solomon equations (involving as many cross-relaxation terms as there exists couples of interacting nuclei) can constitute a good approximation. Nevertheless, in order to illustrate this increasing complexity but also the wealth of information provided by an exhaustive relaxation study, we discuss below the magnetization modes pertaining to an AX_2 spin system [52, 57], recognizing that the ubiquitous $^{13}CH_2$ grouping should in principle be treated accordingly. Without entering the details of the calculations leading to the construction of those modes and to the establishment of the relaxation matrix, some trends can show up. Since a three-spin 1/2 system involves eight eigenstates, seven 'active' modes must be found (excluding the identity which is irrelevant). It turns out that they are divided into four antisymmetric modes and three symmetric modes which are listed in Table 2.1 together with, for some of them, the corresponding spectra which would be obtained after the application of a read-pulse applied either to A (^{13}C) or to X (protons). Let us recall that, in an isotropic medium, and if the A and X nuclei are J-coupled, A appears in the form of a triplet and X in the form of a doublet. Modes not associated with observable quantities, $^{(s)}v_6$ and $^{(s)}v_7$ of Table 2.1, can only be detected in an anisotropic medium where the X spectrum is made of two doublets.

Examination of the various magnetization modes listed in Table 2.1 deserves some comments (i) Except for $^{(a)}v_1$ and $^{(a)}v_2$, which correspond to the longitudinal magnetizations of A and X respectively, equilibrium values of these magnetization modes are zero. This implies that they can be created only through relaxation phenomena, unless a special spin preparation is used. (ii) As for the AX spin system, antisymmetric modes couple into symmetric modes only via *cross-correlation csa-dipolar spectral densities*. (ii) The two 'classical' magnetization modes $^{(a)}v_1$ and $^{(a)}v_2$ couple into $^{(a)}v_3$ and $^{(a)}v_4$ by *cross-correlation dipolar–dipolar spectral densities*. (iv) If all cross-correlation spectral densities can be disregarded, Solomon-like equations are retrieved for the modes $^{(a)}v_1$ and $^{(a)}v_2$:

$$\frac{d}{dt}\left[\frac{1}{\sqrt{2}}(I_{A,z}-I_{A,eq})\right] =$$

$$-R_1^{A'}\left[\frac{1}{\sqrt{2}}(I_{A,z}-I_{A,eq})\right]-\sqrt{2}\sigma_{AX}\left[\frac{1}{2}(I_{X,z}+I_{X',z}-2I_{X,eq})\right]$$

(2.22)

$$\frac{d}{dt}\left[\frac{1}{2}(I_{X,z}+I_{X',z}-2I_{X,eq})\right] =$$

$$-\sqrt{2}\sigma_{AX}\left[\frac{1}{\sqrt{2}}(I_{A,z}-I_{A,eq})\right]-R_1^{X}\left[\frac{1}{2}(I_{X,z}+I_{X',z}-2I_{X,eq})\right]$$

In the previous equations, the specific rate of A, $R_1^{A'}$, differs from the relaxation rate of (2.17) by the fact that the dipolar interaction with X must be multiplied by two.

Under X decoupling conditions, because the X longitudinal magnetization is identically zero, the previous modes, with the exception of $^{(a)}v_1$, are no longer adequate. $^{(a)}v_3$ and $^{(a)}v_4$ must be substituted by $\sqrt{8/3}(I_{A,z}\mathbf{I_X I_{X'}})$ (with $\mathbf{I_X I_{X'}}=I_{X,x}I_{X',x}+I_{X,y}I_{X',y}+I_{X,z}I_{X',z}$), $^{(s)}v_5$ disappears whereas $^{(s)}v_6$ and $^{(s)}v_7$ must be replaced by a single symmetric mode equal to $\sqrt{2/3}(\mathbf{I_X I_{X'}})$. For describing the evolution of A longitudinal magnetization, one is then left with three modes

$$^{(a)}v_1 = (1/\sqrt{2})(I_{A,z}-I_{A,eq})$$

$$^{(a)}v_2' = [(1/\sqrt{3})\,^{(a)}v_3 + \sqrt{2/3}\,^{(a)}v_4 = \sqrt{8/3}(I_{A,z}\mathbf{I_X I_{X'}})$$ (2.23)

$$^{(s)}v_3' = [(1/\sqrt{3})\,^{(s)}v_6 + \sqrt{2/3}\,^{(s)}v_7] = \sqrt{2/3}(\mathbf{I_X I_{X'}})$$

Since $^{(a)}v_2'$ and $^{(s)}v_3'$ are linear combinations of the magnetization modes in the absence of X irradiation, the kinetic equations can be easily constructed, recognizing, however, that the contribution of mode $^{(a)}v_2$ must be retained in the form of a constant $(^{(a)}v_2 \equiv 2I_{X,eq})$. This latter feature leads to stationary state values, especially for $^{(a)}v_1$, resulting in an NOE enhancement for the A magnetization which would be analogous to the one given by equations (2.20) and (2.22) *in the absence of cross-correlation dipolar–dipolar spectral densities*. It can however be pointed out that the latter have a negligible effect if the specific relaxation rate of X is larger than the specific relaxation rate of A, condition which is usually met in practice $(R_1(\text{protons}) \gg R_1(\text{carbon}))$. A similar situation occurs for the evolution of A longitudinal magnetization which, to a good approximation (again if $R_1^X \gg R_1^A$), depends essentially on the specific longitudinal relaxation rate of A. Thus, in most practical situations, the ^{13}C longitudinal magnetization, under proton irradiation, behaves classically even if it is J-coupled and interacting with two or more protons (the above considerations can be extended to a spin system more complex than the CH_2 grouping).

Table 2.1 Magnetization modes of an AX_2 spin system, devised for studying longitudinal relaxation. Superscripts indicate their antisymmetric (a) or symmetric (s) property with respect to total spin inversion

Magnetization mode	$(\pi/2)_A$	$(\pi/2)_X$
Equilibrium		
$^{(a)}v_1 = \dfrac{1}{\sqrt{2}}\,(I_{A,z} - I_{A,eq})$		
$^{(a)}v_2 = \dfrac{1}{2}\,(I_{X,z} + I_{X',z} - 2I_{X,eq})$		
$^{(a)}v_3 = \sqrt{8}\,(I_{A,z}I_{X,z}I_{X'z})$		
$^{(a)}v_4 = 2\,I_{A,z}\,(I_{X,x}I_{X'x} + I_{X,y}I_{X'y})$		
$^{(s)}v_5 = I_{A,z}(I_{X,z} + I_{X'z})$		
$^{(s)}v_6 = \sqrt{2}\,(I_{X,z}I_{X',z})$		
$^{(a)}v_7 = (I_{X,x}I_{X',x} + I_{X,y}I_{X',y})$		

A more general approach for studying nuclear spin relaxation (longitudinal as well as transverse) stems from the Redfield theory [58] which provides an evolution equation for each element of the density matrix.

$$\frac{d}{dt}\,\sigma_{aa'}(t) = i(\omega_{a'} - \omega_a)\sigma_{aa'}(t) + \sum_{b,b'} R_{aa',bb'}\sigma_{bb'}(t) \qquad (2.24)$$

In (2.24), the density matrix is implicitly expressed in the basis of the static Hamiltonian eigenvectors, so that $\omega_{a'}$ and ω_a represent the energy (in rad s^{-1}) of eigenstates $|a'\rangle$ and $|a\rangle$. The first term in the right-hand side of (2.24) stands for the evolution of coherence $\sigma_{aa'}$ under precession, whereas the second term involves the relaxation parameters $R_{aa',bb}$ which are linear combinations of spectral densities. It can be demonstrated that $R_{aa',bb'}$ is significantly different from zero only if $|\omega_a - \omega_{a'}| \approx |\omega_b - \omega_{b'}|$. An immediate consequence is the absence of coupling between coherences ($|b\rangle \neq |b'\rangle$) and populations ($|a\rangle = |a'\rangle$). In other words, *longitudinal relaxation and transverse relaxation are well separated processes.* For a coherence whose frequency ($\omega_{a'} - \omega_a$) is significantly different from all other coherence frequencies (e.g., for one-quantum coherences, this corresponds to a line not overlapping with other lines), the evolution equation is especially simple

$$\frac{d}{dt}\sigma_{aa'} = \left[i(\omega_{a'} - \omega_a) + R_{aa',aa'}\right]\sigma_{aa'} \qquad (2.25)$$

and predicts a monoexponential transverse relaxation which would be of particular value. Unfortunately, field inhomogeneities have not been accounted for in the above equation. In order to circumvent their effects, special experimental procedures described later (spin echo or relaxation in the rotating frame by a radio-frequency spin-locking field) must be devised; they introduce a lot of complications which render transverse relaxation less appealing than it would appear at first sight from equation (2.25).

Returning to longitudinal relaxation amounts to recasting equation (2.24) for populations $P_a = \sigma_{aa}$

$$\frac{dP_a}{dt} = \sum_b W_{ab} P_b \qquad (2.26)$$

where the familiar transition probabilities have been introduced; they possess the following properties $W_{ba} = W_{ab}$ and $W_{aa} = -\Sigma_{b \neq a} W_{ab}$. It can be noticed that equation (2.26) concerns solely longitudinal magnetizations (because they are expressed only according to eigenstate populations). Therefore, the kinetic equations relative to magnetization modes presented above could simply be obtained from equation (2.26) by means of a unitary transformation. It can, however, be recalled that the interest of magnetization modes, with respect to the compact form of equation (2.26), arises from a better physical insight (since they are related to observable quantities) and from possible factorization of the relaxation matrix.

4.2   MEASUREMENTS OF RELAXATION PARAMETERS

For resorting to the relaxation parameters mentioned above and to their subsequent interpretation in terms of molecular motions, the prerequisite is that reliable experimental methods are available. The most commonly used are presented in this section.

4.2.1 Specific relaxation rates under monoexponential decay or recovery

We shall be dealing here with the traditional measurement of the three classical relaxation rates, R_1 (longitudinal relaxation rate), R_2 (transverse relaxation rate) and $R_{1\rho}$ (relaxation rate of magnetization 'locked' by a radio-frequency field), whenever magnetization decay or recovery obeys a simple monoexponential law, that is when Bloch equations apply (see Section 4.1 for the circumstances under which these equations either are strictly valid or constitute a good approximation).

4.2.1.1 Longitudinal relaxation rate R_1

It is by far the most reliable relaxation parameter. Its determination requires an initial perturbation of the longitudinal magnetization, followed by an evolution period of duration τ during which the system recovers exponentially according to R_1 and which ends up by a read-pulse which samples the partially relaxed magnetization as a function of τ. In order to avoid corruption of the measured magnetization, transverse magnetization (if any) must be destroyed prior to the read-pulse. This latter condition is automatically satisfied if one relies on the widely used inversion-recovery method, which will be the only one (with some of its variants) discussed here. It has the general form (Figure 2.1)

$$(\pi) - \tau - (\pi/2) - \text{Acq} \tag{2.27}$$

and is run for a series of τ values. The initial π inverting pulse insures that one is dealing solely with longitudinal magnetization and, from the Bloch equation, the measured magnetization, resulting from the Fourier transform of the free induction decay (fid) following the $\pi/2$ read-pulse, has the form

$$M(\tau) = M_{eq}[1 - 2\exp(-R_1\tau)] \tag{2.28}$$

The method applies obviously to a multiline spectrum since the Fourier transform yields the recovery of each line in the spectrum [59]. Its major

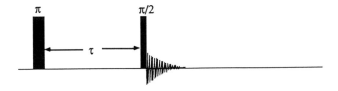

Figure 2.1 Scheme of the basic inversion-recovery sequence

drawback arises from the necessity of waiting complete return to thermal equilibrium before the next measurement; this takes about five times the longest T_1 (in order to retrieve about 99% of the equilibrium magnetization). A minor disadvantage stems from the imperfections of the inverting pulse, which can be accounted for by a parameter k ($k \leqslant 2$) which replaces the factor 2 in front of the exponential. This parameter must be included in the fitting procedure [60] of the evolution equation which becomes

$$M(\tau) = M_{eq}[1 - k \exp(-R_1\tau)] \qquad (2.29)$$

It turns out that this equation holds as well if a waiting time of 5 T_1, for a given τ value, is not respected. This is the basis of the Fast Inversion Recovery Fourier Transform (FIRFT) method [61] which allows a longitudinal relaxation time determination to be run in a reasonable measuring time when a large number of accumulations is required. (This is the case for carbon-13 in natural abundance, for which sensitivity is poor and longitudinal relaxation times are relatively long.) A still more efficient method relies on the measurement of a reference fid and a partially relaxed fid, in non-equilibrium conditions. The method dubbed SUFIR (for SUper Fast Inversion Recovery) [62] can be schematized as follows (Figure 2.2)

$$\left[\left(\frac{\pi}{2}\right)(\tau \text{ including Acq}_1) - (\pi) - (\tau) - \left(\frac{\pi}{2}\right)(\tau \text{ including Acq}_2)\right]_n \qquad (2.30)$$

Both acquisitions are stored in separate blocks (with the relevant signal amplitudes denoted below by S_1 and S_2) and a proper phase cycling avoids any corruption from transverse magnetization. This sort of one-shot T_1 determination is obtained from S_1 and S_2 by the relation

$$T_1 = -\frac{\tau}{\ln\left(1 - \dfrac{S_2}{S_1}\right)} \qquad (2.31)$$

The method is perfectly reliable provided that τ lies in the range 0.5 T_1–3T_1.

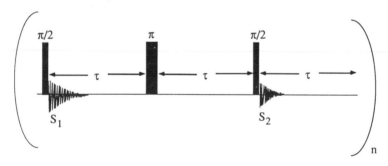

Figure 2.2   Scheme of the super fast inversion-recovery sequence

Problems arise when ^{13}C or ^{15}N relaxation times of large molecules (biomolecules) are to be measured. These problems are twofold: the poor sensitivity (even when one is dealing with enriched material) and overlap of lines. Both difficulties can be circumvented by *two-dimensional inverse detection techniques*. Let us first recall the basic principles of these techniques, known under the acronyms of HSQC (Heteronuclear Single Quantum Correlation) [63] or HMQC (Heteronuclear Multi Quantum Correlation) [64]. They consist essentially in detecting the A nucleus (^{13}C, ^{15}N,) indirectly at the proton measurement frequency (with an obvious sensitivity gain) after that proton magnetization has been transferred by one means or the other to the A nucleus, in order to still enhance the sensitivity of the experiment by a factor equal to γ_H/γ_A. During the evolution period (t_1) magnetization is labelled according to A chemical shifts, whereas the detected fid during t_2 involves only proton chemical shifts. After a double Fourier transform with respect to t_1 and t_2, one obtains cross-peaks indicating correlations between a given A nucleus and the proton to which it is J-coupled. In addition to the correlation information, the two-dimensional map exhibits a spread of cross-peaks such that an effective separation of otherwise overlapping lines is generally obtained. This feature is in fact essential for accessing to individual relaxation parameters of unresolved lines in the one-dimensional spectrum. We shall focus here to the HSQC experiment which will lend itself to the determination of relaxation parameters, better than does the HMQC experiment. Rather than providing all the details of the actual pulse sequence, it may be more instructive to indicate schematically its major steps [65]:

[transfer from H to A]—[evolution under A chemical shift (t_1)]—[back transfer from A to H]—[acquisition according to H chemical shift (t_2) under possible A decoupling: $S(t_1, t_2)$]

It is perfectly clear that the double Fourier transform of $S(t_1, t_2)$ yields A chemical shift information along the ν_1 dimension and H chemical shift information along the dimension ν_2. Both transfer stages are made possible only by the existence of a J-coupling between H and A, limiting in practice the method to A nuclei directly bound to proton(s). They are performed by one of the most common transfer procedures: INEPT, or DEPT [53]. The method for measuring the longitudinal relaxation time of A is usually of the inversion-recovery type and is inserted after the first transfer stage (H to A); this includes an A inverting pulse followed by an evolution interval τ during which A recovers according to T_1. In order to make sure that the recovery is monoexponential, proton decoupling is applied during the τ interval. A magnetization evolves during t_1 with an amplitude dictated by its recovery during τ which is also reflected in the final signal $S(t_1, t_2)$. The whole experiment is repeated for as many τ values as required for a proper determination of the longitudinal relaxation time of each A nucleus.

4.2.1.2   *Transverse relaxation rate R_2 and spin-lock relaxation rate $R_{1\rho}$*

It will be seen in Section 5 that transverse relaxation rates (or, to a certain extent, the spin-lock relaxation rates) provide information complementary to that afforded by longitudinal relaxation rates. As already mentioned, if Bloch equations apply, the transverse relaxation time could be deduced simply from the fid decay or, in the frequency domain, from the linewidth $\Delta\nu_{1/2}$ through

$$T_2 = \frac{1}{\pi\Delta\nu_{1/2}} \tag{2.32}$$

Unfortunately, the measured linewidth is generally corrupted by the inhomogeneity of the static magnetic field B_0, resulting in an apparent transverse relaxation time T_2^*, shorter than the true T_2. Thus, reliable transverse relaxation time determination from linewidths are limited to systems for which field inhomogeneity contributions are negligible, with respect to the natural linewidth (e.g. quadrupolar nuclei) or, possibly, when the so-called 'ultra high resolution' methodology is employed [66]. When this is not the case, one must rely on techniques which annihilate field inhomogeneities. The well-known spin echo technique (or Hahn sequence in Figure 2.3) fulfils in principle this requirement, thanks to the refocusing properties of (π) radio-frequency pulse:

$$\left(\frac{\pi}{2}\right) - \tau - (\pi) - \tau - \text{Acq} \tag{2.33}$$

As explained in more detail in Chapter 7, any precession effect is cancelled by the end of the second τ interval, leading to the formation of an echo. This means that nuclear magnetization is identical to what it was just after the initial $(\pi/2)$, except for an attenuation factor due to relaxation. Since precession arises from chemical shifts as well as from field inhomogeneities, the latter effects are automatically removed. If signal acquisition is started just at the end of the second interval τ, one obtains a fid corresponding to the second half of the echo whose Fourier transform yields a spectrum with line intensities attenuated according to $\exp(-2\tau/T_2)$ where the decay time constant is actually the true T_2. Repeating the experiment for a series of τ values and fitting the amplitude decay versus $\exp(-2\tau/T_2)$ yields in principle the transverse relaxation rate. There are, however, some drawbacks to this simple pulse sequence: (i) imperfections of the central π pulse may yield erroneous results,

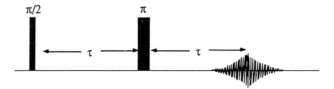

Figure 2.3   Scheme of the spin-echo sequence

(ii) B_0 inhomogeneities, which act as gradients, induce further signal attenuation owing to translational diffusion (see Chapter 7) which can be accounted for by the factor $\exp[(-8/3)(\gamma^2 g^2 D\tau^3)]$, where γ is the gyromagnetic ratio of the considered nucleus, g the gradient amplitude and D the self-diffusion coefficient; (iii) if the spin system involves J couplings, the echo amplitude is modulated according to the quantity $J\tau$ in a more or less complicated manner, depending on the nature of the spin system. Ill-effects corresponding to points (i) and (ii) are suppressed or greatly attenuated by the CPMG pulse train described below. There are few remedies (if any) to point (iii) [67]. Moreover, it can be emphasized that coupled spin systems lead to a lot of additional complications (for instance through the incidence of cross-relaxation or cross-correlation terms [68]) in such a way that spin echo techniques have been seldom considered for determining their transverse relaxation properties. The Carr–Purcell–Meiboom–Gill (CPMG) method [69] consists in applying a train of 180° pulses (instead of a single refocusing pulse) in the following manner

$$\left(\frac{\pi}{2}\right)_x—\tau—\left[(\pi)_y—\tau—(A)—\tau\right]_n \qquad (2.34)$$

Analyzing this sequence leads to the conclusion that a spin echo is formed at each point denoted by (A) in (2.34), whose amplitude decays according to $\exp(-t/T_2)$, where t is the time elapsed from the initial $(\pi/2)$ pulse. The advantages of the method are twofold: (i) the phase of the π pulses (which are applied along the y axis of the rotating frame whereas the initial $\pi/2$ pulse is applied along the x axis) can be shown to compensate for π pulse imperfections (including misadjustment of their length and effects of radio-frequency field inhomogeneities); (ii) for an identical evolution duration t, translational diffusion effects are reduced with respect to the Hahn sequence; in that case, the attenuation factor was given by $\exp[-(\gamma^2 g^2 Dt^3)/3]$ whereas it becomes $\exp[-(\gamma^2 g^2 Dt^3)/3n^2]$ for the CPMG sequence.

As will be seen later, the so-called spin-lock relaxation time (or relaxation time in the rotating frame) is in many instances very close (or even identical) to the transverse relaxation time. In non-coupled spin systems, its measurement is straightforward and not prone to difficulties mentioned above about spin-echo techniques. The sequence is quite simple (Figure 2.4)

$$\left(\frac{\pi}{2}\right)_x \left[(SL)_y\right]_\tau \text{Acq} \qquad (2.35)$$

where $[(SL)_y]_\tau$ means that a radio-frequency field is applied along the y axis of the rotating frame for a time τ. This rf field 'locks' the nuclear magnetization which has been flipped to the y axis of the rotating frame by the initial $(\pi/2)_x$ pulse, avoiding any precession due to B_0 inhomogeneity. Since magnetization is out of equilibrium, it decays according to $\exp(-\tau/T_{1\rho})$ with

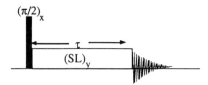

Figure 2.4   Scheme of the spin-lock sequence for measuring $T_{1\rho}$

$R_{1\rho}(=1/T_{1\rho})$ being the relaxation rate in the presence of an rf spin-locking field. The only inconvenient of the method is the necessity of using a strong rf field for preventing offset effects in multiline spectra [70]. Because this rf field is applied for relatively long periods (in order to make the decay observable), the method may be instrumentally demanding.

Finally, it can be noticed that the CPMG pulse train or alternatively a spin-locking field can be inserted in inverse detection schemes [71] in the same manner as the inversion-recovery subsequence (see the previous section). In that case, however, proton decoupling during the evolution interval should be avoided. The extra line broadening caused by a necessarily imperfect decoupling may alter the result of the CPMG method; on the other hand, the application of an rf field to proton transitions could produce unwanted magnetization transfers (of the Hartmann–Hahn type) during the spin-lock interval.

4.2.2   Cross-relaxation rates (systems obeying the Solomon equations)

As we have seen in Section 4.1, when two nuclei interact by dipolar coupling, the evolution of their magnetizations (longitudinal or transverse) can no longer be accounted for by separate kinetic equations. Rather, their longitudinal magnetizations on the one hand, and their transverse spin-locked magnetizations on the other hand, are coupled through a term called cross-relaxation rate. The interest in determining cross-relaxation rates is obvious since they are solely dependent on dipolar interactions, leading directly to structural and/or dynamical information, without worrying about several possible relaxation mechanisms as this is the case for the specific relaxation rates considered in the previous section. Some care must nevertheless be exercised since other coupling terms (cross-correlation terms) may in some instances also interfere and the experimental procedures must be devised in order to cancel their effects.

4.2.2.1   Longitudinal cross-relaxation rates

It was recognized very early that the existence of homonuclear cross-relaxation rates is indicative of a certain proximity of the two relevant nuclei (through their dependence in $1/r^6$, where r is the distance between the two interacting nuclei—see Section 5). The cross-relaxation rate manifests itself through the

Nuclear Overhauser Effect (NOE) which can be easily detected by comparing a reference spectrum (obtained in standard conditions) and a spectrum obtained under saturation conditions (by applying continuously a selective rf field) of a given nucleus (say X). If cross-relaxation between X and A exists, the line intensities of A will be enhanced by the factor given by equation (2.20) so that, in the homonuclear case (identical γ's), the maximum enhancement factor is of 1.5. Another one-dimensional procedure, commonly employed and known as transient NOE, consists in selectivity inverting the X resonances and looking at the build-up of A longitudinal magnetization by cross-relaxation. Referring to Solomon equations (2.18), we can recognize that the initial behaviour yields directly the cross-relaxation rate σ between A and X. Let τ the time elapsed between the X inverting pulse and the read pulse: A magnetization, in the limit of short τ values, can be written as

$$I_{A,z}(\tau) = I_{A,z}(0) + \tau \left(\frac{\mathrm{d}}{\mathrm{d}t} I_{A,z} \right)_{t=0}$$

which, from equations (2.18) yields

$$I_{A,z}(\tau) = I_{eq}(1 + 2\sigma\tau) \tag{2.36}$$

where I_{eq} is the magnetization equilibrium value, common to both A and X nuclei.

The method presents two advantages. First, the factor of two in front of σ enhances the sensitivity to the cross-relaxation rate, provided that τ is kept to a sufficiently short value for fulfilling the initial behaviour conditions. Secondly, if several nuclei are involved, the result of equation (2.36) is not corrupted by cross-relaxation with additional nuclei since the latter would indirectly interfere at longer τ values. This is the so-called spin diffusion phenomenon [72] which can be schematized as follows:

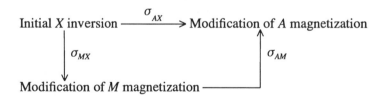

The above representation suggests that spin diffusion implies a two-stage transfer depending roughly on τ^2 which can therefore be minimized (or made negligible) for sufficiently short τ values. It can be recognized that spin diffusion is fully included in the result of the selective saturation method. The selective inversion (or in a more general way, the selective perturbation) method should therefore be preferred. It constitutes in fact the origin of the widely used two-dimensional NOESY (Nuclear Overhauser Effect SpectroscopY) experiment which can be sketched as follows [73]:

$$\left(\frac{\pi}{2}\right) - t_1 - \left(\frac{\pi}{2}\right) - t_m - \left(\frac{\pi}{2}\right) \text{Acq}(t_2) \tag{2.37}$$

In this sequence, the evolution period (t_1, the first dimension) serves as chemical shift labelling, in the sense that longitudinal magnetization after the second ($\pi/2$) pulse is modulated according to $\cos(2\pi\nu_A t_1)$, where ν_A is the resonance frequency of the A spin. This is the initial perturbation of the relevant longitudinal magnetization which replaces the selective inversion of the one-dimensional experiment described above. Then comes a mixing time t_m which is the equivalent of τ in equation (2.36). Any transverse magnetization during t_m is supposed to be eliminated either by a physical means (spoiling gradient pulse) or by a proper phase cycling. The last ($\pi/2$) pulse enables to 'read' longitudinal magnetizations which have evolved under their specific relaxation rates *and* under cross-relaxation rates. As a matter of fact, a double Fourier transform of the signal $S(t_1, t_2)$ provides diagonal peaks whose intensity reflects specific relaxation rates and cross-peaks indicative of cross-relaxation between the two relevant nuclei and which can provide quantitative information about the corresponding rate, thus about the internuclear distance. As everyone knows, this is now the method of choice for obtaining structural information in large biomolecules.

Finally, it can be mentioned that these methods can be transposed to the heteronuclear case. For instance, the popular ^{1}H–^{13}C NOE determination proceeds from the observation of ^{13}C resonances: (i) under continuous proton decoupling which yields I_{stat} (see equation 2.20) including the NOE enhancement, (ii) under decoupling gated on only during acquisition which provides I_{eq}. Comparison of I_{stat} and I_{eq} [74] leads unambiguously to the proton-carbon cross-relaxation rate in the case of protons directly bound to the considered carbon (remote protons have negligible effects due to the $1/r^6$ dependence). Heteronuclear NOE two-dimensional methods are also in use: on the one hand, the HOESY experiment [75] which proceeds from the observation of the heteronucleus (homologous to the NOESY experiment) and, on the other hand, the measurement of the heteronuclear cross-relaxation rate by inverse detection [76] in a manner similar to that described in Section 4.2.1 (insertion of the relevant subsequence in a HSQC experiment). It can be mentioned that the HOESY experiment proves to be very convenient for sorting out different heteronuclear cross-relaxation contributions, for instance in the case where the heteronucleus is not directly bonded to a proton and is therefore subjected to several dipolar interactions of comparable magnitude.

4.2.2.2 *Transverse cross-relaxation rates*

The phenomenon of cross-relaxation for pure transverse magnetizations does not generally exist because resonance frequencies of the two interacting nuclei are different and thus do not fulfil the conditions derived by Redfield [58] (see page 46 of this chapter). Nevertheless, by applying spin-locking procedures (relaxation in the rotating frame), all resonance frequencies of a homonuclear spin system become identical and cross-relaxation can take place with, moreover, no influence from static field inhomogeneities. The cross-relaxation rate, measured in that way and which is denoted by σ_ρ, is actually called the transverse cross-relaxation rate. The interest in σ_ρ lies in the fact that it is always positive (see Section 5); this is in contrast with the longitudinal cross-relaxation rate which may become negative in the case of slow molecular tumbling and which may get close to zero, a situation which precludes the observation of cross-relaxation.

The one-dimensional ROE (Rotating Frame Overhauser Effect) experiment mimics the selective NOE experiment with the two magnetizations of interest in opposite directions along the spin-locking axis (instead of the z axis). A possible sequence [77] for performing such a measurement can be devised as follows:

$$\left(\frac{\pi}{2}\right)_x - \tau' - \left[(SL)_y - \tau\right]\text{Acq} \tag{2.38}$$

If we assume that A is on-resonance (its resonance frequency coinciding with the carrier frequency) and that τ' is set to $1/|v_A - v_X|$, A magnetization remains actually along $+y$ whereas X magnetization is taken to $-y$ by the end of the precession interval τ', so that the relative configuration of A and X magnetizations is optimal for transfer by cross-relaxation in the rotating frame. The homologous two-dimensional experiment is simply obtained by substituting to the fixed interval τ', the incremented time t_1; usually, the spin-locking field application (τ in (2.38)) is denoted by t_m as in the NOESY experiment. This two-dimensional sequence is coined CAMELSPIN [78] or ROESY [79] and is almost systematically employed as a complement to the NOESY experiment. For both sequences, one must worry about possible transfers by J couplings (as in COSY or TOCSY experiments). Remedies have been proposed to circumvent these problems [80].

4.2.3   Cross-correlation spectral densities

Cross-correlation is prone to complicate spin relaxation measurements as soon as more than a single relaxation mechanism (interaction) exists within the spin system. Although, at first sight, this should be the common rule, such an occurrence can fortunately be easily predicted on the basis of the expected strength of the various interactions affecting the spin system. For instance, minor interactions such as spin-rotation or intermolecular dipolar interactions are generally gathered in the so-called 'random field' relaxation mechanism and it can be shown by symmetry considerations that no cross-correlation can occur between random fields and the almost ubiquitous intramolecular dipolar interaction (that is magnetic dipole–magnetic dipole interactions within the spin system). As already mentioned in section 4.1, we shall only consider the two types of cross-correlation between, 'chemical shift anisotropy' (csa) of one spin and a dipolar interaction involving that spin on the one hand, and between two different dipolar interactions sharing a common spin (for instance the AX and AX' dipolar interactions in an AX_2 spin system) on the other hand. It may be worth recalling that, as far as longitudinal relaxation is concerned, csa-dipolar cross-correlation spectral densities couple antisymmetric *and* symmetric magnetization modes whereas dipolar–dipolar cross-correlation appear as coupling terms between the usual longitudinal magnetizations $(I_{A,z}, I_{X,z}...)$ and the other modes required for a complete description of the relaxation in a multispin system.

The general way of determining cross-correlation spectral densities, which may lead to interesting dynamical or structural information, is to analyse thoroughly the complete evolution of the observable quantities after an appropriate initial perturbation has been applied to the spin system. This method relies on the multiexponential character of the recovery or build-up curves (which reflects the fact that several simultaneous differential equations govern the kinetics of the spin system) and requires efficient numerical fitting procedures. A more direct method and maybe more accurate consists in devising a special spin preparation such that the cross-correlation spectral density of interest can be deduced from the initial behaviour of the evolution of an observable. Both methodologies will now be discussed for the csa–dipolar and dipolar–dipolar cross-correlation spectral densities.

4.2.3.1 Chemical shift anisotropy–dipolar cross-correlation spectral densities

In order to explain the experimental procedures aimed at the detection and eventually at the determination of the csa–dipolar cross-correlation spectral density (it may be recalled it is often dubbed interference term) we shall rely on the simple system of two spin 1/2 nuclei (J-coupled) and refer to equation (2.19). There are two experimental manifestations of the existence of an interference term: if the chemical shift anisotropy at nucleus A is not negligible, one may observe a differential broadening of the A doublet (as far as B_0 inhomogeneities do not mask this phenomenon); likewise, a standard inversion-recovery experiment, with however a read-pulse significantly lower than 90°, yields differential recoveries for each line in the doublet [81]. This similar behaviour, for transverse and longitudinal magnetizations, can be understood on the basis of equation (2.19), which only concerns longitudinal magnetization, recognizing that transverse magnetizations obey homologous equations with possibly different frequencies involved in the relevant spectral densities. Owing to the problems inherent to pure transverse magnetization (B_0 field inhomogeneities, modulations by scalar couplings) which can neverthe-less be overcome by relaxation in the rotating frame, we shall rather focus on longitudinal relaxation. The clue of csa–dipolar interference term arises from the coupling of classical longitudinal magnetization, say $I_{A,z}$, with the two-spin order $(2I_{A,z}I_{X,z})$, the coupling term being precisely the relevant (csa–dipolar) spectral density. Of course, one could rely on an accurate analysis of the multiexponential behaviour of $I_{A,z}$ which, in principle, is able to sense the creation of $(2I_{A,z}I_{X,z})$ through the interference term. However, this is a *second-order* effect (since, at thermal equilibrium $(2I_{A,z}I_{X,z})$ is zero), which manifests itself at rather long mixing times and which is necessarily hampered by experimental uncertainties. As a matter of fact, unless the multiexponential character is strongly apparent, the most reliable way to measure a relaxation rate stems from the initial behaviour (initial slope) of a given observable. For instance, one may monitor the build-up of $(2I_{A,z}I_{X,z})$ via the interference term. This is indeed a first-order process, since this build-up starts from zero and is produced via the non-equilibrium value of $I_{A,z}$ (which has for instance been inverted; see equation (2.19)). We must therefore find out a means of converting the non-observable $(2I_{A,z}I_{X,z})$ quantity into an observable quantity. At this point, a distinction between hetero- and homonuclear spin systems must be made. Let us consider first an heteronuclear spin system and let us suppose that we perform first an inversion-recovery experiment, acting only on A

$$(\pi)_x[A] - \tau - (\pi/2)_x[A] - \text{Acq}$$

which allows us to monitor the A longitudinal magnetization as a function of τ. In fact, in addition to $I_{A,y}$ which arises from the perturbed A longitudinal magnetization, the $(\pi/2)_x[A]$ pulse generates the quantity $(2I_{A,y}I_{X,z})$ created by the interference term. The trick [82] is to subtract from this fid, the one obtained by the following sequence

$$(\pi)_x[A] - \tau - (\pi)[X](\pi/2)_x[A] - \text{Acq}$$

This sequence yields again $I_{A,y}$ and now $-(2I_{A,y}I_{X,z})$ because of the X inverting pulse. Owing to the sign of acquisition, the whole experiment provides therefore $(2I_{A,y}I_{X,z})$ which is indeed observable in the form of an antiphase doublet and which reflects the build-up of $(2I_{A,z}I_{X,z})$ by relaxation. Examination of equation (2.19) indicates that *the initial slope* provides directly the coupling term between the A longitudinal magnetization and the longitudinal two-spin order, which is precisely the *csa-dipolar interference term*. A more elaborate method [83], though along the same lines, leads to an in-phase doublet allowing decoupling of the X nucleus which, together with a reference spectrum, makes possible the determination of the csa–dipolar cross-correlation spectral density with quite good accuracy. Two-dimensional experiments in the inverse mode (see Section 4.2.1.1) can be devised for monitoring the creation of $(2I_{A,z}I_{X,z})$ mediated by the interference term [84]. Of course, the analysis of complete recovery or build-up curves obtained under various spin preparations can also yield the interference term. These spin preparations may include the four following combinations: A or X inversion and A or X

observation [85], separated by an evolution interval as in a standard inversion-recovery experiment. Csa–dipolar cross-correlation terms may alter the T_1 and T_2 measurements of heteronuclei in the inverse mode (see Sections 4.2.1.1 and 4.2.2.2). These effects are more pronounced in the case of transverse relaxation determination and can be annihilated by inverting ^{1}H pulses applied during the CPMG sequence [86].

It turns out that the longitudinal interference term becomes negligibly small when the molecular tumbling slows down, as in the case of large biomolecules. In such a situation, one may rely on *transverse* interference terms which, as for standard ROESY experiments, do not vanish. The method, dubbed ortho-ROESY [87], makes use of a spin-locking period applied only to one of the two involved spins. Along the same lines, a two-dimensional experiment has been devised for measuring the interference term between the csa of $X(^1H)$ and the $A-X$ dipolar interaction (A being an heteronucleus); the result is displayed in the ν_1 dimension according to A chemical shifts [88]. During the spin-lock period $2I_{X,x}I_{A,z}$ is created via the interference term from $I_{X,x}$. By means of two 90° pulses applied simultaneously to A and X, $2I_{X,x}I_{A,z}$ is converted into $2I_{X,z}I_{A,y}$ which evolves during t_1 according to A chemical shifts. This antiphase coherence is converted back by a further pair of 90° pulses into $-2I_{X,y}I_{A,z}$ for being observed as a proton signal, that is with an enhanced sensitivity.

The major difference between heteronuclear and homonuclear systems arises from the fact that, in the latter case, non-selective pulses are normally used, which affect both nuclei A and X. Consider for instance a $(\pi/2)_x$ pulse; it will convert the quantity of interest, namely $2I_{A,z}I_{X,z}$ which carries the interference term information, into $2I_{A,y}I_{X,y}$ which is unobservable. Therefore, the standard inversion-recovery experiment $(\pi) - \tau - (\pi/2)$ Acq. does not provide any measurement of csa-dipolar cross-correlation. There are three ways of unravelling this contribution (i) using a selective read-pulse, say a $(\pi/2)_x(A)$ acting solely on A magnetization which will convert $2I_{A,z}I_{X,z}$ into $2I_{A,y}I_{X,z}$ which is observable and which adds to $I_{y,A}$ (coming from the specific relaxation of A and possibly from cross-relaxation with X) as an antiphase doublet. This leads to an A doublet whose lines relax in a different way. (ii) a similar result, although with a reduced sensitivity, can be obtained with a non-selective read-pulse of flip angle θ significantly smaller than 90° which converts $2I_{A,z}I_{X,z}$ into $2(\cos \theta\, I_{A,z} + \sin \theta\, I_{A,y})(\cos \theta\, I_{X,z} + \sin \theta\, I_{X,y})$, the latter expression including the two observable quantities (antiphase doublets) $2I_{A,z}I_{X,y}$ and $2I_{A,y}I_{X,z}$ affected by $\sin \theta \cos \theta$. (iii) In the context of two-dimensional spectroscopy, a double quantum filter can be added at the end of a NOESY type experiment [89, 90]. Again, contributions from $2I_{A,z}I_{X,z}$ are converted into antiphase doublets whose intensities reflect the csa–dipolar cross-correlation term whereas the standard cross-correlation contributions in a NOESY experiment are filtered out since they proceed from one-quantum coherences. As this was already mentioned, interference terms involved in longitudinal relaxation may become vanishingly small in the slow-motion limit. The remedy is thus to turn to relaxation in the rotating frame. Such experiments have indeed been proposed for homonuclear systems where both spin-locking periods and read-pulses are selective [91].

4.2.3.2   Dipolar–dipolar cross-correlation spectral densities

Literature concerning the measurements of dipolar–dipolar cross-correlation spectral density is abundant and can be found in a review article by Grant *et al.* [57]. We shall first outline the methodology generally applied by these authors to a $^{13}CH_2$ grouping. The reader is referred to the AX_2 spin system which has been dealt with in Section 4.1 within the frame of magnetization modes. The following experiments are aimed at disturbing from their thermal equilibrium value a sufficient number of magnetization modes (invariably carbon-13 is observed without proton decoupling at any stage of the sequence):

(i)   a normal inversion-recovery experiment applied to carbon-13 $(^{(a)}v_1$ inverted; see Table 2.1);

(ii)   a carbon-13 recovery experiment with an initial perturbation applied to protons in the form of a hard pulse $(^{(a)}v_2$ inverted);

(iii) a carbon-13 recovery experiment for which the initial perturbation, in the form of an inverting soft pulse, is applied to one of the lines in the proton doublet (creation of the $^{(s)}v_5$ mode);

(iv)   a carbon-13 recovery experiment where the initial perturbation is made of a pulse sequence used for instance in two-dimensional J spectroscopy and which leads to the creation of $^{(a)}v_3$ mode. This initial perturbation can be schematized as follows (A being the carbon-13, X and X' standing for the two protons in the CH_2 grouping) and is depicted in Figure 2.5

$$(\pi/2)_x(A) - 1/4J_{CH} - [(\pi)_y(A); \pi(X)] - 1/4J_{CH} - (\pi/2)_{\pm x}(A)$$

Carbon chemical shift effects are refocused thanks to the central $(\pi)_y(A)$ pulse whereas the simultaneous application of a proton π pulse enables the evolution due to the J_{CH} coupling to continue until a pure antiphase configuration is

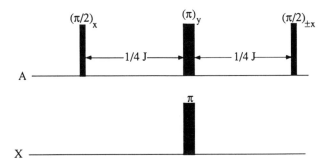

Figure 2.5   Pulse sequence for the creation of the v_3 mode in an AX_2 system

formed (for a total evolution time of $1/2J$). Prior to the last $\pi/2$ carbon pulse, the carbon-13 triplet is thus in an antiphase configuration corresponding to the operator product $4I_{A,y}I_{X,z}I_{X',z}$ (the two outer lines lines of the triplet being in a direction opposite to that of the central line). The last $(\pi/2)_{\pm x}(A)$ pulse yields therefore an initial state, at the beginning of the recovery period, which corresponds to the creation of the mode $^{(a)}v_3$ ($\pm I_{A,z}I_{X,z}I_{X',z}$ omitting the normalization factor). The data relative to each line in the carbon triplet for all these experiments are then analysed by numerical fitting procedures, enabling extraction of all cross-correlation spectral densities of interest (dipolar–dipolar and even csa–dipolar).

Along the same lines, much work has been done for extracting homonuclear dipolar–dipolar cross-correlation spectral densities by analysing multiexponential recovery curves resulting from a single non-selective inversion-recovery experiment [92].

Other methodologies include the direct observation of the three-spin longitudinal order, whose evolution is related to dipolar cross-correlation spectral densities, by dedicated one-dimensional multipulse experiments [93–95] or by using multiple quantum filtration in two-dimensional experiments [91, 96].

5   Relaxation Mechanisms and Applications

Interactions responsible for spin relaxation will be outlined below according to their importance and occurrence, disregarding the quadrupolar contribution which is considered in the Chapter 5 of this book. However, this contribution may be extremely useful for analysing molecular overall motions and will therefore be alluded to if necessary. We shall present below the following relaxation mechanisms: dipolar interaction, chemical shift anisotropy, scalar relaxation of the second kind and spin rotation. The so-called scalar relaxation of the first kind arising from exchange processes is considered in Chapter 3. Other mechanisms, which in some instances may become dominant, as for example dipolar intermolecular interactions or contributions from paramagnetic species, will be gathered in a term denoted by $(R)_{\text{other}}$ and simply added in the relevant relaxation rate. Most nuclear spin relaxation data accumulated for the last twenty years can be found in the exhaustive review of Kowalewski [40]. Hence, only a few illustrative examples will be presented below.

5.1   REDUCED SPECTRAL DENSITIES

Most NMR relaxation parameters, and at least those considered in the present chapter, are linear combinations of *reduced* spectral densities whose expression is deduced from the general form of spectral densities given in Section 1 of this chapter

$$\tilde{J}^{rr'}(\omega) = 4\pi \int_{-\infty}^{+\infty} \left\langle Y_2^0[\beta_r(t)]Y_2^0[\beta_{r'}(0)] \right\rangle \exp\left(-i\omega t\right) \mathrm{d}t \qquad (2.39)$$

Here r and r' may refer to two different directions, relative to the two relevant relaxation mechanisms. Hence, equation (2.39) concerns auto-correlation as well as cross-correlation terms. In the mechanisms we are dealing with, $\tilde{J}^{rr'}(\omega)$ turns out to appear as the Fourier transform of correlation function (see formula (2.16)) of spherical harmonics of rank two and zero projection with the following convention: $Y_2^0(\beta) = \sqrt{5/16\pi}(3 \cos^2\beta - 1)$; for $r = r'$, this corresponds to an autocorrelation function. In general, the angle β_r (or $\beta_{r'}$) indicates the orientation of a molecular vector with respect to the laboratory frame. For instance, this is the internuclear vector in the case of dipolar interaction. The ensemble average, denoted by brackets, implies the knowledge of the individual molecular motion. If the latter can be described by a monoexponential correlation function $\exp(-|t|/\tau_c)$, one has

$$\tilde{J}^{rr'}(\omega) = \left[(3 \cos^2\theta_{rr'} - 1)/2\right] \frac{2\tau_c}{1 + \omega^2\tau_c^2} \qquad (2.40)$$

where $\theta_{rr'}$ is the angle between the two relaxation vectors associated with the interactions r and r' (Figure 2.6). In a more general case, where two rotational diffusion coefficients are equal $D_x = D_y \neq D_z$ (symmetric top molecule), denoting θ and φ the polar angles of the relaxation vectors in the molecular frame (x, y, z) (See Figure 2.6) with $\tau_1^{-1} = 2D_x + 4D_z$, $\tau_2^{-1} = 5D_x + D_z$, $\tau_3^{-1} = 6D_x$, $\tilde{J}^{rr'}(\omega)$ is given by

$$\begin{aligned}
\tilde{J}^{rr'}(\omega) = &\,(3/4)\sin^2\theta_r \sin^2\theta_{r'} \cos 2(\varphi_r - \varphi_{r'})[2\tau_1/(1 + \omega^2\tau_1^2)] \\
&+ (3/4)\sin(2\theta_r)\sin(2\theta_{r'})\cos(\varphi_r - \varphi_{r'})[2\tau_2/(1 + \omega^2\tau_2^2)] \quad (2.41) \\
&+ (1/4)(3 \cos^2\theta_r - 1)(3 \cos^2\theta_{r'} - 1)[2\tau_3/(1 + \omega^2\tau_3^2)]
\end{aligned}$$

This can be extended to the case of a completely anisotropic motion $(D_x \neq D_y \neq D_z)$, leading to a rather complicated algebraic expression [97].

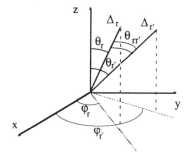

Figure 2.6 Definition of the different polar angles involved in the expressions of cross-correlation spectral densities. Δ_r and $\Delta_{r'}$ represent the direction of the two relaxation vectors

In the case of the 'model free approach' [49] (or two-step model [98]) described in Section 2.2.3, the reduced spectral density $\tilde{J}^{rr'}(\omega)$ is given by

$$\tilde{J}^{rr'}(\omega) = \left[\frac{1}{2} (3 \cos^2 \theta_{rr'} - 1) - S_r S_{r'} \right] \frac{2\tau_f}{1 + \omega^2 \tau_f^2} + S_r S_{r'} \frac{2\tau_s}{1 + \omega^2 \tau_s^2} \quad (2.42)$$

where S_r and $S_{r'}$ are the two order parameters for the directions Δ_r and $\Delta_{r'}$ respectively and with the two correlation times τ_f (fast motion) and τ_s (slow motion) defined in Section 3.2.3. Of course, the autocorrelation functions corresponding to equations (2.40), (2.41) and (2.42) become quite simple ($\theta_{rr'} = 0$, $\theta_{r'} = \theta_r$, $\varphi_{r'} = \varphi_r$ and $S_{r'} = S_r$).

As explained in Section 3.1, in the case of rigid molecules, at most three distinct rotational diffusion coefficients may be extracted from the spectral densities. Defining the correlation time τ_i corresponding to the diffusional coefficient D_i as $\tau_i = 1/6D_i$, this quantity can subsequently be expressed according to $\tau_i = C_i \eta + \tau_i^{FR}$ [99]. τ_i^{FR} is the free rotor correlation time: $\tau_i^{FR} = (2\pi/9)(I_i/k_B T)^{1/2}$, where I_i is the principal component of the inertial tensor relative to the i axis [100]. In the first term of τ_i, η is the solvent viscosity whereas C_i can be expressed according to various hydrodynamic theories which differ essentially by their boundary conditions. Two limiting cases can be considered (i) stick boundary conditions, which imply that the solvent molecule rotates with the molecule under investigation; this is the hypothesis underlying Perrin's calculations discussed in Section 3.1.2 [27], (ii) slip boundary conditions, according to which the solvent is not affected by the solute reorientation [101].

5.2   DIPOLAR INTERACTIONS

The expressions of the relevant relaxation parameters are given in Table 2.2. R_1^A and R_2^A are specific rates obtained by means of selective pulses acting only on the A spin. $(R_1^A)_{like}$ concern identical isotopes A and X subjected to non-selective pulses. σ_e, is the cross-correlation rate involved in longitudinal relaxation. σ_t identified as σ_ρ (see section 4.2.2.2), is the transverse cross-relaxation rate which is given here for identical isotopes since it is the only case of practical interest. To a good approximation, $R_{1\rho} \approx R_2$. Dipolar–dipolar cross-correlation rates include K_D as a multiplying factor and a coefficient depending on the magnetization modes governing the spin system under study.

In general, relaxation rates are the sum of several contributions. However, through the determination of σ which involve only dipolar interactions, it is possible to conclude that the dipolar contribution is the dominant one. For the molecules of concern in this chapter and for the measurement frequencies of the commonly used spectrometers [102] ($\nu(^1H) \leqslant 750$ MHz), the condition of extreme narrowing is fulfilled: $\omega \tau_c \ll 1$ so that relaxation parameters are independent of the measurement frequency. Thus, as far as specific relaxation

Table 2.2 Dipolar relaxation parameters. $K_D = (\mu_0 \gamma_A \gamma_X \hbar / 4\pi r_{AX}^3)^2$; r_{AX}: internuclear distance; γ_A, γ_X magnetogyric ratios of nuclei A and X of resonance frequencies $\omega_A / 2\pi$ and $\omega_X / 2\pi$ respectively. I_X is the spin number of nucleus X

$$R_1^A = (1/15)I_X(I_X+1)K_D[6\tilde{J}(\omega_A+\omega_X)+3\tilde{J}(\omega_A)+\tilde{J}(\omega_A-\omega_X)]$$

$$(R_1^A)_{like} = (1/20)K_D[12\tilde{J}(2\omega_A)+3\tilde{J}(\omega_A)]$$

$$R_2^A = (1/15)I_X(I_X+1)K_D[2\tilde{J}(0)+\tilde{J}(\omega_A-\omega_X)/2+3\tilde{J}(\omega_A+\omega_X)+3\tilde{J}(\omega_X)/2+3\tilde{J}(\omega_A)]$$

$$\sigma_1 = (1/20)K_D[6\tilde{J}(\omega_A+\omega_X)-\tilde{J}(\omega_A-\omega_X)]$$

$$(\sigma_1)_{like} = (1/20)k_D[2\tilde{J}(0)+3\tilde{J}(\omega_A)]$$

rates or cross-relaxation rates are concerned, τ_c/r_{AX}^6 is the only quantity which can be determined. τ_c is an effective correlation time which is related to the reorientation of the r_{AX} direction and which can be expressed as a function of the three diffusion coefficients. Thus *either* dynamical *or* structural information can be extracted from relaxation rates

Concerning structural information, if we have a set of dipolar relaxation rates corresponding to a set of r_{AX} vectors, the knowledge of one r_{AX} distance allows one to derive the effective τ_c value. This τ_c value is then assumed to be appropriate for all the r_{AX} vector reorientations (this corresponds to the assumption of quasi-isotropic motion). Thus, other distances may be obtained with an accuracy which is good enough for deciding between different possible conformations. The most commonly used approach consists of measuring the homonuclear NOE factors (e.g. proton–proton), either in a one-dimensional fashion by selective saturation [56, 103] or by means of the NOESY two-dimensional experiments [73]. Owing to the r^{-6} dependence, it must be pointed out that the method is only appropriate for unknown distances comparable to the reference distance.

Very often, the molecular geometry is known, thus allowing the study of reorientational motion, which, for rigid molecules, depends upon three diffusion coefficients along the three orthogonal axes, assumed to be coincident with the principal axes of the inertia tensor (PAI). Consequently, a good description of the motion is obtained if, along the available relaxation vectors, at least three of them are close to the PAI. It turns out that, in many cases, such a set cannot be found. In fact, three different vectors should be sufficient. However, the $\sin^2$ or $\cos^2$ angular dependencies of autocorrelation spectral densities preclude the use of vectors which have close geometrical dispositions with respect to PAI. As an example, in some simple planar aromatic molecules (e.g. pyridine, quinoline) where carbon-13 longitudinal relaxation is dominated by dipolar interaction, we can find only two sets of usable ^{13}CH vectors: (1, 2, 4, 5) and (3) (see Figure 2.7).

Fortunately, the third piece of information, necessary to fully describe the molecular reorientation, is obtained through ^{14}N relaxation rates which are exclusively quadrupolar [104, 105]. In less symmetrical polyaromatic rings, the CH vectors are of sufficiently different orientations with respect to PAI to provide by themselves the desired information [106].

The above mentioned problems may be circumvented if dipolar cross-correlation spectral densities can be determined from relaxation measurements. This is because the angular dependence is no longer a quadratic form of trigonometric functions (see e.g. equation (2.41)). A first methodology employs proton–proton dipolar cross-correlation spectral densities which can be determined provided that intermolecular contributions are negligible [92]. It allowed, for instance, to detect motional anisotropy in a series of aromatic molecules [107]. In a more recent work [108], both auto- and cross-correlated dipolar spectral densities, extracted from carbon-13 relaxation data, lead to the determination of the three principal elements of the diffusion tensor in ethanol.

As already mentioned in Section 2.3.2, the dipolar relaxation parameters are also sensitive to internal motions. A case very often encountered is the notation of a methyl group. The experiment may lead to a rotational barrier height and/(or) rotation frequencies. A recent discussion of this problem is presented in reference [109].

When the extreme narrowing condition is no longer satisfied, spectral densities cease to be frequency independent (Figure 2.8) and the comparison of T_1 and η (NOE factor) does not allow deduction of the importance of the dipolar contribution. Likewise, the usually accepted equality of T_1 and T_2 is no longer verified. For instance, for an AX spin system, in case of a predominant dipolar contribution, the often invoked relation, indicating a maximum value for the NOE factor $\eta(A) = \frac{1}{2}(\gamma_X/\gamma_A)$ only holds in the extreme narrowing conditions (in the homonuclear case, this corresponds to a 50% increase of the

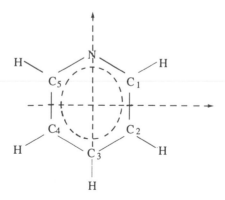

Figure 2.7 Numbering of carbon atoms in the pyridine molecule

signal intensities). Conversely, measurements at different frequencies, whose range nowadays covers almost a decade, enable the determination of additional interesting parameters, depending on the model which is chosen to analyse experimental data. Again the 'model free approach' or 'two-step model' are usually employed especially for studying aggregates or large biomolecules. Those two topics are covered in Chapters 8 and 9 of this book. Because the $\tilde{J}(0)$ term which appears in R_2 and not in R_1, T_2 may become very short (see Table 2.2 and Figure 2.8).

5.3 CHEMICAL SHIFT ANISOTROPY

In an NMR experiment, the static magnetic field $\mathbf{B}$ experienced by a given nucleus is $\mathbf{B} = (1 - \boldsymbol{\sigma} \mathbf{B}_0)\mathbf{B}_0$ is the externally applied magnetic field and $\boldsymbol{\sigma}$ is the second rank shielding tensor, responsible for the chemical shift phenomenon. Thus, owing to molecular motions, $\mathbf{B}$ is time dependent, and prone to contribute to nuclear spin relaxation, giving rise to the CSA (chemical shift anisotropy) mechanism, already largely discussed in this chapter. $\boldsymbol{\sigma}$ may be taken as a symmetrical cartesian tensor, although there is no theoretical reason to assume such a symmetry [110]. Moreover, in the present context and for the sake of simplicity, we shall only consider a $\boldsymbol{\sigma}$ tensor of axial symmetry ($\sigma_{xx} = \sigma_{yy} \neq \sigma_{zz}$ and $\Delta\sigma = \sigma_{zz} - \sigma_{xx}$). In that case, it turns out that the Hamiltonian can be expressed in terms of spherical harmonics of rank two $Y_2^0(\beta)$, where β is the angle between the tensor symmetry axis and the B_0 direction. As far as molecular reorientation is concerned, we notice a behavior similar to the

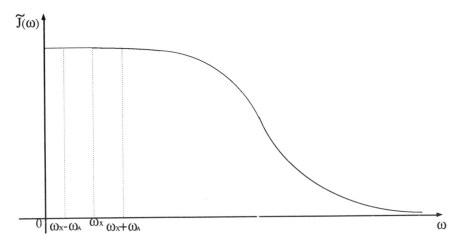

Figure 2.8 Reduced spectral density $\tilde{J}(\omega)$. In the extreme narrowing limit $\tilde{J}(0) \approx \tilde{J}(\omega_A - \omega_X) \approx \tilde{J}(\omega_A + \omega_X)$. For dipolar relaxation, $R_1^A = R_2^A$ (see Table 2.2). Otherwise, spectral densities may decrease more or less depending on the involved frequency

dipolar relaxation mechanism. This also explains the possibility of csa–dipolar cross-correlations whose effects, as already mentioned before, are becoming nowadays more and more visible with the advent of high field spectrometers.

Table 2.3 provides the expression of the relaxation parameters relevant to the CSA mechanism as a function of the reduced spectral densities defined in Section 5.1. In extreme narrowing conditions, $R_1^{CSA}/R_2^{CSA} = 6/7$ and both relaxation rates are quadratic functions of B_0. As a consequence, in case of a large anisotropy (e.g. $^{13}C{=}O$ and $^{31}P{=}O$), this feature makes lines broader at higher fields [102]. Regarding relaxation phenomena, plotting R as a function of B_0 makes it relatively easy to sort out the contribution of CSA (slope) from other contributions (intercept). For instance, using this method, it proved possible to show that the C_{60} fullerene dissolved in a spinless solvent (CS_2) relaxes almost exclusively through the CSA mechanism [111]. Moreover, using the chemical shift anisotropy value measured in the solid state, the correlation time of this spherical molecule is obtained. The result is in perfect agreement with the τ_c value calculated from the Stokes–Einstein equation $\tau_c = V\eta/k_B T$ (V is C_{60} volume; η is solvent viscosity).

Except in favourable cases, the determination of chemical shift anisotropy from conventional relaxation rates is relatively difficult since it implies measurements at different magnetic field values for isolating the CSA contribution. By contrast, the cross-correlation csa–dipolar spectral density affords a convenient means of reaching (or at least estimating) the chemical shift anisotropy value because it appears as the product of two terms, one depending solely on the dipolar interaction (which can be deduced from independent determinations) and the other depending solely on the CSA mechanism. This is reminiscent of cross-relaxation terms which originate from a single relaxation mechanism. To cite a few examples, the method has been successively applied to the determination of phosphorous-31 CSA in PHO_3^{2-}[112] to the estimation of carbon-13 CSA in ethanol [108] and even to the estimation of the anisotropy of proton shielding tensors. The latter, albeit weak, could be detected for amide protons [90] and for anomeric protons in glucose [113].

Table 2.3 CSA mechanism relaxation rates as a function of $\omega_0 = \gamma B_0$. $\mathscr{J}^{csa(A),dip(AX)}$ refers to the coupling term between $I_{A,z}$ and $2I_{A,z}I_{X,z}$ in the evolution equations of longitudinal magnetization in a two-spin system (see Section 4.2.3.1)

$$R_1^{CSA} = (1/15)(\gamma B_0)^2 (\Delta\sigma)^2 \tilde{J}(\omega_0)$$

$$R_2^{CSA} = (1/15)(\gamma B_0)^2 (\Delta\sigma)^2 [2\tilde{J}(0)/3 + \tilde{J}(\omega_0)/2]$$

$$\mathscr{J}^{csa(A),dip(AX)} = (1/15)(\gamma_A B_0)(\Delta\sigma_A)(\mu_0/4\pi)(\gamma_A\gamma_X\hbar/r_{AX})\tilde{J}(\omega_A)$$

5.4   SCALAR RELAXATION OF THE SECOND KIND

This mechanism operates on a slowly relaxing nucleus, itself J-coupled to a fast relaxing nucleus X (generally a quadrupolar nucleus) and can also be involved in exchange phenomena (see Chapter 3). Besides a decoupling effect (the multiplet structure of A disappears if the X relaxation times are sufficiently short), viewed from A, the fast relaxation of X plays a role similar to that of molecular motion through a time modulation of the J coupling interaction. It turns out that the expression of A relaxation rates cannot be derived in the manner employed for dipolar or CSA relaxation because of the special form of the 'space functions' F_r^m. The complete treatment can be found in ref. [21] and leads to

$$R_1^{SC}(A) = [2J_{AX}^2 I_X(I_{X+1})]T_2^X/[1 + (\omega_A - \omega_X)^2(T_2^X)^2] \qquad (2.43)$$

$$R_2^{SC}(A) = [J_{AX}^2 I_X(I_{X+1})/3]\{T_1^X + T_2^X/[1 + (\omega_A - \omega_X)^2(T_2^X)^2]\} \qquad (2.44)$$

In these formulae, J_{AX} is expressed in rad s^{-1}, I_X is the spin number of nucleus X, $1/T_1^X$ and $1/T_2^X$ are its longitudinal and transverse relaxation rates, respectively. A first point to notice is that the contribution of scalar relaxation to the longitudinal relaxation time of A remains negligible unless the two frequencies are very close to each other (e.g. ^{13}C and ^{79}Br) for making small enough the denominator of (2.43). Conversely, due to the factor $1/T_1^X$ in (2.44), scalar relaxation, when it exists, dominates the transverse relaxation rate. With the knowledge of the longitudinal relaxation time of the quadrupolar nucleus, it therefore proves possible to extract J_{AX} from the transverse relaxation rate of the slowly relaxing nucleus (A) (or better from the relaxation rate in the rotating frame). This indirect method is especially valuable since splittings from the J coupling cannot be observed. An example is provided by the determination of C–Cl indirect couplings in the four chloromethanes [114].

5.5   SPIN-ROTATION

The spin-rotation interaction originates from the coupling between the total angular momentum $\mathbf{J}$ and the nuclear spin angular momentum $\mathbf{I}$. It gives rise to a Hamiltonian, $\mathcal{H}_{SR} = \Sigma_k \mathbf{I}_k \mathbf{CJ}$, k is the nucleus index and $\mathbf{C}$ a second rank tensor which has some similarities with the magnetic screening tensor (chemical shielding tensor σ). The spin-rotation contribution to the spin nucleus relaxation rates is seldom significant. It may, however, become important for nuclei belonging to mobile parts of a molecule (e.g. freely rotating methyl groups). However, there exists studies of relaxation times which lead to the conclusion that, in molecules as large as benzene [115], toluene or pyridine [116], the spin rotation interaction contributes significantly to the relaxation processes.

  We shall briefly present the way to derive the spin-rotation interaction. The total molecular Hamiltonian [117] may be split in two parts, (i) the principal contribution $\mathcal{H}_0$ which includes the electron kinetics energy and the electron-electron and electron-nucleus coulombic interaction, (ii) the perturbation contribution $\mathcal{H}_1$, which contains the magnetic interactions: electron-orbital, electron-electron, electron-nucleus... By application of the perturbation theory up to order two, one obtains among other terms an energy contribution $\Delta E = \Sigma_k \mathbf{I}_k \mathbf{CJ}$. This may be interpreted as the interaction of the magnetic moment $\mathbf{I}_k$ with a magnetic field $\mathbf{CJ}$. This magnetic field is generated by the molecular motion, because the electrons do not follow exactly the rotation of the molecular frame. The spin-rotation corresponds to a small spectral line splitting in the microwave region which has been observed and studied only in a few molecules. Moreover, the spin-rotation tensor elements C_{rs} ($r, s = x, y, z$) which appear as an infinite sum of terms (second order perturbation contribution), are difficult to calculate, except for small molecules. Several methods have been proposed for these calculations [117].

As the molecule rotates and changes its angular momentum, there is a magnetic field modulation at each nucleus thus resulting in a contribution to the relaxation rates. T_1 depends on the reduced spectral density $\tilde{J}(\omega_0)$, and T_2 on both $\tilde{J}(0)$ and $\tilde{J}(\omega_0)$; in the extreme narrowing limit $T_{1SR} = T_{2SR}$. The spin-rotation mechanism allows one to define a correlation time τ_J which refers to the changes in the magnitude of the molecular angular momentum vector rather than to changes in the orientation of the molecule (chemical shift anisotropy and quadrupolar contributions) or in the orientation of a vector (dipolar contribution). At room temperature and in the diffusion regime, the Hubbard relation holds $\tau_c \tau_J = I/6kT$, where τ_c corresponds to the overall molecular reorientation, I is the average momentum of inertia about the principal molecular axes. Higher values of T_{SR}^{-1} can occur for small molecules at elevated temperatures. Contrary to the other mechanisms, the contribution of spin-rotation to the relaxation rates increases with the temperature. Thus the study of T_1 as a function of the temperature is a means to analyse the importance of the spin-rotation mechanism. This has been done for instance in a very thorough way with the carbon-13 longitudinal relaxation time of the CS_2 molecule [118], which depends on two mechanisms: spin-rotation and chemical shift anisotropy. The correlation between the spin rotation constant C and the shielding anisotropy $\Delta\sigma$ was verified within experimental error.

6   Conclusion

It can be appreciated that nuclear spin relaxation is capable of providing accurate information about rotational motions. Translational motions may in some instances contribute, but their effect on relaxation parameters are rather difficult to interpret whereas the extraction of rotational correlation times from properly chosen relaxation parameters may be straightforward. In addition to cross-relaxation rates which depend solely on dipolar interactions and which, consequently, are also sensitive to the internuclear distance, one can resort to the accurate measurements of cross-correlation rates. The latter either involve two different dipolar interactions or correlate a dipolar interaction with the CSA mechanism. These measurements yield complementary and refined information in terms of rotational dynamics and/or structure (internuclear distances and shielding tensor). With the advent of sophisticated multi-impulsional techniques dedicated to relaxation measurements and high-field NMR spectrometers, they should become of routine use in the near future.

References

1. G. C. Maitland, M. Rigby, E. B. Smith and W. A. Wakeham, *Intermolecular Forces, their Origin and Determination*, Oxford University Press, Oxford (1981).
2. J. N. Murrell, S. Carter, S. C. Farantos, P. Huxley and A. Varandas, *Molecular Potential Energy Functions*, John Wiley & Sons, New York (1984).
3. M. Rigby, E. B. Smith, W. A. Wakeham and G. C. Maitland, *The Molecular Forces*, Oxford Science Publications (1986).
4. A. D. Buckingham, P. W. Fowler and J. M. Hutson, *Chem. Rev.*, **88**, 963 (1988).
5. A. D. Buckingham, in J. J. C. Teixera-Dias (ed.), *Basic Theory of Intermolecular Forces in Molecular Liquids: New Perspectives in Physics and Chemistry*, Nato Asi. Series, Kluwer Academic Publishers, Dordrecht (1992).
6. A. A. Bothner-By, C. Gayathri, P. C. M. Van Zijl, C. MacLean, J. J. Lai and K. M. Smith, *Magn. Reson. Chem.*, **23**, 935 (1985).
7. C. M. Skoglund, Molecular Orientation Studies using High Field NMR. Thesis Carnegie–Mellon University, Pittsburgh (1987).
8. P. C. M. Van Zijl, B. H. Ruessink, J. Bulthuis and C. MacLean, *Acc. Chem. Research*, **17**, 172 (1984); E.W. Bastiaan and C. MacLean, *Molecular Orientation in High-Field High-Resolution NMR* in *NMR Basic Principles and Progress*, vol. 25, Springer-Verlag, Berlin, Heidelberg, New York, 1991.
9. N. Wax, *Selected Papers on Noise and Stochastic Processes*, Dover, New York (1954).
10. B. J. Berne, in D. Henderson (ed.), *Physical Chemistry and Advanced Treatise*, vol. VIII B, Academic Press, New York, London (1971).
11. W. A. Steele, in I. Prigogine and S. A. Rice (eds), *Advances in Chemical Physics*, vol. XXXIV, p. 1, John Wiley & Sons, New York, London (1976).
12. G. Williams, *Chem. Soc. Rev.*, **7**, 89 (1978).
13. W. G. Rotschild, *Dynamics of Molecular Liquids*, John Wiley & Sons, New York, 1984.
14. P. Madden and D. Kivelson, *Adv. Chem. Phys.*, **56**, 1057 (1984). D. Kivelson and P. Madden, *J. Phys. Chem.*, **88**, 6557 (1984).
15. D. Wei and G. N. Patey, *J. Chem. Phys.*, **91**, 7113 (1989); *J. Chem. Phys.*, **93**, 1399 (1990).
16. R. G. Gordon, in *Advances in Magnetic Resonance*, vol. 3, p. 1, edited by J. S. Waugh, Academic Press, New York (1968).
17. D. A. McQuarrie, *Statistical Mechanics*, Harper & Row, New York (1976).
18. T. Ohba and S. Ikawa, *Molec. Phys.*, **73**, 985 (1991).
19. S. K. Deb, M. L. Bansal and A. P. Roy, *Chem. Phys.*, **110**, 391 (1986).
20. L. Van Hove, *Phys. Rev.*, **95**, 249 (1954).
21. J. C. Lassègues, M. Fouassier and J. L. Viovy, *Molec. Phys.*, **50**, 417 (1983).
22. W. F. Van Gunsteren and J. C. Berendsen, *Molec. Phys.*, **45**, 637 (1982).
23. R. T. Boere and R. G. Kidd, *Annu. Rep. NMR Spectrosc.*, **319** (1982).

24. P. Debye, *Polar Molecules*, Reinhold, New York (1929); M. C. Wang and G. G. Uhlenbeck, *Rev. of. Mod. Physics*, **17**, 323 (1945).
25. A. Abragam, *The Principles of Nuclear Magnetism*. Clarendon Press, Oxford, (1961).
26. G. Williams, *Chem. Soc. Rev.*, **7**, 89 (1978).
27. F. Perrin, *J. Phys. Rad.*, **V**, 33 (1934); *J. Phys. Rad.*, **VII**, 1 (1936).
28. L. D. Favro, *Phys. Rev.*, **119**, 53 (1960).
29. W. T. Huntress, Jr., *Adv. Magn. Reson.*, **4**, 1 (1970).
30. P. A. Egelstaff, *An Introduction to the Liquid State*, Academic Press, London, (1967).
31. R. E. D. McClung, *J. Chem. Phys.*, **55**, 3459 (1969).
32. R. E. D. McClung, *J. Chem. Phys.*, **57**, 5478 (1972).
33. T. E. Bull and W. Egan, *J. Chem. Phys.*, **81**, 3181 (1984).
34. A. G. St. Pierre and W. A. Steele, *J. Chem. Phys.*, **57**, 4638 (1972).
35. J. P. Boon and S. Yip, *Molecular Hydrodynamics*, Dover Inc., New York (1980).
36. M. Fixman and K. Rider, *J. Chem. Phys.*, **51**, 2425 (1969).
37. G. Levi, J. P. Marsault, F. Marsault-Herail and R. E. D. McClung, *J. Chem. Phys.*, **73**, 2443 (1980).
38. G. T. Evans, *J. Chem. Phys.*, **65**, 3030 (1976); *J. Chem. Phys.*, **67**, 2913 (1977).
39. D. H. Lee and R. E. D. McClung, *J. Magn. Reson.*, **73**, 34 (1984); *J. Chem. Phys.*, **112**, 23 (1987).
40. J. Kowalewski, *Annu. Rep. NMR Spectrosc.*, **22**, 306 (1989); **23**, 289 (1991).
41. H. G. Hertz, *Prog. NMR Spectrosc.*, **16**, 115 (1983).
42. D. E. Woessner, *J. Chem. Phys.*, **37**, 647 (1962).
43. D. E. Woessner, B. S. Snowden Jr. and G. H. Meyer, *J. Chem. Phys.*, **50**, 719 (1969).
44. Buu Ban, *J. Magn. Reson.*, **45**, 1981 (1981).
45. D. Wallach, *J. Chem. Phys.*, **47**, 5258 (1967).
46. M. E. Rose, *Elementary Theory of Angular Momentum*, John Wiley & Sons, Inc., New York, London, Sydney (1967).
47. G. C. Levy, A. Kumar and D. Wang, *J. Am. Chem. Soc.*, **83**, 7596 (1983).
48. T. E. Bull, *J. Chem. Phys.*, **65**, 4802 (1976).
49. G. Lipari and A. Szabo, *J. Am. Chem. Soc.*, **104**, 4546 (1982).
50. B. Halle, *Mol. Phys.*, **61**, 693 (1987).
51. L. G. Werbelow and D. M. Grant, *Adv. Magn. Reson.*, **9**, 189 (1977).
52. D. Canet, *Progr. NMR Spectrosc.*, **21**, 237 (1989).
53. R. R. Ernst, G. Bodenhausen and A. Wokaum, *Principles of Nuclear Magnetic Resonance in One and Two Dimensions*, Oxford University Press, Oxford (1987).
54. T. C. Farrar and J. E. Harriman, *Density Matrix and its Applications in NMR Spectroscopy*, The Farragut Press, Madison (1989).
55. M. Goldman, *J. Magn. Reson.*, **60**, 437 (1984).
56. J. H. Noggle and R. E. Schirmer, *The Nuclear Overhauser Effect*, Academic Press, New York (1971).
57. D. M. Grant, C. L. Mayne, F. Lin and T. X. Xiang, *Chem. Rev.*, **91**, 1591 (1991).
58. A. G. Redfield, *Adv. Magn. Reson.*, **1**, 1 (1965).
59. R. L. Vold, J. S. Waugh, M. P. Klein and D. E. Phelps, *J. Chem. Phys.*, **48**, 3831 (1968).
60. E. D. Becker, J. A. Ferretti, R. K. Gupta and G. H. Weiss, *J. Magn. Reson.*, **37**, 381 (1980).
61. D. Canet, G. C. Levy and I. R. Peat, *J. Magn. Reson.*, **18**, 199 (1975).
62. D. Canet, J. Brondeau and K. Elbayed, *J. Magn. Reson.*, **77**, 483 (1988).

63. G. Bodenhausen and D. J. Ruben, *Chem. Phys. Lett.*, **69**, 185 (1980).
64. A. Bax, G. Griffey and D. M. Doddrell, *J. Magn. Reson.*, **55**, 301 (1983).
65. N. R. Nirmala and G. Wagner, *J. Am. Chem. Soc.*, **110**, 7557 (1988).
66. A. Allerhand and S. R. Maple, *Anal. Chem.*, **59**, 441 A (1987).
67. L. Mahi, J. C. Duplan and B. Fenet, *Chem. Phys. Lett.*, **211**, 27 (1993).
68. R. R. Vold and R. L. Vold, *J. Chem. Phys.*, **64**, 320 (1976).
69. S. Meiboom and D. Gill, *Rev. Sci. Instr.*, **69**, 688 (1958).
70. T. K. Leipert, J. H. Noggle, W. J. Freeman and D. L. Dalrymple, *J. Magn. Reson.*, **19**, 208 (1975).
71. N. R. Nirmala and G. Wagner, *J. Magn. Reson.*, **82**, 659 (1989).
72. A. Kalk and H. J. C. Berendsen, *J. Magn. Reson.*, **24**, 343 (1976).
73. J. Jeener, B. H. Meier, P. Bachmann and R. R. Ernst, *J. Chem. Phys.*, **71**, 4546 (1979).
74. R. Freeman, H. D. W. Hill and R. Kaptein, *J. Magn. Reson.*, **7**, 327 (1972).
75. P. L. Rinaldi, *J. Am. Chem. Soc.*, **105**, 5167 (1983); C. Yu and G. C. Levy, *J. Am. Chem. Soc.*, **105**, 6994 (1983).
76. A. G. Palmer III, R. A. Hochstrasser, D. P. Millar, M. Rance and P. E. Wright, *J. Am. Chem. Soc.*, **115**, 6333 (1993).
77. K. Elbayed and D. Canet, *Molec. Phys.*, **71**, 979 (1990).
78. A. A. Bothner-By, R. L. Stephens, J. T. Lee, C. D. Waren and R. W. Jeanloz, *J. Am. Chem. Soc.*, **106**, 811 (1984).
79. A. Bax and D. G. Davis, *J. Magn. Reson.*, **63**, 207 (1985).
80. T. L. Hwang and A. J. Shaka, *J. Magn. Reson.*, **B102**, 155 (1993).
81. H. Néry and D. Canet, *J. Magn. Reson.*, **42**, 370 (1981); M. Guéron, J. L. Leroy and R. H. Griffey, *J. Am. Chem. Soc.*, **105**, 7262 (1983); V. A. Daragan and K. H. Mayo, *Chem. Phys. Lett.*, **206**, 393 (1993).
82. G. Jaccard, S. Wimperis and G. Bodenhausen, *Chem. Phys. Lett.*, **138**, 601 (1987).
83. K. Elbayed and D. Canet, *Molec. Phys.*, **68**, 1033 (1989).
84. J. Boyd, U. Hommel and I. D. Campbell, *Chem. Phys. Lett.*, **175**, 477 (1990).
85. T. C. Farrar and I. C. Locker, *J. Chem. Phys.*, **87**, 3281 (1987).
86. L. E. Kay, L. K. Nicholson, F. Delaglio, A. Bax and D. A. Torchia, *J. Magn. Reson.*, **97**. 359 (1992).
87. R. Brüschweiler and R. R. Ernst, *J. Chem. Phys.*, **96**, 1758 (1992).
88. C. Dalvit, *J. Magn. Reson.*, **97**, 645 (1992).
89. C. Dalvit and G. Bodenhausen, *Chem. Phys. Lett.*, **161**, 554 (1989).
90. C. Dalvit, *J. Magn. Reson.*, **95**, 410 (1991).
91. I. Burghardt, R. Konrat and G. Bodenhausen, *Mol. Phys.*, **75**, 467 (1992).
92. R. L. Vold and R. R. Vold, *Prog. NMR Spectrosc.*, **12**, 79 (1978).
93. J. Brondeau, D. Canet, C. Millot, H. Néry and L. G. Werbelow, *J. Chem. Phys.*, **82**, 2212 (1985).
94. G. Jaccard, S. Wimperis and G. Bodenhausen, *J. Chem. Phys.*, **85**, 6282 (1986).
95. N. Müller, *Chem. Phys. Lett.*, **31**, 218 (1986).
96. C. Dalvit and G. Bodenhausen, *J. Am. Chem. Soc.*, **110**, 7924 (1988).
97. H. W. Spiess, in *NMR Basic Principles and Progress*, vol. 15, Springer-Verlag, Berlin, Heidelberg (1978); J. P. Marchal and D. Canet, *J. Chem. Soc. Faraday Trans. II*, **78**, 435 (1982).
98. H. Wennerström, B. Lindman, O. Söderman, T. Drekenberg and J. B. Rosenholm, *J. Am. Chem. Soc.*, **101**, 6860 (1979).
99. D. Kivelson, *Symp. Faraday Soc.*, **11**, 7 (1979).
100. D. R. Bauer, J. I. Brauman and R. Pecora, *J. Am. Chem. Soc.*, **96**, 6840 (1974).
101. G. K. Youngren and A. Acrivos, *J. Chem. Phys.*, **63**, 3846 (1975).

102. D. Canet and J. B. Robert, in J. B. Robert (guest ed.), *NMR Basic Principles and Progress*, vol. 25, Springer-Verlag, Berlin, Heidelberg (1990).
103. D. Neuhaus and M. Williamson, *The Nuclear Overhauser Effect in Structural and Conformational Analysis*, VCH Publishers, New York (1989).
104. J. P. Kintzinger and J. M. Lehn, *Molec. Phys.*, **22**, 273 (1972).
105. D. Jalabert, J. B. Robert, D. Canet and P. Tekely, *Molec. Phys.*, **79**, 673 (1993).
106. J. J. Delpuech, B. Benayada, D. Nicole and P. Tekely, *J. Magn. Reson.*, **81**, 288 (1989).
107. H. Nery, D. Canet and J. L. Rivail, *Chem. Phys.*, **62**, 123 (1981).
108. Z. Zheng, C. L. Mayne and D. M. Grant, *J. Magn. Reson.*, **A268**, 103 (1993).
109. M. J. Dellwo and J. Wand, *J. Am. Chem. Soc.*, **115**, 1886 (1993) and references therein.
110. R. F. Schneider, *J. Chem. Phys.*, **48**, 4905 (1968).
111. D. Canet, J. B. Robert and P. Tekely, *Chem. Phys. Lett.*, **212**, 483 (1993).
112. T. C. Farrar and I. C. Locker, *J. Chem. Phys.*, **101**, 6860 (1979).
113. L. Poppe and H. Van Halbeek, *Magn. Reson. Chem.*, **31**, 665 (1993).
114. M. Ohuchi, T. Fujito and M. Imanari, *J. Magn. Reson.*, **35**, 415 (1979).
115. A. Dolle, M. A. Suhm and H. Weingartner, *J. Chem. Phys.*, **94**, 3361 (1990).
116. R. D. Brown and M. Head-Gordon, *Chem. Phys.*, **105**, 1 (1986).
117. W. H. Flygare, *Molecular Structure and Dynamics*, Prentice-Hall Inc., Englewood Cliff, New Jersey (1978).
118. H. W. Spiess, D. Schweitzer, U. Haeberlen and K. H. Hausser, *J. Magn. Reson.*, **5**, 101 (1971).

3 TIMESCALES IN NMR: NUCLEAR SITE EXCHANGE AND DYNAMIC NMR

J.-J. Delpuech
Université Henri Poincaré, Nancy I, France

Dynamics of Solutions and Fluid Mixtures by NMR
Edited by J.-J. Delpuech   © 1995 John Wiley & Sons Ltd

1   Introduction

1.1   RELAXATION AND DNMR

It is a common use to classify dynamic effects in NMR as arising from either relaxation or nuclear site exchange processes, themselves originating in the existence of fluctuating magnetic fields. Random fields are described both by their magnitude—depending on the type of magnetic interaction—and by the correlation time(s) responsible for the time dependence. These two features help to clarify the traditional distinction between relaxation and NMR site exchange. Firstly, an NMR site exchange in fluids is equivalent to a longitudinal fluctuation of weak amplitude—by about a few hertz to up to a few hundred hertz—essentially involving the isotropic frequency shifts described by the chemical shift Hamiltonian $\mathcal{H}_{CS}$ (equation (1.7)). It should be noted in this respect that the anisotropic part $\mathcal{H}_{CSA}$ of the full chemical shift Hamiltonian, which accounts for one important relaxation mechanism (see Chapter 2), is considered to be averaged to zero on the NMR timescale of the exchange, this means that the exchange is assumed to be clearly slower than molecular tumbling (see below). Isotropic chemical shift fluctuations are not covered, indeed, under the headings of the various relaxation mechanisms as mentioned in Chapter 2, and thus require being dealt with in the present chapter.

Secondly, correlation times accounting for nuclear relaxation are directly related to Brownian motion in fluids. Relaxation mechanisms can be described as random walk processes occurring by small steps with a fast repetition time τ_c. Very low energy barriers are involved in these diffusional processes and all successive steps are consequently equiprobable. On the contrary, molecular processes over higher energy barriers may be envisaged to take place by jumps carrying one given nucleus among a set of chemical species standing in equilibrium, each one characterized by the resonant frequency ν_A, ν_B ... of the observed nucleus. In this case, the effects induced on the NMR patterns are said to result from a nuclear exchange between sites A, B... . Detailed representations of NMR site exchanges are likely to be very complicated, as are the descriptions of reaction mechanisms themselves. Fortunately, as far as the modifications of the NMR patterns are concerned, they can be accounted for by using a simplified model ignoring what precisely occurs during the act of exchange. This model assumes that jumps are occurring very suddenly, that is in a time much less than the Larmor period, so that (i) the phase and the spin state of each nucleus is preserved during the transfer from one site to another, and (ii) the duration of each jump is negligible with respect to the average time elapsed between two jumps, the so called exchange time τ_{ex}. Such a picture applies as well to any intra- or intermolecular process, involving or not bond-breaking and bond-forming steps. The chemical process producing the nuclear site exchange is often called *chemical exchange*; it should involve forward and reverse reactions standing in equilibrium. Of course, NMR spectroscopy may also be used for kinetic purposes merely as a titration procedure to perform NMR spectra as a function of time in the case of reactions going slowly to equilibrium (or completion). Equilibration studies are not included under the denomination of DNMR; and they will be quoted in this presentation only in a few examples where DNMR measurements are simultaneously possible, for the sake of comparison. It may even happen that chemical exchange is not accompanied by any net reaction, e.g. isomerization between identical conformers (as in DMF, see Chapter 1, Section 4) or ligand exchanges about symmetrically coordinated metal ions. In this case, the exchange is said to be *degenerate*, or else the sites are said to be in *mutual* or *self-exchange*.

Nuclear site exchanges are not restricted to NMR lines which are chemically shifted from each other, but apply as well to spin–spin coupling multiplets. The equivalence between these two phenomena is especially clear for weakly coupled nuclei where the full spin–spin coupling Hamiltonian $\mathcal{H}_s$ (equation (1.8)) can be truncated to its diagonal part

$$\mathcal{H}_s = J\mathbf{I}.\mathbf{S} \sim JI_z S_z \tag{3.1}$$

The exchange is occurring if either the coupling constant J or the spin state of one of the two coupled nuclei, say S_z in the following, is jumping at random among a series of discrete values in the course of time. Under these conditions, the time-dependent Hamiltonian $\mathcal{H}_s(t)$ represents a longitudinal fluctuation which may generate coalescence phenomena between the components of spin multiplets in the same manner as that described in Chapter 1 for chemical shift variations. The NMR timescale of the exchange is determined in this case by the value of the coupling constant J, still representing the frequency shift between the multiplet components. The two types of exchange, involving variations of either J or S_z, are said to be of the first or second kind, respectively, a denomination already used in the case of scalar relaxation considered to represent the fast exchange regime. First kind exchanges are brought about by chemical processes suddenly changing the value of J, e.g. in the case of a series of conformers rapidly interconverting into each other. In exchanges of the second kind, the NMR pattern of spins $\mathbf{I}$ is observed as the spin state S_z is modulated by either a true chemical exchange—e.g. the transfer of a coupled acidic proton—or the longitudinal relaxation of spins **S**. The latter possibility applies in principle to any pair of coupled nuclei. It is, in fact, observed only if the relaxation time T_{1s} of spins **S** fits the NMR timescale of the exchange, i.e., according to equation (1.52), if

$$2\pi J T_{1s} \sim 1 \tag{3.2}$$

This requires in practice to observe a spin 1/2 nucleus **I** coupled to a nucleus **S** undergoing moderately fast quadrupolar or electronic relaxation

The distinction between relaxation and DNMR processes tends to vanish in the case of fast exchange. Nuclear site exchange then becomes a relaxation mechanism where the time-dependent magnetic interaction is either a chemical shift difference ($\Delta\omega$) or a coupling constant (J) modulated with a characteristic correlation time τ_c taken as the exchange lifetime τ_{ex}. However, the magnitude of this relaxation estimated from the mean square of fluctuations ($\Delta\omega^2$ or J^2) is weak (10^2–10^6 square hertz at most) as compared to that of other modes of relaxation, e.g. dipolar or quadrupolar relaxations (10^8–10^{12} square hertz). It thus appears that the effects of chemical exchange cannot be detected at higher rates, and that the exchange lifetimes cannot be extracted in this case from the NMR patterns.

There is an important exception to this rule when the chemical exchange can be considered as a time-dependent process which contributes to modulate the interaction actually controlling NMR linewidths, i.e., in most cases, dipolar (D), or quadrupolar (Q), or chemical shift anisotropy (CSA) relaxation. There is in this case at least one dynamical process, besides chemical exchange, modulating D, Q or CSA interactions, namely molecular tumbling with a rotational correlation time τ_r. In the probable event of uncorrelated dynamical processes, the time-dependent Hamiltonian $\mathcal{H}_D(t)$ or $\mathcal{H}_Q(t)$ or $\mathcal{H}_{CSA}(t)$, fluctuates with a characteristic correlation time τ_c where:

$$1/\tau_c = 1/\tau_r + 1/\tau_{ex} \tag{3.3}$$

NMR lifetimes may thus be deduced as correlation times from relaxation time measurements, provided that nuclear site exchange is faster than molecular tumbling, i.e. when $\tau_{ex} < \tau_r = 10^{-9}$–$10^{-11}$ s in most fluids. This is typically the case of (i) internal motions in flexible molecules when they can be described as discrete jumps over small energy barriers between a set of conformers (see Chapters 2 and 9), (ii) fast proton exchanges on coordinated water molecules around metal ions or at active sites of enzymes, (iii) fast exchange of metal ions bound to non-equivalent sites in polypeptide solutions (see below, Section 5).

Scalar relaxation of the second kind deserves a special mention in this respect, since this interaction is not modulated by molecular tumbling, but by the interplay of the chemical exchange (if any) and of the relaxation of the coupled spin **S** (important only in the case of quadrupolar or electronic relaxation, see above). The NMR timescale in this case is thus limited by either the quadrupolar relaxation time $T_{1Q} \sim 10^{-1}$–10^{-4} s, or the electronic relaxation time $T_{1e} \sim 10^{-6}$ to 10^{-12} s in most cases, respectively. In this respect, there are privileged quadrupolar nuclei, e.g. ^2H, ^{14}N, ^{23}Na, $^{43}Ca\ldots$, or paramagnetic centres, e.g. transition metal ions Cu^{2+}, Mn^{2+}, $Co^{2+}\ldots$, which can be used as probes to study the dynamic properties of solutions. The nature and the use of such probes are two important topics treated in depth through the next two chapters.

1.2 HISTORICAL OVERVIEW

In earlier observations of nuclear magnetic resonance, it was soon recognized that molecular rate processes might have an effect on NMR spectra. Thus, exchanges in the fast exchange limit were set forth to account for the presence of a single line in cases when a set of chemically shifted resonances, or else a spin multiplet, was expected instead. An early example was found with the hydroxylic resonance in ethanol whose frequency was shown to be concentration and temperature dependent, this was properly accounted for by the presence of variable quantities of free ethanol molecules and of hydrogen-bonded aggregates standing at equilibrium and rapidly converting into each other on the NMR timescale [1, 2]. Spin-echo measurements were also used at this time to detect indirect nuclear spin–spin couplings (J) from the existence of beats at a frequency of J in the amplitude of echoes. Anomalous damping of echoes was, however, observed in the case of methanol when a very small percentage of water ($\sim 1\%$) was added, this was assigned to direct interchange of protons between H_2O and CH_3OH molecules [3] and permitted a rough estimate of protons lifetimes $\tau_{ex} \sim 1$ s at $-4\,°C$.

A stepping-stone in the history of DNMR was brought by Gutowsky, McCall and Slichter, hereafter designated as GMS, who gave in 1953 the first theory of NMR lineshapes in the presence of chemical exchange, using for this purpose modified Bloch equations [4]. There was no experimental evidence at that time to support the theory in the coalescence region. Applications still concerned the fast exchange limit, e.g. intramolecular rearrangement between equatorial and apical positions of fluorine atoms in PF_5, BrF_5 or IF_5 molecules [4], or else fast proton transfers between water and strong electrolytes [5], or between water and ammonia [6]. Three years later, two series of experiments nicely confirmed the lineshapes predicted by the GMS theory. The first one, performed again by the group of Gutowsky, described a two-site exchange between two lines of equal intensities on the example of N,N-dimethylformamide [7] already presented in Chapter 1. The second experiment re-examined the hydroxylic triplet of ethanol, which was shown to suffer a progressive coalescence by adding traces of either acid or basic compounds (themselves not detected in the spectrum). A successful simulation of coalescence patterns was given for this three-site exchange, which, however, involved a single exchange time, namely the average lifetime of one hydroxylic proton on an individual ethanol molecule between two successive transfers [8].

An unrestricted treatment of multisite exchange between weakly coupled sites was given first by Kubo [9], Sack [10], and Anderson [11]—using the picture of a random walk of the phase of the transverse processing magnetization—and then by McConnell [12] using the more simple picture of *transverse magnetization transfers* among the nuclear sites. Both treatments led to a common matrix formalism with the introduction of the so called *exchange matrix*. A last step of development concerned NMR patterns involving strongly coupled lines, i.e. second-order spectra. This was achieved by using the *density matrix* formalism, as in the case of relaxation theories. The fundamental equation 2.10 giving the evolution of the density matrix ρ as a function of time, was added with terms expressing exchanges as transfers of density matrix elements [13, 14]. The computation of ρ permitted in turn the expectation value of nuclear magnetization, and then the NMR signals in the presence of exchange to be derived.

As far as chemical applications are concerned, another key step was the pioneering investigations of Grunwald and coworkers on proton transfers in first a series of ammonium salts [15], and then of alcohols [16]. These examples showed how to deal with complex reaction mechanisms accounting for one or several NMR site exchanges, and were later used as a model by many investigators.

Up to now, NMR site exchanges have been considered to involve a transfer of transverse magnetization. An important step further in the development of DNMR was taken in 1956 by Solomon and Bloembergen [17] who pointed out the possibility of also observing *longitudinal magnetization* transfers from the modification of the apparent longitudinal relaxation rate T_1 or of the transient or steady-state nuclear Overhauser effect. Experiments using T_1 measurements or saturation transfers to derive exchange rates were described a few years later by Forsen and Hoffmann [18, 19]. Transfers of longitudinal magnetization are the basis of exchange-transferred NOE experiments [20] and of pulse sequences used in 2D-NMR to study exchange phenomena [21], e.g. in the so called exchange spectroscopy (EXSY).

Another important contribution of Solomon and Bloemberger [17] was to show the possibility of measuring NMR lifetimes well beyond the fast exchange limit when they are so short as to take part in the modulation of relaxation interactions, as already explained in the previous section. The NMR timescale $\omega_0 \tau_{ex} \sim 1$ is determined by the Larmor frequency $\omega_0 \sim 10^9$ rad s^{-1}. The method is therefore appropriate to rate processes with lifetimes shorter than about 10^{-8} s. This leaves a wide range of unexplored exchange rates, since, on the other hand, lineshape analysis is restricted to lifetimes greater than about $10^{-4} - 10^{-5}$ s. The NMR timescale can, however, be moved into the intermediate range of the microsecond by measuring relaxation times $T_{1\rho}$ in the rotating frame (see Chapter 2). Measurements of rotating frame relaxation have been known for many years, but the first application of the $T_{1\rho}$ method to rate processes seems to have been reported by Deverell *et al.* in 1970 in a study of the chair-to-chair inversion of cyclohexane [22].

All these pioneering studies gave rise to innumerable applications in various fields of chemistry and biochemistry; they constituted the basic framework for all investigations using NMR spectroscopy to study kinetic processes.

1.3   TOPICS IN CURRENT DNMR

From the foregoing it appears that problems in dynamic nuclear magnetic resonance can be presented from different points of view. Matters of interest may first be categorized into either the methodologies or the applications of DNMR. *Experimental procedures* may in turn be classified according to the mode of observation, using either *transverse* or *longitudinal* components of nuclear magnetization. *Transverse magnetization* is used in at least two variants of DNMR: lineshape analysis and T_2 measurements.

Lineshape analysis offers the advantages of a wide range of validity throughout the NMR window, a reasonable degree of accuracy and the existence of a well-established methodology. Its drawbacks are first those of one-dimensional NMR, namely the impossibility to assign by inspection the sets of exchanging lines in complex spectra. Another limitation is the small range of measurable rate constants. Exchange rates τ_{ex}^{-1} cannot be measured beyond the slow side or the fast side of the NMR window, i.e. when $\tau_{ex}^{-1} < \sim\Delta\omega_{1/2}$ or $\tau_{ex}^{-1} > \Delta\omega^2/\Delta\omega_{1/2}$, respectively, according to equations (1.49) and (1.47).

This limitation arises in part from the linewidths $\Delta\omega_{1/2}$ pre-existing in the absence of exchange. This means that sharper lines in a given NMR pattern should bring about some extension of the NMR window. Apparent linewidths are mainly determined by residual heterogeneities of the magnetic field $\mathbf{B}_0$; they may be decreased to the natural width due to transverse relaxation by using a Carr–Purcell–Meiboom–Gill (CPMG) train of spin echoes (see Chapter 2). T_2 *measurements* have been used as a complementary method to lineshape analysis to increase the NMR window by about one order of magnitude towards fast exchange rates.

In a second category of DNMR experiments, *selective* transfers of *longitudinal magnetization* are observed among exchanging sites. The selectivity of transfers involves observing individual sites as uncoalesced lines, thus restricting the method to slow transfers. The method has two specific advantages. (i) The range of measurable exchange rates in the slow exchange region is limited by 'natural' longitudinal relaxation times in the absence of exchange, several seconds or more for spin 1/2 nuclei in diamagnetic solutions; the NMR window can thus be extended by about one order of magnitude beyond the slow exchange limit. (ii) The selectivity of transfers can provide mapping of exchange pathways.

These transfers are equivalent to *cross-relaxation* processes, and their effects are intrinsically superimposed to those of NOE, if any. Experimental methods may then use T_1 measurements and steady-state or transient NOE to derive exchange rates. Recent and important variants use two-dimensional NMR spectroscopy which allows NMR encoding of nuclear sites during the evolution period of the pulse sequence. In this way, exchanging sites are detected by cross-peaks in two-dimensional spectra. Quantitative estimations of rate constants are also possible, but with a lesser degree of accuracy than in one-dimensional lineshape analysis, at least at the present time. Another important possibility is to observe the nuclear magnetization spin-locked along the RF field $\mathbf{B}_1$ (see Chapter 2) and to perform either $T_{1\rho}$ *measurements* or *magnetization transfer experiments in the rotating frame.*

Besides all of these methods which are specific to DNMR, let us recall the possibility of using T_1, T_2 and NOE measurements if NMR lifetimes are so short as to play the role of correlation times in nuclear relaxation.

All these procedures have widely different timescales, this determines the type of chemical applications for each of them. The so called NMR timescale, $\Delta\omega\tau_{ex} \sim 1$ (equation (1.52)), actually refers to lineshape analysis. This is an important limitation even for all the other procedures in DNMR, since it controls the possibility of distinguishing separate resonances for each site, this means that, beyond coalescence, the effects of rate processes can be observed only on a weighted mean over all sites simultaneously. As mentioned above, both the slow and fast limits on either side of the NMR timescale—the so called NMR or kinetic window—are determined by the apparent linewidth in the absence of exchange $\Delta\omega_{1/2} \sim 1$ rad s^{-1} in most cases, and by the frequency shift between exchanging sites, $\Delta\omega = 1$ to 10^2 rad s^{-1} in current practice. Measurable exchange rates should consequently be comprised between $\Delta\omega_{1/2}$ and $\Delta\omega^2/\Delta\omega_{1/2}$, i.e. between about 1 and 10^4 s^{-1}. The kinetic window may be somewhat enlarged on either side by about one order of magnitude by measuring apparent T_1, T_2 relaxation times. One-dimensional lineshape analysis is likely to cover more than 90% of applications described in the literature, its use is only challenged by the upsurge of two-dimensional methods.

Quite different timescales are, however, available from $T_{1\rho}$ and NMR correlation times, corresponding to lifetimes of about one microsecond and one nanosecond, respectively. These two methods, however, have severe intrinsic limitations and are much less familiar to the NMR users than lineshape analysis which requires no additional device to be performed. Nevertheless, rotating-frame methods are now routinely implemented in most NMR spectrometers, mainly because of the necessity to perform ROESY experiments, so that $T_{1\rho}$ measurements may enter in current practice in the future.

Applications of DNMR to chemical kinetics should consequently involve fast reversible reactions at equilibrium which fit the above NMR timescales;

this means that pseudo first-order rate constants for magnetic site exchange should usually lie between 10^{-1} and $10^4\,s^{-1}$. From a chemical point of view, exchanges can be classified as *intramolecular* or *intermolecular* processes, the latter being much less studied than the former in this way.

The term *chemical reorganization* [23] has been proposed to cover the large variety of *internal molecular processes* studied by DNMR. Another denomination pointing at the objective of these investigations is *stereodynamics*. The term of *stereochemically nonrigid* [24] has been coined to characterize all the molecules that undergo intramolecular rearrangements rapidly enough to influence NMR lineshapes at temperatures within the practical range of applications. Another specificity of DNMR which attracted many investigators is the possibility to detect a degenerate rearrangement in which all of the interconverting species that are observable are chemically and structurally equivalent. This was named an identity, or invisible process, in which the forward and reverse reactions are identical in a so called *isodynamic* equilibrium [25]. Sterochemically nonrigid molecules liable to undergo degenerate rearrangements are named *fluxional* molecules [24]. Typical examples are the chair-to-chair inversion of unsubstituted cyclohexane, Berry pseudo-rotations about identically penta-substituted central atoms (as in the compounds mentioned above, PF_5, BrF_5 . . .), or degenerate valence isomerizations.

In *intermolecular* processes, several species, denoted symbolically as BE, CE, DE . . . have in common a structural unit E which can be interchanged from one molecule to the next. It is, for example, the case of protons ($E = H^+$) resulting from the ionization of a mixture of acidic compounds: BH^+, CH^+, DH^+ . . . , and then reattached in the reverse reaction to the conjugate bases B, C, D . . .. The chemical bonds between the exchanging unit E and the rest of the parent molecules, B, C, D . . . should be fragile enough to permit fast and reversible bond cleavage. Moreover, the molecular fragments as well as the parent molecules should be able to exist as stable species. These considerations restrict the range of possible applications, as compared to the case of intramolecular processes. Another distinctive feature is that the relative spin states of structural fragments, such as B and E, associated in a given parent molecule, say BE, are continuously changing in the course of time; this allows the study of *degenerate* exchanges between chemically identical molecules provided that nuclei from one fragment are scalarly coupled to nuclei in the other fragment. This is the case of proton transfers between ethanol molecules which can be monitored, as already mentioned above, by the coalescence of the hydroxylic triplet into a singlet (NMR pattern of the exchanging fragment $E = H$) or of the methylenic eight lines multiplet into a quadruplet (NMR pattern of the complementary fragment $B = CH_3CH_2O$).

The various types of intramolecular and intermolecular exchanges mentioned in the above paragraphs show that areas of applications concern organic chemistry as well as inorganic and organometallic chemistry, and biochemical investigations. The methods and applications of DNMR have been reviewed in a great number of journals and books in the course of years. Earlier reports could review the field of DNMR in its entirety [25–28]. More recent reviews are restricted to specialized aspects of DNMR, such as lineshape analysis [29], density matrix formalism [23], $T_{1\rho}$ measurements [30], exchange spectroscopy [31, 32] (EXSY), experimental procedures [33], effects of chemical exchange on the NOE [20, 34], applications to conformational rearrangements and stereodynamics [25, 29, 36], to inorganic and organometallic chemistry [37], to organic chemistry [38] and to biology [39–43].

1.4 PLAN OF THE CHAPTER

Topics of interest in DNMR may thus be conveniently classified after the mode of observation. The first two sections try bring together a variety of DNMR procedures—lineshape analysis, transient methods, inversion and saturation transfer, two-dimensional exchange spectroscopy—under the unifying themes of transverse or longitudinal magnetization transfer experiments. Determinations of exchange rates as correlation times are displayed in a further section of smaller importance. These methodological sections are preceded by mechanistic considerations relating the magnetic site exchange to chemical rate constants. Typical examples of reactions liable to DNMR measurements are given throughout this chapter and are to be found in the next chapters as well. Main fields of interest in DNMR are briefly reviewed in two sections where they are classified according to the type of reaction (intra or intermolecular) rather than to the fields of chemistry and biochemistry involved.

2 Nuclear Sites and NMR Site Exchanges

Any problem in DNMR is made up of two elements. There is first an NMR description to assign the exchanging sites and denote the unknown probabilities of magnetization transfer from one site to the next. The second step of a DNMR treatment is to devise a chemical model accounting for site exchange. The term *chemical exchange* which is commonly used in this type of investigation is somewhat ambiguous. It should be understood as standing for the chemical process generating the NMR site exchange. In fact, an additional feature is implied in the abridged denomination of chemical exchange, namely the exchange of some part of the examined structure, i.e. the exchange of substituents or bond rearrangements in intramolecular processes, or else the transfer of molecules or molecular fragments in intermolecular processes. Whatever the exact meaning of chemical exchange may be, the NMR site exchange and the associated molecular rate process should be carefully distinguished from each other. There are several reasons for that. The first is that there is no univocal connection between both phenomena; examples will thus be given below where the same chemical event can give rise to several types of NMR exchanges, and, conversely, several chemical processes may contribute to the same site exchange.

Another distinctive feature is that the magnetic effects of chemical exchange are not restricted to nuclei which are effectively transferred into a non-equivalent position as the result of bond distortion or bond cleavage, as in the case of the two *N*-methyl groups of DMF (see above) or of the hydroxylic protons of methanol. Non-exchanging parts of the examined molecules may also suffer an NMR site exchange. This is the case of the methylic protons of methanol which, although not directly involved in the transfer of hydroxylic protons, suffer an NMR exchange between the two components of their representative doublet. Such a possibility also exists in the first example above using hindered rotation about an amide bond, if we consider an unsymmetrically *N*-substituted amide such as a 2-acetamidothiophene derivative, the *N*-acetyl,*N*-ethoxycarbonylmethyl-2-amino–5-nitrothiophene **1** [44]. Detailed studies of substituent effects in the thiophene ring and quantum mechanical calculations have shown that the planes of the thiophene ring and of the amide unit are perpendicular to each other, with the presence of two isomers depending on the *cis* (Z) or *trans* (E) position of the carbonyl group (Figure 3.1). The interconversion of the *cis* into the *trans* isomer carries the methylic 2-acetamido protons into the vicinity of the thiophene unit where they are shielded by ring current effects. On the other hand, the non-exchanging protons of the thiophene ring, H_3, and at a much lesser extent H_4, are located within the anisotropy cone of the carbonyl group in the *cis* isomer, where they are shielded as compared to the *trans* isomer. These chemical shift variations allow measuring the rate of *cis-trans* interconversion from the NMR pattern of the exchanging 2-acetamido as well as of the nonexchanging thiophenic protons. This example is also an illustration of two independently observed NMR site exchanges monitored by the same rate process, hindered rotation about the amide bond.

These examples clearly show that, although if they are closely linked together, chemical and NMR site exchanges are not identical to each other. A

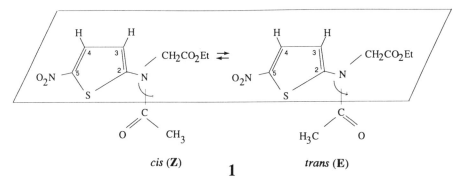

Figure 3.1 The interconversion of *cis* and *trans* isomers of compound **1**

third distinctive feature in this respect concerns the kinetic parameters describing either the NMR site exchange or the molecular rate processes, which are seldom found to be equal together. This is in line with the fact that NMR timescales limitations strictly apply to NMR site exchanges; rate constants for the parent chemical processes may reach values well beyond these limitations, especially in the case of concentration-dependent kinetics.

After a re-examination of the concept of nuclear sites, we shall first study in this section the general formalism used to describe the kinetics of NMR site exchanges, and then give selected examples of how to relate them to chemical kinetics.

2.1 NUCLEAR SITES

In the simple event of an uncoupled nucleus, or else of a set of isochronous nuclei uncoupled to the rest of the molecule, nuclear sites are defined by the set of chemical shifts corresponding to each position successively occupied by the observed nucleus. Nuclear sites are in this case represented by a set of chemically shifted lines, each one denoted either by a letter, A, B, C..., or more conveniently, by numerical indexes: $1, 2, \ldots, n_S$. Each site i is characterized by three NMR parameters: the resonant frequency ν_i, the relaxation rate $1/T_{2i}$ or the linewidth $\Delta\nu_{1/2}^i$ or $\Delta\omega_{1/2}^i$, and the relative intensity p_i.

A self-evidence is that sites should have different NMR parameters to be distinguished from each other. Thus, degenerate exchanges between like molecules in which one nucleus may occupy either of several equivalent positions are not accompanied by coalescence phenomena. This is true as well for inter- as for intramolecular processes. Thus, hindered rotation about the amide bond of N,N-dimethylacetamide **2** exchanges the *cis* and *trans* positions of N-methyl substituents, while the acetamido protons occupy equivalent positions and therefore are not subject to coalescence phenomena. Another example is proton transfers between water molecules which do not alter the water line. On the contrary, proton transfers in aqueous solution of an acid AH will give rise to exchanges between lines of water and acid protons.

2

Great care should be exercised in such cases to ascertain whether the two positions are really non-equivalent. A matter of concern in this view for organic chemists is the case of pairs of protons H_A, H_B borne by one methylenic carbon, which may be or not mirror images of each other, and are thus liable to generate or not NMR site exchanges in conformational isomerization processes. Such protons are said to be *enantiotopic* or *diastereotopic* [45], respectively, and the interchange of diastereotopic protons is called *topomerization* [46]. The presence of an adjacent asymmetric carbon is often quoted as a criterion to observe diastereotopic protons. In spite of the fact that couples of enantiomers give rise to identical spectra and cannot be distinguished from each other (at least in non-optically active solvents), protons H_A and H_B are effectively non-equivalent in this case due to the disymmetry of their chemical environment. Thus 1-methylenic protons in 2-methyl-butyllithium **3** are anisochronous (even on assuming fast rotation about the C_1–C_2 bond) and are thus liable to topomerization [38, 46] when the Li–C bond is cleaved and inversion of the carbonium ion takes place on carbon 1. This is in sharp contrast with the case of unsubstituted butyllithium in which protons H_A, H_B are equivalent and thus not liable to topomerization effects.

$$CH_3 - CH_2 - \overset{*}{C}H - \underset{\underset{\displaystyle H_B}{|}}{\overset{\overset{\displaystyle H_A}{|}}{C}} - Li$$
$$\underset{\displaystyle CH_3}{|}$$
3

It should be stressed, however, that the presence of a chiral centre is not required to observe topomerization. Thus, the 1-methylenic protons of 3–3-dimethyl-butylmagnesium chloride **4** are magnetically non-equivalent and exchange their magnetic sites with one another in the inversion of the carbonium ion accompanying the ionization of the Grignard reagent [25].

$$H_3C - \underset{\underset{\displaystyle CH_3}{|}}{\overset{\overset{\displaystyle CH_3}{|}}{C}} - CH_2 - \underset{\underset{\displaystyle H_B}{|}}{\overset{\overset{\displaystyle H_A}{|}}{C}} - MgCl$$
4

The same is true for nitrogen inversion of optically inactive methyl-diethyl(or dibenzyl) amine **5** [48, 49]. In spite of overall molecular symmetry, methylenic protons H_A, H_B occupy diastereopic positions, as shown in their Newman projection along one N–CH_2 bond, and are interchanged by nitrogen inversion.

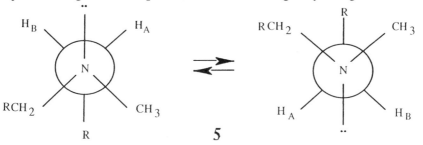

5

R = CH_3 or C_6H_5 (one conformation only is represented)

Another important example is that of cyclohexane **6** [50] where each carbon bears two non-equivalent methylenic protons H_A, H_B, in axial (a) and equatorial (e) positions, which are interchanged in chair-to-chair inversion, an exchange often symbolized as $ae \Leftrightarrow ea$, or else $a \Leftrightarrow e$.

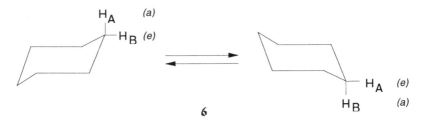

6

Topomerizations are not restricted to methylenic protons; diastereotopic positions may be as well occupied by other nuclei, e.g. two fluorine atoms or two identical substituents. An important example is that of stereochemically nonrigid structures about a metal ion M in tris-chelated complexes [51, 52] $M(LL)_3$, where LL is a symmetrical bidentate ligand, such as nonamethylimidodiphosphoramide (NIPA) **7** [53]. Such structures are existing in indifferentiated couples of enantiomers Δ, Λ because of chirality at the metal centre. The two N-methyl groups borne by each terminal nitrogen atom of the three coordinated NIPA molecules are diastereotopic. This may be visualized on a Newman projection along the C_3 axis of the molecule, which shows that no symmetry operation of the D_3 symmetry groups $(C_3, 3C_2)$ can interchange the two nitrogen substituents (Figure 3.2). Two singlets are effectively obtained by $^1H \{^{31}P\}$ or $^{31}C \{^1H\}$ NMR at low temperature, which are coalescing into a single line at higher temperatures as the result of the fast racemization process $\Delta \Leftrightarrow \Lambda$ through internal rearrangement within the coordination sphere.

$$\left[(CH_3)_2N \right]_2 - \underset{\underset{O}{\overset{\|}{}}{P}}{} - \underset{}{N} - \underset{\underset{O}{\overset{\|}{}}{P}}{} - \left[N(CH_3)_2 \right]_2 \qquad (NIPA)$$

$$\overset{CH_3}{\underset{}{|}}$$

7

It may be noted that, at higher temperatures, diastereotopic nuclei become dynamically equivalent and can be considered as a set of isochronous nuclei described by a single site. This site may in turn be engaged in another exchange which is slow as compared to the previous one. This is typically the case with tris-chelates of NIPA, where the coordinated NIPA molecules may suffer a slow intermolecular exchange between the coordination sphere and the bulk solution at temperatures where the above internal rearrangement is fast. Along the same lines, in the hindered rotation about the amide bond of DMF, the three protons in each N-methyl substituent are assigned a single averaged resonant frequency

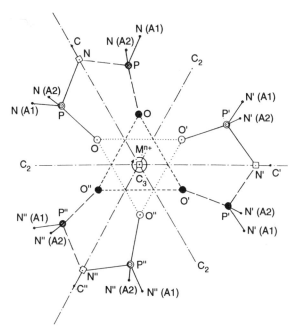

Figure 3.2 A projection of the tris-bidentate NIPA complex of a metal ion M^{n+} along the C_3 rotation axis showing the non-equivalence of the terminal nitrogen atoms (A1 and A2). Oxygen and phosphorus atoms of the chelate rings are drawn in either of two ways [(1) open and circled circles; (2) solid and shaded circles] depending on whether they are located above or below the mean molecular plane defined by the central metal ion M^{n+} and the three bridging nitrogen atoms (open squares) [53]. Reprinted with permission from Rubini *et al.*, *Inorg. Chem.*, **18**, 2964. Copyright (1979) American Chemical Society

on account of fast internal rotation about the $N-CH_3$ bond. To be valid, such considerations require that the two NMR windows for both fast and slow processes do not overlap and are far from each other on the temperature and/or concentration scale. In this way, the slower process can be considered to occur between nuclear sites in smaller number due to averaging effects from the faster process. This greatly simplifies the analysis of DNMR patterns where several exchanges are processing at quite different rates. This simplification is not restricted to topomerization and conformational processes. Let us give as a final example in this respect the case of the hydroxylic protons in aqueous solutions of strong acids or bases, represented by a single line standing for H_2O, OH^- and H_3O^+ species, which may in turn be engaged in slow transfers with weaker acids or bases.

Up to now we have neglected any line splitting from spin–spin coupling phenomena. In most *intramolecular processes*, spin–spin splitting is a source of complication in NMR patterns without actually bringing any new information. A simple example is the topomerization of two methylenic protons H_A, H_B, which are in fact unavoidably coupled to each other. If we assume weak coupling and first-order analysis, each nucleus has its own spin state α or β; the magnetic state of the set of nuclei is given by product wave functions $\alpha\alpha$, $\alpha\beta$, $\beta\alpha$, $\beta\beta$; four transitions are obtained as $\alpha\alpha \to \beta\alpha$ (site 1) and $\alpha\beta \to \beta\beta$ (site 3) for doublet A, and $\alpha\alpha \to \alpha\beta$ (site 2) and $\beta\alpha \to \beta\beta$ (site 4) for doublet B (Figure 3.3). Topomerization carries each nucleus with the same spin state from position A (chemical shift ν_A) to position B (chemical shift ν_B), and vice versa. The exchange is said of the type AB $\Leftrightarrow$ BA. Transitions such as $\alpha\alpha \to \beta\alpha$ are thus transformed into $\alpha\alpha \to \alpha\beta$. This means that lines 1 and 2 on the one hand, 3 and 4 on the other , are exchanging together, as shown by the arrows (Figure 3.3). The NMR exchange is thus the superposition of two two-site exchanges between pairs of chemically shifted lines. No additional information is gained as compared to the case of uncoupled nuclei. Spin–spin decoupling can be used in such instances to suppress spin multiplicity whenever it is possible, this was the case when observing diastereotopic protons in NIPA tris-chelates, spin-decoupled from phosphorus nuclei.

In the case of *intermolecular processes*, each chemical species, such as BE (Section 1.3), exchanges a molecular fragment E with the bulk liquid. The degree of information brought to DNMR depends on whether the coupled nuclei belong to the same part, E or B, of the molecule, or not. In the first event, there is no additional information from spin-coupling effects, as for intramolecular processes. In the second event, spin multiplicity can afford an independent way for kinetic measurements. This is especially true for *degenerate* processes, where spin multiplicity is the unique source of kinetic information, e.g. proton transfers between like molecules, as in liquid methanol. The three methylic protons of methanol are represented by a doublet, each component of which corresponds to a given magnetic state, α or β, of the coupled hydroxylic proton (X). In each transfer, an individual hydroxylic proton is substituted by another proton from the bulk liquid, incoming with

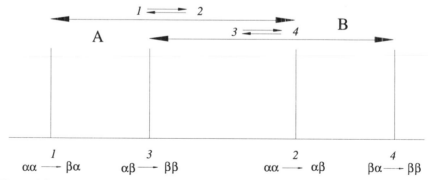

Figure 3.3   NMR site exchange of the type AB $\Leftrightarrow$ BA

either the same or the opposite spin state, α or β, as that of the leaving nucleus X, with equal probabilities. Proton transfer (the chemical exchange) is thus accompanied by a two-site NMR exchange in the chemically non-exchanging part of the molecule, with an NMR rate constant half that of proton transfer. The same information can be obtained from the coalescence of the quadruplet of the coupled nucleus X in the exchanging part of the methanol molecule. The four components of the quadruplet correspond to the four possible magnetic states of the set of methylic protons: $\alpha\alpha\alpha$, $\beta\alpha\alpha$ (indistinguishable from $\alpha\beta\alpha$ and $\alpha\alpha\beta$), $\alpha\beta\beta$ (and, similarly, $\beta\alpha\beta$ and $\beta\beta\alpha$), and $\beta\beta\beta$. A hydroxylic proton leaves a companion methyl group and is carried into another methanol molecule where its state will be, at random, any of the four possible states mentioned above, with probabilities equal to the relative intensities of the quadruplet components, namely 1/8, 3/8, 3/8 and 1/8, respectively. In the limit of fast exchange, the hydroxylic and methylenic multiplets will be transformed into two singlets exactly as if they were spin-decoupled from each other by double irradiation; this effect is called *chemical spin-decoupling*. This example is also an illustration of the fact that the same chemical event can result in several independent NMR site exchanges, and that NMR rate constants may be different from chemical rate constants. Along the same lines, protons transfers between like water molecules are amenable to DNMR measurements by using oxygen-17 enriched samples [54], in which the exchanging protons are spread over six sites, equally spaced and intense, each corresponding to one of the six magnetic states of the non-exchanging ^{17}O nucleus ($I = 5/2$). More complex examples may involve both chemical shift and spin–spin coupling effects.

At this point, we may draw the conclusion that, in the first-order approximation, chemical exchanges are accompanied by NMR site exchanges from line to line in the NMR pattern of the observed nuclei. A general formal description of NMR site exchanges may thus be given in the next subsection. However, this conclusion does not hold as such in the case of strong coupling and second-order analysis. This can be easily shown in the above exchange of diastereotopic nuclei AB $\Leftrightarrow$ BA. Second-order analysis mixes the nuclear states of individual nuclei to obtain four wave functions:

$\varphi_1 = \alpha\alpha$; $\varphi_2 = \alpha\beta \cdot \cos \theta + \beta\alpha \cdot \sin \theta$; $\varphi_3 = -\alpha\beta \cdot \sin \theta + \beta\alpha \cdot \cos \theta$; $\varphi_4 = \beta\beta$ (where $\theta = 0.5 \tan^{-1} (J/\delta)$). Along the same line of reasoning as that used in first-order analysis, it cannot be said that line 1, corresponding in this case to the transition $\varphi_1 \rightarrow \varphi_3$, is carried into line 3, corresponding to $\varphi_1 \rightarrow \varphi_2$, since the permutation of the nuclear states α and β of A and B transfomms φ_3 into $\alpha\beta \cdot \cos \theta - \beta\alpha \cdot \sin \theta$ which is not identical to φ_2. The term of nuclear site exchange is inappropriate is this case. It should rather be said that chemical exchange of the type AB $\Leftrightarrow$ BA brings about the coalescence of the AB-type NMR spectrum. A quantum-mechanical treatment is necessary in this case to compute the NMR patterns from the expectation value of transverse nuclear magnetization, using the density matrix formalism for this purpose.

Fortunately, with the advent of high-field spectrometers and of heteronuclear NMR, most NMR spectra are of the first-order type. NMR site exchanges will thus be examined below in the first-order approximation. Some indications on second-order NMR patterns will be given in a special section.

2.2   THE EXCHANGE MATRIX

NMR site exchanges can be envisioned as magnetization transfers from site i to site j, in just the same way as chemical reactions may be formally considered as transfers of matter from one species to another. Such transfers are characterized by means of pseudo first-order rate constants k_{ij} (in s^{-1}). Let us first examine a two-site exchange, denoted as

$$1 \underset{k_{21}}{\overset{k_{12}}{\rightleftharpoons}} 2, \quad \text{or else} \quad A \underset{k_B}{\overset{k_A}{\rightleftharpoons}} B \tag{3.4}$$

The number F of nuclei transferred per second from site 1 to site 2 is $k_{12}p_1$, where p_1 is the population of site 1, expressed as the molar fraction, or the number of nuclei (generally in moles), or the area of the absorption curve, or any proportional number. In steady-state conditions, there should be an equal number of nuclei transferred from site 2 to site 1, so that

$$k_{12}p_1 = k_{21}p_2 \tag{3.5}$$

Let us call $r = p_1/p_2$ the time-independent ratio of the populations of the two sites. This ratio is equivalent, but not identical, to a chemical equilibrium constant. This ratio is generally known, either a priori, or from preliminary NMR measurements performed in the absence of exchange. This reduces the number of independent unknown kinetic parameters to one, e.g. to k_{12}, with $k_{21} = rk_{12}$.

NMR mean lifetimes τ_1, τ_2 are also used in this case instead of rate constants k_1, k_2. The average time spent by one nucleus in site 1 (or 2) is given by the ratio of population p_1 (or p_2) to the flow of nuclei transferred per second in either sense:

$$\tau_1 = p_1/F = 1/k_{12} \quad \text{and} \quad \tau_2 = 1/k_{21} \tag{3.6}$$

Conservation of matter is expressed with these notations as

$$p_1/\tau_1 = p_2/\tau_2 \quad \text{or} \quad p_{1,2} = \tau_{1,2}/(\tau_1 + \tau_2) \tag{3.7}$$

In the case of multi-site exchanges, the array of exchange rate constants k_{ij} is displayed for convenience as the off-diagonal elements P_{ij} of a square matrix $\mathbf{P}$ of order n_S, the total number of lines in the NMR pattern:

$$\mathbf{P} = \begin{bmatrix} \diagdown & k_{12} & k_{13} & \cdots & k_{1n} \\ k_{21} & \diagdown & k_{23} & \cdots & k_{2n} \\ & \text{Diagonal elements} & & \\ k_{31} & k_{32} & \diagdown & \cdots & k_{3n} \\ \vdots & \vdots & \vdots & & \vdots \\ k_{n1} & k_{n2} & \cdots & \cdots & \diagdown \end{bmatrix} \tag{3.8}$$

To facilitate further calculations of lineshapes, diagonal elements P_{ii} are conventionally defined as the opposite sum of off-diagonal elements k_{ij} in the ith row:

$$P_{ii} = - \sum_{\substack{j=1,n_s}}^{j \neq i} k_{ij} \tag{3.9}$$

This is the so called *exchange matrix* **P** which summarizes all the unknown kinetic parameters. For example, in the above example of the intramolecular degenerate exchange of the type AB $\Leftrightarrow$ BA, the exchange matrix:

$$\mathbf{P} = \left[\begin{array}{cc:cc} -k_{12} & k_{12} & 0 & 0 \\ k_{21} & -k_{21} & 0 & 0 \\ \hdashline 0 & 0 & -k_{34} & k_{34} \\ 0 & 0 & k_{43} & -k_{43} \end{array} \right] \tag{3.10}$$

has four zero elements and is partitioned into two submatrices (as shown by the broken lines), this expresses the overall NMR exchange to result from the superposition of two independent two-site exchanges, $1 \Leftrightarrow 2$ and $3 \Leftrightarrow 4$.

The flows of nuclei, F_{+i} and F_{-i}, incoming to or leaving each site i per second, respectively, are:

$$F_{+i} = \sum_{\substack{j=1,n_s}}^{j \neq i} p_j P_{ji} \quad \text{and} \quad F_{-i} = -p_i P_{ii} \tag{3.11}$$

Conservation of matter requires that $F_{+i} = F_{-i}$ for each site, so that we can write n_s equations of the type

$$p_i P_{ii} + \sum_{j \neq i} p_j P_{ji} = 0 \tag{3.12}$$

Mean lifetimes on each site i is found by dividing the steady-state concentration p_i by the flow of nuclei F_{-i}:

$$\tau_i = p_i / F_{-i} = -1/P_{ii} \quad or \quad 1/\tau_i = -P_{ii} \tag{3.13}$$

The overall exchange rate from site i can be divided into n_s components, each one from the transfer from site i to site j:

$$1/\tau_i = \sum_{j=1,n_s}^{j \neq i} 1/\tau_{ij} \qquad \text{where } 1/\tau_{ij} = k_{ij} \qquad (3.14)$$

This notation is cumbersome and reserved to the simple case of two-site exchanges. Another special case in this respect is when the number of unknown parameters k_{ij} is reduced to unity, this means that all NMR site exchanges arise from a single chemical event. Under this condition, the lifetime associated to this single event may be conveniently defined as the exchange lifetime $\tau_{ex} = 1/k_{ex}$, and all the rate constants k_{ij} can be expressed as

$$k_{ij} = k_{ex}\Pi_{ij} \qquad (3.15)$$

where k_{ex} is the unique exchange rate, and Π_{ij} an array of numerical constants, often determined on statistical grounds, displayed in the so-called probability matrix Π. Thus, in the above example of proton transfers between like methanol molecules, τ_{ex} indicates the mean residence time of a hydroxylic proton in an individual molecule. The exchange matrices associated with the methylic doublet and the hydroxylic quadruplet are respectively:

$$\mathbf{P} = \frac{k_{ex}}{2} \begin{bmatrix} -1 & 1 \\ 1 & -1 \end{bmatrix} \qquad (3.16)$$

and

$$\mathbf{P} = \frac{k_{ex}}{8} \begin{bmatrix} -7 & 3 & 3 & 1 \\ 1 & -5 & 3 & 1 \\ 1 & 3 & -5 & 1 \\ 1 & 3 & 3 & -7 \end{bmatrix} \qquad (3.17)$$

2.3   RELATIONSHIPS TO CHEMICAL KINETICS

The most difficult task left to the discretion of DNMR users is to set-up an exchange matrix on the basis of a plausible kinetic scheme. This permits in turn to compute theoretical lineshapes by standard procedures (see next section). The kinetic scheme is then adopted or rejected depending on whether the simulated curves tightly fit experimental spectra or not. This task remains relatively simple as long as the whole set of NMR rate constants k_{ij} can be reduced to one unknown parameter, taken as the exchange rate $k_{ex} = 1/\tau_{ex}$, as explained above.

Let us first consider the very simple case of hindered rotation in DMF [7]. The chemical exchange of N-methylic protons between cis and trans positions with rate constant k results in an NMR exchange ($1 \Leftrightarrow 2$) between two equally populated sites ($p_1 = p_2$), with both forward and reverse rate constants equal to the chemical rate ($k_{12} = k_{21} = k$ or k_{ex}). The exchange lifetime $\tau_{ex} = 1/k$ thus represents the mean lifetime of each conformer: $\tau_1 = \tau_2 = \tau_{ex}$. The exchange matrix is written as

$$\mathbf{P} = k_{ex} \begin{bmatrix} -1 & 1 \\ 1 & -1 \end{bmatrix} \tag{3.18}$$

where the matrix between brackets is the associated probability matrix Π. Lineshapes obtained in this case are represented in Figure 1.4.

A related example of higher complexity is hindered rotation in the unsymmetrically N-substituted amide 1. Two sets of lines x and y, all in the same intensity ratio r, are observed, corresponding to the *cis* and *trans* isomers (Figure 3.4). Sites are unequally populated and the ratio r is identical to the *cis* to *trans* equilibrium constant K ($K > 1$). The two N-methylic singlets of the amide bond give rise to a two-site exchange described by the population vector $\mathbf{p} = [r, 1]$ and the exchange matrix:

$$\mathbf{P} = k_{ex} \begin{bmatrix} -1 & 1 \\ r & -r \end{bmatrix} \tag{3.19}$$

in which k_{ex} is the rate of *cis* to *trans* conversion. The NMR patterns [55] obtained for different values of k_{ex} are shown on the right-hand side of Figure 3.5. An equivalent information can be obtained from the two quadruplets of coupled sets of protons H_3, H_4 in each isomer. Each line of the *cis* quadruplet is exchanged with the corresponding line of the *trans* quadruplet, since the spin state of nuclei H_3, H_4 is preserved during internal rotation. There are thus four independent exchanges, all with the same exchange matrix (3.19). There is in

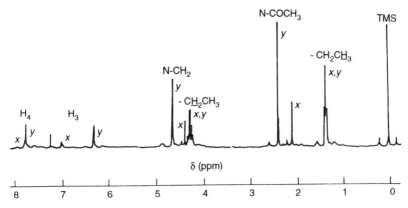

Figure 3.4 The ^1H spectrum of compound 1 at $-10\,°C$ and 250 MHz in $CDCl_3$ [55]. Reprinted from Nicole *et al.*, *Tetrahedron*, **36**, 3234. Copyright (1977), with kind permission from Elsevier Science Ltd, The Boulevard, Langford Lane, Kidlington, OX5 1GB, UK3

fact a strong overlap of H_3 lines from both isomers, and theoretical lineshapes are obtained by adding all the four individual NMR patterns together. Such an addition requires specially adapted computer programming. If standard DNMR programs only are available, an equivalent procedure should be preferred, which considers overall exchanges over the eight sites with a population vector $\mathbf{p} = [r, r, r, r, 1, 1, 1, 1]$ and a matrix $\mathbf{P}$ composed of four diagonal submatrices (3.19):

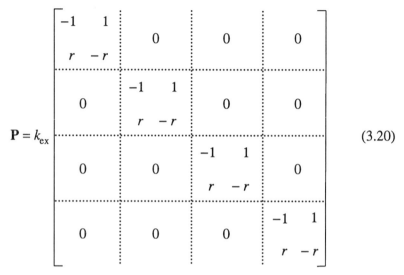

$$
\mathbf{P} = k_{\text{ex}}
\begin{bmatrix}
\begin{matrix} -1 & 1 \\ r & -r \end{matrix} & 0 & 0 & 0 \\
0 & \begin{matrix} -1 & 1 \\ r & -r \end{matrix} & 0 & 0 \\
0 & 0 & \begin{matrix} -1 & 1 \\ r & -r \end{matrix} & 0 \\
0 & 0 & 0 & \begin{matrix} -1 & 1 \\ r & -r \end{matrix}
\end{bmatrix}
\tag{3.20}
$$

Lineshapes computed in this way are compared to the corresponding set of experimental spectra in Figure 3.5.

Examples of multisite exchanges which can be considered to result from the superposition of two-site exchanges are also found for intermolecular rate processes. Let us consider the transfer of ligand A (or solvent) molecules between the coordination (or solvation) sphere about a metal ion and the bulk solution. This results in an NMR exchange between the so-called bound and free sites (b and f, respectively):

$$
A_b \underset{rk_{\text{ex}}}{\overset{k_{\text{ex}}}{\rightleftharpoons}} A_f, \qquad \text{where } r = [A_b]/[A_f]
\tag{3.21}
$$

This may be a simple two-site exchange if the observed nuclei are uncoupled to the rest of the molecule, e.g. when observing the methylic protons of coordinated dimethylsulfoxide $O = S(CH_3)_2$ (DMSO). If a set of coupled nuclei is observed instead, as in the case of the methylic protons of trimethyl-phosphate $O = P(OCH_3)_3$ (TMPA) coupled to ^{31}P nucleus, two NMR patterns, two doublets in the present case (Figure 3.6), are observed for the bound lines (1 and 3, enunciated in the sequence of decreasing frequencies) and free lines (2 and 4), chemically shifted from each other by 0.3 ppm in the example of the

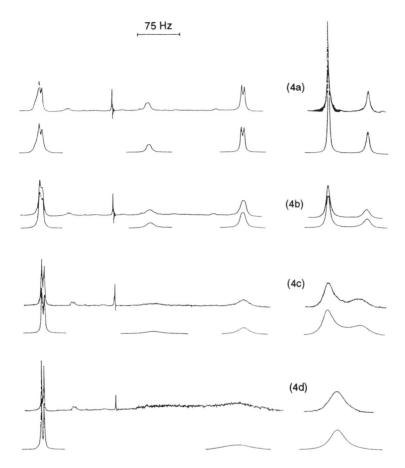

Figure 3.5 Experimental and theoretical curves (above and below respectively) for the exchange of the methylic protons of the amide group (right) and the ring protons H_3, H_4 (left) as a function of the temperature: 15 °C (4a), 25 °C (4b), 35 °C (4c), 45 °C (4d) [55]. Reprinted from Nicole *et al.*, *Tetrahedron*, **36**, 3237. Copyright (1977), with kind permission from Elsevier Science Ltd, The Boulevard, Langford Lane, Kidlington, OX5 1GB, UK3

complex ion Be(TMPA)$_4^{2+}$ dissolved (as its perchlorate) in nitromethane [56], an inert cosolvent. The solution is added with variable concentrations C_f of ligand in excess so as to provide the free site molecule. As the spin states of nuclei in an individual ligand molecule are preserved in the transfer, exchanges occur exclusively between lines 1 and 2 on the one hand, and 3 and 4 on the other, with a common exchange rate k_{ex}. The DNMR problem may be treated as the superposition of two two-site coalescence patterns using the exchange matrix (3.19). A four-site treatment is generally preferred, with a population vector $\mathbf{p} = [r, r, 1, 1]$ and an exchange matrix $\mathbf{P}$, again partitioned into two diagonal submatrices:

$$\mathbf{P} = k_{ex} \begin{bmatrix} -1 & 1 & \vdots & 0 & 0 \\ r & -r & \vdots & 0 & 0 \\ \cdots & \cdots & \vdots & \cdots & \cdots \\ 0 & 0 & \vdots & -1 & 1 \\ 0 & 0 & \vdots & r & -r \end{bmatrix} \qquad (3.22)$$

The relationship to chemical kinetics is less straightforward than in the above examples. Ligand exchange about metal ion (M) complexes MA_n (charges omitted) may proceed through either a dissociative or associative mechanism. In the first event, the rate-determining step is the dissociation of the complex into a transient $(n-1)$-coordinated species, rapidly followed by the reattachment of a ligand molecule from the bulk solution (A^*):

$$MA_n \xrightarrow[\text{(slow)}]{k_d} MA_{n-1}, \quad \text{and} \quad MA_{n-1} + A^* \xrightarrow[\text{(fast)}]{} MA_{n-1}A^* \qquad (3.23)$$

in which case $k_{ex} = k_d$. In the second event, the incoming molecule A^* binds to the complex to form a $(n+1)$-coordinated transition state. This first step is rapidly followed by the removal of one ligand molecule A:

$$MA_n + A^* \xrightarrow[\text{(slow)}]{k_a} MA_nA^* \xrightarrow[\text{(fast)}]{} MA_{n-1}A^* + A \qquad (3.24)$$

The kinetic law is first order with respect to both the bound (C_b) and excess free ligand (C_f) concentrations, so that:

$$\tau_{ex} = C_b/(k_a C_b C_f) = 1/(k_a C_f), \quad \text{or} \quad k_{ex} = k_a C_f \qquad (3.25)$$

This is an example where the NMR rate constants (k_{ex} and rk_{ex}) are different from the chemical rate constant k_a, and are concentration-dependent.

Additional information can be obtained by observing the metal nucleus itself. The ^{9}Be spectrum of Be(TMPA)$_4^{2+}$ is a quintuplet of equally spaced lines, of relative intensities $1 : 4 : 6 : 4 : 1$, as the A part of an AX$_4$-type pattern, in which each ^{9}Be nucleus is coupled to four equivalent ^{31}P nuclei. Lines are sharp and the quintuplet is well-resolved because of slow quadrupolar relaxation of ^{9}Be nuclei in a tetrahedral cubic-symmetrical ligand arrangement. Each of these lines corresponds to a given magnetic state of the set of four coordinated ^{31}P nuclei, respectively: $\alpha\alpha\alpha\alpha$, $\beta\alpha\alpha\alpha$ (and three more degenerate functions by permuting all α and β states), $\beta\beta\alpha\alpha$ (sixfold degenerate), $\beta\beta\beta\alpha$ (fourfold degenerate), $\beta\beta\beta\beta$. The state of one set may be changed whenever *any* of the four coordinated TMPA molecules leaves the complex, this occurs with a frequency four times as large as that for an individual molecule, i.e. the exchange rate among ^{9}Be sites is $4k_{ex}$. Similar information is thus expected from ^{9}Be and ^{1}H NMR, however the range of measurable rates will be extended towards lower k_{ex} values by using ^9Be instead of ^1H NMR (Figure 3.6). The population vector to be used in this case is simply $\mathbf{p} = [1, 4, 6, 4, 1]$. Exchanges may occur only between adjacent lines of the spin multiplet, since one ^{31}P nucleus at most can change its spin state from α to β during the exchange of *one* ligand molecule. Non-zero elements of the probability matrix are those subscripted with two consecutive digits, such as Π_{12}, Π_{21}, Π_{23}, etc. ..... Their values are determined on statistical considerations as usual, e.g. a phosphorus nucleus leaving a complex ion in the ^{31}P magnetic state $\alpha\alpha\alpha\alpha$ (site 1) is substituted by another ^{31}P nucleus either in the α or β state, at

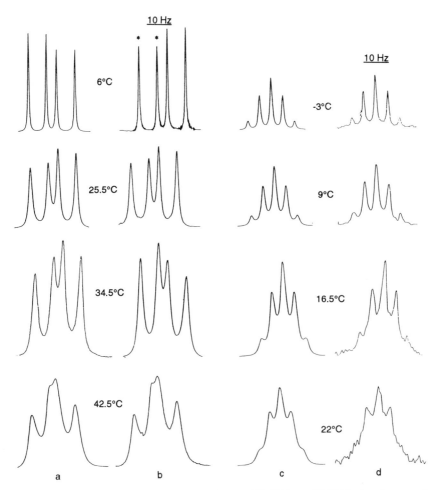

Figure 3.6 DNMR spectra for the exchange $(TMPA)_b \rightarrow (TMPA)_f$ in the complex $Be(TMPA)_2^{2+}$ at different temperatures: a and b, or c and d, experimental and theoretical curves for ^1H and ^9Be nuclei, respectively [56]. Reprinted with permission from Delpuech *et al.*, *New J. Chem.*, **1**, 133. Copyright (1977) Gauthier-Villars

random, with equal probabilities of $1/2$. This means that the only non-zero element in the first row of the Π matrix is $\Pi_{12} = 1/2$. The full matrix is obtained by inspection along the same lines as:

$$\mathbf{P} = 4k_{ex} \begin{bmatrix} -1/2 & 1/2 & 0 & 0 & 0 \\ 1/8 & -1/2 & 3/8 & 0 & 0 \\ 0 & 1/4 & -1/2 & 1/4 & 0 \\ 0 & 0 & 3/8 & -1/2 & 1/8 \\ 0 & 0 & 0 & 1/2 & -1/2 \end{bmatrix} \qquad (3.26)$$

Ligand exchange is accompanied by chemical spin-decoupling of the ^{9}Be quintuplet, in a similar way as described for degenerate proton transfers between like methanol molecules.

In all the above examples, one chemical reaction only is accounting for NMR site exchanges. Other examples illustrate the possible occurrence of several chemical processes simultaneously. If we again consider proton transfers in methanol [16, 57], the NMR exchange rate was found to be strongly dependent upon the purity of methanol and clearly increased by traces of acids or bases. These species, although in too small amounts to be detected by NMR, actually determine the exchange kinetics. In *acidic medium*, an individual methanol molecule is transformed into the conjugate methyloxonium ion. This is the rate-determining step, which is rapidly followed by the opposite deprotonation of the methyloxonium ion, in which either the initial (marked H) or the newly income proton (H*) is carried into the bulk solution, with equal relative probabilities of $1/2$.

$$CH_3OH + H^* \xrightarrow{k_1} CH_3\overset{+}{O}\begin{smallmatrix} H \\ \diagup \\ \diagdown \\ H \end{smallmatrix} \xrightarrow{fast} \begin{smallmatrix} \longrightarrow CH_3OH + H^* \\ \\ \longrightarrow CH_3OH^* + H \end{smallmatrix} \qquad (3.27)$$

The first event only is accompanied by an NMR site exchange, so that the exchange rate k_{ex} (or $1/\tau_{OH}$, where τ_{OH} is the mean lifetime of one hydroxylic proton in methanol) is equal to $k_1 [H^+]/2$. In *basic medium*, a small fraction of methanol molecules are converted into methoxide ions CH_3O^-. Proton transfers are then occurring between neutral methanol and its conjugate anion (not detected by NMR):

$$CH_3OH + CH_3^*O^- \xrightarrow{k_2} CH_3O^- + CH_3^*OH \qquad (3.28)$$

in which case: $k_{ex} = k_2[CH_3O^-] = k_2 K_{CH_3OH}/[H^+]$
where K_{CH_3OH} is the self-ionization constant of methanol. In *neutral media* both processes (3.27) and (3.28) are contributing to chemical exchange, so that

$$k_{ex} (\text{or } 1/\tau_{OH}) = k_1 [H^+]/2 + k_2 K_{CH_3OH}/[H^+] \qquad (3.29)$$

This kinetic law also holds for any other alcohol than methanol. The validity of the scheme is established by studying the variations of k_{ex}, deduced from the

NMR coalescence patterns by using the appropriate exchange matrices (3.18 in the present example), as a function of pH, at constant temperature. It is then seen that exchange rates are accelerated either at higher or lower pH (Figure 3.7), with a minimum value—which may be or not beyond the slow limit of the NMR window—when $[H^+] = (2k_2 K_{CH_3OH}/k_1)^{1/2}$.

The DNMR problem in the above example is easily set because (i) there is one unknown NMR rate constant k_{ex} (ii) the chemical processes result into a common probability matrix. The treatment is more complex in problems where these two constraints are raised. There are further possibilities of simplification permitting an easy approach in some cases. First the chemical processes may have quite different rates so that the NMR timescale for each of them is reached at clearly different temperatures, or at different concentrations of reactants if the reaction is concentration-dependent as mentioned above for the intermolecular and intramolecular processes in NIPA tris-chelates [53].

A second factor of simplification, which may facilitate the set up of complex exchange matrices, is the presence of several independent sets of exchanging lines, each one possibly submitted to different rate processes among those participating to the overall kinetic scheme. This is especially the case when there is one set of lines (or several) subject to one chemical process only. An example is provided by proton transfer investigations on 1-cis-2,6-trimethylpiperidine (P) in acidic solutions [58]. Piperidine P exists in two different geometric isomers A and B, depending on whether the N-methyl substituent is in the equatorial or axial position, respectively. In acidic solutions (pH 0 to 8), the isomeric amines are almost fully protonated (by more than 99%), and the ammonium cations AH^+ and BH^+ only are observed by NMR.

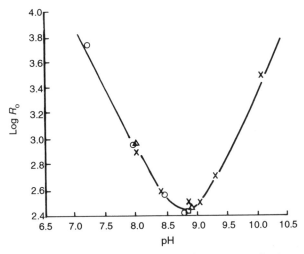

Figure 3.7 The dependence of the rate $R_0 = k_{ex}$ of proton transfer in methanol on the pH at 25 °C [57]. Reprinted with permission from Grunwald *et al.*, *J. Am. Chem. Soc.*, **84**, 4667. Copyright (1962) American Chemical Society

Lines of interest are those of the N-methyl (a doublet for each isomer through coupling to the acidic exchangeable N–H proton) and C-methyl (a doublet for each isomer through coupling to the tertiary vicinal non-exchangeable C–H proton) substituents (Figure 3.8). Two NMR exchanges are observed in this compound resulting from two different pH-dependent rate processes:

(i) Proton transfers E_{HA} (or E_{HB}) in which each isomeric ion AH^+ (or BH^+) is exchanging its acidic proton with those from the bulk solution:

$$AH^+ + H^{*+} \underset{k_{HA}}{\overset{k_{HA}}{\rightleftharpoons}} AH^{*+} + H^+ \qquad (E_{HA})$$

$$BH^+ + H^{*+} \underset{k_{HB}}{\overset{k_{HB}}{\rightleftharpoons}} BH^{*+} + H^+ \qquad (E_{HB})$$

(ii) Nitrogen inversion E_N carrying isomers AH^+ and BH^+ into one another:

$$AH^+ + \underset{k_{NB}}{\overset{k_{NA}}{\rightleftharpoons}} BH^+ \qquad (E_N)$$

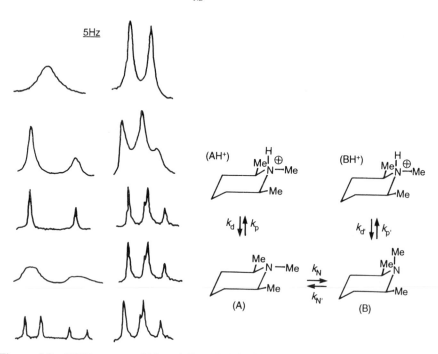

Figure 3.8 NMR spectra of N- and C-methyl doublets (left and middle, respectively) of 0.4 molar aqueous 1-cis-2,6-trimethylpiperidinium ion at 33 °C and pH = 9.33, 8.15, 7.30, 4.20, 0.00 (from top to bottom), and the kinetic scheme for competitive protonation [58]. Reprinted with permission from Delpuech *et al.*, *New J. Chem.*, **2**, 563. Copyright (1978) Gauthier-Villars

The overall kinetic scheme proposed to account for these processes is depicted in Figure 3.8, where k_N (and $k_{N'}$), k_d (and $k_{d'}$), k_p (and $k_{p'}$) are pseudo first-order microscopic rate constants for nitrogen inversion (on the free amines A and B), deprotonation and reprotonation processes, respectively. The two latter processes result from different mechanisms, involving either the solvent or the free amines A and B, so that rate constants k_d, k_p are written as the sum of several contributions, some of which are pH-dependent. NMR exchange rates k_{HA}, k_{HB}, k_{NA} and $k_{NB} = k_{NA} p_{AH}/p_{BH}$ (where p_{AH} and p_{BH} are the fractions of AH$^+$ and BH$^+$ in the mixture of isomers) can in turn be formally expressed as a function of all the above microscopic constants.

The C-methylic doublets offer the example of a set of exchanging lines through one rate process only, *nitrogen inversion* E_N, while the four N-methylic lines are subject to both *proton transfer and nitrogen inversion* processes. The coalescence of the *C-methylic lines* is thus selectively providing information on NMR site exchange E_N. As all spin states are preserved in this intramolecular process, the exchange between the two C-methylic doublets results again from the superposition of two two-site exchanges, treated by using the exchange matrix 3.22. The *N-methylic lines*, denoted as 1,2 (AH$^+$) and 3,4 (BH$^+$) in the sense of decreasing frequencies, suffer three types of exchanges: $1 \Leftrightarrow 2$ (E_{HA}), $3 \Leftrightarrow 4$ (E_{HB}); (1 and 2) $\Leftrightarrow$ (3 and 4) (E_N), so that the exchange matrix is written (omitting diagonal elements) as

$$
\mathbf{P} = \begin{bmatrix}
 & k_{HA} & k_{NA}/2 & k_{NA}/2 \\
k_{HA} & & k_{NA}/2 & k_{NA}/2 \\
k_{NB}/2 & k_{NB}/2 & & k_{HB} \\
k_{NB}/2 & k_{NB}/2 & k_{HB} &
\end{bmatrix} \tag{3.30}
$$

The prior knowledge of k_{NA}, k_{NB} (from the C-methylic lines) will greatly facilitate the adjustment of the two parameters k_{HA}, k_{HB} left unknown in matrix (3.30).

In aqueous solution, a coalescence of each N-methylic doublet into a singlet is observed when the pH is raised from about 3 to 5, while C-methylic lines remain unaltered. This means that the deprotonation of the piperidinium cations is increased by elevation of the pH, and that the free amines obtained A or B are quickly protonated before nitrogen inversion can occur, so that k_{NA}, $k_{NB} \sim 0$, and the exchange matrix (3.30) is partitioned into two independent diagonal submatrices. NMR rate constants k_{HA}, k_{HB} are then obtained from the coalescence of each N-methylic doublet individually, using in each case the simple exchange matrix (3.18) with $k_{ex} = k_{HA}$ or k_{HB}, respectively. The concentration and pH dependence of k_{HA}, k_{HB} yields in turn the chemical rate constants for elementary proton transfers. At pH 7, the two N-methylic doublets are averaged to two sharp singlets of relative intensities p_{AH} and p_{BH}. From pH 7 to 9, a second coalescence occurs between these two singlets, resulting from a two-site exchange which can be studied with matrix (3.20), where $k_{ex} = k_{NA}$ and $r = p_{AH}/p_{BH}$. This second process is brought about by nitrogen inversion on the free amines A and B which are in sufficient amount in this pH range, this view is corroborated by the simultaneous coalescence of the C-methylic doublets (Figure 3.8). Both C-methylic and N-methylic patterns yields k_{NA}, k_{NB} as a function of pH, and, in turn, the nitrogen inversion rates k_A, $k_B \sim 10^3$ s^{-1} at 25 °C.

Similar investigations in DMSO as a solvent [59] instead of water show a simultaneous coalescence of the four N-methylic lines on the one hand and of the four C-methylic lines on the other hand within the same pH range, pH 6 to 7.5. The study of C-methylic lines yields as above the values of k_{NA}, k_{NB} which are then reported into matrix (3.30). The best fit between experimental and computed N-methylic patterns is obtained when parameters k_{HA}, k_{HB} in matrix (3.30) are adjusted so that $k_{HA} = k_{NB}/2$ and $k_{HB} = k_{NA}/2$. Under these conditions, matrix (3.30) can be recast as

$$\mathbf{P} = k_{NT} \begin{bmatrix} \diagdown & P_{AH}/2 & P_{BH}/2 & P_{BH}/2 \\ P_{AH}/2 & \diagdown & P_{BH}/2 & P_{BH}/2 \\ P_{AH}/2 & P_{AH}/2 & \diagdown & P_{BH}/2 \\ P_{AH}/2 & P_{AH}/2 & P_{BH}/2 & \diagdown \end{bmatrix} \tag{3.31}$$

where $k_{NT} = k_{NA} + k_{NB}$. These results, in sharp contrast with those in aqueous solution, were interpreted by a slow reprotonation of the free amines A and B, which then suffer a great number of nitrogen inversions before their reprotonation, this means that $k_p \ll k_A$ in DMSO instead of $k_p \gg k_A$ in water. The protons released from the piperidinium cations with an overall rate constant $k_d = k_{NT}$ are then redistributed at random on A and B molecules, either in the α or β states, in the statistical ratios p_{AH} (or $p_{AH}/2$ for each component of the AH doublet) and p_{BH} (or $p_{BH}/2$), respectively. These ratios are just the elements of the probability matrix (3.31). Deprotonation rates only, at the exclusion of nitrogen inversion rates, can be obtained under these conditions, at least in this pH range. This interpretation was nicely confirmed by ^{1}H$\{^{14}$N$\}$ NMR of the acidic exchangeable proton itself. The NMR spectrum consists of a dozen lines for each isomer and coalescence patterns can effectively be accounted for by using the above k_d value as an exchange rate and a probability matrix whose elements Π_{ij} are, as in matrix (3.31), equal to the relative populations p_j of sites j.

This last example illustrates well the wealth of kinetic information and the potential of NMR to assign complex molecular mechanisms. It is also an illustration of the difficulties encountered by NMR users in the very first steps of DNMR experiments, essentially in setting up exchange matrices by inspection from a kinetic scheme to be devised in each case. Fortunately, the ensuing NMR methodology is more straightforward, using elaborated standard procedures, as we shall see in the next section.

3   Transverse Magnetization Transfer Experiments

Many theoretical developments have been proposed to account for NMR coalescence patterns. They may be classified along three lines of argument which originate in the pioneering work of Gutowsky, McCall and Slichter (GMS) [4], or Anderson [11], Kubo [9] and Sack [10], or else Kaplan [13] and Alexander [14], respectively. The first two methods apply to loosely coupled sites only, when the exchange can be considered as the fluctuations of local magnetic fields parallel to the z axis, generated either by chemical shift Hamiltonians, or by truncated scalar coupling Hamiltonians (see above Section 2.1). These theories rely on classical dynamics of spin motions, and are often called 'adiabatic' to indicate that no transition is induced during exchange. A third approach, based on a non-adiabatic quantummechanical treatment, should be used when the scalar coupling Hamiltonian cannot be reduced to its diagonal part, i.e. in the event of strong couplings. This section will mainly deal with the adiabatic case, which is most important to NMR users because of its generality and simplicity.

Two parallel ways have been used for this purpose, either Bloch equations modified for chemical exchange in steady-state regime (CW-NMR) [4], or stochastic theory for Markoffian random modulation of the processing transverse magnetization $M_{xy}(t) = M_0 \exp i\omega t$ following a 90° RF pulse (FT-NMR), represented by a correlation function [11]:

$$g(t) = \left\langle \exp\left[i \int_0^t \omega'(t')\, dt' \right] \right\rangle$$

where $\omega'(t')$ is a random variable, taking any of the frequency values ω_i of the exchanging sites i. The brackets $\langle \, \rangle$ mean an average over all sequences of ω_i values from time zero to t. The signal in the frequency domain is obtained as expected by Fourier transformation of $g(t)$. The theory, applied to multisite exchanges, led to the most useful expressions in DNMR. However, an equivalent result may be obtained much more economically by an extension of the alternative method using Bloch equations, which will thus be the only one to be developed in the following. A merit of the parallel Anderson–Kubo–Sack theory is to make clear the equivalence of CW-NMR and FT-NMR experiments in the obtaining of coalescence patterns.

3.1   LINESHAPE ANALYSIS FOR TWO WEAKLY COUPLED SITES

A simple but very instructive step in DNMR theory concerns the two-site exchange, which has been the subject of successive improvements and extensions since the early GMS treatment. The most fruitful view [12] is to consider opposite transfers of transverse magnetization with rate constants k_{12} and k_{21} as pointed out in Section 2.1. In the absence of exchange, transverse magnetizations G_1 and G_2 in each site obey individual Bloch equations (1.29):

$$dG_1/dt + a_1 G_1 = i\gamma B_1 M_o p_1$$
$$dG_2/dt + a_2 G_2 = i\gamma B_1 M_o p_2 \tag{3.32}$$

where M_o is the total magnetization over both sites, p_1 and p_2 are the fractional populations in each site, and a_j ($j = 1$ or 2) are complex quantities:

$$a_j = 1/T_{2j} + i(\omega - \omega_j) \quad \text{or} \quad \pi\Delta\nu_{1/2}^j + 2\pi i(\nu - \nu_j) \tag{3.33}$$

NMR exchange between sites 1 and 2 couples the set of equations (3.32):

$$dG_1/dt + a_1 G_1 = i\gamma B_1 M_o p_1 - k_{12} G_1 + k_{21} G_2$$
$$dG_2/dt + a_2 G_2 G_2 = i\gamma B_1 M_o p_2 + k_{12} G_1 - k_{21} G_2 \tag{3.34}$$

Steady-state solutions of equations (3.34) are obtained by putting $dG_1/dt = dG_2/dt = 0$. The overall transverse magnetization G is then

$$G = G_1 + G_2 = i\gamma B_1 M_o \frac{\tau_1 + \tau_2 + \tau_1\tau_2(a_1 p_2 + a_2 p_1)}{(1 + a_1\tau_1)(1 + a_2\tau_2) - 1} \tag{3.35}$$

with the notations given in Section 2.2. Another expression of G is often given as

$$G = i\gamma B_1 M_o[1 + \tau(a_1 p_2 + a_2 p_1)]/[(a_1 p_1 + a_2 p_2 + \tau a_1 a_2)] \tag{3.36}$$

where τ is a kinetic parameter defined as

$$\tau = \tau_1\tau_2/(\tau_1 + \tau_2) = (k_{12} + k_{21})^{-1} = p_1\tau_2 = p_2\tau_1 \tag{3.37}$$

The NMR lineshape $v(\nu)$ is obtained as the v component of $G = u + i\cdot v$, i.e. as the real part of $G/i\gamma B_1 M_o$. Programs have been built to compute G and v over a set of equally spaced values of the RF frequency ν.

The simple case when the two sites are equally populated has been so far the most widely studied. In this case, two systems of notations have been simultaneously employed: $\tau_1 = \tau_2 = \tau_{ex}$, where τ_{ex} is the exchange lifetime used in the exchange matrix (3.18), and $\tau_1 = \tau_2 = 2\tau$ where τ is the kinetic parameter in equation (3.37). This dual notation has been the cause of much confusion in the literature.

3.2   NMR TIMESCALES FOR LINESHAPE ANALYSIS

Two extreme situations are of great importance in lineshape analysis, the so called *fast and slow exchange limits*. These limits, already introduced in Chapter 1 on other grounds, are given by the conditions:

$$\tau|\omega_1 - \omega_2| \ll 1 \quad \text{or} \quad \tau|\omega_1 - \omega_2| \gg 1 \tag{3.38}$$

respectively. Let us discuss these conditions in the light of magnetization transfers and of equation (3.36). To do so, it should be realized that the modulus of complex quantities a_j ($j = 1$ or 2) is mainly determined by their imaginary component $(\omega - \omega_j)$, at least in the case when sharp lines are observed in conditions of no exchange. This means that $|a_j|$ is roughly increasing from $\pi \Delta v_{1/2}^j \sim 0$ to $|\omega_2 - \omega_1|$ in the frequency range of interest. In the vicinity of each site, e.g. site 1, these considerations result in $|a_1| \ll |a_2|$, while the condition of *slow exchange* (3.38) becomes equivalent to $\tau |a_2| \gg 1$. If the two sites are not too unequally populated, i.e. in the absence of one predominant line (cf. Section 3.5), the above conditions may be extended by including p_1, p_2 as multiplicative factors, namely: $|a_1 p_2| \ll |a_2 p_1|$, or $|a_1 p_1| \ll |a_2 p_2|$, and $|\tau a_2 p_1| \gg 1$. In the vicinity of line 1, equation (3.36) can thus be simplified into

$$G = i\gamma B_1 M_0 [\tau p_1 / (p_2 + \tau a_1)]$$

The complex steady-state magnetization can be rearranged into

$$G = i\gamma B_1 M_o / [(1/T_{21} + 1/\tau_1) + i(\omega - \omega_1)] \tag{3.39}$$

and, compared to the similar equation (1.31), is representing a Lorentzian absorption curve, still centered at frequency v_1, with an unchanged relative intensity p_1, and broadened by a quantity $\delta \Delta v_{1/2}^1$ such that:

$$\pi \delta \Delta_{1/2}^1 = 1/\tau_1 \tag{3.40}$$

A similar formula holds for site 2 on changing subscript 1 into 2. This shows that in the slow exchange region an individual line is still observed for each site, and that extra line-broadening induced by chemical exchange is liable to experimental detection as long as it clearly surpasses the natural linewidth $\Delta v_{1/2}^1 \sim ca$ 0.1 to 1 Hz. These conclusions are in accord with those of Section 1.4 (equation (1.49) and Figure 1.4), established on other lines in the archetypal case of two equally populated sites where $\tau_1 = \tau_2 = \tau_{ex}$.

The opposite situation of *fast exchange* when $|\tau a_2| \ll 1$ results in the cancellation of all the terms containing τ in equation (3.36), which is thus reduced to

$$G = i\gamma B_1 M_o / (a_1 p_1 + a_2 p_2) \tag{3.41}$$

The imaginary part of G represents an averaged Lorentzian curve, of intensity $(p_1 + p_2)$, centred at the frequency $v_{av.} = p_1 v_1 + p_2 v_2$, and of linewidth $\Delta v_{1/2}^{av.} = p_1 \Delta v_{1/2}^1$. These conclusions are those already introduced in Chapter 1, and they can be extended in fact to any type of NMR exchange.

In between these two regions of fast and slow exchange, *coalescence* is reached when line-broadening is so large as to be of the same order of magnitude as the frequency separation $|v_1 - v_2|$ between the exchanging sites, in which case both lines have merged into a unique broad resonance. This occurs when:

$$\tau \delta \omega \quad \text{or} \quad 2\pi \tau \, \delta v \sim 1 \tag{3.42}$$

in full analogy with equation (1.52). This is the so-called NMR timescale which the slow and fast limits should be referred to. The dimensionless quantity $\tau \, \delta\omega$ is correspondingly large or small by comparison to unity, and should not vary from this value by more than about plus or minus an order of magnitude for accurate measurements. The corresponding range of τ values is called the *NMR window* for lineshape analysis. It should be observed that NMR timescales depend not only on a kinetic parameter (τ), but also on a purely nuclear parameter $(\delta\omega)$. The NMR timescales are thus determined by chemical shift variations or J scalar couplings. This means that exchange lifetimes τ_{ex} should be comprised between about 0.1 ms to 1 s, and the corresponding exchange rates k_{ex} between 10^4 and 10^0 s^{-1}. In fact, several different NMR timescales may be associated to a same chemical process depending on the nucleus observed. This was clear on some of the examples presented in the previous section, e.g. the ligand exchange on the complex ion $Be(TMPA)_4^{2+}$ observed either by ^1H or ^9Be NMR (Figure 3.6). Finally, NMR timescales can be modified to a certain extent by magnetically changing the nuclear parameter $\delta\omega$, when it represents a chemical shift variation, in either of two ways: (i) Modification of the working frequency of the spectrometer: the NMR timescale may thus be varied by about one order of magnitude in this way; (ii) Modification of resonance frequencies by means of shift reagents: an increase of the difference $\delta\nu = \nu_1 - \nu_2$ obtained in this way results in a displacement of the NMR timescale towards faster exchange rates, which may be monitored to some degree by varying the concentration of the reagent; however, attention should be paid to possible interactions of lanthanide complexes, which often prevent reliable comparisons with results obtained in the absence of reagents.

Exact coalescence is obtained on increasing the NMR rate constant when the two maxima on each individual line have just disappeared (Figure 1.4) and the resulting lineshape is perfectly flat, this occurs for an exchange rate close to but not identical to that given by condition 3.42, depending on the nuclear parameters p_j and $\Delta\nu_{1/2}^j$ on each site. For example, the coalescence of two equal and infinitely sharp lines ($p_1 = p_2 = 0.5$ and $1/T_1 = 1/T_2 = 0$) is obtained from equation (1.50) when

$$\delta\omega . \tau_{ex}/2 \quad \text{or} \quad \pi(\nu_1 - \nu_2)\tau_{ex} \quad \text{or} \quad 2\pi(\nu_1 - \nu_2)\tau = \sqrt{2} \qquad (3.43)$$

3.3   EXPERIMENTAL WINDOWS FOR KINETIC MEASUREMENTS

As NMR exchange rates are related to one or several chemical processes, there are consequently experimental windows for kinetic measurements. Kinetic windows are determined in most investigations by the temperature range over which kinetic rate constants are brought to an order of magnitude resulting in NMR site exchanges within the NMR timescale. This may be called the *temperature window*, which extends on either side of the *coalescence temperature* by plus or minus a few ten degrees, depending on the magnitude of the associated activation enthalpy $\Delta H^{\neq}$. Temperature windows amount to about $\pm 20°$, $\pm 30°$ and $\pm 90\,°C$ if $\Delta H^{\neq} = 60$, 40, or 20 kJ mol^{-1} respectively, for rate constants increasing or decreasing by one order of magnitude around coalescence.

Concentration-dependent reactions may be characterized as well by *concentration windows*, extending (for bimolecular mechanisms) over an order of magnitude on either side of the concentration at coalescence, i.e. over about two units on a logarithmic scale. A peculiar case, encountered in most proton transfer studies, involves pH-dependent reaction rates, which can be measured in most cases only over pH windows of about two units. It should be observed that chemical rate constants are different from NMR exchange rates in such cases. Ultra-fast proton transfers may thus be brought to NMR timescale by simply choosing an appropriate pH window. The required temperature or concentration windows may be located beyond experimental possibilities, in which case the NMR exchange is immeasurably fast or slow, e.g. nitrogen inversion in piperidine could be brought to the NMR timescale at $-140\,°C$ only in 1977 by an appropriate choice of solvent systems ($CHFCl_2$–CHF_2Cl mixture) [60].

A molecular rate process may be assigned one or several kinetic windows, depending on the number of nuclei simultaneously observed, e.g. the aforementioned ligand exchange on Be(TMPA)$^{2+}$ has two overlapping temperature windows, 25–55 °C and 5–35 °C, for ^1H or ^9Be NMR, respectively [56]. The existence of several temperature windows extends the overall temperature range studied and thus improves the reliability of computed activation parameters. Another possibility in this respect is to combine DNMR data with kinetic data obtained by other methods, e.g. by chemical relaxation on the fast side of the DNMR window, or by equilibration methods on the slow side. Thus, measurements of proton transfer rates on piperidinium salts in aqueous acidic solutions were carried out at pH 3 to 5, and pH -3 to -1, by DNMR and hydrogen-deuterium isotopic substitution, respectively [61]. Another example, mentioned in previous sections, concerns pH controlled nitrogen inversion on 1-*cis*-2,6-trimethylpiperidinium ion in aqueous solution [58], which may be measured either by DNMR from pH 7.5 to 9, or from the slow conversion of the *cis* to the *trans* N-Me isomer in the pH range 3 to 5. Diagrams showing the continuity between plots of kinetic rates within both pH windows are particularly convincing to assess the validity of the mechanisms accounting for chemical exchange (Figure 3.9 and 3.10).

If several exchanges are simultaneously occurring in a given system, there are consequently several associated kinetic windows. Thus, two rate processes are simultaneously present in the piperidinium salt mentioned above, namely either proton transfer or nitrogen inversion, each one occurring in a well separated pH window, at pH 3–5 and 7.5–9 respectively. Another example is afforded by metal ion tris-chelates, where internal rearrangement and intermolecular exchange of ligands are observed at low and high temperature, respectively. A third example concerns unsubstituted neutral piperidine, where

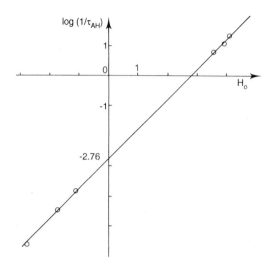

Figure 3.9 Plots of proton transfer rate $1/\tau_{AH}$ on 0.4 molar aqueous 1-*cis*-2,6-trimethylpiperidinium ion as a function of pH at 33 °C. Reproduced by permission of John Wiley & Sons from Ref. 61

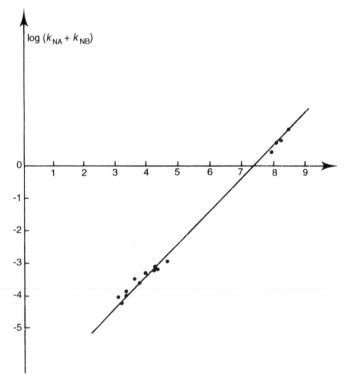

Figure 3.10 Plots of the overall NMR site exchange rate $k_{NH} = k_{NA} - k_{NB}$ owing to nitrogen inversion in 0.4 molar aqueous 1-*cis*-2,6-trimethylpiperidinium ion as a function of pH at 27 °C [58]. Reprinted with permission from Delpuech *et al.*, *New J. Chem.*, **2**, 563. Copyright (1978) Gauthier-Villars

two conformational processes are possible, namely cyclohexane-like ring reversal (I_R) and nitrogen inversion (I_N) which interchange CH_2 and NH protons, respectively, between equatorial and axial positions. Two coalescence temperatures were effectively observed, one at $-60\,°C$ and the other at $-140°\,C$ which were assigned to I_R and I_N processes, respectively [60]. The existence of two rate processes was in fact recognized in all six-membered N-heterocycles, and the assignment of the low and high temperature windows was the matter of many errors and controversies in the past [62].

3.4   EVALUATION OF RATE CONSTANTS

There is a constant interest in the computation of rate constants in two-site exchanges [63–66], since this case remains by far the most familiar to chemists. Three steps can be distinguished in this evaluation: the extraction of NMR rate constants from NMR patterns, the kinetics of the chemical processes involved and, eventually, the derivation of the associated activation parameters through operations at *variable temperature*.

NMR exchange rates k_{12} and k_{21} are easily obtained at the two ends of the NMR window. In the *slow exchange* region, the extra linebroadenings $\delta\Delta_{1/2}^i$, independently measured on each line i, yield the exchange lifetimes $\tau_1 = 1/k_{12}$ and $\tau_2 = 1/k_{21}$ according to equation (3.40). In fact, the knowledge of the ratio of populations $r = p_1/p_2$ may restrain these measurements to one site by using the conservation law (3.5); this is necessary when the other site is too small and/or too broad for accurate measurements (Section 3.5). Measurements are reliable as long as the ratio $\Delta v_{1/2}^j/\delta v$ is close to zero, i.e. when lines are sharp compared to the frequency interval δv in conditions of no exchange. In the opposite situation, theoretical simulations are required to improve the accuracy. In any case, kinetic parameters become fully uncertain when $\delta\Delta v_{1/2}^i$ approaches the non-exchanging linewidth $\Delta v_{1/2}^i$.

When exchange rates are located above the coalescence point while remaining moderately fast, there is one line centred at the average frequency $v_{av.}$ (equation (3.41)). This line is, however, broader than the average linewidth $\Delta v_{1/2}^{av.}$, there is an extra linebroadening $\Delta v_{1/2}^{av.}$ due to incomplete averaging, which may be computed from equation (3.36) as:

$$\pi\Delta v_{1/2}^{av.} = 4\pi^2 p_1^2 p_2^2\,\delta v^2(\tau_1 + \tau_2) \quad \text{or} \quad 4\pi^2 p_1 p_2 \tau\,\delta v^2 \tag{3.44}$$

For equally populated sites, equation (3.44) simplifies to:

$$\pi\Delta v_{1/2}^{av.} = \tau_{ex}\,\delta\omega^2/8 \tag{3.45}$$

which is identical to equation (1.45) where $\Delta\omega = \delta\omega/2$ and $\tau_1 = \tau_2 = \tau_{ex}$. Limitations in the usefulness of this formula are of the same nature as for slow exchange rates: (i) The extra line-broadening should be substantially larger than the original linewidths; (ii) Theoretical simulations are necessary as a last step of computation, mainly in the case of two closely spaced NMR lines.

The maximum sensitivity in lineshape analysis is expected not at the extremities of the NMR window, but in its central part near the coalescence point. The accuracy of measurements in this region is, however, limited by a subsequent decrease of the signal-to-noise ratio S/N, by a factor of *ca.* $\delta v / \Delta v_{1/2}$, owing to the fact that the overall intensity is spread over the frequency interval δv. High-performance spectrometers and concentrated samples are required for a good accuracy at coalescence. In the case of widely separate resonances, no signal at all can even be detected at coalescence, and DNMR measurements should then be performed on the contrary at the two ends of the NMR window, where now they have precisely a good degree of accuracy due to the high value of the ratio $\delta v / \Delta v_{1/2}$.

Rate constants at coalescence have an order of magnitude given by equation (3.42). Exact values, however, depend on the static line intensities p_i, linewidths $\Delta v_{1/2}^i$, and the frequency interval δv, they should be preferably determined by theoretical simulation of NMR patterns and adjustment of kinetic parameters by comparison through trial and error procedures starting their first iteration with the above approximate value. The simulation can be performed with home-made computer programs built on the basis of equation (3.36), or else standard programs using matrix formalism (Sections 2.2 and 3.6). The comparison of calculated and experimental curves can be made either visually or automatically by least squares procedures in the so called *total lineshape analysis.*

The case of two equally populated sites has been extensively studied in the past. There exists in this case the simple formula (3.43) for the exchange rate at coalescence. This formula is valid only in the case of infinitely sharp lines, i.e. when the ratio $\delta v / \Delta v_{1/2}$ is larger than about 10. Under the same assumptions, exchange rates k_{ex} just below the coalescence point can be obtained by measuring the shifted positions of the two peaks with a line separation $\delta v'$ such that, according to equation (1.51):

$$k_{ex} = \pi [(\delta v^2 - \delta v'^2)/2]^{1/2} \qquad (3.46)$$

In all other cases [64, 65] and for rate constants outside coalescence, abacus have been drawn to determine τ as a function of dimensionless parameters characterizing NMR profiles for various ratios $\delta v / \Delta v_{1/2}$. Total lineshape analysis should be used, however, whenever reliable values are necessary, this is especially true for further computations of activation parameters (see below).

Owing to computing facilities presently available, uncertainties from the DNMR methodology itself may be generally considered to be negligible as compared to other sources of error, namely those on the temperature and the static parameters δv, $\Delta v_{1/2}^i$ and p_i. Temperature gradients within the sample tube and errors in the temperature scale are commonly encountered in DNMR, in spite of technological progress to eliminate them [33]. NMR parameters should be determined effectively in the absence of exchange. Fractional populations p_i (at the exception of degenerate exchanges where p_i values are determined statistically) are measured from line intensities at temperatures or concentrations well below those corresponding to the NMR window. In the more common event of concentration-independent equilibria (K), this requires extrapolating p_i values up to the temperature window by measuring the free energy $\Delta G°$, the enthalpy $\Delta H°$ and entropy $\Delta S°$ of equilibrium at lower temperatures [63]. Chemical shifts contained in δv may also depend on the temperature (while scalar couplings J are much less sensitive to thermal effects), and should be extrapolated by some empirically determined continuity law up to the NMR window. Linewidths $\Delta v_{1/2}$ depend both on the setting of the spectrometer (through B_0 inhomogeneities) and on the temperature (through correlation times appearing in T_2). Estimations of linewidths $\Delta v_{1/2}$ are performed in either of two ways (i) Measurement of $\Delta v_{1/2}$ at low temperatures and extrapolation to the NMR window using semiempirical Arrhenius-like plots, or (ii) Measurements of linewidths on resonances not involved in the rate process. For all these reasons, DNMR measurements are suspected to be less accurate than those from classical kinetics. Comparisons have been made on a few systems where both types of measurements are simultaneously possible [38, 58, 63]. However, even in this exceptional case, comparisons are not straightforward, since the experimental windows in both types of experiments are generally quite different, especially for non-overlapping temperature windows, in which case the comparison can be made only on activation enthalpies and entropies (see below).

A second step is to derive chemical rate constants from exchange rates using an appropriate kinetic scheme for this purpose, as explained in Section 2.3. *Concentration-independent* exchange rates are usually assigned to some first order or pseudo first-order rate-determining step. This is the case for most chemical reorganization processes which are generally assumed to proceed along a reaction path with one transition state (TS) neglecting all possible intermediate states. Under these circumstances, NMR rate constants k_{12} and k_{21} are identified to forward and return rate constants k_1, k_2 for the configurational or conformational equilibrium:

$$1 \underset{k_2}{\overset{k_1}{\rightleftarrows}} 2 \qquad\qquad (3.47)$$

where $K = k_1/k_2 = k_{12}/k_{21}$, if sites 1 and 2 are assigned to isomers **1** and **2**. The equilibrium (3.47) should not be too biased in either sense, otherwise one isomer cannot be detected, depending on the sensitivity of the spectrometer. This limits K values between ca 10^{-2} and 10^{+2}, and, consequently, standard free energy barriers between ca ± 10 kJ mol^{-1}. This explains why nitrogen inversion can be studied by DNMR on unsubstituted piperidine, and not on N-methylpiperidine, due to a standard free energy $\Delta G°_{e \to a}$ between the N-axial and N-equatorial positions of 1.7 and 11.3 kJ mol^{-1}, respectively. The best situation in this respect is that of *degenerate equilibria*, where $K \equiv 1$ and $\Delta G° = 0$.

Standard free energies are themselves the difference $\Delta G_1^\ddagger - \Delta G_2^\ddagger$ between half-barriers $\Delta G_1^\ddagger$ and $\Delta G_2^\ddagger$ for forward and reverse rate processes, respectively. Half-free energy barriers are always positive, while standard free energies may be positive, zero, or negative, depending on the relative magnitudes of $\Delta G_1^\ddagger$ and $\Delta G_2^\ddagger$. Each half-free energy barrier is defined as accompanying either of the two fractional reaction paths $\mathbf{1} \to TS$ or $\mathbf{2} \to TS$, according to the *Eyring rate equation*:

$$k_i = \kappa(k_B T/h)\exp(-\Delta G_i^\ddagger/RT) \tag{3.48}$$

where the transmission coefficient κ is generally taken as unity, k_B is the Boltzmann constant, and h is the Planck's constant. Standard and activation free energies are often quoted in conformational analysis in place of equilibrium and rate constants, from which they can be computed (in $J\ mol^{-1}$) as:

$$\Delta G° = -19.143T \log K \tag{3.49}$$

$$\Delta G_i^\ddagger = 19.143T(10.318 - \log k_i/T) \tag{3.50}$$

Comparisons of standard and activation free energies are often made for structurally related sets of compounds. These comparisons should indicate which is the half-reaction being considered ($\mathbf{1} \to \mathbf{2}$, or $\mathbf{2} \to \mathbf{1}$), (except for degenerate isomerizations), and they should be preferably performed for a common temperature, except if the entropic contribution $T\Delta S$ to ΔG may be assumed to be negligible compared to the enthalpic contribution ΔH, itself considered as temperature independent. This assumption is commonly accepted for conformational equilibria, this permits to compare $\Delta G°$ and $\Delta G^\ddagger$ values taken at the coalescence temperature T_c of each compound.

If it is not the case, activation enthalpies and entropies should be determined through operations at variable temperature. $\Delta H_i^\ddagger$ and $\Delta S_i^\ddagger$ are determined from the least squares slope and intercept of a linear plot of $\log(k_i/T)$ versus $1/T$:

$$\log(k_i/T) = 10.318 - \Delta H_i^\ddagger/(19.143\ T) + \Delta S_i^\ddagger/19.143 \tag{3.51}$$

A variant is to use an Arrhenius plot of $\log k_i$ versus $1/T$:

$$\log k_i = \log A_i - E_i^\ddagger/(19.143\ T) \tag{3.52}$$

in which case:

$$\Delta H_i^\ddagger\ (J\ mol^{-1}) = E_i^\ddagger - 8.314T$$

$$\Delta S_i^\ddagger\ (J\ K^{-1}\ mol^{-1}) = 19.143 \log(A_i/T) - 205.8 \tag{3.53}$$

Standard free energies themselves are also varying with the temperature. If their variations cannot be neglected over the temperature range of interest, standard enthalpy and entropy, $\Delta H°$ and $\Delta S°$, should be computed from a linear plot of log k versus $1/T$ (obtained at temperatures below the NMR window):

$$\log K = -\Delta H°/(19.143\ T) + \Delta S°/19.143 \qquad (3.54)$$

Other examples of concentration-independent kinetics are the dissociative intermolecular exchange of ligands around metal ions, e.g. in the complex ion $Be(HMPA)_4^{2+}$, where $HMPA = OP[N(CH_3)_2]_3$, or proton transfers between an acid AH (charges omitted) and the solvent SH. In the first example, the metal ion Be^{2+} is assumed to be totally coordinated. The kinetic scheme involves the dissociation of the complex to yield an undetectable tricoordinated intermediate $Be(HMPA)_3^{2+}$; this is the rate-determining step, quickly followed by the reattachment of a ligand molecule ($HMPA^*$) from the bulk nitromethane solution [56]:

$$Be(HMPA)_4^{2+} \underset{\text{fast}}{\overset{k_d}{\rightleftharpoons}} HMPA + Be(HMPA)_3^{2+} \underset{\text{fast}}{\overset{k_d}{\rightleftharpoons}} Be(HMPA)_3(HMPA^*)$$

$$\Big\Updownarrow \text{fast}$$

$$HMPA^* \text{ (free)}$$

Free HMPA in excess plays only the role of a reservoir allowing ligand molecules to change their resonant frequency in the course of the dissociation-reassociation process. The overall chemical reaction equivalent to the above sequence:

$$Be(HMPA)_4^{2+} + HMPA^* \underset{k_d}{\overset{k_d}{\rightleftharpoons}} Be(HMPA)_3(HMPA^*)^{2+} + HMPA$$

is degenerate, while the NMR site exchange is not so, being of the type:

$$(HMPA)_{\text{bound}} \underset{k_{21}}{\overset{k_{12}}{\rightleftharpoons}} (HMPA)_{\text{free}}$$

It should be clear that, in sharp contrast to the above examples of conformational equilibria, the population ratio in the two sites has no chemical significance. In particular this ratio should not be identified with the equilibrium constant for the formation of the complex. The unique information brought by DNMR is the activation barrier between the four-coordinated complex in the ground state and the assumed three-coordinate intermediate:

$$\Delta G^{\neq} = -RT \log k_d \qquad (3.55)$$

Operations at variable temperature reveal the enthalpic and entropic contribution to $\Delta G^{\neq}$.

In the second example, the proton transfer is an intermolecular process, usually assumed to be bimolecular:

$$AH + SH \underset{k_2'}{\overset{k_1'}{\rightleftharpoons}} A + SH_2 \qquad (3.56)$$

The kinetics of deprotonation is, however, of pseudo first-order because of the large and constant excess of solvent molecules SH . The reverse reprotonation is assumed to be very fast, so as to deal with negligible and undetected amounts of species A and SH_2. This is equivalent to a degenerate proton transfer between like molecules AH, promoted by the presence of the solvent. DNMR investigations are possible only in the case when the acidic proton is coupled to some nucleus in the non-exchanging part A of the molecule AH (see Section 2.1). NMR studies can only yield in this context the rate constant for deprotonation

$$k_{ex} = k_{12} (\text{or } k_{21}) = k_0 \quad \text{with} \quad k_0 = k'_1 [S]$$

Other possibilities of transfers have been devised in the case of hydrogen-bonded solvents, typically water, involving solvent molecular clusters [16].

Examples of *concentration-dependent kinetics* have been given in Section 2.3. Complex kinetics may involve several reactions simultaneously, each one of any order, or even with no order at all. Under these conditions, activation parameters computed from overall NMR exchange rates k_{ex} are meaningless. Kinetic investigations should therefore first separate the contributions of each reaction through operations at variable concentrations and the simultaneous observation of several NMR patterns. Activation parameters are determined in a second step through operations at variable temperature. A last example will illustrate this point. It has been observed in many cases that reaction (3.56) contributes to a negligible extent to proton transfers in solution. Proton transfers are predominantly effected by the conjugate bases of either the substrate or the solvent [15, 16]:

$$AH + A \underset{k_1}{\overset{k_1}{\rightleftharpoons}} A + AH \qquad (3.57)$$

$$AH + S \underset{fast}{\overset{k_2}{\rightleftharpoons}} A + SH \qquad (3.58)$$

In either reaction, the rate-determining step is the deprotonation of the acidic substrate AH, so that the exchange rate k_{ex} on AH is

$$k_{ex}{}^{(AH)} = k_0 + k_1 [A] + k_2 [S]$$

or, else, by introducing the ionization constants K_{AH} and K_{SH} of the substrate and the solvent:

$$k_{ex}{}^{(AH)} = k_0 + (k_1 K_{AH} [AH] + k_2 K_{SH}) / [H^+] \qquad (3.59)$$

The three rate constants k_0, k_1, k_2 at one temperature have to be determined separately, this is achieved (i) by analysing the NMR patterns obtained at constant substrate concentration [AH] and variable pH, this yields k_0 and the sum $(k_1 K_{AH}[AH] + k_2 K_{SH})$, and then (ii) by repeating this cycle of operations for a set of different concentrations [AH] in order to separate in turn the two terms in the above sum. The way to such studies was opened by the pioneering work of Grunwald and coworkers [15] on methylammonium salts $(CH_3)_n$ NH^+_{4-n} ($n = 0$ to 3) in aqueous acidic solution (where the conjugate base A is the corresponding free amine in undetectable amounts); their conclusions were a negligible contribution of reaction (3.56) except in very acidic medium, and variable relative contributions of reactions (3.57) and (3.58) depending on steric hindrance around the nitrogen atom, with orders of magnitude for k_1 and k_2 of ca 10^8 and 10^{10} mol^{-1} dm^3 s^{-1}, respectively, close to the diffusion-limited values. Another interesting conclusion was obtained by simultaneously measuring exchange rates $k_{ex}^{(H_2O)}$ of protons in water molecules (see the next Section 3.5), and comparing the values thus obtained to those computed from equation (3.58) as

$$k_{ex}^{(H_2O)} = k_2[S], \qquad \text{where } S = OH^- \qquad (3.60)$$

A large discrepancy was observed between these two evaluations, which were unexpectedly found to be also dependent on the salt concentration. This was accounted for by devising a third reaction of transfer implying one water molecule in a concerted mechanism, whose efficiency was demonstrated later on by theoretical computations [67]:

$$AH + O\!\!-\!\!H + A \underset{k_3}{\overset{k_3}{\rightleftharpoons}} A + H\!\!-\!\!O + HA \qquad (3.61)$$
$$\qquad | \qquad\qquad\qquad | $$
$$\qquad H \qquad\qquad\qquad H$$

The rate law (3.60) was then completed into:

$$k_{ex}^{(H_2O)} = k_2[S] + k_3[A] = (k_2 K_{SH} + k_3 K_{AH}[AH])/[H^+] \qquad (3.62)$$

while k_1 in equation (3.59) was replaced by the sum $(k_1 + k_3)$. Measurements of both exchange rates $k_{ex}^{(AH)}$ and $k_{ex}^{(H_2O)}$ at variable pH and concentrations [AH] led to the determination of the four rate constants k_0, k_1, k_2, k_3.

3.5   EXCHANGES INVOLVING ONE PREDOMINANT SITE

Exchanges involving a predominant species **1** are of great importance in a few specific cases: (i) The dilute species **2** is paramagnetic and present in small concentration so as not to perturb the observation of the diamagnetic species **1**; (ii) Solute **2** is a diamagnetic species either standing in a biased equilibrium with species **1**, as e.g. in certain conformational equilibria, or (iii) added in small amount as a chemical reagent, e.g. traces of solvated hydronium ion H^+ to catalyse proton transfers on the predominant species **1**.

In such situations, one line only is observed at frequency v_1 for the abundant species, while the minor species may either be observed as a small line at frequency v_2, or even vanish into the baseline noise. DNMR phenomena may be detected, provided that $p_2 \delta v$ is greater than a few hertz (see below), as a weak line-broadening $\delta \Delta v_{1/2}$ of the major site, which passes by a maximum at the coalescence point, and then is followed by a resharpening of lines above coalescence. Simultaneously, the intense line is slightly shifted from its initial position at frequency v_1 by $\Delta \omega_{obs}$ (or Δv_{obs} in Hz). A general deceptive feature is the disappearance of the smaller line even before the larger line be somewhat altered by exchange. This is a consequence of a much smaller lifetime τ_2 in the minor site since:

$$p_2 \ll p_1; p_1 \pm 1 \quad \text{and} \quad \tau_2 \sim p_2 \tau_1 (= \tau) \tag{3.63}$$

Formulas for $\Delta \omega_{obs}$ and $\delta \Delta v_{1/2}$ have been given by Swift and Connick [68] in the case of a paramagnetic species **2**, but they may be applied to any two-site exchange involving one predominant site. A noteworthy and often ignored feature of these formulas is their validity throughout the NMR window, as well in the fast as in the slow exchange regions and at coalescence. They were established from modified Bloch equations (3.35) under the simplifying conditions (3.63):

$$\pi \delta \Delta v_{1/2} = \frac{p_2}{\tau_2} \frac{1/T_{22} (1/T_{22} + 1/\tau_2) + \delta \omega^2}{(1/T_{22} + 1/\tau_2)^2 + \delta \omega^2} \tag{3.64}$$

$$\Delta \omega_{obs} = \frac{p_2 \delta \omega}{(\tau_2 / T_{22} + 1)^2 + \tau_2^2 \delta \omega^2} \tag{3.65}$$

The first example of application is ligand exchange in labile complexes of paramagnetic ions (see also Chapters 4 and 6). The free ligand in excess (often the pure solvent itself) is called the *diamagnetic site D* (site 1 with the above notations), while ligand molecules in the first coordination sphere constitute the *paramagnetic site M*, with $p_D \pm 1$ and $p_M = n_M C_S / C_L \ll 1$, where C_S and C_L are the analytical concentrations of metal ion and ligand, and n_M is the coordination number in the complex. In this specific example, τ_2, T_{22} and $\delta \omega$ are denoted as τ_M, T_{2M}, and $\Delta \omega_M = \omega_M - \omega_D$; in equation (3.64), line-broadening $\pi \delta \Delta_{1/2}$ is rather expressed as the difference in relaxation rates $(1/T_2^{obs} - 1//T_{2D})$. Formulas (3.64) and (3.65) show that quantities $(1/T_2^{obs} - 1/T_{2D})$ and $\Delta \omega_{obs}$ should be proportional to p_M at constant temperature. This is the very first step to control the applicability of Swift and Connick's relationships. The proportionality coefficients may be called the *specific* relaxation rate $1/T_{2r}$ and chemical shift variation $\Delta \omega_r$, so that equation (3.64) and (3.65) are recast under the operational forms:

$$1/T_{2r} = \frac{1}{\tau_M} \left[\frac{1/T_{2M} (1/T_{2M} + 1/\tau_M) + \Delta \omega_M^2}{(1/T_{2M} + 1/\tau_M)^2 + \Delta \omega_M^2} \right] \tag{3.66}$$

$$\Delta\omega_r = \frac{\Delta\omega_M}{(\tau_M/T_{2M} + 1)^2 + \tau_M^2 \Delta\omega_M^2} \tag{3.67}$$

$\Delta\omega_M/\omega$ is the paramagnetic or Knight shift, itself proportional to the reciprocal of temperature:

$$\Delta\omega_M/\omega = 2\pi A\mu_M[S(S+1)]^{1/2}/(3k_B T\gamma) \tag{3.68}$$

where A is the hyperfine coupling constant (in hertz), μ_M and S are the electronic magnetic moment and spin number of the cation, γ is the gyromagnetic ratio of the observed nucleus.

As the paramagnetic site cannot be observed in most cases, three unknowns T_M, $\Delta\omega_M$ and $1/T_{2M}$ are appearing in equations (3.66) and (3.67). The apparent undeterminacy is raised by relating T_{2M} to T_{1M}, the longitudinal relaxation time in the bound site, itself determined by T_1 measurements (see below, Section 4.2), e.g. by putting $T_{2M} = T_{1M}$ if motional narrowing conditions are fulfilled. The system of equations (3.66) and (3.67) may then yield $1/\tau_M$ and $\Delta\omega_M$ through some simple algebra, provided that exchange rates $1/\tau_M$ are within the NMR window.

The NMR timescale is determined by the frequency shift $\Delta\omega_M$ between the two sites, generally much higher than those usually found in diamagnetic systems, typically 10^3–10^4 Hz, so that the NMR exchange rate $k_M = 1/\tau_M$ (from the bound to the free site) at coalescence is of the order of $\Delta\omega_M \sim 10^4$–10^5 s^{-1}. Values of the chemical rate constants themselves depend on the kinetic law for ligand exchange as explained in the previous section. In the slow or fast exchange regions, equation (3.67) simplifies to $\Delta\omega_r = 0$ and $\Delta\omega_r = \Delta\omega_M$, respectively. This shows that $\Delta\omega_M$ can be determined in practice at high temperatures where $\Delta\omega_M = \Delta\omega_r$, and then extrapolated to any temperature by means of equation (3.68).

The kinetic information is then extracted from linewidths measurements and equation (3.66). Two extreme situations are usually met, where either $\Delta\omega_M^2 \gg (1/T_{2M})^2$ or $\Delta\omega_M^2 \ll (1/T_{2M})^2$ depending on whether the paramagnetic linewidth is small or large compared to the line separation $A\omega_M$. In the first event, equation (3.66) simplifies to:

$$1/T_{2r} = \Delta\omega_M^2 \tau_M/(1 + \tau_M^2 \Delta\omega_M^2) \tag{3.69}$$

Two limiting laws are obtained in the slow or fast exchange regions, when $\tau_M^2 \Delta\omega_M^2$ is (a) either much larger or (b) much smaller than unity, respectively:

$$\text{(a) } 1T_{2r} = 1/\tau_M, \quad \text{or} \quad \text{(b) } 1/T_{2r} = \Delta\omega_M^2 \tau_M \tag{3.70}$$

In a doubly logarithmic plot of $1/T_{2r}$ versus $1/\tau_M$, the limiting laws (3.70) are represented by two straight lines (a) and (b) of opposite slopes connected to each other by a curved line passing by a maximum (A) at coalescence when $1/\tau_M = \Delta\omega_M$ (Figure 3.11). The straight line in the fast exchange region is limited to its intersection (B) with the asymptotic horizontal line $1/T_{2r} = 1/\tau_M$

(c) obtained for infinitely fast exchanges. This point represents the fast exchange limit when $1/\tau_M = \Delta\omega_M^2/T_{2M}$, typically $1/\tau_M = 10^4 - 10^5 \text{ s}^{-1}$ if $\Delta\omega_M = 10^3 - 10^4 \text{ Hz}$ and $T_{2M} = 10^{-2} - 10^{-3} \text{ s}$. The slow exchange limit on the other hand is determined by the smallest fraction p_M^{min} of bound ligand such that the line-broadening in the diamagnetic site just exceeds the static linewidth $\Delta\nu_{1/2}^D$, this corresponds to

$$1/\tau_M(\text{min}) = \pi\Delta\nu_{1/2}^D/p_M^{min} \qquad (3.71)$$

typically $1/\tau_M(\text{min.}) = 10^2 - 10^3 \text{ s}^{-1}$ if $\Delta\nu_{1/2}^D \sim 1 \text{Hz}$ and $p_M^{min} = 10^{-2} - 10^{-3}$. The temperature window corresponding to the NMR window should be examined in each case, since the above parameters $\Delta\omega_M$, $1/T_{2M}$ and p_M^{min} are themselves depending on the temperature and on the nature of the system examined, e.g. the graph in Figure 3.11 refers to the example of DMSO exchange on the hexasolvate $Co(DMSO)_6^{2+}$ at 25 °C, observed by ^{13}C NMR at 22.63 MHz [69].

In the alternative event when the Knight shift is smaller than the linewidth in the paramagnetic site $(\Delta\omega_M^2 \ll (1/T_{2M})^2)$, equation (3.66) simplifies to

$$1/T_{2r} = 1/(T_{2M} + \tau_M) \qquad (3.72)$$

In this case, $1/T_{2r}$ is continuously increasing from the slow to the fast exchange region, reaching an asymptotic value of $1/T_{2M}$ when the NMR exchange rate $1/\tau_M$ is larger than $1/T_{2M}$. There is no singularity at the coalescence point. Measurable exchange rates $1/\tau_M$ should be brought to an NMR window limited by the values $\pi\Delta\nu^D_{1/2}/p_M^{min}$ and $1/T_{2M}$ at the slow and fast ends, respectively. An example of this situation was found for DMSO

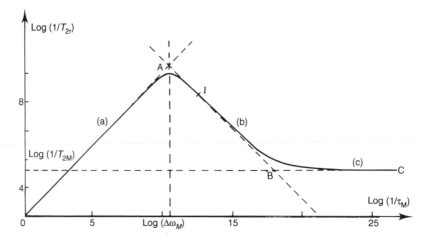

Figure 3.11 Theoretical dependence of the specific relaxation rate $1/T_{2r}$ on the exchange rate $1/\tau_M$ for constant values $\Delta\omega_M = 6580 \text{ Hz}$ and $1/T_{2M} = 26 \text{ s}^{-1}$, in the example of the $Co^{2+}/$ DMSO system observed by ^{13}C NMR at 22.63 MHz and 25 °C

exchange [70] on the hexasolvate $Fe(DMSO)_6^{3+}$ in pure DMSO at 95°–135 °C, observed by ^1H NMR at 14.0 MHz, where $\Delta\omega_M \sim 10^3$ Hz and $1/T_{2M} \sim 10^4$ s^{-1}.

The results of this discussion apply as well to fully diamagnetic as to paramagnetic systems. Two equations are obtained, with the notations of equations (3.64), in full analogy to equations (3.69) and (3.72) when the line in site 2 is sharp or broad compared to line separation:

$$\pi\delta\Delta\nu_{1/2} = p_2\delta\omega^2\tau_2/(1 + \tau_2^2\delta\omega^2) \quad \text{if} \quad \delta\nu \gg \Delta\nu_{1/2}^2 \qquad (3.73)$$

$$\pi\delta\Delta\nu_{1/2} = p_2/(T_{22} + \tau_2) \quad \text{if} \quad \delta\nu \ll \Delta\nu_{1/2}^2 \qquad (3.74)$$

Both equations have the noticeable property of being valid throughout the NMR window.

An important application of equation (3.74) is found in the binding of small ligands to macromolecules. The bound ligand has a reduced mobility and consequently a large relaxation rate $1/T_{22}$. The concentration of the macro-molecule is generally small, and thus that of the bound ligand, assumed to be exchanging with free ligand in large excess, so that all the conditions are met for a successful use of equation (3.74). Some enzyme-substrate, inhibitor-enzyme, or protein-drug complexes have been studied in this way [39–43].

Formula 3.73 has been typically applied to proton transfers [54] between a predominant species, usually the solvent, and a substrate in smaller amount, as in aqueous solutions of methylammonium salts examined in Section 3.4, equations (3.57) to (3.62). Equation (3.73) is the clue to DNMR studies when the acidic proton in the predominant site is not coupled to the rest of the molecule, thus precluding chemical spin-decoupling effects. This is the case for proton transfers in aqueous solution, except if isotopic $H_2^{17}O$ is used in the place of $H_2^{16}O$. Line separations are usually much smaller in diamagnetic than in paramagnetic systems, so that DNMR effects can be detected only in the region of coalescence, where equation (3.73) predicts a maximum broadening

$$\delta\Delta\nu_{1/2}(\max) = p_2|\nu_1 - \nu_2| \qquad (3.75)$$

$$\text{when} \quad 1/\tau_2 = \delta\omega \quad \text{or} \quad k_{21} = 2\pi|\nu_1 - \nu_2| \qquad (3.76)$$

The extension of the method towards smaller and smaller populations p_2 of the minor species is of course limited by the accuracy in linewidth measurements, i.e. to $ca.$ 1 hertz. Therefore, the fractional population of the minor species should be approximately greater than $\delta\omega^{-1}$. In this respect, ^{13}C NMR is better than ^1H NMR because line separations are intrinsically greater. A clever application of equation (3.75) has been devised to study biased equilibria between two conformers 1 and 2, where the second is in too small quantity to be detected [71]. The fractional population p_2, and consequently the conforma-tional equilibrium constant $K = p_2/p_1 \sim p_2$ and standard free energy $\Delta G°$, can be computed by observing the maximum line-broadening and using equation (3.75), if the line separation $|\nu_1 - \nu_2|$ is estimated by comparison with structurally related compounds. The method was first applied to nitrogen inversion in 1,2,2,6-tetramethylpiperidine **8** which carries the *N*-methyl substituent from the N-equatorial (e) to the N-axial (a) position (predominant and minor conformers, respectively).

Variable-temperature ^{13}C NMR spectra indeed show the broadening and then the resharpening of lines in the range 307–97 K. A maximum line-broadening of 4 Hz was observed at $-60\,°$C for axial C(2)-methyl group. Application of formula (3.75) yielded $p_a = 0.01$, $\Delta G°_{e \to a} = 7.9$ kJ mol^{-1} and k_{21} $(e \to a) = 2000$ s^{-1} at $-60\,°$C. An estimated frequency shift $\delta\omega = 5$ ppm was used on the basis of an additional γ effect for the N-axial conformer. Many other examples followed this pioneering experiment, mainly concerning the stereodynamics of polyheterocyclic systems [62].

3.6   MULTISITE EXCHANGES

The so called Kubo–Sack matrix equations [9, 10] for multisite exchanges have been established using Fourier transformation of appropriate autocorrelation functions introduced by Anderson [11]. A much simpler presentation, in the continuity of the theory used for two-site exchanges, is to write modified Bloch equations (3.34) at the rate of one equation per site i, where $i = 1$ to n_s, the total number of exchanging lines [72a]:

$$dG_i/dt + a_i G_i = -i\gamma B_1 M_o p_i + P_{ii} G_i + \sum_{j \neq i} P_{ji} G_j \qquad (3.77)$$

Steady-state nuclear magnetizations G_i in each site i are obtained as the solutions of a set of n_s linear equations:

$$-a_i G_i + P_{ii} + \sum_{j \neq i} P_{ji} G_j = iC p_i \qquad (3.78)$$

where $C = \gamma B_1 M_o$ is a constant which will determine the relative intensities of lines, and which can thus be equated arbitrarily to unity. *In matrix notation*, the system of linear equations (3.78) may be written as

$$\mathbf{GQ} = i\mathbf{p}$$

G and **p** are the row vectors for nuclear magnetizations and populations

$$\mathbf{G} = [G_1, G_2 \ldots, G_i, \ldots G_{n,}]$$
$$\mathbf{p} = [p_1, p_2 \ldots, p_i, \ldots p_{n,}] \tag{3.79}$$

Q is the matrix for the coefficients of the unknowns G_i in each equation (3.78). It is obtained by subtracting the diagonal matrix $\mathbf{\alpha} = diag(\alpha_1, \alpha_2, \ldots, \alpha_i, \ldots, \alpha_{ns})$ from the exchange matrix **P**:

$$\mathbf{Q} = \mathbf{P} - \mathbf{\alpha} \tag{3.80}$$

This is a complex matrix of the form

$$\mathbf{Q} = \mathbf{A} + i\mathbf{B} \tag{3.81}$$

A and **B** are two real matrices computed from equations (1.30) and (3.79) as

$$\mathbf{A} = \mathbf{P} - \mathbf{R} \tag{3.82}$$

$$\mathbf{B} = \mathbf{\Omega} \tag{3.83}$$

where **R** and $\mathbf{\Omega}$ are two diagonal matrices

$$\mathbf{R} = diag(1/T_{21}, \ldots, 1/T_{2i}, \ldots) = diag(\pi \Delta v_{1/2}^1, \ldots) \tag{3.84}$$

$$\mathbf{\Omega} = diag(2\pi(\omega - \omega_1), \ldots) \tag{3.85}$$

The solutions of equations (3.78) are obtained as

$$\mathbf{G} = i\mathbf{p}\mathbf{Q}^{-1}$$

In matrix notation the total transverse magnetization $G = \Sigma G_i$ may be considered to result from the matrix product

$$\mathbf{G} = \mathbf{G} \cdot \mathbf{1} = i\mathbf{p}\mathbf{Q}^{-1}\mathbf{1}$$

where **1** is a column vector with all n_s elements equal to unity. The NMR signal $v(v)$ is obtained as the imaginary par of G, i.e. as the real part of a matrix product

$$v(v) = real \ (\mathbf{p}\mathbf{Q}^{-1}\mathbf{1})$$

or else as:

$$v(v) = \mathbf{p} \cdot real \ (\mathbf{Q}^{-1}\mathbf{1})$$

real $\mathbf{Q}^{-1}$ is the real part of the inverse of the complex matrix **Q** which is obtained from the algorithm

$$real \ \mathbf{Q}^{-1} = (\mathbf{A} + \mathbf{B}\mathbf{A}^{-1}\mathbf{B})^{-1}$$

The NMR signal, a real number, is obtained by a sequence of matrix operations

$$v(v) = \mathbf{p}(\mathbf{A} + \mathbf{B}\mathbf{A}^{-1}\mathbf{B})^{-1}\mathbf{1} \tag{3.86}$$

where **p**, **A** and **B** are defined by equations (3.79) and (3.82) to (3.85). Matrix operations are standard multiplication, addition, and inversion of real matrices. There are simple and efficient programs to perform such calculations in NMR textbook [33]. Such programs may cover over 90% of problems in current DNMR, and are used as well for two-site exchanges in the place of those specifically written on the basis of equation (3.36).

The simulation of coalescence patterns not only yields the NMR rate constants k_{ij}, but may help to assess the exchange mechanism. Thus, in the study of simultaneous proton transfer and nitrogen inversion on 1-*cis*-2,6-trimethylpiperidinium ion in DMSO explained in Section 2.3, an alternative assumption to account for NMR patterns is a Walden inversion on nitrogen brought about by a proton substitution. In this event, components of each C-methylic doublet for the two isomers AH^+ and BH^+ do not exchange together and the exchange matrix (3.31) should be modified by putting $P_{12} = P_{21} = 0$. It was then observed that no good fit between experimental spectra of the N-methylic protons could be obtained at any exchange rate k_{NT} when using the modified matrix (3.31), this permitted to rule out this assumption.

As for two-site exchanges, a simple formula is obtained in the case of *slow exchanges*. In this event, each individual line j is only broadened. The transverse magnetization G in its vicinity can be reduced to G_i:

$$G_i = -iCp_i/(\alpha_i - P_{ii})$$

leading to a line-broadening given by

$$\delta\Delta v^i_{1/2} = -P_{ii} = 1/\tau_i = \Sigma k_{ji} \qquad (3.87)$$

where τ_i is the average residence time in site i according to equation (3.14).

Owing to the multiplicity of lines, coalescence phenomena are progressive, occurring first between neighbouring sites and then between remote sites, roughly in the increasing order of line separations between pairs of exchanging lines, if we try to refer to the NMR timescale (3.38). An example is given in studies of base-catalysed proton transfers on DMSO. The proton-undecoupled ^{13}C NMR spectrum of $^{13}CH_3 S(O)^{12}CH_3$ is composed of sixteen lines displayed in a quadruplet of four quadruplets, denoted A, B, C, D. The spacing between each quadruplet is equal to the scalar coupling $^1J_{13_{C-H}} = 138$ Hz to the three protons directly bonded to each carbon-13, and the closely spaced components in each multiplet originate in long-range couplings $^4J_{13_{C-H}} = 4$ Hz with the second methylic group $^{12}CH_3$. In conditions where the exchange rate k_{ex}, is varied from 4 to 17 s^{-1}, it is seen that the components of each multiplet are collapsing together, while the four averaged singlets, A, B, C, D are still observed individually. It would be, however, erroneous to consider overall coalescence patterns as the simple superposition of pairwise exchanges, since all exchanges are in fact contributing simultaneously to modify lineshapes. Thus, in the present example, the exchange matrix contains nonzero elements connecting pairs of lines not only within each multiplet, but also between adjacent multiplets [72b].

Exchanges involving one predominant line [54] give rise to a simple treatment as in equation (3.64), when rate constants k_{ji} from minor sites j to the abundant site i have a common value $k_{ex} = 1/\tau$. Exchanges between minor species themselves are assumed to have a negligible contribution, so that τ represents the residence time in all the minor sites. As in the case of two-site exchanges (equation (3.37)), this amounts to saying that each elementary chemical event, occurring with an average repetition time τ, is followed by exchanges from site i to site j with a probability $k_{ij} = p_j/\tau$, proportional to the population of site j. The single dominant line i is then broadened by

$$\pi\delta\Delta v_{1/2}^i = \frac{1}{\tau}\sum_{j \neq i} p_j \frac{1/T_{2j}(1/T_{2j} + 1/\tau) + \delta\omega_j^2}{(1/T_{2j} + 1/\tau)^2 + \delta\omega_j^2} \tag{3.88}$$

where $\delta\omega_j = \omega_j - \omega_i$. The predominant line i is slightly shifted from its resonance frequency ω_i by the quantity

$$\Delta\omega_{obs} = \sum_{j \neq i} \frac{p_j\delta\omega_j}{(\tau/T_{2j} + 1)^2 + \tau^2\delta\omega_j^2} \tag{3.89}$$

Formula (3.88) simplifies to two limiting laws in analogy to equation (3.73) and (3.74)

$$\pi\delta\Delta v_{1/2}^i = \tau\sum_{j \neq i} p_j\delta\omega_j^2/(1 + \tau^2\delta\omega_j^2) \quad \text{if} \quad \delta v_j \gg \Delta v_{1/2}^j \tag{3.90}$$

$$\pi\delta\Delta v_{1/2}^i = \sum p_j/(T_{2j} + \tau) \quad \text{if} \quad \delta v_j \ll \Delta v_{1/2}^j \tag{3.91}$$

These equations again cover the entire range from fast to slow exchange rates and are valid as well for diamagnetic sites j. They were first applied to a study of proton transfers in water where the predominant line was that of $H_2^{16}O$ molecules and the minor sites were the six equally spaced lines of isotopic $H_2^{17}O$ molecules, cleverly used to overcome the impossibility of observing exchanges between like $H_2^{16}O$ molecules (Section 2.1).

3.7   SCALAR RELAXATION AS AN EXCHANGE PROCESS

It has been pointed out that slow NMR relaxation of spin $1/2$ nuclei coupled to a quadrupolar nucleus Q $(I_Q \geqslant 1)$ can be described in analogy to chemical exchange. Jumps of nucleus Q between its $(2I_Q + 1$) magnetic states as the result of quadrupolar relaxation are accompanied by an exchange between the corresponding lines of the multiplet. This exchange occurs at a rate Π_{ij} $(1/T_{1Q})$, where T_{1Q} is the quadrupolar relaxation time and Π_{ij} is the matrix element giving the relative probability of a jump between magnetic states i and

j. These probabilities depend only on the value of the spin number I_Q, and can be computed according to Abragam [73], e.g. in the case of spin 1 nuclei:

$$\mathbf{\Pi} = \begin{bmatrix} -0.6 & 0.2 & 0.4 \\ 0.2 & -0.4 & 0.2 \\ 0.4 & 0.2 & -0.6 \end{bmatrix} \tag{3.92}$$

Line-shapes for the multiplet can thus be simulated by using standard multisite exchange programs, in which the only parameter to be fitted is the relaxation time of the coupled nucleus Q. A noteworthy feature of coalescence patterns is their dependence on the temperature which is opposite to that observed in chemically exchanging systems. Chemical exchange times τ_{ex} are decreasing, while quadrupolar relaxation times are, indeed, increasing on elevation of the temperature. Slow exchange spectra are thus obtained at higher temperatures in this case. Thus the NMR spectrum of the two methoxylic protons coupled to one deuterium nucleus in 1-(monodeuterated methoxy)-2-tertiobutylbenzene appears as a triplet ($I_D = 1$) at 35 °C, coalescing into a broad singlet at −54 °C (Figure 3.12) [74].

Figure 3.12 Temperature dependence of ^{1}NMR site exchange induced by coupled deuterium quadrupolar relaxation in monodeuterated 2-*tert*-butylanisole. Reproduced by permission of the author from Ref. 74

If the slow exchange region cannot be reached, two parameter have to be fitted simultaneously, T_{1Q} and J. This was the case of Al^{3+}, Ga^{3+} and In^{3+} octahedral complexes with organophosphorus ligands such as trimethylphosphate (TMPA). The $^{31}P\{^1H\}$ spectrum is a broad band, even in conditions when ligand exchange is well beyond the slow exchange limit. This was assigned to quadrupolar relaxation of ^{27}Al, ^{71}Ga and ^{115}In nuclei ($I = 5/2$, $3/2$ and $9/2$, respectively), coupled to the phosphorus nuclei in the six coordinated TMPA molecules. Theoretical simulations of NMR spectra using the appropriate probability matrices yielded adjusted values $J = 20$, 33 and 48 Hz, and $1/T_{1Q} = 10$, 280 and 850 s^{-1} at 25 °C [75].

3.8 TRANSIENT METHODS: T_2 MEASUREMENTS

Transient methods for the measurement of rates include fast passage effects, multiple resonance experiments and pulse methods. The importance of fast passage methods has declined since the advent of multipulse NMR. Multiple resonance experiments mainly concern longitudinal magnetization transfers. We are then left with pulse methods which aim at measuring transverse relaxation rates $1/T_2$. Transverse magnetization may be examined as a time-domain signal after a 90° read pulse about the x axis of the rotating frame, the FID signal. If the FID contains one transition only, the decay of oscillations yields the relaxation time T_2. The difference in relaxation rates measured in the presence or absence of exchange, $(1/T_2)_{obs}$ and $(1/T_2^0)$, permits kinetic parameters to be deduced by means of equations such as (3.44) or (3.64).

No further advantage seems to be brought by pulsed NMR in this presentation as compared to CW lineshapes. Let us say at this point that the equivalence of dynamic information obtained by CWN-MR or FT-NMR of chemically exchanging systems has been the matter of at least three demonstrations, using Fourier Transformation of an autocorrelation function (Kubo–Sack–Anderson, see above), or integration of modified Bloch equations [76], or else density matrix formalism [77, 78].

In fact, relaxation times T_2 are measured by the *spin-echo sequence* (Chapter 2) so as to rule out B_0 inhomogeneity effects. A first advantage is to reduce the relaxation rate $1/T_2^0$ to the natural linewidth; this increases in turn the precision and the accuracy in the measurement of small line-broadenings induced by exchange phenomena at both ends of the NMR window, or even throughout the NMR window in the case of one predominant line. Relaxation rates measured by the spin echo method are widely used to determine ligand exchange and proton transfer rates where the exchanging predominant line is that of the solvent.

However, it soon appeared that if a train of pulses is used in place of the simple two-pulse spin echo sequence, another parameter could monitor the influence of rate processes upon relaxation rates, namely the *pulse repetition rate*. The Carr–Purcell–Meiboom–Gill (CPMG) sequence [79] used in this case (Chapter 2) involves n 180° pulses at times $t_p, 3t_p, ..., (2n-1)t_p$ with an echo at $2t_p, 4t_p, ..., 2nt_p$. The peak amplitudes of the successive echoes are modulated by the pulse rate $a = 1/2t_p$ (s^{-1}) in the case of kinetically influenced lines, while non-exchanging lines decay with their natural spin–spin relaxation time according to $\exp(-t/T_2^0)$, where $t = 2nt_p$ denotes the total time, ignoring the duration of each pulse. Spin-echo sequences may be combined with phase detection and Fourier transform in order to give a series of high-resolution spectra corresponding to different values of n. The intensities of each line from these spectra can be plotted as a function of the operation time $t = nt_p$ so as to reconstruct the decay of successive echoes and to extract apparent relaxation times $T_2(a)$ whenever an exponential decay is actually obtained. A last option is to use several working frequencies ω_0. All these operations are delicate and time-consuming, this explains in part the small development of this methodology in spite of its relevance to chemical kinetics.

Multiexponential decays are in fact expected with a number of components equal to the number of sites [80–83], this explains why computing programs have been set up almost exclusively in the case of two-site exchanges. Simulations of DNMR spectra using the CPMG sequence however show a predominantly single-exponential decay, this greatly facilitates quantitative evaluations of kinetic and NMR parameters. This is especially true in fast exchange conditions, where only one peak with a single relaxation time $T_2(a)$ appears. In slow exchange conditions, single exponentials with identical time constants are obtained for both sites 1 and 2 if the first few echoes are discarded [80].

The apparent relaxation times $T_2(a)$ sharply depend on the pulse repetition rate a. If the pulse repetition is slow $(a \rightarrow 0)$, the observed spectrum is just that described in the above sections with values of (T_2^{-1}) $a \rightarrow 0$ equal to those previously denoted as $(1/T_2)_{\text{obs}}$. On the other hand, if the mean time between spin transfers τ_{ex} is much greater than the duration $2t_p$ between two pulses, the apparent (T_2^{-1}) $a \rightarrow \infty$ describes the natural transverse relaxation rate $(1/T_2^0)$ in each site, without any contribution from exchange. In between these two extremes, where $\tau_{\text{ex}} \leqslant 2t_p$, there is an exchange contribution to the apparent (T_2^{-1}) a value, which is not equal to $(1/T_2)_{\text{obs}}$. It thus appears that there is a step in the log (T_2^{-1}) versus log a curve from a higher $((1/T_2)_{\text{obs}})$ to a smaller $(1/T_2^0)$ value of (T_2^{-1}) by going from slow to fast pulse rates. The mid-point of the step is approximately $1/\tau_{\text{ex}}$ or $\Delta\omega/2$, depending on whether the exchange rate is fast or slow on the NMR timescale [83]. These two limiting cases may be distinguished from one another, if necessary, by noting that the position of the mid-point in the slow exchange conditions $(\Delta\omega/2)$ depends on the field strength and not upon the temperature of operation, with opposite conclusions for the fast exchange conditions. There is thus a possibility of measuring independently the DNMR parameters $\Delta\omega$, τ_{ex} and T_{21}^0, T_{22}^0 in a two-site exchange [84, 85]. Consideration of the full range of pulse rates and of exchange rates in principle permit the adjustment of populations p_1, p_2 besides the above static and dynamic parameters. A noteworthy feature is also the obtaining of a spectrum free from exchange alterations at high pulse rate $(a \rightarrow \infty)$, this allows *kinetic difference spectroscopy* by subtracting this spectrum from spectra obtained at any pulse rate a so as to get rid of all the non-exchanging lines.

The treatment of experimental data may use simulation of the decay of echoes with adjustment of parameters, or closed analytical formulas relating the apparent relaxation rates $(T_2^{-1})_a$ to kinetic and NMR parameters. Luz and Meiboom [84] were the first to derive such a formula for two-site exchanges assuming equal transverse relaxation times in both sites $(T_{21}^0 = T_{22}^0 = T_2^0)$ and fast exchange conditions (with the notations of Section 3.1):

$$(1/T_2)_a = 1/T_2^0 + p_1 p_2 \delta\omega^2 \tau [1 - 2a\tau \tanh(1/2a\tau)] \qquad (3.93)$$

This formula has two limiting values as mentioned above:

$$(1/T_2)_{a \rightarrow 0} = 1/T_2^0 \quad \text{and} \quad (1/T_2)_{a \rightarrow v} = 1/T_2^0 + p_1 p_2 \delta\omega^2 \tau$$

The apparent line-broadening

$$(\delta \Delta \nu_{1/2})_a = \pi[(1/T_2)_a - 1/T_2^0]$$

is thus equal to that expected from lineshape analysis (equation 3.44) decreased by the factor $[1 - 2a\tau \tanh (1/2a\tau)]$ itself very close to $(1 - 2a\tau)$ as soon as $2a\tau \leqslant 0.5$. A 50% decrease is thus observed when $\tau^{-1} = 4a$. This is the midpoint of the step function mentioned above [86]. It is then apparent that exchange rates and natural relaxation rates can be obtained from this step function, independently from the exact values of p_1, p_2, and $\delta\omega$. The prior knowledge of NMR exchanging sites is not even required in this respect. The determination of faster and faster exchange rates in this way remains possible as long as the instrumental requirements for high pulse repetition rates are fulfilled. The time $2\tau_p$ between two 180° pulses should be clearly longer than the pulse width t_w, e.g. $2\tau_p \geqslant 10 t_w$. With pulse widths of typically 10 to 50 μs, pulse repetition rates are limited to $ca.\ 2 \times 10^4\,\mathrm{s}^{-1}$ and exchange rates $1/\tau$ to $ca.$ $10^5\,\mathrm{s}^{-1}$. An erroneous belief is the possibility of extending the NMR window towards faster exchange rates. DNMR effects are indeed decreased by spin-echo techniques as compared to lineshape analysis, so that resolution limitations are still more stringent in the spin-echo method. This is another reason why the method remains confined to a very few experimental applications, a third reason being the possibility of erroneous measurements in the case of homonuclear spin–spin couplings.

An attractive feature of the method is the possibility of independent determinations of $1/T_2^0$, τ and $p_1 p_2\ \delta\omega^2$. This means that, if the populations in each site are known a priori as is the case in degenerate chemical exchanges, the frequency shift $\delta\omega$ can be computed even if the rate process cannot be frozen out on the NMR timescale. Conversely the populations p_1, p_2 in the two sites may be evaluated if the frequency shift $\delta\omega$ is estimated from structural considerations (cf. Section 3.5), e.g. the number of solvent molecules exchanging between the bulk and bound sites in solutions of paramagnetic ions could be determined unequivocally in this way [86].

Formula (3.93) applies as well to the case of one predominant line i quickly exchanging with one or several minor species j with a common rate constant $k_{ji} = 1/\tau$, in which case the term $p_1 p_2 \delta\omega^2$ should be replaced by $\Sigma p_j \delta\omega_j^2$ $(j \neq i)$, respectively. Examples of application, already mentioned in Section 3.6, concern proton transfers in ^{17}O enriched water containing various acidic, or basic solutes [84, 87]. The formula giving the apparent relaxation rate of the water line was recast under the operational form [84]:

$$(1/T_2)_a = (1/T_2^0) + (1 - 2a\tau)\tau\left(\frac{35}{12}\right)pJ^2 \qquad (3.94)$$

where p is the atom fraction of ^{17}O, J is the ^{17}O–^{1}H spin–spin coupling, and τ is the mean time of residence of protons in H_2O molecules.

The restricting assumptions made in the above treatment: fast exchange, two-site exchanges, sites with equal relaxation time, were raised in more elaborate developments [80–82]. Monoexponential decays and closed formulas for apparent relaxation rates $(1/T_2)_a$ are available in the case of two-site exchanges over the whole range from slow to fast exchange conditions [80, 81]. The general formula for two-site exchanges in its more refined version is [80]

$$(1/T_2)_a = (1/T_{21}^0 + 1/T_{22}^0)/2 + 1/2\tau - 0.5a \, \mathrm{Log}[F + (F^2 - 1)^{1/2}] \quad (3.95)$$

where

$$F = \frac{p^2 + \Delta\omega^2}{p^2 + q^2} \cosh \frac{2p}{a} + \frac{q^2 - \Delta\omega^2}{p^2 + q^2} \cos \frac{2q}{a}$$

$(p + iq)$ is a complex number defined as:

$$(p + iq^2)1 = (i \, \Delta\omega + s + \sigma)^2 + 1/\tau_1 \tau_2$$

and:

$$2s = 1/T_{21}^0 - 1/T_{22}^0; \quad 2\sigma = 1/\tau_1 - 1/\tau_2; \quad 2\Delta\omega = \omega_2 - \omega_1$$

There are simplified versions of equation (3.95) in the special case of two equally populated sites, which have been applied to the study of degenerate hindered rotations in a few N,N-dimethylamides [83, 85]. A good agreement was found in this case with data obtained in parallel investigations using lineshape analysis.

On the whole, this methodology appears complementary to lineshape analysis. Its main advantage is to offer an independent determination of exchange rates. However, it does not extend fundamentally the NMR timescale and the NMR window; small improvements in this respect are, however, appreciated in the search of accurate activation parameters. The complexity in the treatment of data restricts experimental investigations to two-site exchanges, and explains the small number of applications found in the literature.

3.9   EXCHANGES AMONG STRONGLY COUPLED LINES

When the spins in a molecule interact with each other through a strong scalar coupling, NMR transitions no longer represent changes of the spin state of individual nuclei, and the concept of nuclear site becomes meaningless. The set of coupled spins should then be described by a nuclear wave function , and the transverse magnetization G is taken as proportional to the ensemble average of the expectation value of the operator, $\hat{I}^+ = \hat{I}_x + i\hat{I}_y$:

$$\langle \hat{I}^+ \rangle = \mathrm{Tr}. \, \rho \hat{I}^+ = \Sigma \rho_{kl} I_{kl}^+ \quad (3.96)$$

where ρ is the density matrix. The NMR signal $v(\omega)$ is then computed as the imaginary part of $\langle \hat{I}^+ \rangle$ [13, 14, 23].

Let us consider the simple case of an intramolecular degenerate exchange where there is only one species, and consequently a single set of spin wavefunctions and one density matrix ρ. The time evolution for the density matrix (equation (2.10)) is then completed to include exchange effects by adding a term on the model of modified Bloch equations (3.34):

$$d\rho/dt = i[\rho, \mathcal{H}] + [\rho \ (after \ exchange) - \rho]/\tau_{ex} \qquad (3.97)$$

Following Kaplan [13] and Alexander [14], we introduce the exchange operator $\hat{E}$ defined so that the exchange process can be represented by the change of spin wave function: $\psi \rightarrow \hat{E}\psi$. In the basis of product wave functions, the operation consists in the interchange of the indices of the exchanging spins in the initial wave function. According to this definition, $\hat{E}^2 = 1$, and the associated matrix is real, symmetric and unitary. Thus in the exchange $AB \Leftrightarrow BA$, the basis set: $\psi_1 = \alpha\alpha$, $\psi_2 = \alpha\beta$, $\psi_3 = \beta\alpha$ and $\psi_4 = \beta\beta$ is transformed into:

$$\hat{E} \begin{vmatrix} \alpha\alpha \\ \alpha\beta \\ \beta\alpha \\ \beta\beta \end{vmatrix} = \begin{vmatrix} \alpha\alpha \\ \beta\alpha \\ \alpha\beta \\ \beta\beta \end{vmatrix} \quad whence \quad E = \begin{bmatrix} 1 & 0 & 0 & 0 \\ 0 & 0 & 1 & 0 \\ 0 & 1 & 0 & 0 \\ 0 & 0 & 0 & 1 \end{bmatrix} \qquad (3.98)$$

The density matrix ρ is then transformed into, $\hat{E}\rho\hat{E}$, and the time-evolution of π is given by the master equation:

$$d\rho/dt = i[\rho, \mathcal{H}] + (\hat{E}\pi\hat{E} - \rho)/\tau_{ex} - \rho/T_2^0 \qquad (3.99)$$

The Hamiltonian $\mathcal{H}$ includes contributions from Zeeman effect, chemical shift, spin–spin coupling and driving RF field. Relaxation terms $-\rho/T_2^0$ are inserted to provide the proper linewidths. The matrix equation (3.99) should be developed in full at the rate of one coupled differential equation for each of its elements. However, the useful elements of the density matrix are those which correspond to the non-zero elements of the I^+ matrix, themselves equal to unity for all one-quantum transitions and to zero in all other cases.

Thus, in the above $AB \Leftrightarrow BA$ example, there are four transitions and four matrix elements to consider: $\rho_{12}, \rho_{13}, \rho_{24}, \rho_{34}$. The contribution of exchange to the time derivatives of these four elements is respectively:

$$(\rho_{13} - \rho_{12})/\tau_{ex}; \quad (\rho_{12} - \rho_{13})/\tau_{ex}; \quad (\rho_{34} - \rho_{24})/\tau_{ex} \quad and \quad (\rho_{24} - \rho_{34})/\tau_{ex}$$

The steady-state values are obtained by setting all the derivatives in equation (3.99) equal to zero and computing the ρ_{ij} elements as the solutions of a system of four linear equations for a range of values of τ. The NMR lineshape is then given according to equation (3.96) by

$$v(\omega) = imag(\rho_{12} + \rho_{13} + \rho_{24} + \rho_{34}) \qquad (3.100)$$

An example is given in Figure 3.13 for the intramolecular rearrangement of NIPA molecules coordinated to the linear uranyl cation UO_2^{2+} in the equatorial

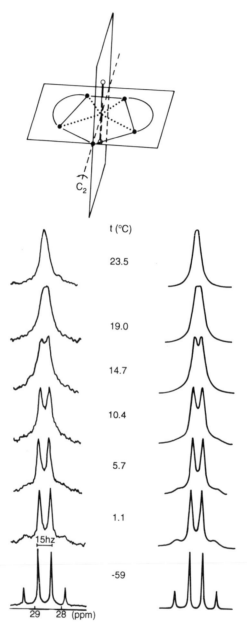

Figure 3.13 Experimental and calculated $^{31}P\{^1H\}$ (32.44 MHz) lineshapes (on the left- and right-hand sides of the figure) of a 0.1 molar solution of $[UO_2(NIPA)_2 EtOH]^{2+}$ at various temperatures; best fit exchange rates $k_{ex} = 105, 78, 54, 38, 26, 18, 0\ s^{-1}$ (from top to bottom)l [88]. Reprinted with permission from Rodehüser *et al. Inorg. Chem.*, **21**, 1062. Copyright (1982) American Chemical Society

plane of the complex $[UO_2(NIPA)_2 X]^{2+}$, where $X = C_2H_5OH$. The two phosphorus nuclei occupy non-equivalent positions P_A, P_B in each ligand molecule and can exchange their positions through a dissociative mechanism in which one NIPA molecule is detached by one end only and is reattached to either site A or B. Experimental spectra can be compared to the theoretical curves obtained for best-fitting τ_{ex} values [88].

The treatment of more complicated spin systems is not fundamentally different. However the number n_T of coupled linear equations deriving from the matrix equation (3.99) rapidly increases with the number of coupled spins: $n_T = 4, 15, 56, 210, 792 \dots$ for systems, of 2, 3, 4, 5, 6 ... spins, respectively. This number still increases for non-degenerate exchanges involving at least two species 1 and 2, each one with its own density matrix ρ^1 and ρ^2. In this case the exchange consists in the permutation of indices on the joint density matrix ρ^{12} for the combined system of interacting molecules $\rho^{12} \to \hat{E}\rho^{12}\hat{E}$. The joint density matrix is obtained as the direct product of the density matrices for the individual molecules: $\rho^{12} = \rho^1 \times \rho^2$. The next step is to separate ρ^1 (and ρ^2) from the joint density matrix by summing over the indices in $\hat{E}\rho^{12}\hat{E}$ which pertain to ρ^2 (and ρ^1, respectively). This operation is denoted by Alexander as

$$\rho^1 \ (after \ exchange) = \Omega^2(\hat{E}\rho^{12}\hat{E}) \tag{3.101}$$

These values are then reported in two coupled matrix equations identical to equation (3.99), each one giving the time evolution $d\rho^1/dt$ and $d\rho^2/dt$, respectively. These two matrix equations are again developed into one set of coupled linear equations in number $n_{T1} + n_{T2}$, where n_{T1} and n_{T2} are the numbers of transitions for each species. The NMR signal $v(\omega)$ results from a weighted mean of the expectation value $\langle \hat{I}^+ \rangle$ over all species, i.e. in the present case:

$$v(\omega) = imag(p_1\langle \hat{I}_1^+ \rangle + p_2\langle \hat{I}_2^+ \rangle) \tag{3.102}$$

Specific programs can be written for the more simple cases, e.g. $AB \Leftrightarrow BA$; $AA'BB' \Leftrightarrow BB'AA'$; $ABCD \Leftrightarrow BADC$ (intramolecular, degenerate); $AB \Leftrightarrow A'B'$ (intramolecular, non-degenerate); $AB \Leftrightarrow A^*B$; ABC (or AB_2) $\Leftrightarrow A^*BC$ (or A^*B_2); $ABCD \Leftrightarrow A^*BCD$ (intermolecular, degenerate).... A general methodology encompassing all types of exchanges over up to six spins has been devised by Binsch [89] with the production of several standard programs for the simulation of spectra and the adjustment of kinetic parameters either by comparison (DNMR3) or by automatic least squares procedures (DNMR5, DAVINS) [90].

3.10 APPLICATIONS OF DNMR: THE INTERNAL DYNAMICS OF MOLECULES AND AGGREGATES

Stereochemically non-rigid molecules have been extensively studied in the early times of DNMR. Molecular motions may not, however, be completely free so that the exchange rates do not exceed the fast exchange limit.

Conformational changes can be classified for convenience as restricted rotations, pyramidal inversion and ring reversal.

Internal rotations about single bonds are generally immeasurably fast by DNMR except in the case of sterically crowded single bonds or of single bonds with partial double bond character. The first situation is met in heavily substituted ethanes where the free rotation from a staggered conformation to another one is slowed down by halogen substituents, or else by the presence of a tertiary butyl group. Fluorine substituents have been especially investigated, since conformational changes induce large ^{19}F chemical shifts. The rotation of a tertiary butyl group attached to a cyclic unit as in *tert*-butylcycloalkanes is so much hindered at low temperatures that the three methyl groups may become non-equivalent. Typical compounds containing single bonds with partial double bond character are amides and peptides, thioamides, carbamates, nitrosamines, nitriles, aldehydes and ketones. A few examples are given in the course of this book to illustrate the potential of DNMR in this area. Exhaustive reviews and books on these topics can be consulted for further reading [91–95, 25, 35, 36, 38].

Hindered rotations about carbon–carbon bonds [38] in organic molecules can be conveniently classified according to the types of carbon atoms. The internal rotations mentioned above can thus be named as rotations about an sp^3–sp^3 bond. The barrier to rotation about a normal double bond is, of course, too high to be measured by DNMR. Rotations about sp^2–sp^2 bonds are thus normally unobservable except if there is some reduction in the degree of double bond character. This was found to be the case (i) in sterically hindered olefins, (ii) in olefins having an electron-withdrawing substituent on one end and an electron-donating substituent at the other end of the double bond, the so called push-pull ethylenes [96], as in enamino ketones, enamino acid derivatives and ketene aminals, (iii) in polyenes with a high electronic delocalization, e.g. in cumulenes, fulvenes..., (iv) in conjugated chiral polyenes with steric hindrance to the coplanarity of π-bonds, as in highly congested biphenyls (including fused rings attached to O,O'-carbons), 1,3-butadienes or acid anilides, (v) in systems involving a sp^2–sp^2 carbon to heteroatom bond, e.g. the C=N bond in iminium salts and *N*-arylimines, or C=O$^+$ and CS$^+$ bonds in uronium or thiouronium salts. Restricted rotations may also occur about a trigonal to tetrahedral bond, e.g. about olefin or aryl to alkyl bond, especially in the case of bulky *sec*-alkyl or *tert*-alkyl substituents, or else in cyclophanes and related compounds.

Pyramidal inversion refers to inversion of configuration of an atom bonded to three substituents in a pyramidal geometry [91, 94, 97, 98]. The pyramidal atom bears a lone pair of electrons, such as nitrogen in amines and imines, carbon in carbonium or carbenium ions, and oxygen in oxonium or ketonium salts [38]. The process of inversion normally involves a planar transition state and the interchange of two equivalent configurations or invertomers. The most investigated example is that of nitrogen inversion. The barrier to inversion in ammonia and in many amines is so low that it is difficult to measure inversion rates by DNMR, although they can be detected in some cases from the pH dependence of the NMR spectra [62]. Ring strain or electronegative substituents may bring nitrogen inversion to the NMR timescale. Ambiguities are often difficult to raise as to the exact nature of the molecular motion, e.g. (i) nitrogen or ring inversion in N-heterocyclic compounds, or (ii) N-inversion or hindered rotation about the C—N or C=N bond. Stereodynamics of cyclic [99] and open chain [100] organonitrogen compounds has been exhaustively reviewed in two recent books.

Concerted configurational or conformational changes have been investigated in a variety of processes such as (i) *ring inversion* in cyclohexane derivatives, six-membered heterocycles, and larger rings [99, 101], (ii) *correlated rotations in molecular gears* [38, 102, 103]: triarylmethane and related compounds Ar$_3$ZX (**9**), where Z is a central atom such as carbon, nitrogen or boron, and Ar is a substituted phenyl ring attached to Z by an aromatic carbon C$_{ar}$, assume propeller-like conformations in the ground state; changes of helicity of the aryl ring may involve up to three ring flips about the C$_{ar}$–Z bond, simulating the motion observed in a gear; other examples of molecular gears are displayed by 9-triptycene derivatives [104], such as 9-benzyltriptycene **10**, and (iii) Berry [105] pseudo-rotational interchange of axial and equatorial substituents in trigonal bipyramidal molecules such as PF$_5$, BrF$_5$, IF$_5$ [51].

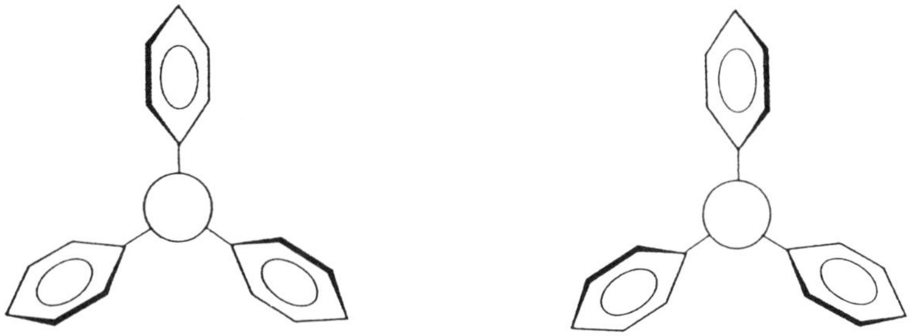

9

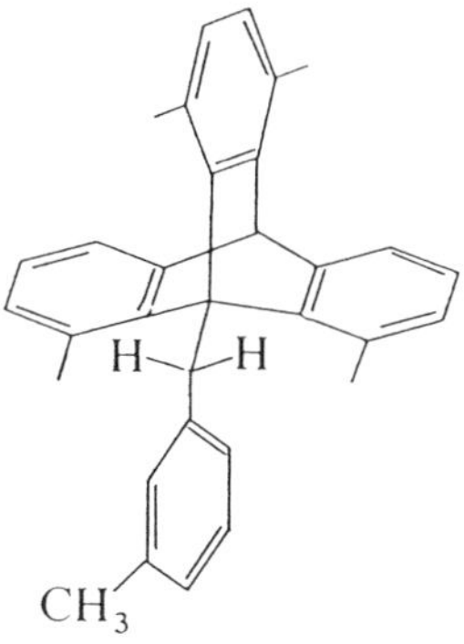

10

In the foregoing, internal processes were implicitly assumed to involve only bond distortions. Intramolecular exchanges, degenerate or not, may as well involve bond-breaking and bond-forming steps. A first example is *valence isomerizations*, also called *valence tautomerisms*, that interconvert ρ and π bonds in certain molecules with fluxional structures. Typical cases amenable to DNMR are (i) rapid Cope rearrangements in structures such as 3,4-homotropilidene [106] **11**   or bullvalene [107] **12**, (ii) rapid equilibration between valence isomers, such as cycloheptatriene-norcaradiene derivatives, e.g. oxepine-benzene oxide [108] **13**, (iii) the rearrangement of benzofuroxan [109] **14**.

A second example of intramolecular rearrangement involving a dissociation and then a recombination of the molecular fragments is 1,2-hydride or alkide shifts in carbocations. Thus, hexamethylbenzene in very acidic medium forms the σ-complexed cation [110]. At room temperature, there is a rapid migration of the acidic proton from one carbon to the next, as shown by the presence of a doublet for all the methylic protons and a nineteen lines multiplet for the tertiary proton, due to an average coupling ($J = 2.1$ Hz) to the eighteen methyl protons made dynamically equivalent. The frozen structure **15** was obtained at low temperature, below -105 °C. The proton migration was shown to be purely intramolecular by comparison of experimental and calculated lineshapes. A related example is the heptamethylbenzenonium ion [111] **16**, where the exchange process was shown, again on the basis of DNMR calculations, to involve 1,2-methyl shifts, and not random intramolecular or intermolecular methyl shifts.

Intramolecular exchanges are observed not only in single molecules, but in molecular aggregates, typically in stereochemically nonrigid metal complexes. A great number of studies have been reported in organometallic and solution chemistry, involving either rearrangements of ligands without bond rupture (pseudo-rotations, twist mechanisms...), or with bond rupture. The latter event is found in a great number of labile metal chelates, in bridge-terminal exchanges of ligands in polynuclear complexes and in fluxional valence isomerizations which interconvert σ- and π-bonded ligands. Some examples are given throughout this chapter, and there are several reviews in the field for further reading [24, 37, 51, 52, 112–114]. Inorganic coordination chemistry is in fact a vast domain of application of NMR, due to the variety of resonant nuclei possibly observed and to the large range of chemical shifts and coupling constants for each of them. Another factor of development is the simplicity of structures, which can be described in many cases in terms of idealized polyhedra the vertices of which correspond to the ligand positions [29, 51]. Exchange phenomena are, however, difficult to interpret owing to the multiplicity of coordination sites. A simplifying assumption is then to assume transition states with again an idealized polyhedral geometry; intramolecular ligand rearrangements are said to be *polytopal* in this case. Polytopal rearrangements can be given a topological representation by examining all the permutational possibilities to interchange the ligand and by using permutational analysis to classify them according to their effect on NMR spectra [29, 51, 115]. This permits in turn the corresponding exchange matrices to be set up for the simulation of DNMR spectra. The comparison of experimental and theoretical spectra is the final criterion used to assess the mechanism of exchange. These efficient but lengthy procedures are being presently superseded by two dimensional exchange spectroscopy, which establishes connectivities between exchanging sites and can thus map out possible pathways for ligand rearrangements (see Section 4.4).

3.11 APPLICATIONS OF DNMR: INTERMOLECULAR PROCESSES

A number of intermolecular processes have already been described in the course of this chapter, pertaining to two essential types of exchanges: proton transfers and ligand exchanges. A major difficulty in these studies is to select chemical reactions whose rates are fitting the NMR timescale. *Proton transfers* can thus be conveniently studied in aqueous solutions using acid-base pairs where the basic site is nitrogen in amines [15, 116] or amides [117–119], oxygen in alcohols and phenols [16, 116], sulphur in mercaptans [120], phosphorus in phosphonium salts [121], or carbon in acetylenic compounds [122]. Transfers on oxygen in carboxylic acids is too fast, and those on sp^3 or sp^2 carbons too slow to be amenable to DNMR. Most transfer reactions have been studied in aqueous solution. A few additional protogenic media have also been investigated, e.g. methanol and carboxylic acids, with the objective of studying the proton abnormal conductance through Grotthus chains of hydrogen-bonded solvent molecules [116] . A more simple situation in this respect can be hoped for in poorly ionizing solvents, as are dipolar aprotic solvents such as dimethylsulfoxide [123]. However, this advantage is counterbalanced by the necessity to establish pH and pK scales in those novel solvents before any DNMR experiment. Protonation and deprotonation phenomena can in turn control the occurrence of conformational changes, such as pyramidal inversion on nitrogen in amines [62, 124] or on carbon in carbonium ions [38, 111]. An example of pH-dependent nitrogen inversion has been given in Section 2.3. Related topics are (i) exchanges of Lewis acids, such as boranes, boron trifluoride [125–127], alanes [128] or gallanes [129], on Lewis bases such as amines, ethers or carbonyl compounds, and (ii) dissociation of neutral species into ion pairs or free ions, e.g. carbon acids into their associated carbonium ions, or organometallic compounds into ion pairs, with the possibility of inversion of configuration at the carbon or metal centre [38, 111].

Ligand exchanges in solutions of diamagnetic or paramagnetic ions may occur either between coordinated and bulk solvent molecules, or between various ligands competing for the coordinated metal ion [114, 130]. To fit the NMR timescale, ligand coordination to the metal should be neither too weak nor too strong; this depends, in fact, on the couple ion/ligand investigated. Ions with a small charge to radius ratio, typically those from the alkaline metals, require strong ionophores such are multidentate ligands: polyamides, polyethers, amino-acids, cryptands [131], crown ethers [132], calixarenes [133].... Ions with large charge to radius ratio, from the alkaline earths, trivalent normal or transition metals, or from the lanthanides and actinides, require on the contrary moderately coordinating ligands, otherwise the exchange is immeasurably slow on the NMR timescale. Related reactions are the scrambling of organic ligands about metal in organometallic compounds, e.g. organolithium solutions [134], or halogen exchanges between polyhalo-genated metals, such as Me_2SnBr_2, Me_2SnBrI, Me_2SnI_2 [135], or polyhalogenated metal ions in solution. Another important field of development is bioinorganic chemistry with the metal or metal ion coordination to amino acids and peptides [39, 136], and the association of large biomolecules, proteins, carbohydrates, nucleic acids, to small substrates considered as ligands: water molecules, metal ions, drugs and inhibitors [39–43]. Several of these topics will be developed in Chapters 4, 5, 6 and 9.

Electron self-exchange reactions are also amenable to DNMR measure-ments. These experiments use a redox pair in which one species (**D**) is diamagnetic and the other (**P**) is paramagnetic. To fit the NMR timescale, the NMR exchange rate should be of the same order as the hyperfine coupling constant A in the paramagnetic species, in most cases about one megahertz. Electron exchange rates may be still higher if the reaction is concentration-dependent. An example is the electron exchange between Cu(I) and Cu(II) according to the reaction [137]

$$Cu(I)(D) + Cu(II)(P) \overset{k}{\underset{k}{\rightleftharpoons}} Cu(II) + Cu(I)$$

with $k_{NMR} = k\,[Cu^{2+}]$ and $k = 0.5 \times 10^8\,mol^{-1}\,dm^3\,s^{-1}$ in concentrated HCl. The experiments were performed by ^{63}Cu or ^{65}Cu NMR using a large excess of Cu(I), which is the only species to be detected, and observing the line-broadening induced by the addition of small quantities of Cu(II). The electron transfer was found much slower $(k \sim 10^4\,mol^{-1}\,dm^3\,s^{-1})$ when the copper ions are sterically hindered through complexation within macrocyclic tetrathiaether structures [138]. Other examples are electron exchanges between (i) Co(I) (D) and Co(II) (P) [51] in the complex $Co(CNR)_5^+$ $(k = 10^6\,s^{-1})$ (ii) tetra-phenylenediaamine (D) and its cation radical (P):

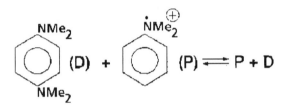

with $k = 2.5 \times 10^4$ and $10^8\,mol^{-1}\,dm^3\,s^{-1}$ in aqueous acidic solution [139] and in acetonitrile [140], respectively, and (iii) cyclooctatetraene and its anion radical (very slow), and then between the anion radical and the dianion $(k \sim 10^9\,mol^{-1}\,dm^3\,s^{-1})$ [141]. In these examples, one site only is observed, corresponding to the diamagnetic species D standing in large excess, with applications of formulas of Section 3.5. This operating procedure is required because the observed nucleus is quite close to the electronic spin with the effect of extreme line-broadening preventing the detection of the paramagnetic site. This constraint can be raised if the observed nucleus is far from the lone electron, this implies using large molecular structures. An example is the work on electron self-exchange between ferro- and ferri-cytochromes [142], which was further extended to several metalloproteins [143], myoglobins and chemically modified cytochromes c [144].

Intermolecular exchange processes can also be investigated by utilizing the special features of *quadrupolar nuclei*. A pioneering example was the study of the very fast reaction between I^- and I_2:

$$I^- + I_2 \overset{k_1}{\underset{k_2}{\rightleftharpoons}} I_3^-$$

The ^{127}I nucleus has a large quadrupole moment and its resonance is sharp enough for detection only in species where the field gradient is zero at the observed nucleus. This is the case for the iodide ion and not for the iodine molecule or the adduct I_3^-. Analysis of linewidth variations of the iodide ion after addition of small quantities of iodine yielded reaction rates $k_1 = 4.1 \times 10^{10}$ mol dm^3 s^{-1} and $k_2 = 7.6 \times 10^7$ s^{-1}. Other examples using quadrupolar probes are given throughout Chapter 5.

In conclusion, this short review aimed at emphasizing the main fields of application of DNMR: stereodynamics of organic, inorganic and biochemical substrates, proton transfers, ligand exchanges, paramagnetic and quadrupolar probes. It must be understood that all the above applications have been worked out by using the whole set of DNMR methodologies, which are further reviewed in the next sections.

4 Longitudinal Magnetization Transfer Experiments

This category of experiments involves the selective introduction of a non-equilibrium spin polarization into an ensemble of nuclei which exchange with other unperturbed spin ensembles. The perturbation is carried out in either of two ways, in which the longitudinal magnetization is put to zero by selectively saturating one signal or is inverted by a 180° RF pulse. These two experiments are respectively called *transfer of saturation* and *inversion transfer*. Chemical exchange as well as longitudinal relaxation processes are then working together to restore Boltzmann equilibrium. The longitudinal magnetization transferred in this process between the spin ensembles may yield information on the rate of chemical exchange provided that it contributes more than longitudinal relaxation to the overall flux of magnetization [146]. This means that exchange rates $k_{ex} = 1/\tau_{ex}$ should be larger than relaxation rates $R_1 = 1/T_1$; this introduces another NMR timescale, the *relaxation time* T_1 timescale such that $k_{ex} > R_1$. The *NMR timescale* for lineshape analysis, defined as a *frequency shift timescale*, however, keeps its relevance in this type of experiment, since the two (or more) signals exchanging non-equilibrium magnetization should remain separate so as to permit their relative intensities to be measured at any time. Consequently, the exchange should be slow on the NMR timescale ($k_{ex} < \delta v$), and, at the same time, fast on the T_1 timescale ($k_{ex} > R_1$). Fortunately, faster exchange rates are usually necessary to cause coalescence $k_{ex} \leqslant 10$ to 10^4 s^{-1} than those required by T_1 timescale, $k_{ex} > 1$ to 10 s^{-1} for relaxation times T_1 of *ca* 1 to 0.1 s. There will often exist the possibility of a moderately wide range of exchange rates in which transfer of magnetization occurs between separate signals below their coalescence point. In this respect, longitudinal transfer experiments are enabling the extension of the NMR window for kinetic measurements towards slow exchange rates by about one order of magnitude.

Another point deserving notice is that transfers of polarization from site to site may also be carried out by dipole–dipole cross-relaxation, the so-called nuclear Overhauser effect or NOE (Chapter 2). There is conversely an effect of chemical exchange on the NOE [20, 34]. This analogy between NOE and longitudinal magnetization transfers also suggests that the two-dimensional

NOE spectroscopy (NOESY) may be used to detect signals which are correlated to each other as well through dipole–dipole relaxation as through chemical exchange. This is effectively the case and the corresponding two-dimensional spectroscopy is called **E**xchange **S**pectroscop**Y** or EXSY.

The following points: saturation transfer, inversion transfer, effects of exchange on the NOE, EXSY, are briefly developed in the following sections. All these experiments are assumed to involve uncoupled or weakly coupled nuclei. The basic time-evolution for longitudinal magnetizations M_{zi} in sites i will be described by using modified Bloch equations (1.22) written in the absence of the RF field $\mathbf{B}_1$ as [18]:

$$dM_{zi}/dt = -R_{1i}(M_{zi} - M_{0i}) - \left(\sum_{j \neq i} k_{ij}\right)M_{zi} + \sum_{j \neq i} k_{ji}M_{zj} \qquad (3.103)$$

or else, after introducing the conservation law (3.12), as:

$$d(M_{zi} - M_{0i})/dt = -\left(R_{1i} + \sum_{j \neq i} k_{ij}\right)(M_{zi} - M_{0i}) + \sum_{j \neq i} k_{ji}(M_{zj} - M_{0j}) \qquad (3.104)$$

A matrix formulation [147] may be given for the set of equations (3.104) as:

$$d\mathbf{M}/dt = \mathbf{A'M} \qquad (3.105)$$

where $\mathbf{M}$ is a column vector of elements $(M_{zi} - M_{0i})$ and $\mathbf{A'} = \mathbf{P'} - \mathbf{R}$ is defined as the transpose of matrix $\mathbf{A}$ (equation (3.82)). In fact, most discussions will deal with the archetypal two-site example, where:

$$dM_{z1}/dt = -R_{11}(M_{z1} - M_{01}) - k_{12}M_{z1} + k_{21}M_{z} \qquad (3.106)$$

$$dM_{z2}/dt = -R_{12}(M_{z2} - M_{02}) - k_{21}M_{z2} + k_{12}M_{z1} \qquad (3.107)$$

As in the classical inversion-recovery experiment, a final 90° read pulse is used after a total time t to obtain an FID of initial intensity proportional to $M_z(t)$. The intensities of the frequency-domain signals S_i after Fourier transformation of the FID are thus reflecting the state of longitudinal magnetizations $M_{zi}(t)$ at the end of the evolution period. Equations (3.103) tot (3.107) may be rewritten in an equivalent form using as variables the intensities of the signals $S_i(t)$ and S_{0i} corresponding to magnetizations M_{zi} and M_{0i} at time t and at equilibrium, respectively.

4.1   SATURATION TRANSFER

In this experiment [146, 148], a radio-frequency field $\mathbf{B}_2$ is applied at exactly the resonance frequency of one (or more) exchanging signal with an amplitude sufficiently strong to saturate the signal and however not too strong to saturate other resonances, i.e. such that $\gamma^2 B_2^2 T_1 T_2 \gg 1$ and $\gamma B_2/2\pi < \delta\nu$ respectively. This requires a FT spectrometer equipped with a low-power channel allowing selective homonuclear spin decoupling or selective presaturation pulse sequences. In its more simple variant, the experiment, first described by Forsen and Hoffmann [18], consists in saturating line 2 from time $t = 0$ $(M_{z2}(0) = 0)$, and observing the time evolution of the magnetization $M_{z1}(t)$ in site 1, assumed to be in slow exchange with site 2. Equation (3.106) integrates, for the initial condition $M_{z1}(0) = M_{01}$, into:

$$M_{z1}(t) = M_{01}[\tau_{11}/T_{11} + (\tau_{11}/\tau_1)\exp(-t/\tau_{11})] \qquad (3.108)$$

where:

$$1/\tau_{1i} = 1/T_{1i} + 1/\tau_i = R_{1i} + k_{ij} \qquad (i = 1 \text{ or } 2, j = 2 \text{ or } 1) \qquad (3.109)$$

Thus, from the moment at which site 2 is irradiated, we observe an exponential decrease of magnetization in site 1 from its equilibrium value up to a steady-state value:

$$M_{z1}{}^{\infty} = (\tau_{11}/T_{11})M_{01} = [R_{11}/(R_{11} + k_{12})]M_{01} \qquad (3.110)$$

This amounts to a partial *transfer of saturation* from site 2 to site 1, with a fractional change in signal intensity:

$$f_1 = (M_{z1}{}^{\infty} - M_{01})/M_{01} = -k_{12}/(R_{11} + k_{12}) \qquad (3.111)$$

This fraction may assume values comprised between 0 and 1 depending on whether the exchange rate k_{12} from the observed site to the irradiated site is much slower or faster than the relaxation rate R_{11}, respectively.

The steady-state fractional intensity change f_1 may yield the rate constant $k_{12} = -f_1 R_{11}/(1 + f_1)$ provided that (i) f_1 is not too small and not too large, say between *ca* 0.1 and 10, and (ii) R_{11} is determined independently by an inversion recovery experiment performed on site 1 while continuously irradiating site 2 (see Section 4.2). An equivalent method consists in measuring the time evolution of $M_{z1}(t)$ according to equation (3.108), a *temporal magnetization transfer* experiment [146]: a semi-logarithmic plot of Log $(M_{z1} - M_{z1}{}^{\infty})$ against time is a straight line of slope $-1/\tau_{11}$, whence: $k_{12} = f_1/\tau_{11}$. An example where this technique was applied to extend the temperature window for lineshape analysis towards lower temperatures and thus to improve the accuracy of activation parameters is the ring inversion of cyclohexane $-d_{11}$. Two ^{1}H{D} signals are observed for the lone proton in the two interconverting chair conformations corresponding to either the equatorial (H_e) or axial (H_a) position [101]. Saturation of the low field resonance (H_e) causes the intensity of high field resonance (H_a) to decrease by 88% at $-97\,°C$. The exponential decay of intensity has a time constant $\tau_{11} = 3.0$ s, whence: $f_1 = 0.88$; $k_{12} = k_{21} = k_{inv.} = 0.29$ s^{-1}; $T_{11} = T_{12} = 25$ s. Inversion rates as low as 0.004 s^{-1} were measured in this way at temperatures decreased down to $-117\,°C$. Similar experiments were carried out to measure the rate of interconversion of α- to β-glucopyranosyl derivatives through transfer of saturation on the anomeric proton from one conformer to the other [149].

This shows the capability of measuring rates occurring in the range between 10^{-3} and $1\,s^{-1}$ by saturation transfer, when the relaxation times T_1 are sufficiently long. This range of rates is difficult to explore by other means, being too fast for isotope flux experiments and too slow for lineshape analysis. A typical application in this respect is the study of slow proton transfers on weakly acidic or basic substrates, e.g. phenols [150], alcohols [151], amides [118] or carbonium ions [152]. Saturation transfer from water to NH hydrogens in peptides has been used in many investigations to characterize the accessibility and consequently the vicinity, of NH hydrogens to the surrounding solvent molecules [153]. Another advantage in the measurement of slow exchange rates is when unstable species are studied for which the higher temperatures necessary for lineshape analysis would result in fast thermal decomposition, e.g. in the study of the barrier to rotation in dimethylcyclopropylcarbinyl cation **17** between $-49°$ and $-21\,°C$ by transfer of saturation between the two methyl resonances [154].

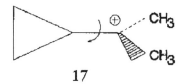

17

Additional advantages of the spin saturation methods are (i) to specifically demonstrate which sites are involved in exchange (ii) to detect small or hidden resonances by observing changes of intensities at exchanging sites with a high population. A few examples will illustrate these two points. (i) There are fluxional organometallic compounds in which a transition metal is bound to a fragment of a cyclic polyene system. Metal migrations may occur from one bond to another through 1–2, 1–3, or random shifts. The distinction between these possibilities was made clear by applying spin saturation techniques to the ^{13}C resonances of the polyene system [155]. Another example in this respect concerns the exchange of phosphate units among phosphorus metabolites (inorganic phosphate, AMP, ADP, ATP, phosphocreatine) or with enzyme-bound intermediates, in the presence of enzymes or of living cells or organisms. The exchanging units and the kinetics of exchange can be determined on the basis of ^{31}P saturation transfer experiments [146]. (ii) Restricted rotation in certain sulfonium ylides such as $(CH_3)_2S^+\,CH{=}CO^-\,Ph$ in the presence of traces of water has been shown to proceed after reprotonation to the keto form $(CH_3)_2S^+\underline{CH_2}C(O)Ph$. Evidence for this mechanism is the transfer of saturation from the weak resonance of water to the methylene protons (underlined) [156].

An interesting observation is that the reverse rate constant k_{21}—and consequently the fractional populations p_1, p_2—can be obtained independently from the forward rate constant k_{12} by observing in turn site 2 after sudden irradiation of site 1. Variants of the saturation transfer method are *saturation-*

recovery experiments [18] which consist in measuring the time course recovery of line 1 upon sudden removal of the saturation at line 2, or at line 1, or at both lines 1 and 2. These variants yield the same information as above, e.g. the last variant results into the following law for the time evolution of magnetization in site 1:

$$M_{z1} = M_{01} \frac{\tau_{11}}{T_{11}} \, [1 - \exp\left(- t/\tau_{11}\right)] \qquad (3.112)$$

The above treatments may be generalized to multisite exchanges in loosely coupled systems. Thus equation (3.111) applies to any individual site i when all other exchangeable sites are saturated simultaneously, if we define τ_{1i} such that $1/\tau_{1i} = R_i - P_{ii} = 1/T_{1i} + 1/\tau_i$. In the same way, the recovery of the ith magnetization follows the rate law (3.112) if the saturating field is applied to all the exchanging sites and then is selectively removed at site i. Under these conditions, steady-state and temporal magnetization measurements can yield for each site i saturation factors f_i and time constants τ_{1i}, from which relaxation rates R_{1i} and residence times τ_i are easily computed. Individual rate constants k_{ij} and k_{ji} connecting a pair of sites i and j are obtained in a second type of experiments, in which all the resonances with the exception of those for the ith and jth sites are saturated. Rate constants are obtained from the steady-state values of magnetizations M_{zi} and M_{zj} measured in these conditions, according to modified Bloch equations recast under the form [28]:

$$0 = M_{oi}/T_{1i} - M_{zi}/\tau_{1i} + k_{ji}M_{zj}$$
$$0 = M_{oj}/T_{1j} - M_{zj}/\tau_{1j} + k_{ij}M_{zi} \qquad (3.113)$$

This methodology was first applied to a three-site system involving the keto-enol equilibrium of acetylacetone:

$$CH_3COCH_2COCH_3 \rightleftharpoons CH_3CO—CH\!=\!\!\!=\!C(OH)—CH_3$$

Sites 1 and 2 are taken as the methyne and hydroxyl protons in the enol, and site 3 as the methylene protons in acetylacetone. The series of multiple resonance experiments outlined above allowed Forsen and Hoffmann [18] to obtain the six unknown rate constants: $k_{12} = k_{21} = 0.00$; $k_{13} = 0.08$; $k_{31} = 0.11$; $k_{23} = 0.06$; $k_{32} = 0.10 \text{ s}^{-1}$ at room temperature. This confirmed the suggestion that the base-catalysed exchange of —OH and $=$CH protons takes place through the keto form as an intermediate.

A variant of this methodology applied to three sites i, j, k uses the measurement of (a) six saturation factors $f_i(j)$ in the site i when site j is saturated, and (b) three apparent relaxation rates R_{1i} $(j, k) = = -(R_{1i} + k_{ij} + k_{ji})$ in site i when sites j and k are simultaneously saturated. Individual rate constants are then computed as

$$k_{ji} = R_{1i}(j,k) \, \frac{f_i(j) + f_i(k)f_k(j)}{1 - f_j(k)f_k(j)} \, \frac{p_i}{p_j} \qquad (3.114)$$

This methodology was applied to primary amide hydrogen exchanges [157]. These compounds have two non-equivalent hydrogens, in *cis* Z (site 1) or *trans* E (site 2) position, which can be interchanged through internal rotation ($1 \Leftrightarrow 2$), or be exchanged with the solvent (site 3). The forward and reverse rate constants for the three site exchanges: $1 \Leftrightarrow 2$; $2 \Leftrightarrow 3$ and $1 \Leftrightarrow 3$ were determined in various solvents, e.g. H_2O, H_2SO_4. The data demonstrated that $k_{12} = k_{21}$ as expected for mutual exchange, and that $k_{23} = k_{13}$ or not, depending on the mechanism of proton transfers to the solvent.

Finally, it should be stressed that unwanted saturation transfers may occur if a presaturation sequence is used as a means to suppress the solvent line, in which case the exchangeable nuclei in the dissolved compounds, e.g. the NH hydrogens of peptides in water, may be suppressed from the NMR spectrum.

4.2   INVERSION TRANSFER

Besides saturation transfer and saturation-recovery experiments, a third category of temporal magnetization transfer uses selective inversion of a resonance followed by observation of the exchanging resonances. The observed magnetizations start from and return to their equilibrium values after passing by a minimum, more or less pronounced, depending on the magnitude of chemical rates compared to relaxation rates and on the fractional populations [158]. A large panel of procedures is presently available to achieve selective 180° inversions using either soft pulses or combinations of hard pulses [159]. Magnetization recovery no longer follows pure monoexponential behaviour. The set of coupled equations (3.104) result in multiexponential recovery [147, 158, 160] of magnetizations M_{zi}:

$$M_{zi}(t) = M_{zi}(0) + \Sigma \; C_{ij} \exp(-\lambda_j t) \tag{3.115}$$

where the time constants λ_j are the eigenvalues of matrix **A'** in equation (3.105), and the coefficients C_{ij} are determined from the initial conditions $M_{zi}(0) = \pm M_{oi}$ and fractional populations p_i. Explicit expressions are existing in the simple case of two or three equally populated sites with a common relaxation rate and a single exchange rate [158, 161, 162]. In all other cases, the time course of magnetization changes should be simulated until a best fit is obtained with experimental data by visual or least-squares adjustment of parameters R_i, p_i and k_{ij}. Other combinations of inverted lines can be envisaged, depending on experimental possibilities, i.e. one or several lines, or even all the resonances can be inverted and the time course recovery of magnetization observed in all exchanging sites. The selective pulse(s) used to initiate the experiment may not be 180° pulses, but at the expense of a loss in overall sensitivity. The first experiment was performed by adiabatic passage through one of the components of the proton triplet signal from NH_4^+ in saturated ammonium nitrate solution. The non-equilibrium magnetization thus introduced was then transferred to the other components of the triplet by proton exchange.

The first quantitative experiment concerned a degenerate conformational flipping in a derivative of paracyclophane [158]. The same range of applications can be envisaged as for saturation transfer experiments, e.g. hindered rotations in amides and peptides, or enzyme-catalysed phosphate exchange between ATP, AMP and ADP. Another type of application, also possible through lineshape analysis at higher exchange rates, is a permeation experiment [146]: the transport of ions through the membrane of vesicles, in which the outside ions are chemically shifted from the inside ions by the presence of a shift reagent in the outside solution.

As pointed out in Section 3.5, an important type of two-site exchange is met in situations where the first site is much more populated than the second. In this case one line only is observed and the usual inversion-recovery experiment applied to this line has a monoexponential behaviour with an apparent longitudinal relaxation time given as:

$$(1/T_{11})_{obs} = 1/T_{11} + p_2/(T_{12} + \tau_2) \tag{3.116}$$

In full analogy to T_2 measurements described previously, we define a specific relaxation rate:

$$1/T_{1r} = [(1/T_{11})_{obs} - 1/T_{11}]/p_2 = 1/(T_{12} + \tau_2) \tag{3.117}$$

which contains all the experimental observations. This formula is also valid if the minor site is paramagnetic, in which case the subscript M is used to denote the minor site and equation (3.11) is recast under the form:

$$1/T_{1r} = 1/(T_{1M} + \tau_M) \tag{3.118}$$

This formula holds throughout the NMR window. In sharp contrast to T_2 measurements (equation (3.66)), the timescale attached to equation (3.118) is only determined by the relaxation rate in the minor site. In the *slow exchange* region, the specific relaxation rate $1/T_{1r}$ is equal to the exchange rate $1/\tau_2$ or $1/\tau_M$; no additional information is obtained in this case with respect to that from T_2 measurements. In the *fast exchange* region, T_1 measurements are yielding the longitudinal relaxation rate $1/T_{12}$ in the minor site; this datum constitutes a necessary complement to T_2 measurements in the determination of the four unknown parameters: T_{12}, T_{22}, τ_2 and $\Delta\omega_2$, as explained in Section 3.5.

4.3　THE EFFECTS OF EXCHANGE ON THE NOE

Dipole–dipole cross-relaxation may also contribute to longitudinal magnetization transfers, this is the nuclear Overhauser effect between pairs of nuclei I and S which are in close spatial proximity [20, 34] (Chapter 2). Cases of practical interest may be divided into two categories. In the first type of problems, the two nuclei I and S are in mutual exchange:

$$I \underset{k_{21}}{\overset{k_{12}}{\rightleftharpoons}} S$$

An example is the topomerization AB $\Leftrightarrow$ BA of methylenic protons H_A, H_B in valium [163] **18**, as the consequence of a conformational flip between two degenerate boat conformations:

18

The rate of exchange can be determined through lineshape analysis [164] at temperatures between 340 and 400 K. At lower temperatures, the conformational inversion was studied by a combination of saturation transfer and inversion-recovery experiments. Nuclei I and S are scalarly coupled to each other in this example; we shall consider in the following each doublet taken as a whole to represent a single site, i.e. sites 1 and 2 for I and S (or H_A and H_B) nuclei, respectively. The basic equations (3.104) to (3.107) for magnetization transfer should be completed to include cross-relaxation contributions on the model of Solomon's equations:

$$dM_{z1}/dt = -R_{11}(M_{z1} - M_{01}) - \sigma(M_{z2} - M_{02}) - k_{12}M_{z1} + k_{21}M_{z2} \quad (3.119)$$

A similar equation holds for the second site. σ is the cross-relaxation rate, from which the NOE factor η in the absence of chemical exchange is computed as $\eta = (\sigma/R_{11})(M_{02}/M_{01})$. The time evolution of magnetization is again biexponential with two time constants λ_1, λ_2 taken as the eigenvalues of the modified A' matrix of equations (3.105). Parameters σ, R_{11}, R_{12}, k_{12} and k_{21} are obtained by fitting experimental data to the above equations; in the present example: $\sigma = 0.18$ s^{-1}; $R_{11} = R_{12} = 0.60$ s^{-1} and $k_{12} = k_{21} = k = 5.4$ s^{-1} at 320 K. The saturation factor f_{1N} in the presence of the NOE is easily obtained from equation (3.119) by putting dM_{z1}/dt and dM_{z2}/dt equal to zero:

$$f_{1N} = (R_{11}\eta - k_{12})/(R_{11} + k_{12}) \quad (3.120)$$

This value differs from that obtained in the absence of the NOE by the term $R_{11}\eta$ (equation (3.111)). The influence of the NOE on saturation transfer experiments will be weak if $|R_{11}\eta| \ll k_{12}$. This is the case in the present example [163], where $\eta = 0.3$; $f_1 = -0.90$ and $f_{1N} = -0.87$, and also in all the saturation transfer experiments mentioned in the previous section [165]. Proton transfers on primary amides remain exceptional in this respect, since there is a possibility of strong NOE between the *cis* and *trans* amide hydrogens [157, 166, 167].

In a second type of problem, the interacting nuclei I and S are both carried through chemical exchange from one chemical environment to another, in which the NOE factor is clearly different. The NOE in one site is said to be transferred to the other site through chemical exchange. This is the so called exchange-transferred NOE (ET-NOE), which was first devised to study a small molecule of biological interest, called the ligand L, reversibly bound to a macromolecule E, generally a protein:

$$E + L \underset{k_{-1}}{\overset{k_1}{\rightleftharpoons}} EL$$

where EL is a protein–ligand complex (or association). In this case, the NOE factors η_F and η_B for nuclei in the free and bound ligands are expected to be quite different and even of opposite signs, with limiting values of 0.5 and -1.0, respectively, on the basis of small or large reorientational correlation times (Chapter 2). The transferred negative NOE from the bound to the free ligand nuclei, is prone to have measurable effects on the resonances in the free ligand molecule, with the decisive advantage of being easily detectable. Four sites have to be considered, which are denoted as I_F and S_F in the free ligand, and I_B and S_B in the bound ligand. These four sites are transferring longitudinal magnetization by either cross-relaxation or chemical exchange, according to the following scheme [168]:

$$\left(I_F \underset{\sigma_F}{\overset{\sigma_F}{\rightleftharpoons}} S_F \right) \underset{k_{BF}}{\overset{k_{FB}}{\rightleftharpoons}} \left(I_B \underset{\sigma_B}{\overset{\sigma_B}{\rightleftharpoons}} S_B \right)$$

where $k_{FB} = k_1\,[\text{E}]$ and $k_{BF} = k_{-1}$. Four equations (3.119) may be written for the magnetizations in each site. These equations were solved for steady-state saturation transfer experiments, in which I and S are the observed and irradiated nuclei, respectively. The resonances in the free and bound sites for each nucleus are either separate or dynamically averaged to single lines I_{av} and S_{av}, according to whether the exchange is slow or fast on the *chemical shift timescale*. In the first event, irradiation can be performed on resonances S_F or S_B, while observing either I_F or I_B resonances: four saturation factors $f_{I_{F(or\,B)}}(S_{F(or\,B)})$ may be obtained in this way. In the second event, irradiation and observation are performed on the average line so as to determine one saturation factor $f_{Iav.}(S_{av.})$. In practical applications, the ligand is chosen to be in fast exchange on the *relaxation T_1 timescale*, the kinetic information is lost, and attention is focused on cross-relaxation rates. Under these conditions, a single ET-NOE $f_I(S)$ is observed, irrespective of the exchange regions on the chemical shift scale:

$$f_I(S) = (p_F \sigma_F + p_B \sigma_B)/(p_F R_{11_F} + p_B R_{11_B}) \tag{3.121}$$

where p_B and p_F, σ_F, R_{11_B} and R_{11_F} are the fractional populations, the cross-relaxation and longitudinal relaxation rates of the bound and free ligand respectively. Attention should be drawn on that the ET-NOE factor $f_I(S)$ does not represent a weighted average over the NOE factors $\eta_F = \sigma_F/R_{11_F}$ and $\nu_B = \sigma_B/R_{11_B}$ in the free and bound ligand, but rather involves weighted means over relaxation rates. A negative ET-NOE will be observed on either the resonances I_F or I_{av} provided that $|p_B\sigma_B| > |p_F\sigma_F|$. In this way, information concerning cross-relaxation between two bound nuclei is transferred from the bound to the free state. Knowledge of the NOEs in the bound state enables information to be obtained in turn on the spatial proximity, qualitatively and also quantitatively in favourable instances, between observed nuclei, and consequently on the conformation of the bound ligand. An example concerns the study of nucleotides bound to protein receptors [34, 168], which can be shown in this way to be either in the *syn* or the *anti* conformation about the glycosidic bond depending on the nature of the nucleotide-protein pair.

Applications are by no means restricted to proton–proton nuclear Overhauser effect—e.g. the ^{31}P {^{1}H} NOE of phosphorus-containing metabolites in fast exchange between free and macromolecular bound states [169]—nor to the example of two interacting spins. Multispin systems are advantageously studied by two-dimensional NOESY, observing the decrease of cross-peaks intensities for pairs of nuclei submitted to the ET-NOE; this spectroscopy is called the ET-NOESY [170]. In this case the transfer of the NOE may be indirect, occurring over more than three spins; another difficulty is the possibility of spillover saturation due to spin diffusion [171].

4.4 TWO-DIMENSIONAL EXCHANGE SPECTROSCOPY

For multisite systems, two-dimensional exchange spectroscopy—EXSY or 2D EXSY-offers many advantages. All sites are simultaneously explored, so that exchanges are clearly delineated. Site-to-site rate constants can be obtained individually, without the parallel measurement of spin relaxation rates, this yields information on reaction mechanisms in complex systems. No selective pulses are required in these experiments, this permits investigations even when the exchanging resonances have a low chemical shift separation. On a fundamental point of view however, these methods use the same principles as one-dimensional longitudinal magnetization transfers. They have consequently the same limitations: rate constants should be slow on the NMR timescale and fast on the relaxation time T_1 timescale: nuclear Overhauser effects and scalar spin-couplings tend to obscure magnetization transfer phenomena.

4.4.1 The pulse sequence (EXSY)

The EXSY experiment, pioneered by Jeener *et al.* [21], uses the same pulse sequence as the NOESY technique (equation 2.37). The only difference is that the phenomenon responsible for transfer of z magnetization between spins during the mixing time t_m is chemical exchange in the place of cross-relaxation.

An EXSY spectrum has cross-peaks which show which NMR lines exchange with which other lines. During the labelling period t_1, nuclei in site i, initially aligned along the y axis in the rotating frame [31, 32, 172] after the first 90°_x pulse, precess by an angle $\omega_i t_1$. The magnetization that the second 90° pulse rotates into xz plane has a z component $M_{oi} \cos \omega_i t_1$. (The useless x component can be eliminated by means of a homospoil pulse or a field gradient.) During the mixing time t_m, magnetization transfers between sites i and all other sites j take place in conformance to equation (3.105). This means that after the last 90° pulse, which is an observing pulse, a nucleus initially precessing with frequency ω_i has a given probability Λ_{ij} of precessing with frequency ω_j during time t_2. The time-domain signal will thus contain terms of the form: $\exp(i\omega_i t_1)\exp(i\omega_j t_2)$. Double Fourier transformation will yield two types of peaks along the frequency axes Ω_1 and Ω_2: *cross-peaks* at $\Omega_1 = \omega_i$ and $\Omega_2 = \omega_j$, connecting together exchanging sites i and j, and diagonal peaks at $\Omega_1 = \Omega_2 = \omega_i$ for nuclei which are still in a site i after the mixing period. The intensity I_{ij} of cross-peaks is proportional to Λ_{ij}; this value may be computed through integration of equation (3.105) as

$$I_{ij} = p_i (\exp \mathbf{A}' t_m)_{ij} \tag{3.122}$$

Taking the exponential of matrix $\mathbf{A}'$ first requires its diagonalization, so that

$$\exp(\mathbf{A}' t_m) = \mathbf{X} \, diag(\lambda_1 t_m, \lambda_2 t_m \dots) \mathbf{X}^{-1} \tag{3.123}$$

where $\lambda_1, \lambda_2, \dots$ are the eigenvalues and $\mathbf{X}$ the matrix of eigenvectors of matrix $\mathbf{A}'$. Cross-peak intensities thus depend in a complex manner upon the NMR rate constants k_{ij}, which should be adjusted by fitting the relative intensities I_{ij} of cross-peaks to equation (3.122) and (3.123).

This explains why 2D EXSY has been used mainly for qualitative applications at the present time. Another reason for that is the necessity to select pure absorption mode spectra in both dimensions, free of dispersion components, for quantitative purposes (instead of the more usual absolute value transform). This is obtained by ingenious phase cycling procedures over 16 accumulations, according to either the method of States *et al.* [173], or to the time-proportional phase incrementation (TPPI) of Redfield and Kunz [174] and Marion and Wüthrich [175]. Two cases of simplification are, however, possible. For two-site systems [21], the exchange rate $k_{ex} = 1/\tau$ is given by a closed formula

$$1/\tau = t_m^{-1} \log[(1 + r)/(1 - r)]$$

where

$$r = 4p_1 p_2 (I_{11} + I_{22})/[(I_{12} + I_{21}) - (p_1 - p_2)^2] \tag{3.124}$$

Another possibility is to use short mixing times t_m so that the exponential can be reduced to the first terms of its series development [176]:

$$I_{ij} = p_i k_{ij} t_m \tag{3.125}$$

The unwanted diagonal peaks can also be removed from the spectrum by additional phase cycling [174, 175]. The 2D spectrum becomes an approximate graphic display of the exchange matrix **P**, providing a clear picture of the exchange pathways. According to equation (3.125), all cross-peaks should have intensities with the *same positive sign* under appropriate phase detection, while cross-peaks arising from the NOE may be either positive or negative [177–179], depending on the sign of the cross-relaxation constant σ (see above). Intensities I_{ij} and I_{ji} of cross-peaks which are symmetrical about the diagonal are equal [31, 32] since balanced flows of nuclei i and j should connect the two exchanging sites at equilibrium so that

$$k_{ij}p_i = k_{ji}p_j \qquad \text{and therefore } I_{ij} = I_{ji} \qquad (3.126)$$

In all these experiments, the mixing time t_m is constant and chosen close to the average exchange lifetime, usually a few tenths of second. Since a two-dimensional operation requires much instrument time, it is not usual to repeat the experiment with a series of mixing times t_m. This amounts to performing one-point kinetics without statistical control of the precision. However, it may be necessary to monitor the build-up of magnetization transfers through variations of t_m so as to assign indirect transfers in multisite exchanges (see below).

4.4.2 Specific problems

Timescales for longitudinal magnetization transfers are restricting EXSY experiments to a rather narrow range of rate constants and to corresponding narrow experimental windows. To however obtain accurate activation parameters, EXSY can be combined with lineshape analysis at higher temperature. In the event of short relaxation times T_1, experiments are only feasible if they are accompanied by large chemical shift variations $\Delta\omega$. This is often the case of paramagnetic complexes, where relaxation times of ligand nuclei are short, around 1 to 100 ms, and paramagnetic shifts are large, up to 10–100 ppm. Quadrupolar nuclei in diamagnetic non-cubic-symmetrical substrates are often characterized by short relaxation times and small chemical shift variations; this explains why most EXSY experiments involve spin 1/2 nuclei: ^{1}H, ^{13}C, ^{31}P, ^{19}F, ^{195}Pt, ^{119}Sn, ^{205}Tl, ^{29}Si ..., or nuclei endowed with relatively small quadrupolar constants: ^{2}H, ^{6}Li, ^{51}V, ^{23}Na.... In the latter case, large chemical shifts were, in fact, obtained through addition of dysprosium ion as a shift reagent to a Na$^+$ solution [180].

Two-dimensional exchange NMR offers the unique feature of graphically visualizing pairs of sites in mutual exchange. Attention should be paid, however, to the possibility of magnetization transfers arising from sequential processes in multisite systems. This means that cross-peaks between sites i and k may result from two successive exchanges $i \Leftrightarrow j$ and $j \Leftrightarrow k$, without being indicative of any real exchange between sites i and k. This is called an *indirect* transfer, which can be only evidenced by fully analysing intensities according to equation (3.120) or by repeating the EXSY experiment for a series of mixing times t_m. Indirect cross-peaks do not appear at short t_m and are usually recognized by an induction period in the build-up of their intensities. An example of that, among many others, is offered by the rates of exchange among the monomeric (19a) and the *cis* and *trans* dimeric nitrosobenzene (19b, 19c). The 2D^{13}C spectrum in CDCl$_3$ at 10 °C shows cross-peaks between the ortho carbon resonances of all pairs of species (Figure 3.14). However, quantitative analysis shows that the rate constant for direct *cis/trans* interconversion is zero; this excludes, as expected, internal rotation about the N–N double bond in the dimers. Cross-peaks between *cis* and *trans* dimers arise only indirectly and the *cis/trans* interconversion occurs exclusively through dissociation to the monomer [181].

19a   19b   19c

Scalar coupling introduces artefacts in 2D exchange spectroscopy. Cross-peaks called *J* cross-peaks are appearing in this case even in the absence of chemical exchange [182]. This is actually the basis of the 2D COSY method. Uncoupled systems are thus best fitted to 2D exchange spectroscopy. Decoupling is another possibility in the case of heteronuclear coupling, e.g. with ^{13}C{^{1}H} or ^{31}P{^{1}H} NMR. The effects of homonuclear coupling can, however, be removed by appropriate phase cycling [182] or by introducing variations δt_m in the mixing time t_m [183].

Finally, it should be stressed that any phenomenon which contributes to longitudinal magnetization transfer during the mixing time will modify the intensity of cross-peaks, thus interfering with chemical exchange. We may quote at least three possibilities of this type. The most usual one is cross-relaxation which adds a contribution $-\sigma_{ij}$ to the rate constant k_{ij}. This effect is indeed the basis of another 2D NMR method, called NOESY, which uses exactly the same pulse sequence (Chapter 2). Several features fortunately can help to separate cross-relaxation from chemical exchange. Firstly, cross-relaxation can be eliminated if the exchanging nuclei are well apart from each other. Secondly, homonuclear intramolecular cross-relaxation rates σ_{ij} are positive for small molecules in the motional narrowing limit; the corresponding cross-peaks are thus negative in pure absorption mode in contrast to those arising from chemical exchange [177]. However, the intramolecular NOE has an opposite sign, and the contribution to cross-peaks is positive, in the case of

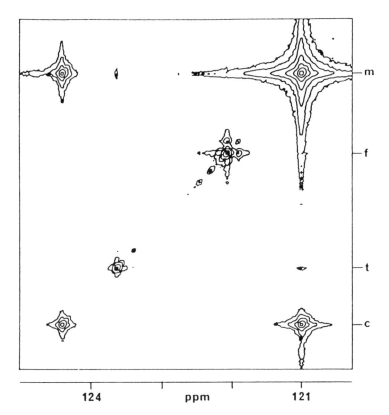

Figure 3.14 Carbon-13 2D EXSY spectrum of nitrosobenzene in CDCl$_3$ at 10 °C showing the ortho carbons region only. Assignments are given: m = monomer; c = *cis* dimer; t = *trans* dimer (f = fold-back signal). Reproduced by permission of John Wiley & Sons from Ref. 181

large molecules, this is the basis of the ET-NOE examined in the previous section. In this case, the differentiation may come from temperature (and eventually concentration) effects. Cross-relaxation rates are indeed governed by correlation times τ_c (Chapter 2) which obey Arrhenius-type laws with very low activation energies, as compared to those usually found for most reaction rates. This suggests the need to evaluate the contribution of cross-relaxation at lower temperature and/or in the absence of coreactants and catalyst, and to perform some extrapolation to the temperatures used for kinetic measurements. Relaxation reagents may also be used to suppress the NOE. Finally, specific methods also exist in the case of scalar couplings, using modified pulse sequences [184, 185].

The combined effects of indirect exchange and cross-relaxation may result in unexpected connectivities. Unexpectedly positive cross-peaks were observed between some non-labile protons in biomolecules and water. Direct intermolecular NOE would result in cross-peaks of opposite sign, since fast translational diffusion ensures motional narrowing conditions and, consequently, positive cross-relaxation rates. This phenomenon was accounted for by considering a three-site model, in which the observed connectivities result from the combined effect of cross-relaxation between a non-exchangeable proton and an exchangeable one and of chemical exchange of the latter with water. This was called a 'chemically relayed NOE' [186].

A second possibility of non-chemical magnetization transfer is quadrupolar relaxation [187] in scalarly coupled spin systems described in Section 3.7. Thus the ^{13}C spectrum of benzene-d_6 exhibits a $1:1:1$ triplet owing to coupling to deuterium with spin $I = 1$. Cross-peaks are connecting each pair of components of this triplet in a ^{13}C EXSY experiment, whose intensities are reflecting the off-diagonal elements of matrix (3.92).

A third possibility is to induce a *fast chemical reaction* during the mixing time t_m, e.g. a photochemical reaction of the type:

$$A \xrightarrow{\text{h}\nu} B.$$

In this way, a nucleus resonating at frequency ω_i in molecule A is carried into molecule B, where it resonates at frequency ω_j, by means of a laser pulse of shorter duration (a few nanoseconds). A mapping is then obtained in which each resonance i in the photoproduct is coupled to the corresponding resonance j in the starting compound. The transfer may involve longitudinal magnetization if the three 90° pulse sequences of EXSY are preserved, or transverse magnetization if the laser pulse is applied at a time t_1 after the first 90° pulse and is immediately followed by acquisition of the FID during subsequent time t_2. This is the SCOTCH experiment [188–190] (Spin COherence Transfer in CHemical reactions), first described on the example of the photocyclization of compound 20 into 21. Compound 20 is partially converted only into 21, hence the presence of both diagonal and off-diagonal peaks. This sequence in fact does not fulfil all the conditions for a DNMR experiment, since the photochemical reaction is irreversible. This explains why the spectrum is asymmetric, with only the cross-peaks $i \rightarrow j$ corresponding to the forward reaction, all located on the same side of the diagonal. Another feature of this experiment was to use the cross-peak pattern of the methyl region of the starting material to reassign the product resonances.

20   21

4.4.3   Qualitative and quantitative applications

The essential feature of 2D EXSY experiments is to provide a graphic display of the exchange process and to map out the exchange pathways by inspection of the NMR sites in mutual exchange. Another feature is the relationship between the intensity of a cross-peak and the rate constants for chemical exchange. Quantitative analysis is required to recognize indirect cross-peaks and to derive accurate kinetic parameters. In this respect, the two-dimensional method has the advantage over lineshape analysis that the temperature dependence of the chemical shift and the relaxation times need be known only approximately. Another advantage over saturation or inversion transfer experiments is to avoid incomplete selective saturation or spillover effects when the NMR peaks are too close to each other, especially in the case of complex spectra in large molecules. The analysis of the propagation of experimental errors to computed parameters is difficult in two-dimensional spectroscopy [31, 191, 192]. The accuracy depends on the signal-to-noise ratio, the digital resolution, the quality of phase corrections and of the integration procedure in the estimation of the volumes of cross-peaks. Comparisons between the results from various NMR techniques have been performed in a number of examples, e.g. for the hindered rotation in N-trifluoroacetyl-N-methylbenzylamide 22[193]. It was found that, although the scatter of data for two-dimensional experiments is larger than that for the other techniques, the rate constants remain comparable to those obtained from the other methods; the root-mean-squared error over the barrier to rotation was found equal to 8.5 instead of 3.3 or 10.8 kJ mol^{-1} for lineshape analysis or saturation transfer, respectively; the average value of the barrier for overall data was 91.0 kJ mol^{-1}.

22

Two-dimensional spectroscopy is an invaluable tool to study multisite and/or multimechanism exchanges. Even in the case of a single chemical exchange, a high number of exchanging sites generally results in inextricably overlapping coalescence patterns in one-dimensional NMR. An example is the intermolecular electron transfer reaction between the reduced and oxidized forms of cytochrome c [194], the rate of which fits the timescales for 2D EXSY. Cross-peaks were used in this study for signal assignments rather than for kinetic purposes. The Fe(III) oxidized form is indeed paramagnetic, with a well-dispersed ^{1}H NMR spectrum, permitting the identification of structural units such as the methyl groups of leucine and isoleucine residues in the 15–30 ppm region. Cross-correlation of these signals enabled the signals of these structural units in the Fe(II) reduced form to be assigned. This study illustrates the wealth of information which can be gained from 2D EXSY in the realm of *biochemistry*. Other typical examples in this respect are the multiple proton transfers between water and peptide bonds [195], the interconversion of saccharides [196–197], or the enzyme-catalysed phosphate exchanges in phosphorus-containing metabolites [198–200].

The *chemistry of solutions* offers many examples of complex mixtures of species, e.g. metallic or non-metallic oxyanions such as silicates, phosphates, stannates, vanadates …, or still the multiple complexes of certain metal ions. The identification of these species and the pathways for exchanges between them can be greatly clarified by the cross-correlations obtained from ^{29}Si, ^{31}P, ^{119}Sn, ^{51}V … 2D EXSY, Another example of complexity is offered by isotopomerization equilibria, e.g. the hydrogen–deuterium exchange between NH_4^+ and D_2O, where up to four isotopomers were simultaneously observed by proton NMR with deuterium decoupling [31]. The experiments were carried out with ^{15}N-labeled ammonium nitrate, so that eight sites were observed, giving rise to 56 cross-peaks. The full array of corresponding pseudo first-order NMR rate constants was determined by quantitative EXSY, from which an isotope kinetic effect $k_H/k_D = 1.56$ was calculated for the chemical reaction predominantly accounting for the exchange.

Another feature of isotopomers is to allow the distinction between inter and intramolecular processes, e.g. the ligand exchange [201] in the square-planar complex **23**: $(R_3P)_2PtH(SiR'_3)$, where R = cyclohexyl and R' = p-CF$_3$C$_6$H$_4$. This requires the presence of an isotopically abundant nucleus coupled to another nucleus of low abundance, ^{31}P and ^{195}Pt in the present example (in natural abundance of 100 and 33.7%, respectively). The ^{31}P {^{1}H} spectrum distinguishes the two non-equivalent phosphorus nuclei contained in each isotopomer according to their NMR pattern: two small doublets for the ^{195}Pt isotopomer and two intense singlets for the NMR-inactive platinum complex. ^{31}P {^{1}H} EXSY showed connectivities within each isotopomer, but not between the two isotopomers, this excludes the possibility of an intermolecular mechanism with bond rupture for the ligand rearrangement.

23

The latter example illustrates the potential of EXSY NMR to make the distinction between several possible mechanisms. This is especially valuable in the case of degenerate intramolecular rearrangements which cannot be studied by any other method. An archetypal example in this respect is again that of 1,2 methyl shifts in heptamethylbenzenonium ion **16** which were shown first by lineshape analysis and then by the saturation transfer method. ^{13}C {^1H} EXSY brings a particularly elegant demonstration of the existence of 1,2-methyl migrations in the carbonium ion [202]. Connectivities were indeed observed between methylenic carbons C_1 and C_2, between C_2 and C_3, and between C_3 and C_4, to the exclusion of any other combination. This also precludes as well intermolecular methyl exchange as random methyl migration to any position.

In fluxional systems allowing several types of molecular motion, connectivities observed by 2D EXSY distinguish between the possible routes in conformational rearrangements, e.g. between ring inversion and bond-shifting in the stereoisomerization of 1,3-bridged cyclooctatetraenes [203], between synchronous and non-synchronous double sulphur inversion in alkyldisulfides used as chelating agents [191], or between concerted or non-concerted ring flips in propeller molecules [204] .... From this point of view, organometallic systems offer a great variety of dynamical stereochemical problems, whose description requires recourse to permutational and group-theoretical analysis [29]. Investigations in this area are quoted in two reviews, by Orrell and Sik [37] (1987) and by Willem [32] (1987). On the other hand, a recent review (1990) by Perrin and Dwyer [31] provides detailed information on all the aspects of two-dimensional exchange spectroscopy with an exhaustive list of references.

4.5 MAGNETIZATION TRANSFERS IN THE ROTATING FRAME

Magnetization transfers, as well as longitudinal relaxation times $T_{1\rho}$, can be measured in the *rotating frame*. In this method, the nuclear magnetization, after an initial 90°_{x} pulse, is spin-locked along the y axis (Chapter 2) during a period t_{SL} which is analogous to the mixing time t_m used in NOESY or EXSY. Cross-relaxation and chemical exchange are acting on the spin-locked magnetization M_{yi} in exactly the same way as on magnetizations M_{zi} in the inversion transfer or EXSY experiments. Magnetization transfers in the rotating frame are essentially of the same nature as longitudinal magnetization transfers in the laboratory frame, since in both cases the temporal magnetization is aligned along the magnetic field B_{eff} defined in equation (1.13), in the presence or in the absence of the RF field B_1, respectively. The set of equations (3.105) to (3.107) can be used to give the evolution of $M_{yi}(t)$ during spin-lock, where M_{zi}, M_{oi} and R_{1i} should be replaced by M_{yi}, 0 and $1/T_{1\rho}$, respectively. The method still applies to exchange rates slow on the NMR timescale and fast on the relaxation time $T_{1\rho}$ timescale. The principle of these transfers was first enunciated as a one-dimensional experiment by Hennig and Limbach [205, 206] as soon as 1982. The two-dimensional version was described two years later by Bothner-By and coworkers [207], and took the name of ROESY CAMELSPIN [207–209] (see Chapter 2), with the main purpose of observing the NOE in large molecules. Its interest as a tool to study chemical exchange or to separate chemical exchange and cross-relaxation effects in two-dimensional NMR spectra, was recognized later [210–212].

The pulse sequence for these experiments is (see Figure 2.4):

$$90^\circ_x - t_1 - [Spin\text{-}lock)]_y - Acquire\ (t_2)$$

The first pulse is a hard non-selective pulse. After the t_1 delay, the phase of the RF field $\mathbf{B}_1$ is changed by 90°, and the power output is decreased to obtain selective spin-locking. Two 90°-phase shifted RF channels may be used for the high and low power respectively. The *one-dimensional version* [205] has been described for a two-site exchange only. The delay time t_1 is then given a constant value τ and the RF wave is adjusted at the frequency of line 1. Two experiments are performed successively: (i) There is no delay between the two pulses ($\tau = 0$) and the strength of B_1 is sufficient for spin-locking the magnetizations of both sites, i.e. $B_1 \gg 2\pi\Delta\nu/\gamma$, where $\Delta\nu$ is as usual the frequency shift between the two resonances. (ii) A delay $\tau = (2\Delta\nu)^{-1}$ is imposed so that the magnetization of line 2 is dephased by 180° along the $-y$ direction during this time. Both magnetizations $M_{y1}(0) = M_{01}$ and $M_y(0) = -M_{02}$ are then spin-locked by a low-power RF field $\mathbf{B}_1$ applied along the y axis. This is equivalent to a selective inversion-recovery experiment, in which the magnetizations $M_{y1}(t)$ and $M_{y2}(t)$ evolve according to equations analogous to (3.106):

$$dM_{yi}/dt = -\rho_i M_{yi} - \sigma' M_{yj} - k_{ij} M_{yi} + k_{ji} M_{yj} \tag{3.127}$$

with $i = 1$ or 2 and $j = 2$ or 1. $\rho_i = 1/T^i_{1\rho}$ and σ' are the usual relaxation constants of a dipolar-coupled system, taken in the rotating frame. The inversion-recovery experiment assumes a negligible contribution from cross-relaxation ($\sigma' \ll k_{ij}$). Closed formulas were derived in the case of two equally ($k_{12} = k_{21} = k$) or unequally populated sites. An example was given for the H_1/H_2 protons of meso-tetraphenylporphine dissolved in tetrahydrofuran-d_8 at $-70\,^\circ$C, in slow interchange owing to an intramolecular hydrogen migration. Spectra were recorded for a series of spin-lock times t_{SL} (Figure 3.15). Two monoexponential behaviours were obtained for both experiments (i) and (ii), with apparent relaxation rates of (i) $1/T^\circ_{1\rho} = (\rho_1 + \rho_2)/2$ and (ii) $1/T^\circ_{1\rho} + 2k$. The slopes of straight lines representing the logarithmic intensities of signals 1 and 2 as a function of t_{SL} yielded $1/T^\circ_{1\rho} = 2.6\ \mathrm{s}^{-1}$ and, by difference, $k = 2.9\ \mathrm{s}^{-1}$. The advantages of this method over selective inversion-recovery in the laboratory frame is that no selective pulses are required and that the magnetization decay curves are simplified owing to the nearly zero equilibrium magnetizations in the rotating frame.

The *two-dimensional* version of the experiment is obtained by varying the delay time t_1, while the spin-locking time t_{SL} is given a constant value. Compared to EXSY, time t_1 is the labelling period in both spectroscopies, and the transfer of magnetization along the quantification axis, z or y in the laboratory or the rotating frame, respectively, takes place during either the

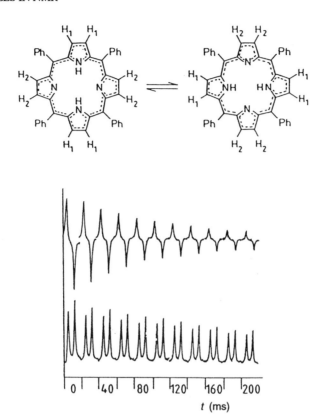

Figure 3.15 Magnetization transfer experiments (i) ($\tau = 0$) and (ii) ($\tau = (2\Delta \nu)^{-1} = 0.02$ s) carried out at $-70\,^\circ$C and 90 MHz on the H$_1$/H$_2$ signals of tetraphenylporphine dissolved in tetrahydrofuran-d_8. Reproduced by permission of Academic Press Inc. from Ref. 205

contact time t_m or the spin-lock time t_{SL}. This is followed by data acquisition. The sequence is repeated after a relaxation delay of at least $5T_1$, and the accumulated FIDs are Fourier-transformed. The main advantage of ROESY over EXSY is that motional narrowing conditions are respected for cross-relaxation and, consequently, the spin-locked NOEs are always positive, even for macromolecules with long correlation times. In phase-sensitive experiments, cross-peaks arising from chemical exchange are in phase with the diagonal peaks, whereas cross-peaks due to NOE have opposite phases. This property can be taken to advantage for the separation of chemical exchange and cross-relaxation effects in macromolecules [210–212], especially in biomolecules where the NOE is an essential tool for structure determinations (see Chapter 9). No quantitative determinations of rate constants seem to have been performed, however, using this two-dimensional exchange spectroscopy.

5   Short Exchange Lifetimes as NMR Correlation Times

As pointed out in the introductory section, the fluctuating chemical shifts induced by fast chemical exchange are generally of too small amplitude to contribute significantly to the relaxation rate in the averaged lines. Moreover, they are directed along the z axis of the laboratory frame and can only participate to transverse relaxation; this participation in fact is that obtained in the fast exchange region and is immeasurably small under the present conditions where we assume that $\tau_{ex} \blacktriangleleft \Delta\omega^{-2}$. There are two possibilities to overcome this difficulty: (i) The chemical exchange modulation $1/\tau_{ex}$ is introduced into a dipolar, or quadrupolar, or scalar interaction, which is representing a major mechanism for T_1, T_2 relaxations of the observed nucleus: the determination of the apparent correlation time τ_c may then yield information on the exchange lifetime τ_{ex}, (ii) the nuclear magnetization is spin-locked along the y axis so as to measure $T_{1\rho}$ relaxation times in the rotating frame (see Chapter 2); chemical shift fluctuations are now orthogonal to the effective magnetic field $\mathbf{B}_{eff}$ and can consequently bring a contribution $(1/T_{1\rho})_{ex}$ to the overall relaxation rate:

$$(1/T_{1\rho}) = (1/T_{1\rho})_{ex} + 1/T_{1\rho}^{\circ} \qquad (3.128)$$

where $1/T_1$ is the value observed in (or extrapolated to) conditions of no exchange. The non-secular contribution $(1/T_{1\rho})_{ex}$ obeys a law similar to that in equation (1.36) with a correlation time equal to τ_{ex}: the measurement of $(1/T_{1\rho})_{ex}$ can thus bring information on chemical dynamics. Let us see these two possibilities in the following.

5.1   CORRELATION TIMES IN THE LABORATORY FRAME

Nuclear spin–nuclear spin dipolar relaxation can be modulated by chemical exchange which modifies the nature, the distance and direction of interacting nuclei. However, the tumbling rate $1/\tau_r$ is generally much larger than the exchange rate $1/\tau_{ex}$ in the expression of $1/\tau_c$ (see equation (3.3)), this makes the contribution of chemical exchange negligibly small. An opposite conclusion is reached when the exchange rate happens to be greater than the tumbling rate. A typical example is the presence of internal motions which are faster than the overall tumbling, specially in large molecules. Ultrafast internal motions are usually limited to unhindered rotations of small substituents about single bonds, e.g. methyl rotations or side-chain motions, or to segmental backbone motions (see Chapter 9). For the sake of convenience, these ultrafast internal motions are usually treated together with overall molecular tumbling (see Chapter 2). We are then left with two other major examples, those of nuclei close to a paramagnetic centre and of quadrupolar nuclei.

5.1.1 Paramagnetic relaxation

Chemical exchanges between a paramagnetic (M) and a diamagnetic (D) site have been outlined in the previous sections. Typical examples are complexes of transition or lanthanide metal ions in the presence of a great excess of free ligand. Beyond the NMR fast exchange limit, one averaged line only is observed. The only information which can be extracted in this case is the contact shift $\Delta\omega_M = \Delta\omega_r$ and the relaxation times, $T_{1M} = T_{1r}$ and $T_{2M} = T_{2r}$ in the paramagnetic site. Exchange rates may, however, be obtained from T_{1M}, T_{2M} if they are sufficiently rapid to compete with the other mechanisms controlling the modulation of paramagnetic interactions [213], namely (i) the rotational correlation time τ_r for Brownian motion of the complex, usually $\tau_r = 10^{-9}-10^{-11}$s, and (ii) the electron spin relaxation times T_{1e}, T_{2e} which span a range of values from $ca.$ $10^{-6}-10^{-9}$ s in certain organic radicals and in a few transition metal ions (Mn^{2+}, Cu^{2+}) up to $10^{-12}-10^{-13}$ s in lanthanide complexes used as shift reagents. τ_M should be of the same order of magnitude as the shortest correlation time, while it remains, however, larger than diffusion-limited values, $\tau_M > 10^{-9}$ to 10^{-10}s, this explains why metal ions with long electronic relaxation times, such as Mn^{2+}, Cu^{2+}, can be used as T_{1M}, T_{2M} probes for kinetic purposes.

Theories of paramagnetic relaxation [17, 213–215] allow experimental T_{1M}, T_{2M} relaxation times to be related to correlation times τ_r, T_{1e} and τ_M (see Chapter 4). In their more simple version, where the electronic spin is assumed to be fully centred on the metal, and the g tensor and the rotational diffusion to be isotropic, the Solomon–Bloembergen equations [17] describe the paramagnetic contributions $1/T_{1M}$, T_{2M} to the spin–lattice and spin–spin relaxation rates of spin 1/2 nuclei in the coordination sphere of paramagnetic ions. Paramagnetic relaxation essentially arises from dipole–dipole coupling and scalar coupling between nuclei and electron spins. Relaxation rates $1/T_{iM}$ ($i = 1$ or 2) are thus the sum of two contributions: (i) the *dipolar* contribution $(1/T_{iM})_{DD}$ which depends on the electron to nucleus distance r and on overall correlation times τ_{ci} defined as

$$1/\tau_{ci} = 1/\tau_r + 1/T_{ie} + 1/\tau_M \tag{3.129}$$

and (ii) the *scalar* contribution $(1/T_{iM})_{RS}$ which depends on the hyperfine coupling constant A (in Hz) and on overall correlation times τ_{ei} defined as

$$1/\tau_{ei} = 1/T_{ie} + 1/\tau_M \tag{3.130}$$

These contributions are given by the Solomon and Bloembergen equations:

$$(1/T_{1M})_{DD} = K_{DD}[3\tau_{c1}/(1 + \omega_I^2\tau_{c1}^2) + 7\tau_{c2}/(1 + \omega_s^2\tau_{c2}^2)] \tag{3.131}$$

$$(1/T_{1M})_{RS} = K_{RS}[\tau_{e2}/(1 + \omega_s^2\tau_{e2}^2)] \tag{3.132}$$

$$(1/T_{2M})_{DD} = K_{DD}[4\tau_{c1} + 3\tau_{c1}/(1 + \omega_I^2\tau_{c1}^2) + 13\tau_{c2}/(1 + \omega_s^2\tau_{c2}^2)] \tag{3.133}$$

$$(1/T_{2M})_{RS} = K_{RS}[\tau_{e1} + \tau_{e2}/(1 + \omega_s^2\tau_{e2}^2)]/2 \tag{3.134}$$

where:

$$K_{DD} = \left(\frac{2 \times 10^{-14}}{15}\right) \frac{\mu_M^2 \gamma_I^2}{r^6} \quad \text{and} \quad K_{RS} = 8\pi^2 A^2 S(S+1)/3 \qquad (3.135)$$

in MKSA units). S and μ_M are the electronic spin number and magnetic moment for the paramagnetic cation, γ_I is the gyromagnetic ratio of the observed nucleus I, ω_I and ω_S are the nuclear and electronic Larmor frequencies. The Larmor frequency ω_S is much larger than ω_I (e.g. $\omega_S \sim 658\, \omega_H$) so that usually: $\omega_S^2 \tau_c^2 \gg 1$, and equations (3.131) to (3.14) simplify to

$$(1/T_{1M})_{DD} = K_{DD}[3\tau_{c1}/(1 + \omega_I^2 \tau_{c1}^2)] \quad \text{and} \quad (1/T_{1M})_{RS} = 0 \qquad (3.136)$$

$$(1/T_{2M})_{DD} = K_{DD}[4\tau_{c1} + 3\tau_{c1}/(1 + \omega_I^2 \tau_{c1}^2)] \quad \text{and} \quad (1/T_{2M})_{RS} = K_{RS}\tau_{el} \qquad (3.137)$$

A study of $1/T_{1M}$ as a function of the working frequency ω_I may then yield K_{DD} and τ_{c1}. This in turn allows evaluation of $(1/T_{2M})_{DD}$ and thus to deduce $(1/T_{2M})_{RS}$ from the difference $(1/T_{2M}) - (1/T_{2M})_{DD}$. Correlation times τ_{el} are obtained from $(1/T_{2M})_{RS}$ when the hyperfine coupling constants have been determined from the paramagnetic shift $\Delta\omega_M$ according to equation (3.68).

The knowledge of K_{DD}, and consequently of the electron to nucleus distance r, is important for structural determination, since this allows the paramagnetic cation to be positioned with respect to a set of interacting nuclei, e.g. the binding site of a paramagnetic cation to an enzyme or a polynucleotide [39, 42, 43]. The knowledge of overall correlation times τ_{c1} and τ_{el} can yield the tumbling, electronic or exchange rates, $1/\tau_r$, $1/T_{1e}$ and $1/\tau_M$ respectively, provided that their contributions to equations (3.129) and (3.130) can be determined separately. The tumbling rate is usually predominating over the exchange rate in equation ((3.129), so that, in most cases, kinetic information is provided on the sole basis of equation (3.130). This means that ligand exchange rates in the fast exchange region can be determined only if two conditions are fulfilled (i) Scalar contribution should contribute to a high extent to the overall relaxation rate (ii) The electronic relaxation time T_{1e} should not be clearly shorter than the lifetime τ_M of a bound ligand molecule, this obliges to use well-defined paramagnetic cations (Mn^{2+}, Cu^{2+}...), or else spin-labelled molecules, the so called *paramagnetic probes* (see Chapter 4). Under these conditions, the distinction between T_{1e} and τ_M in equation (3.130) is possible on the basis of different behaviour towards frequency or temperature variations: τ_M is frequency-independent and strongly temperature-dependent, while T_{1e} has opposite properties. Thus, for ligand exchange on the complex $Mn(DMSO)_6^{2+}$ dissolved in pure DMSO [216], the contributions of exchange and relaxation rates to $1/\tau_{el}$ are respectively: 0.46 and 5.52 ($\times 10^7$ s^{-1}) at 35 °C, and 45.2 and 3.68 ($\times 10^7$s^{-1}) at 165 °C, for a working frequency $\omega_H = 60$ MHz. Delicate stepwise adjustments of parameters, including theoretical expressions for electronic relaxation rates, are necessary to extract the required information.

It is interesting to note that, in such systems, two treatments are possible, using either the paramagnetic relaxation theories, or the Swift and Connick equations. This depends on whether the fast exchange limit on the chemical shift timescale is overpassed or not, i.e. whether the quantity $p_M \Delta \omega_M^2$ is clearly smaller or larger than unity, respectively. Thus for the $Mn^{2+}/DMSO$ system mentioned above, where $p_M = 0.01$ and $1/\tau_M = 0.46 \times 10^7 \, s^{-1}$ at 35 °C, paramagnetic shifts $\Delta \omega_M$ of 723 and 6700 Hz are observed for ^1H and ^{13}C nuclei, at 60 and 22.63 MHz, respectively [216, 217]. The corresponding values of $p_M \Delta \omega_M^2 \tau_M$ are respectively 0.04 and 3.85. This permitted two independent determinations of the ligand exchange rate, using either ^1H NMR and the relaxation method, or ^{13}C NMR and lineshape analysis, with: $1/\tau_M$ (35 °C) = (4.6 ± 2.0) or (3.2 ± 2.5) × $10^6 \, s^{-1}$, respectively.

The whole methodology, in spite of its difficulty and its limited precision, has found important applications in the fields of solution chemistry and of biochemistry. Well-documented topics in these fields are (i) ionic solvation and complexation of paramagnetic cations (see Chapters 4 and 6), and (ii) the study of labile water molecules binding to Mn(II)–protein complexes or to spin-labelled nucleotides or DNA, or of substrate–protein–Mn(II) complexes [39, 42. 43].

5.1.2 Quadrupolar relaxation

A similar methodology [39, 42, 43] has been applied to quadrupolar nuclei used as chemical probes. Quadrupolar relaxation rates $1/T_{1Q}, 1/T_{2Q}$ are determined by the quadrupolar coupling constant $\chi = e^2 qQ/h$ (see Chapters 2 and 5) and the rotational correlation time τ_r, e.g. for spin 3/2 nuclei such as ^{35}Cl, ^{23}Na, ^{39}K, ^{81}Br, in an axially symmetric electrical field gradient and in extreme narrowing conditions:

$$1/T_{1Q} = 1/T_{2Q} = \frac{2\pi^2}{5} \chi \tau_r \qquad (3.138)$$

Relaxation rates may be widely different from an exchanging site to the next, with sharp lines in cubic- or nearly cubic-symmetrical environments, e.g. the chloride ion in aqueous solution (ca. 10 Hz) or broad lines in the presence of a macromolecular liganding group (ca. 10 kHz). Line-broadenings in the latter case may be due as well to a loss of cubic symmetry ($\chi^2 \gg 0$) or to slow molecular tumbling. To bring the observed linewidth into a measurable range (<ca. 100 Hz), an excess of 'free' chloride ion is used as in the case of paramagnetic probes. ^{35}Cl NMR then allows the detection of the exchange of freely solvated chloride (site A) and chloride ions bound to the macromolecule (the site B). As in the case of paramagnetic probes, two methodologies are possible: either the exchange is slow on the chemical shift timescale and is

liable to Swift and Connick equations (see the above sections), or the exchange is beyond the fast exchange limit. In this case, an averaged line is observed with a linewidth

$$(1/T_2)_{obs} = p_A/T_{2Q}^A + p_B/T_{2Q}^B \qquad (3.139)$$

where $p_A \gg p_B$ is close to unity, and T_{2Q}^A and T_{2Q}^B are the relaxation rates in the free and bound sites. This expression predicts a linear variation of the linewidth as a function of the fraction of bound nuclei, from which T_{2Q}^B may be obtained if p_B is known, or vice versa. The method was introduced by Stengle and Baldeschwieler [218] as early as 1966, using $^{35}Cl^-$ as an indirect probe of the interactions of mercury with the sulphydryl groups of haemoglobin. The haemoglobin/Hg(II) complex is, indeed, capable of accepting Cl^- as a ligand (site B), rapidly exchanging with aqueous chloride ion in large excess.

The point of interest here is to know whether chemical exchange may compete with molecular tumbling, so as to modify the correlation time τ_r in equation (3.138). The calculation of NMR relaxation times for quadrupolar nuclei in the presence of ultrafast chemical exchanges between two sites A and B was examined by Marshall [219] in the frame of time-dependent perturbation theories. As expected from the foregoing, it was proposed to add NMR exchange rates k_{AB} or k_{BA} to the rotational rates $1/\tau_r^A$ or $1/_r^B$ in sites A or B, where, however, $k_{AB} \ll k_{BA}$ is usually neglected. It thus seems that fast exchange rates can be deduced from relaxation data as in the case of paramagnetic probes. This is generally not the case because of two additional features of quadrupolar relaxation (i) quadrupolar couplings χ cannot be measured independently from these experiments, as was the case for the hyperfine coupling constants; they remain largely unknown and are often deduced from relaxation data, using independent determinations of correlation times, e.g. from the observation of other nuclei in the molecule (ii) exchanges between chemically identical sites, i.e. of the type

$$A \xrightarrow{k_{AA}} B \quad or \quad B \xrightarrow{k_{BB}} B$$

may contribute to quadrupolar relaxation since the mechanism of relaxation involves a geometrical factor, the interaction-tensor principal axis direction, which is changed after any site exchange. Under these conditions, it was shown that equation (3.139) is still valid when correlation times τ_c^A and τ_c^B are introduced into T_{2Q}^A and T_{2Q}^B:

$$1/\tau_c^A = 1/\tau_r^A + k_{AB} + k_{AA} \sim 1/\tau_r^A + k_{AA} \qquad (3.140)$$

$$1/\tau_c^B = 1/\tau_r^B + k_{BA} + k_{BB} \qquad (3.141)$$

It should be noted however that exchanges between equivalent free halide sites involve cubic symmetry in the solvated ion, this suppresses the presence of k_{AA} in equation (3.140). Moreover, exchanges between protein-bound halide sites are presumably very slow, so that k_{BB} can be ignored in equation (3.141).

This means that the halide probe technique based on equation (3.139) is generally valid and that the overall correlation time in the bound site contains a measurable contribution from exchange rates only in the exceptional situation of rotational diffusion rate (including eventual internal rotations) slower than the exchange rate, e.g. when $\tau_r^B = 10^{-7}$–10^{-8} s and $\tau_{ex} = 10^{-8}$–10^{-10} s. Typical examples using quadrupolar probes are described in Chapter 5.

5.2 CORRELATION TIMES IN THE ROTATING FRAME

The contribution $(1/T_{1\rho})_{ex}$ to the overall relaxation rate $(1/T_{1\rho})$ is related to the exchange time by an equation analogous to equation (1.36) expressing the spectral density of the chemical shift variations:

$$(1/T_{1\rho})_{ex} = p_1 p_2 \delta\omega^2 \tau / (1 + \omega_r^2 \tau^2) \tag{3.142}$$

for a two-site exchange with the notations of Section 3.1 and $\omega_r = |\gamma B_1|$. A plot of $(T_{1\rho})_{ex}$ versus ω_r^2 should be a straight line with a slope $a = \tau/(p_1 p_2 \delta\omega^2)$ and intercept $b = 1/(p_1 p_2 \delta\omega^2 \tau)$. The lifetime τ is given as $(a/b)^{1/2}$. The value of $\delta\omega$ is also obtained if the molar fractions p_1, p_2 are known. This is especially the case when p_1 and p_2 have fixed values on symmetry grounds, e.g. the axial (a) and equatorial (e) sites of hydrogen in cyclohexane-d_{11}, where $p_1 = p_2 = 1/2$ and $p_1 p_2 \delta\omega^2 = \delta\omega^2/4$ with $\delta\omega = \omega_a - \omega_e$. Rotating-frame relaxation rates measured as a function of the RF field strength ω_r may then give the exchange lifetimes and the chemical shift between sites; the latter possibility is appreciated when the rate process is too fast to be brought to the chemical shift timescale.

Conversely, the populations in the two sites can be obtained if the frequency shift $\delta\omega$ is known independently. This is the case in the above example of ligand exchange in solutions of paramagnetic ions, where the shift $\Delta\omega_M$ between the two sites can be determined in conditions of fast exchange from the observed chemical shift of the averaged line. Rotating-frame measurements are then combined with bulk ligand shifts to determine or to reconfirm the number of coordinated ligand molecules [220].

The timescale of the experiment is determined by the magnitude of the RF field expressed in hertz, a few kilohertz, instead of that of B_0, hundreds of megahertz for experiments in the laboratory frame. Moderately fast exchange rates, $k_{ex} = 10^4$–10^6 s^{-1}, can be measured in this way. This range of values contributes to bridge the gap between the relatively slow NMR exchange rates measurable by lineshape analysis and the ultrafast exchange rates deduced from correlation times in the laboratory frame. The combined use of both methodologies allows the kinetic window to be enlarged, and consequently the temperature (or concentration) window, thus improving the accuracy in the determination of activation parameters. The complementarity of the two methodologies has been clearly illustrated [30] with the example of the chair-to-chair interconversion in cyclohexane-d_{11}. The data obtained from $1/T_{1\rho}$

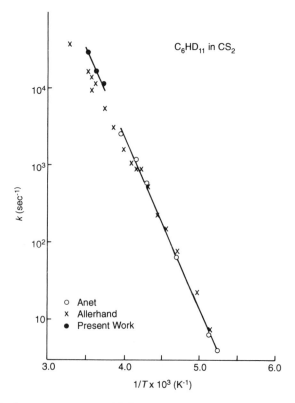

Figure 3.16 Exchange rate versus $1/T$ for chair-to-chair isomerization in cyclohexane-d_{11} ($\times$) Allerhand *et al.* [223]; ($\circ$) Anet and Bourn [221]; ($\bullet$) Wang [30]. Reproduced by permission of NMR Concepts Corporation from Ref. 30

measurement at temperatures of $-9°$ to $5\,°C$ and reported in an Arrhenius plot are in close continuity with those obtained from complete lineshape analysis [221], inversion transfer [222] and spin echo [223] methods at temperatures of -117 to $-24\,°C$ (Figure 3.16).

In spite of some attractive features, this method has not yet received all the attention it deserves for kinetic applications. A recent review by Wang [30] illustrates the potential of $1/T_{1\rho}$ measurements in conformational analysis of small molecules and of biomolecules such as peptides, cyclopeptides and DNA oligomers, and in proton transfers on amides.

6   Conclusions

In this chapter, we have tried to give an account of all the methodologies used in dynamic NMR and of the types of chemical kinetics amenable to DNMR.

Exhaustive descriptions of experimental results obtained in this way are to be found in the references given in the text to specialized books, reviews and publications.

A main point of concern in dynamic applications is the existence of timescales. A positive feature of DNMR is the possibility, for a given system under definite experimental conditions (temperature, concentrations), of changing the chemical shift timescale by variations of the magnetic field or of the observed nucleus. However, the range of exchange rates accessible to lineshape analysis remains relatively narrow, from $ca. 10^{-1}$–$10^4 \, \text{s}^{-1}$. The NMR window can be extended by one to two orders of magnitude towards slow rates by using the panel of longitudinal magnetization transfer methods, and towards faster rates by using transient methods. Other possibilities which still remain to be thoroughly explored are experiments in the rotating frame which can decrease exchange lifetimes down to the microsecond domain. Ultra-fast rate processes may be considered as correlation times in relaxation interactions, and can be measured as such, provided that they compete successfully with molecular tumbling and internal motions or with electronic relaxation times in paramagnetic species. However, it should be realized that these fast exchange methodologies have severe limitations owing to the fact that experiments are performed on averaged lines, with the loss of the primary information on the number and nature of exchanging sites.

The determination of exchanging sites and the connectivity between them is in most cases as important to the chemist and the biochemist as the quantitative measurement of reaction rates, since this provides insight into mechanisms for reactions of interest. From this point of view, two-dimensional exchange spectroscopy is bringing invaluable information in the study of multisite kinetics.

One-dimensional methods still remain at the present time the preferred way to accurate rate determinations, using if necessary the values obtained by two-dimensional NMR in the slow exchange region as trial values for least squares total lineshape analysis. Highly accurate kinetic measurements are a difficult task, as well by DNMR as by any other method. This results from multiple sources for errors, not only due to the method of detection, but also to the control of temperature, concentrations and, in some experiments, of the pressure (see Chapter 6). A great deal of technological achievement has been carried out in the last twenty years, and kinetic results obtained nowadays by DNMR can be compared conclusively to results from other methods whenever this is possible. A few examples of cross-checkings between various DNMR data or between DNMR results and equilibration kinetic data have been given in this chapter, others may be found in the literature [38].

In spite of some shortcomings, it is clear that the advent of nuclear relaxation and DNMR methodologies has changed our vision of molecular dynamics. This is especially true in some advanced fields of chemistry and biochemistry: organic, inorganic and bioinorganic coordination in solution; the

physico-chemistry of fluid inclusions in heterogeneous and microheterogeneous materials; and the dynamics of large molecules, synthetic polymers or
biomolecules.

References

1. J. T. Arnold and M. E. Packard, *J. Chem. Phys.*, **19**, 1608 (1951).
2. U. Liddel and N. F. Ramsey, *J. Chem. Phys.*, **19**, 1608 (1951).
3. E. L. Hahn and D. E. Maxwell, *Phys. Rev.*, **88**, 1070 (1952).
4. H. S. Gutowsky, D. W. McCall and C. P. Slichter, *J. Chem. Phys.*, **21**, 279 (1953).
5. H. S. Gutowsky and A. Saika, *J. Chem. Phys.*, **21**, 1688 (1953).
6. R. A. Ogg, Jr., *Discussions Faraday Soc.*, **17**, 215 (1954).
7. H. S. Gutowsky and R. C. Holm, *J. Chem. Phys.*, **25**, 1228 (1956).
8. J. T. Arnold, *Phys. Rev.*, **102**, 136 (1956).
9. R. Kubo, *J. Phys. Soc. Jap.*, **9**, 935 (1954).
10. R. A. Sack, *Molec. Phys.*, **1**, 163 (1958).
11. P. W. Anderson, *J. Phys. Soc. Jap.*, **9**, 316 (1954).
12. H. M. McConnell, *J. Chem. Phys.*, **28**, 430 (1958).
13. J. I. Kaplan, *J. Chem. Phys.*, **28**, 278 (1958); **28**, 462 (1958).
14. S. Alexander, *J. Chem. Phys.*, **37**, 967 (1954); **38**, 1787 (1963).
15. E. Grunwald, A. Loewenstein and S. Meiboom, *J. Chem. Phys.*, **27**, 630 (1957).
16. E. Grunwald and S. Meiboom, *J. Am. Chem. Soc.*, **85**, 2047 (1963); E. Grunwald and C.F. Jumper, *J. Am. Chem. Soc.*, **85**, 2051 (1963).
17. I. Solomon and N. Bloembergen, *J. Chem. Phys.*, **25**, 261 (1956).
18. S. Forsen and R. A. Hoffmann, *J. Chem. Phys.*, **39**, 2892 (1963); **40**, 1189 (1964).
19. S. Forsen and R. A. Hoffmann, *Prog. NMR Spectrosc.*, **1**, 173 (1966).
20. J. H. Noggle and R. E. Schirmer, *The Nuclear Overhauser Effect*, Academic Press, New York, 1971, p. 125.
21. J. Jeener, B. H. Meier, P. Bachman and R. R. Ernst, *J. Chem. Phys*, **71**, 4546 (1979).
22. C. Deverell, R. E. Morgan and J. H. Strange, *Molec. Phys.*, **18**, 553 (1970).
23. J. I. Kaplan and G. Fraenkel, *NMR of Chemically Exchanging Systems*, Academic Press, New York, 1980.
24. F. A. Cotton, 'Stereochemical Nonrigidity in Organometallic Compunds' in L. M. Jackman and F. A. Cotton (eds), *Dynamic Nuclear Magnetic Resonance Spectroscopy*, Academic Press, New York, 1975, ch. 10.
25. H. Günther, *NMR Spectroscopy*, John Wiley, Chichester, 1980, ch. 8.
26. A. Loewenstein and T. M. Connor, *Ber. Bunsenges. Physik. Chem.*, **67**, 280 (1963).
27. J.-J. Delpuech, *Bull. Soc. Chim. France*, 2697 (1964).
28. C. S. Johnson, Jr., *Adv. Magn. Reson.*, **1**, 33 (1965).
29. A. Steigel, 'Mechanistic Studies of Rearrangements and Exchange Reactions by Dynamic NMR Spectroscopy', in P. Diehl, E. Fluck and R. Klosfeld (eds), *NMR Basic Principles and Progress*, Springer-Verlag, Berlin, 1982, vol. 15.
30. Y. S. Wang, *Concepts in Magnetic Resonance*, **4**, 327 (1992); **5**, 1 (1993).
31. C. L. Perrin and T. M. Dwyer, *Chem. Rev.*, **90**, 935 (1990).
32. R. Willem, *Prog. Nucl. Magn. Reson. Spectrosc.*, **20**, 1 (1987).
33. M. L. Martin, J.-J. Delpuech and G. J. Martin, *Practical NMR Spectroscopy*, Heyden–Wiley, London, 1980.
34. D. Nauhaus and M. Williamson, *The NOE in Structural and Conformational Analysis*, VCH, New York, 1989, ch. 5.

35. G. Binsch, *Top. Stereochem.*, **3**, 97 (1968).
36. I. O. Sutherland, *Annu. Rep. NMR Spectrosc.*, **4**, 71 (1971).
37. K. G. Orrell and V. Sik, *Annu. Rep. NMR Spectrosc.*, **19**, 79 (1987).
38. M. Oki, *Applications of Dynamic NMR Spectroscopy to Organic Chemistry*, VCH, Deerfield Beach, Fl., 1985.
39. K. Wüthrich, *NMR in Biological Research: Peptides an Proteins*, North-Holland, Amsterdam, 1976.
40. L. J. Berliner and J. Reuben (eds), *Biological Magnetic Resonance*, Vols 1 and 2, Plenum Press, New York, 1978.
41. S. J. Opella and P. Lu, *NMR and Biochemistry*, Marcel Dekker, New York, 1979.
42. R. A. Dwek, *Nuclear Magnetic Resonance in Biochemistry*, Clarendon Press, Oxford, 1973.
43. T. L. James, *Nuclear Magnetic Resonance in Biochemistry*, Academic Press, New York, 1975.
44. P. Andriamadio, D. Nicole, A. Cartier, M. Wierzbicki and G. Kirsch, *J. Chim. Phys.*, **88**, 689 (1991).
45. K. Mislow and M. Raban, *Top. Stereochem*, **1**, 1 (1967).
46. G. Binsch, E. L. Eliel and H. Kessler, *Angew. Chem. Int. Ed. Engl*, **10**, 570 (1971).
47. G. Fraenkel, E. Beckenbaugh and P. P. Yang, *J. Am. Chem. Soc.*, **98**, 6878 (1976).
48. M. Saunders and F. Yamada, *J. Am. Chem. Soc.*, **85**, 1882 (1963).
49. C. H. Bushweller and J. W. O'Neil, *J. Am. Chem. Soc.*, **92**, 2159 (1970).
50. F. R. Jensen, D. S. Noyce, C. H. Sederholm and A. J. Berlin, *J. Am. Chem. Soc.*, **82**, 1256 (1960).
51. J. P. Jesson and E. L. Muetterties, 'Dynamic Molecular Processes in Inorganic and Organometallic Compounds', in L. M. Jackman and F. A. Cotton (eds), *Dynamic Nuclear Magnetic Resonance Spectroscopy*, Academic Press, New York, 1975, ch 8.
52. R. H. Holm, 'Stereochemically Nonrigid Metal Chelate Complexes' in *Dynamic Nuclear Magnetic Resonance Spectroscopy*.
53. P. R. Rubini, L. Rodehüser and J.-J. Delpuech, *Inorg. Chem.*, **18**, 2962 (1979).
54. S. Meiboom, *J. Chem. Phys.*, **34**, 375 (1961).
55. D. Nicole, J.-J. Delpuech, M. Wierzbicki and D. Cagniant, *Tetrahedron*, **36**, 3233 (1980).
56. J.-J. Delpuech, A. Peguy, P. R. Rubini and J. Steinmetz, *New J. Chem.*, **1**, 133 (1977).
57. E. Grunwald, C. F. Jumper and S. Meiboom, *J. Am. Chem. Soc.*, **84**, 4664 (1962).
58. J.-J. Delpuech and M. N. Deschamps, *New J. Chem.*, **2**, 563 (1978).
59. J. J. Delpuech and B. Bianchin, *J. Am. Chem. Soc.*, **101**, 383 (1979).
60. F. A. L. Anet and I. Yavari, *J. Am. Chem. Soc.*, **99**, 2794 (1977).
61. J.-J. Delpuech, M. N. Deschamps and F. Siriex, *Org. Magn. Reson.*, **4**, 651 (1972).
62. J.-J. Delpuech, 'Six-Membered Rings' in J. B. Lambert and Y. Tateuchi (eds), *Cyclic Organonitrogen Stereodynamics*, VCH, New York, 1992, ch 7.
63. A. Jaeschke, H. Muensch, H. G. Schmid, H. Friebolin and A. Mannschreck, *J. Mol. Spectrosc.*, **31**, 14 (1969).
64. R. Küspert, *J. Magn. Reson.*, **47**, 91 (1982).
65. C. L. Perrin, *Magn Reson. Chem.*, **26**, 224 (1988).
66. R. Laatikainen, *J. Magn. Reson.*, **64**, 375 (1985).
67. J.-J. Delpuech, G. Serratrice, A. Strich and A. Veillard, *Molec. Phys.*, **29**, 849 (1975).
68. T. J. Swift and R. E. Connick, *J. Chem. Phys.*, **37**, 307 (1962).

69. J.-C. Boubel, K. Bouatouch and J.-J. Delpuech, *C.R. Acad. Sci.*, **284B**, 393 (1977).
70. J.-C. Boubel, J.-J. Delpuech and G. Mathis, *Molec. Phys.*, **33**, 1729 (1977).
71. F. A. L. Anet and V. J. Basus, *J. Magn. Reson.*, **32**, 339 (1978).
72. (a) J.-J. Delpuech, 'RMN dynamique, Echanges entre sites faiblement couplés, Exemples', in M. Rico (ed.), *Curso de Resonancia Magnetica Nuclear*, University of Zaragoza, 1980, p. 305.
    (b) J. Chrisment, J.-J. Delpuech and P. Rubini, *Molec. Phys.*, **27**, 1663 (1974).
73. A. Abragam, *The Principles of Nuclear Magnetism*, Oxford University Press, London, 1961.
74. J. P. Kintzinger, Thesis, University of Strasbourg, 1970.
75. L. Rodehüser, P. R. Rubini and J.-J. Delpuech, *Inorg. Chem.*, **16**, 2837 (1977).
76. R. K. Gupta, T. P. Pitner and R. Wasylishen, *J. Magn. Reson.*, **13**, 383 (1974).
77. J. I. Kaplan, *J. Chem. Phys.*, **57**, 5615 (1972); **59**, 990 (1973).
78. R. R. Ernst, *J. Chem. Phys.*, **59**, 989 (1973).
79. S. Meiboom and D. Gill, *Rev. Sci. Instrum.*, **29**, 688 (1958).
80. J. Jen, *J. Magn. Reson.*, **30**, 111 (1978).
81. H. S. Gutowsky, R. L. Vold. and E. J. Welles, *J. Chem. Phys.*, **43**, 4107 (1965).
82. A. Allerhand and E. Thiele, *J. Chem. Phys.*, **45**, 902 (1966).
83. J. Frahm, *J. Magn. Reson.*, **47**, 209 (1982).
84. Z. Luz and S. Meiboom, *J. Chem. Phys.*, **39**, 366 (1963).
85. A. Allerhand and H. M. Gutowsky, *J. Chem. Phys.*, **41**, 2115 (1964); **42**, 1587 (1965).
86. J. P. Carver and R. E. Richards, *J. Magn. Reson.*, **6**, 89 (1972).
87. D. Lankhorst, J. Schriever and J. C. Leyte, *Chem. Phys.*, **77**, 319 (1989).
88. L. Rodehüser, P. R. Rubini, K. Bokolo, and J.-J. Delpuech, *Inorg. Chem.*, **21**, 1061 (1982).
89. G. Binsch, *J. Am Chem. Soc.*, **91**, 1304 (1969); *J. Magn. Reson.*, **3**, 146 (1970).
90. D. S. Stephenson and G. Binsch, *J. Magn. Reson.*, **32**, 145 (1978); **37**, 395 (1979).
91. J. A. Pople, W. G. Schneider and H. J. Bernstein, *High Resolution Nuclear Magnetic Resonance*, McGraw-Hill, New York, 1959.
92. S. Sternhell, 'Rotation about Single Bonds in Organic Molecules', in L. M. Jackman and F. A. Cotton (eds), *Dynamic Nuclear Magnetic Resonance Spectroscopy*, Academic Press, New York, 1975, ch. 6.
93. L. M. Jackman, 'Rotation about Partial Double Bonds in Organic Molecules', ibid., ch. 7.
94. H. Kessler, 'Detection of Hindered Rotation and Inversion by NMR Spectroscopy' *Angew. Chem. Internat. Edit.*, **9**, 219 (1970).
95. W. E. Stewart and T. H. Sidall, III, 'NMR Studies of Amides', *Chem. Rev.*, **70**, 517 (1970).
96. J. Sandström, *Top. Stereochem.*, **14**, 83 (1983).
97. J. B.Lambert, *Top. Stereochem.*, **6**, 129 (1971).
98. J. M. Lehn, *Fortschr. Chem. Forsch.*, **15**, 311 (1970).
99. J. B. Lambert and Y. Takeuchi (eds), *Cyclic Organonitrogen Stereodynamics*, VCH, New York, 1992.
100. J. B. Lambert and Y. Takeuchi (eds), *Acyclic Organonitrogen Stereodynamics*, VCH, New York, 1992.
101. F. A. L. Anet and R. Anet, 'Conformational Processes in Rings', in L. M. Jackman and F. A. Cotton (eds), *Dynamic Nuclear Magnetic Resonance Spectroscopy*, Academic Press, New York, 1975, ch. 14.
102. K. Mislow, *Acc. Chem. Res.*, **9**, 26 (1976).
103. R. Willem, M. Gielen, C. Hoogzand and H. Pepermans, *Advances in Dynamic*

Stereochemistry (M. Gielen, ed.), Free University of Brussels, Belgium, pp. 207–285.

104. M. Oki, *Top. Stereochem.*, **14**, 1 (1983).
105. R. S. Berry, *J. Chem. Phys.*, **32**, 933 (1960).
106. Von Doering and W. R. Roth, *Tetrahedron*, **19**, 715 (1963).
107. M. Saunders, *Tetrahedron Lett.*, 1699 (1963).
108. E. Vogel and H. Günther, *Ang. Chem. Int. Ed. Engl.*, **6**, 385 (1967).
109. L. W. Reeves, E. A. Allan and K. O. Stromme, *Can. J Chem.*, **38**, 1249 (1960).
110. E. L. Mackor and C. MacLean, *Pure and Appl. Chem.*, **8**, 393 (1964).
111. L. A. Telkowski and M. Saunders, 'Dynamic NMR Studies of Carbonium Ion Rearrangements', in L. M. Jackman and F. A. Cotton (eds), *Dynamic Nuclear Magnetic Resonance Spectroscopy*, Academic Press, New York, 1975, ch. 13.
112. K. Vrieze, 'Fluxional Allyl Complexes', in L. M. Jackman and F. A. Cotton (eds), *Dynamic Nuclear Magnetic Resonance Spectroscopy*, Academic Press, New York, 1975, ch. 11.
113. R. D. Adams and F. A. Cotton, 'Nonrigidity in Metal Carbonyl Compounds, in L. M. Jackman and F. A. Cotton (eds), *Dynamic Nuclear Magnetic Resonance Spectroscoy*, Academic Press, New York, 1975, ch. 12.
114. J.-J. Delpuech, *Analyst*, **117**, 267 (1992).
115. J. Brocas, M. Gielen and R. Willemen, *The Permutational Approach to Dynamic Stereochemistry*, McGraw-Hill, New York, 1983.
116. E. Grunwald and E. K. Ralph, 'Proton Transfer Processes', in L. M. Jackman and F. A. Cotton (eds), *Dynamic Nuclear Magnetic Resonance Spectroscopy*, Academic Press, New York, 1975, ch. 15.
117. A. Berger, A. Loewenstein and S. Meiboom, *J. Am. Chem. Soc.*, **81**, 62 (1959).
118. C. L. Perrin, *J. Am. Chem. Soc.*, **96**, 5628 (1974).
119. J. Chrisment, J.-J. Delpuech, W. Rajerison and C. Selve, *Tetrahedron*, **42**, 4743 (1986).
120. M. Sheinblatt and Z. Luz, *J. Phys. Chem.*, **66**, 1535 (1962).
121. B. Silver and Z. Luz, *J. Am. Chem. Soc.*, **83**, 786 (1961).
122. H. B. Charman, D. R. Vinard and M. M. Kreevoy, *J. Am. Chem. Soc.*, **84**, 347 (1962).
123. B. Bianchin, J. Chrisment, J.-J. Delpucch, M. N. Deschamps, D. Nicole and G. Serratrice, 'An NMR Study of Proton Transfers to and from Nitrogen, Oxygen, Sulphur, Carbon and Phosphorus Atoms', in P. Laszlo (ed.), *Protons and Ions Involved in Fast Dynamic Phenomena*, Elsevier, Amsterdam, 1978, pp. 265–285.
124. M. Saunders, 'Measurement of Rates of Fast Reactions using Magnetic Resonance', in S. Ehrenberg *et al.* (eds), *Magnetic Resonance in Biological Systems*, Pergamon Press, London, 1962, pp. 85–99.
125. A. H. Cowley and J. L. Mills, *J. Am. Chem. Soc.*, **91**, 2911 (1969).
126. A. Rutenberg, A. A. Palko and J. S. Drury, *J. Am. Chem. Soc.*, **85**, 2702 (1963).
127. J. S. Hartman, P. Stilbs and S. Forsen, *Tetrahedron Lett.*, 3497 (1975).
128. Z. Buczkowski, A. Gryff-Keller and P. Szczecinski, *Tetrahedron Lett.*, 607 (1971).
129. J. B. DeRoos and J. P. Oliva, *Inorg. Chem.*, **4**, 1741 (1965).
130. For a review, see (a) C. Deverell, *Prog. Nucl. Magn. Reson. Spectrosc.*, **4**, 325 (1969); (b) J.-J. Delpuech, A. Peguy and M. R. Khaddar, *J. Electroanal. Chem.*, **39**, 31 (1971); (c) J. F. Hinton and S. Amis, *Chem. Rev.*, **71**, 627 (1971); (d) A. Fratiello, *Inorganic Reaction Mechanims*, Part 2, J. O. Edwards (ed.), Wiley-Interscience, New York, 1972; (e) J. Burgess, *Metal Ions in Solution*, Ellis Horwood, Chichester, 1978, ch. 2.

131. J. M. Lehn, J. P. Sauvage and B. Dietrich, *J. Am. Chem. Soc.*, **92**, 2916 (1970).
132. K. H. Wong, G. Konizer and J. Smid, *J. Am. Chem. Soc.*, **92**, 666 (1970).
133. J. Vicens and V. Böhmer (eds), 'Calixarenes a Versatile Class of Macrocyclic Compounds', *Topics in Inclusion Science*, Vol. 3, Kluwer Academic Publishers, Dordrecht, 1990.
134. G. Fraenkel, H. Hsu and B. M. Su, *Lithium: Current Applications in Science, Medicine and Technology*, R. O. Bach (ed.), John Wiley, New York, 1985, p. 273.
135. S. O. Chan and L. W. Reeves, *J. Am. Chem. Soc.*, **95**, 673 (1973).
136. H. Sigel (ed.), *Metal Ions in Biological Systems*, Marcel Dekker, New York, 1979, Vol. 9.
137. H. M. McConnell and H. E. Wawer, *J. Chem. Phys.*, **25**, 307 (1956).
138. G. H. Leggett, B. C. Dunn, A. M. Q. Vande Linde, L.A. Ochrymowycz and D.B. Rorabacher, *Inorg. Chem.*, **32**, 5911 (1993).
139. C. R. Bruce, R. E. Norberg and S. I. Weissman, *J. Chem. Phys.*, **24**, 473 (1956).
140. C. S. Johnson, Jr., *J. Chem. Phys.*, **39**, 2111 (1963).
141. T. J. Katz, *J. Am. Chem. Soc.*, **82**, 3785 (1960).
142. R. K. Gupta and A. G. Redfield, *Science*, **169**, 1204 (1970).
143. D. W. Dixon, X. Hong, S. C. Woehler, A. G. Mauk and B. P. Shihta, *J. Am. Chem. Soc.*, **112**, 1082 (1990).
144. N. Legrand, A. Bondon, G. Simonneaux and A. Schejter, *Magn. Reson. Chem.*, **31**, 523 (1993).
145. O. E. Myers, *J. Chem Phys.*, **28**, 1027 (1958).
146. J. R. Alger and R. G. Shulman, *Quart. Rev. Biophys.*, **17**, 83 (1984).
147. J. Schotland and J. S. Leigh, *J. Magn. Reson.*, **51**, 48 (1983).
148. J. W. Faller, 'Spin Saturation Labeling', in F. C. Nachod and J. J. Zuckerman (eds), *Determination of Organic Structures by Physical Methods*, Academic Press, New York, 1971, Vol. 5, ch. 2.
149. C. L. Perrin and K. A. Armstrong, *J. Am. Chem Soc.*, **115**, 6825 (1993).
150. J. Feeney and A. Heinrich, *J. Chem. Soc. Chem. Commun.*, 295 (1966).
151. B. M. Fung, *J. Chem. Phys.*, **47**, 1409 (1967).
152. J. I. Brauman, D. F. McMillen and Y. Kanazawa, *J. Am. Chem. Soc.*, **89**, 1728 (1967).
153. N. R. Krishna, D. H. Huang, J. D. Glickson, R. Rowan and R. Walter, *Biophys. J.*, **26**, 345 (1979).
154. D. S. Kabaloff and E. Namanworth, *J. Am. Chem. Soc.*, **90**, 3234 (1970).
155. B. E. Mann, *Prog. NMR Spectrosc.*, **11**, 95 (1977).
156. S. H. Smallcombe, R. J. Holland, R. H. Fisch and M. C. Casserio, *Tetrahedron Lett.*, 5987 (1968).
157. C. L. Perrin and E. R. Johnston, *J. Magn. Reson.*, **33**, 619 (1979).
158. F. W. Dahlquist, K. J. Longmuir and R. B. Du Vernet, *J. Magn. Reson.*, **17**, 406 (1975).
159. E. Gaggelli and G. Valensin, *Concepts in Magnetic Resonance*, **4**, 339 (1992); **5**, 19 (1993).
160. A. D. Bain and J. A. Cramer, *J. Phys. Chem.*, **97**, 2884 (1993).
161. S. V. S. Mariappan and D. L. Rabenstein, *J. Magn. Reson.*, **100**, 183 (1992).
162. J. J. Led and H. Gesmar, *J. Magn. Reson.*, **49**, 444 (1982).
163. I. D. Campbell, C. M. Dobson, R. G. Ratcliffe and R. J. P. Williams, *J. Magn. Reson.*, **29**, 397 (1978).
164. P. Linscheid and J. M. Lehn, *Bull. Soc. Chim. France*, 992 (1967).
165. I. D. Campbell, C. M. Dobson and R. G. Ratcliffe, *J. Magn. Reson.*, **27**, 455 (1977).
166. A. G. Redfield and S. Waelder, *J. Am. Chem. Soc.*, **101**, 6151 (1979).

167. N. R. Krishna, K. P. Sarathy, D. H Huang, R. L. Stephens, J. D. Glickson, C. W. Smith and R. Walter, *J. Am. Chem. Soc.*, **104**, 5051 (1982).
168. G. M. Clore and A. M. Gronenborn, *J. Magn. Reson.*, **48**, 402 (1982).
169. S. X. Wang, G. H. Caines and T. Sleich, *J. Magn. Reson.*, **B102**, 47 (1993).
170. G. M. Clore and A. M. Gronenborn, *J. Magn. Reson.*, **53**, 423 (1983).
171. R. G. S. Spencer, A. Horska, J. A. Ferretti and G. H. Weiss, *J. Magn. Reson.*, **B101**, 294 (1993).
172. S. Macura and R. R. Ernst, *Molec. Phys.*, **41**, 95 (1980).
173. D. J. States, R.A. Haberkorn and D. J. Ruben, *J. Magn. Reson.*, **66**, 240 (1986).
174. A. G. Redfield and S. Kunz, *J. Magn. Reson.*, **19**, 250 (1975).
175. D. Marion and K. Wüthrich, *Biochem. Biophys. Res. Comnun.*, **113**, 967 (1983).
176. A. Kumar, G. Wagner, R. R. Ernst and K. Wüthrich, *J. Am. Chem. Soc.*, **103**, 3654 (1981).
177. P. Mirau and F. Bovey, *J. Am. Chem. Soc.*, **108**, 3654 (1981).
178. J. N. Scarsdale, R. K. Yu and J. H. Prestegard, *J. Am. Chem. Soc.*, **108**, 6778 (1986).
179. S. W. Fesik, T. J. O'Donnell, R. T. Gampe, Jr. and E. T. Olejniczak, *J. Am. Chem. Soc.*, **108**, 3165 (1986).
180. D. C. Shungu and R. W. Briggs, *J. Magn. Reson.*, **77**, 491 (1988).
181. K. G. Orrell, V. Sik and D. Stephenson, *Magn. Reson. Chem.*, **25**, 1007 (1987).
182. S. Macura, Y. Huang, D. Suter and R. R. Ernst, *J. Magn. Reson.*, **43**, 259 (1981).
183. S. Macura, K. Wüthrich and R. R. Ernst, *J. Magn. Reson.*, **46**, 269 (1982).
184. G. Wagner, G. Bodenhausen, N. Muller, M. Rance, O. Sorrensen, R. R. Ernst and K. Wüthrich, *J. Am. Chem. Soc.*, **107**, 6440 (1985).
185. G. T. Montelione and G. Wagner, *J. Am. Chem. Soc.*, **111**, 3096 (1989).
186. F. J. M. Van de Ven, H. G. J. M. Janssen, A. Gräslund and C. W. Hilbers, *J. Magn. Reson.*, **79**, 221 (1988).
187. G. Bodenhausen and R. R. Ernst, *Molec. Phys.*, **47**, 319 (1982).
188. J. Kemmink, G. W. Vuister, R. Boelens, K. Dijkstra and R. Kaptein, *J. Am. Chem. Soc.*, **108**, 5631 (1986).
189. P. J. W. Pouwels and R. Kaptein, *J. Magn. Reson.*, **A 101**, 337 (1993).
190. P. J. W. Pouwels and R. Kaptein, *J. Phys. Chem.*, **97**, 13318 (1993).
191. E. W. Abel, T. P. J. Coston, K. Orrell, V. Sik and D. Stephenson, *J. Magn. Reson.*, **70**, 34 (1986).
192. P. W. Kuchel, B. T. Bulliman, B. E. Chapman and G. Mendz, *J. Magn. Reson.*, **76**, 136 (1988).
193. P. Baine, *Magn. Reson. Chem.*, **24**, 304 (1986).
194. H. Santos, D. L. Turner, A. V. Xavier and J. Le Gall, *J. Magn. Reson.*, **59**, 177 (1984).
195. C. M. Dobson, L. Y. Lian, C. Redfied and M. D. Topping, *J. Magn. Reson.*, **69**, 201 (1986).
196. R. S. Balaban and J. A. Ferretti, *Proc. Natl. Acad. Sci. U.S.A.*, **80**, 1241 (1983).
197. J. R. Snyder, E. R. Johnston and A. S. Seriani, *J. Am. Chem. Soc.*, **111**, 2681 (1989).
198. G. L. Mendz, G. Robinson and P. W. Kuchel, *J. Am. Chem. Soc.*, **108**, 169 (1986).
199. H. L. Kantor, J. A. Ferretti and R. S. Balaban, *Biochim. Biophys. Acta*, **789**, 128 (1984).
200. J. Boyd, K. M. Brindle, I. D. Campbell and G. K. Raddla, *J. Magn. Reson.*, **60**, 149 (1984).
201 M. J. Hampden-Smith and H. Ruegger, *Magn. Reson. Chem.*, **27**, 1107 (1989).
202. B. H. Meier and R. R. Ernst, *J. Am. Chem Soc.*, **101**, 6441 (1979).

203. L. A. Paquette, T.-Z. Wang, J. Luo, C. E. Cottrell, A. E. Clough and L. B. Anderson, *J. Am. Chem. Soc.*, **112**, 239 (1990).
204. S. Biali and Z. Rappoport, *J. Org. Chem.*, **51**, 2245 (1986).
205. J. Hennig and H. H. Limbach, *J. Magn. Reson.*, **49**, 322 (1982).
206. J. Hennig and H. H. Limbach, *J. Chem. Soc. Faraday Trans. 2*, **75**, 752 (1979).
207. A. A. Bothner-By, R. L. Stephens, J. Lee, C. D. Warren and R. W. Jeanloz, *J. Am. Chem. Soc.*, **106**, 811 (1984).
208. A. Bax and D. G. Davis, *J. Magn. Reson.*, **63**, 207 (1985).
209. D. Marion, *FEBS Lett.*, **192**, 99 (1985).
210. H. Bleich and J. Wilde, *J. Magn. Reson.*, **56**, 149 (1984).
211. D. G. Davis and A. Bax, *J. Magn. Reson.*, **64**, 533 (1985).
212. J. Belleney and M. Delepierre, *Magn. Reson. Chem.*, **27**, 491 (1967).
213. N. Bloembergen and L. O. Morgan, *J. Chem. Phys.*, **34**, 842 (1961).
214. D. R. Eaton and D. W. Philipps, 'Nuclear Magnetic Resonance of Paramagnetic Molecules', in G. N. La Mar, W. D. W. Horrocks, R. H. Holm (eds), *NMR of Paramagnetic Molecules, Principles and Applications*, Academic Press, New York, 1973, pp. 103–148.
215. A. Carrington and A. D. McLachlan, *Introduction to Magnetic Resonance*, Harper & Row, New York, 1965, ch. 13.
216. J. C. Boubel and J.-J. Delpuech, *Adv. Mol. Rel. Int. Proc.*, **7**, 209 (1975).
217. J. C. Boubel, J. Brondeau and J.-J. Delpuech, *Adv. Mol. Rel. Int. Proc.*, **11**, 323 (1977).
218. T. R. Stengle and J. D. Baldeschwieler, *Proc. Natl. Acad. Sci. U.S.A.*, **55**, 1020 (1966).
219. A. G. Marshall, *J. Chem. Phys.*, **52**, 2527 (1970).
220. S. Chapra, R. E. D. McClung and R. B. Jordan, *J. Magn. Reson.*, **59**, 361 (1984).
221. F. A. L. Anet and A. J. R. Bourn, *J. Am. Chem. Soc.*, **89**, 760 (1967).
222. P. T. Inglefield, E. Krakower, L. W. Reeves and R. Stewart, *Molec. Phys.*, **15**, 65 (1968).
223. A. Allerhand, F. Chen and H. S. Gutowsky, *J. Chem. Phys.*, **42**, 3040 (1965)

4 NUCLEAR PARAMAGNETIC SPIN RELAXATION THEORY. PARAMAGNETIC SPIN PROBES IN HOMOGENEOUS AND MICROHETEROGENEOUS SOLUTIONS

Per-Olof Westlund
University of Umeå, Sweden

Dynamics of Solutions and Fluid Mixtures by NMR
Edited by J.-J. Delpuech © 1995 John Wiley & Sons Ltd

1 Introduction

The presence of paramagnetic species has a profound influence on the NMR spectra of solutions. First, the nuclear spin relaxation rates are enhanced and secondly, the NMR signal may be shifted. In this chapter we will focus on the former and present a general theory which is capable of treating paramagnetically enhanced nuclear spin relaxation in both 'simple' homogeneous and complex heterogeneous solutions. A couple of experimental studies have been selected to present different aspects of using paramagnetic spin probes in heterogeneous systems and to shed light on theoretical problems in extracting molecular information of the system studied. We do not discuss the increasing use of paramagnetic probes in two-dimensional NMR spectroscopy e.g. to cancel cross-peaks and thus obtaining structural information.

It all started almost fifty years ago with the first measurements of solvent water proton spin relaxation rates in the presence of $Fe(NO_3)_3$. The water proton spin relaxation rates increase considerably in solutions containing small concentrations of the hydrated paramagnetic ions. This effect, which is often linearly dependent on the concentration of the paramagnetic ion, is denoted *Proton Relaxation Enhancement* (PRE) or more generally *Paramagnetically Enhanced Nuclear Spin Relaxation* (PER) [1–3].

During five decades PRE and PER studies have been reported on a large variety of systems ranging from 'simple' homogeneous aqueous solutions of transition metal ions to heterogeneous biochemical systems. However, the theoretical tool available in analysing the relaxation data, developed about 1960, only accounted for solvent proton relaxation of transition metal hexa-aquo complexes. For some of the hydrated ions such as Ni(II) and Co(II), the electron spin relaxation rates were estimated to be very fast (about 1–2 ps), this indicated that the simple second-order perturbation theory was not valid. A second problem with this theory may appear when the paramagnetic ion forms a low-symmetry complex with a solute macromolecule. We will focus on PER studies of biochemical or heterogeneous systems and its implications on the theoretical framework. One of our aims is to present some recent theoretical results directed to understand the frequency dependence of experimental nuclear spin relaxation data of paramagnetic complexes in microheterogeneous systems.

In experimental studies, the enhanced nuclear spin-lattice relaxation rates $(1/T_{1E})$ and $(1/T_{2E})$ are monitored. We may split the relaxation rates into three different contributions: the background rate in the absence of paramagnetic ions $(1/T_{i,w}, i = 1, 2)$; a contribution from exchange of solvent nuclei with the first coordination shell $(1/T_{ip})$; and the outer sphere contribution $(1/T_{io})$.

$$\frac{1}{T_{iE}} - \frac{1}{T_{iw}} = \frac{1}{T_{ip}} + \frac{1}{T_{io}} \qquad (4.1)$$

The main contribution to the measured excess relaxation rates (l.h.s of equation (4.1)) of water protons comes from the exchangeable water molecules of the first coordination shell. This contribution to the relaxation rate is given by [4–5] (see also Chapter 3, equations (3.66) and (3.118)):

$$\frac{1}{T_{1p}} = Pq \frac{1}{T_{1M} + \tau_M} \qquad (4.2)$$

$$\frac{1}{T_{2p}} = \frac{Pq}{\tau_M} \left[\frac{(1/T_{2M})^2 + 1/(T_{2M}\,\tau_M) + \Delta\omega_M^2}{(1/T_{2M} + 1/\tau_M)^2 + \Delta\omega_M^2} \right] \qquad (4.3)$$

where τ_M is the mean residence lifetime of q equivalent nuclei of the paramagnetic complex. P is the fraction of exchanging nuclei which are present in the paramagnetic complex. In aqueous solution of transition metal ions q (12 in most aquo complexes) is the number of equivalent and exchanging water protons of the first hydration shell. Then $P = [M]/55.5$, where [M] is the molar concentration of paramagnetic ions and 55.5 is the molar concentration of water $[(1-P) \approx 1]$. Under fast-exchanging condition on the T_1 timescale, $\tau_M \ll T_{1M}$, and assuming $T_{1M} \ll T_{1w}$, where T_{1M} is the spin-lattice relaxation time of these q ligand nuclei, structure and dynamical information of the metal ion complex is transferred to the relaxation behaviour of bulk water protons. It is important to guarantee fast-exchange conditions which are indicated by an increase in $(1/T_{1p})$ with decreasing temperature. Under these conditions one may obtain microscopic information from a theoretical analysis of the frequency dependence of T_{1M}. In theory T_{1M} is expressed in terms of molecular quantities such as the distance between the electron spin and the nuclear spin under study, r_{IS}, the reorientation correlation time of the complex, τ_R; the coordination number q, and parameters (interaction strength and characteristic correlation times) which describe the dynamics of the electron spin system.

The second term on the r.h.s. of equation (4.1) refers to the outer sphere contribution and is dominant for second sphere coordination complexes. Here a translational diffusion modulated dipole–dipole coupling is usually the most important relaxation path. Among complexes acting as relaxation agents of magnetic resonance imaging MRI and in translational diffusion studies of counterions among paramagnetic ions this contribution is important and will be discussed in detail in Section 4.

1.1   THE T_1-NMRD CURVE

The most powerful experiment in studying PER effects is to measure the excess spin-lattice relaxation rate over a wide range of frequencies and at different temperatures. Recent development in PER theory is directed to describe these (NMRD) dispersion profiles [12], where for instance the excess water proton spin lattice relaxation rate is displayed from 0.01 to 100 MHz and sometimes complemented with data measured at some field strengths of a superconducting magnet. Solvent water proton ^{1}H T_1-NMRD profiles have been published for a large variety of systems ranging from hexa-aquo transition metal complexes, aqueous solution with different paramagnetic relaxation agents, low-symmetric complexes and complexes formed with macromolecules in biochemical systems. The relaxivity [12] $R_1 = (1/T_{1p}[M])(mM\ s)^{-1})$ is the quantity most often displayed as a function of the Larmor frequency. Hence the complete nuclear spin-lattice coupling correlation function is often probed. Therefore, we believe NMRD experiments are powerful enough to distinguish between different dynamic models although the number of model parameters may become quite large. One complication which has already been mentioned is due to the very low field region of a NMRD curve starting at 0.001 Tesla. To cover this low field limit special low-field theories have been suggested. For many paramagnetic spin systems the low-field region also introduces an additional conceptual difficulty since the electron spin may be quantized in a molecular fixed frame rather in the laboratory frame.

1.2   SYSTEMS OF INTEREST

One class of transition metal species being of fundamental importance are the aqueous transition-metal ion complexes $M(H_2O)_6^{n+}$. They constitute the natural reference systems for studies of aqueous solutions. In aqueous solution there are two principal different situations of *intra*molecular relaxation to be distinguished: (i) transition metal complexes in homogeneous aqueous solution have an averaged cubic symmetry which is only perturbed by ligand water molecules motions (Figure 4.1(a)). (ii) The hydrated metal ion has an averaged low symmetry configuration (Figure 4.1(b)). For instance, in aqueous solution of transition metal ions interacting with polyelectrolytes or different amphiphilic molecules forming liquid crystalline lamellar or micellar phases. Biochemical systems are of particular importance since most of the applications of the simple SBM theory are found among metalloproteins. Distances, between the electron spin and nuclear spin and coordination numbers, are usually extracted. This is accomplished with, at least from a theoretical point of view, an inadequate theory. A number of NMRD profiles refer to the solvent protons spin lattice relaxation in solution with paramagnetic ions (Cu^{2+}, Fe^{3+}, Co^{2+}, Mn^{2+}) interacting with different proteins (pyruvate kinase, ATP-phosphofructokinase, carboxypeptidase, carbonic anhydrase, concanavaline

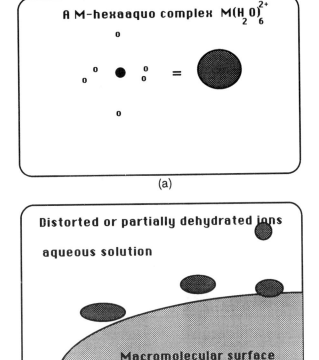

(a)

(b)

Figure 4.1 Two principal different situations. (a) The hexa-aquo complex is pictured as a spherical hydrated 'ball' emphasizing the averaged cubic symmetry of the ligand field. In (b) the hydrated metal ion is perturbed by the macromolecule surface resulting in a low-symmetry complex

A). The majority of the experimental profiles are discussed in qualitative terms and a couple of proteins have been analysed in detail using the SBM theory.

1.2.1   Homogeneous systems

In *homogeneous* systems we have only one type of paramagnetic species present. We may refer to organic or aqueous solution of paramagnetic complexes. Solutions of *low symmetry* complexes such as many contrast agents used in MRI is one example. *Hexa-aquo complexes* of transition metal ions where only a transient zero-field splitting interaction (ZFS) is active represents another example of a homogeneous system.

Let us for a while focus on transition metal-aqueous ions. The spectroscopic properties vary substantially owing to the differences in the d-electron configuration at the metal ion. This sensitivity to the electron configuration is reflected in the electron spin relaxation. For complex ions such as $Mn(H_2O)_6^{2+}$, $Cr((H_2O)_6^{2+}$, and $Fe((H_2O)_6^{3+}$ an observable ESR signal may be recorded but for $Ni(H_2O)_6^{2+}$ and $Co(H_2O)_6^{2+}$ the electron spin relaxation is too fast and the electron spin properties can only be inferred indirectly from the proton spin relaxation properties. In Figure 4.1(a) an hydrated metal ion of a homogeneous system is schematically drawn as a 'sphere'.

1.2.2   Microheterogeneous systems

In figure 4.1(b) a hydrated ion of a microheterogeneous system is schematically pictured as a deformed 'sphere'. In *microheterogeneous* systems we usually have more than one type of paramagnetic species. We may have new complexes formed and at different concentration in the solution as for instance when a transition metal complex $M(H_2O)_6^{n+}$ interacts with a charged molecular interface as illustrated in Figure 4.1(b). The local concentration may then be much larger than the bulk concentration. Then some of the water molecules in the first coordination shell may be replaced by other ligand atoms thus creating a complex of the form $M(H_2O)_{6-k}^{n+}(L)_k$. Generally, the deformed hydration shell has a reduced coordination number q compared to the homogeneous solution. For transition metal ions with electron spin quantum number $S \geqslant 1$ we may expect that the deformation of the hydration shell appears by the creation of a static zero-field splitting (ZFS) interaction which, of course, is depending on the type of transition metal ion. The complex formed is then characterized by both a *transient* and a *static* ZFS interaction. The former is modulated by intramolecular motions and the latter by the reorientation of the whole complex. A static ZFS interaction may be defined as a partial average of the ZFS interaction in a time $t < \tau$ taken over all configurations generated by the orientation of the water molecules in the first coordination shell. The characteristic time of this local fluctuation is similar to the characteristic correlation time of a transient ZFS interaction, τ_v. This correlation time is expected to be in the range of 2–10 ps which is shorter than the reorientational correlation time τ_R of a hexa-aquo complex [10]. A *static* ZFS interaction is defined by

$$< H_{SL}^{ZFS}(t) >_{\text{Static}} = \lim_{t > \tau} \frac{1}{t} \int_0^t H_{SL}^{ZFS}(s) \, ds \qquad (4.4)$$

In Figure 4.2 the limits of large static ZFS interaction and large static magnetic field (the Zeeman limit) are schematically illustrated. In the limit of ZFS interaction we have for a electron spin quantum number $S = 1$ a degenerated excited level $|1>$, $|-1>$ and a level splitting of D (the axial ZFS parameter) in the absence of a static magnetic field B. In the other limit the spin levels are split owing to the static magnetic field perhaps perturbed by the ZFS.

The *transient* ZFS interaction with a zero average is defined

$$\Delta H_{SI}^{ZFS}(t) = H_{SI}^{ZFS}(t) - < H_{SI}^{ZFS}(t) >_{\text{Static}} \qquad (4.5)$$

A very important question may now be raised. Can we decide, using an adequate theoretical description of PER, if a certain deformations of the first coordination sphere (cf. Figure 4.1(b) is transferred to the bulk water proton spin-lattice relaxation rate of the NMRD profiles. Or if we put it another way: What can we decide is 'seen' or not 'seen' by the solvent water proton spin concerning the structure and dynamics of the paramagnetic spin system? An answer which is, of course, also depending on the particular paramagnetic spin probe used.

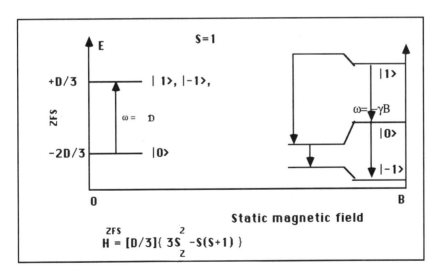

Figure 4.2 The ZFS limit is illustrated for $S = 1$ displaying a spin level splitting of D, where D is the axial ZFS parameter in the principal frame of the ZFS interaction. On the right hand side the Zeeman limit is displayed with spin level splitting proportional to the static magnetic field B weakly perturbed by a static ZFS interaction. (We have assumed only a cylindrical symmetric ZFS tensor in the molecular fixed frame)

Summarizing the main objective of a PER theory, we may say that our goal is to supply a theoretical framework (if possible by expressions in closed analytical form) which connects the time evolution of the macroscopic spill magnetization vectors with the microscopic or molecular level, without introducing phenomenological electron spin relaxation rate constants. The experimental data available at very low static field and studies of fast relaxing or strongly coupled electron spin systems requires a general theoretical approach beyond the SBM theory [1]. In this chapter we present a general theoretical framework [6–8] valid far beyond the traditional Solomon–Bloembergen–Morgan (SBM) theory [9,10] and not treating the electron spin system only within the Redfield–Bloch–Wangsness (RBW) theory [11]. However, we have only partly reached our goal with the development of the general theoretical framework presented in the following sections. The main problem in the application of this general theory still remains, that is to find reasonable simple dynamic models of the lattice (cf. Section 2.2).

This chapter is organized as follows. In the theoretical Sections 2 and 3 we present a general theoretical framework in detail. This theory treats slow-motion problems of the electron spin system, low field as well as high field limits, low- and high-(cubic) symmetric complexes. In Section 3 a number of theoretical NMRD curves are presented) that is, the field dependence of nuclear spin-lattice relaxation rates. In Section 4 we turn to some experimental NMRD studies of different systems. First, aqueous solutions of transition metal ions are presented. Then, some examples of relaxation studies using paramagnetic species in microheterogeneous aqueous solutions are presented: the liquid crystalline lamellar phase, the micellar phase of a lyotropic system, polyelectrolyte solutions and metallo-protein systems.

2 A General theory of Paramagnetically Enhanced Nuclear Spin Relaxation (PER)

In this section we review the general nuclear paramagnetic theory developed by the Stockholm group in the beginning of 1980 [6–8]. This theoretical approach is valid irrespectively of the strength of the electron spin-lattice coupling thus including the slow-motion regime of the electron spin system. It also covers the whole range of static magnetic field strengths from very low field where the Zeeman interaction is smaller than the ZFS interaction to the high field regime where the reverse is true. Within this framework we derive the Solomon–Bloembergen (SB) [9], Solomon–Bloembergen–Morgan (SBM) equations [2] as special cases. Furthermore, we present the theoretical approaches given in the phenomenological theories of Lindner [13], Bertini et al. [14] and Sharp [15].

2.1 NUCLEAR PARAMAGNETIC SPIN RELAXATION THEORY

The basic physical idea behind spin relaxation theory in the perturbation regime is the division of the total system into a pseudo-isolated spin-system comprising the relevant nuclear spin degrees of freedom (I) and a lattice (L) which contains the remaining degrees of freedom. The coupling between the two sub-systems is assumed to be weak and the lattice merely acts as a thermal reservoir. Schematically we illustrate the sub-systems in Figure 4.3.

The isolated system as schematically drawn in Figure 4.3 is defined in terms of spin Hamiltonians. The lattice degrees of freedom are defined by the spin-lattice couplings $H_{IL}(t)$, $H^{SC}(t)$, $H^{DD}(t)$ which comprise classical degrees of freedom (as indicated by the explicit time dependence) describing orientational and translational motions together with quantum mechanical electron spin degrees of freedom and sometimes enlarged with vibrational degrees of freedom. Without any further specification of the lattice degrees of freedom we may apply the RBW perturbation theory [11] and derive the expressions describing the nuclear spin lattice and spin–spin relaxation rates in terms of a Fourier–Laplace transform of time-correlation functions defined by the perturbation Hamiltonian $H_{IL}(t)$. The time independent Hamiltonian is taken in the laboratory frame and given by the Zeeman interaction (rad s^{-1}):

$$H_z = -\gamma_I B_z I_z \tag{4.6}$$

where $-\gamma_I B_z = \omega_I$ defines the Larmor frequency. The nuclear spin vector operator $\mathbf{I}$ with components (I_z, I_1^1, I_{-1}^1) couples to a stochastic time dependent lattice vector operator $T_{-n}^1(t)$ and the nuclear spin-lattice coupling ($H_{IL}(t)$) is

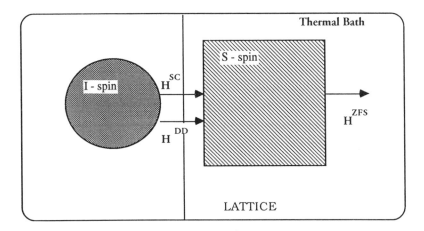

Figure 4.3 The nuclear spin system (I), the lattice (L) comprising the electron spin system (S) and the thermal bath (T). The nuclear spin-lattice interactions, H^{SC} and H^{DD}, are referring to the scalar and nuclear spin-electron spin dipole–dipole interactions. H^{ZFS} is the zero field spitting interaction and represents the electron spin thermal bath coupling.

written most conveniently as a scalar product of irreducible spherical tensor operators:

$$H_{IL}(t) = \sum_n (-1)^n I_n^1 T_{-n}^1(t) \tag{4.7}$$

We obtain, by straightforward application of the RBW theory, expressions which relate the nuclear spin relaxation rates to spectral density $K_n(n\omega_I)$;

$$\frac{1}{T_{1M}} = -2Re\{K_1(\omega_I)\} \tag{4.8}$$

$$\frac{1}{T_{2M}} = -2Re\{K_1(\omega_I)\} + K_0(0) \tag{4.9}$$

The spectral densities of equation (4.8) and (4.9) are defined by the Fourier–Laplace transform at zero and at the Larmor frequency of the time correlation function formed by use of the lattice vector operator $T_{-n}^1(t)$,

$$K_n(\omega_I) = \int_0^\infty \mathrm{tr}_L \{T_n^1 e^{iL_\tau} T_{-n}^1 \sigma_L^{Teq}\} e^{-i\omega_I \tau} \, d\tau \tag{4.10}$$

where $\sigma_L^{T,eq}$ is the lattice equilibrium density operator. The lattice operator T_{-n}^1 may refer to the nuclear spin dipole–electron spin dipole interaction (DD) or scalar interaction (SC) (see Figure 4.3). In the former case it takes the form (see Chapter 2):

$$T_n^{1(DD)}(t) = (-1)^{1+n} \sqrt{C^{DD}} \frac{1}{r_{IS}^3(t)} \sum_{q \in (-1, 0, 1)} \begin{pmatrix} 2 & 1 & 1 \\ n-q & q & -n \end{pmatrix}$$
$$S_q^1 D_{0n-q}^2 [\alpha_{LM}(t), \beta_{LM}(t)] \tag{4.11}$$

where the standard electron spin operator $\mathbf{S}$ (S_Z, S_1^1, S_{-1}^1) is expressed in the laboratory fixed frame. $D_{0n-q}^2[\Omega_{LM}(t)]$ are second rank Wigner rotation matrix elements (see Appendix C). $\Omega_{LM} = (\alpha_{LM}, \beta_{LM})$ is the solid angle between the intermolecular vector $\mathbf{r}_{IS}$ and the laboratory coordinate frame where the z_L-axis is defined by the static magnetic field B_Z and z_M-axis along the $\mathbf{r}_{IS}$ vector. The strength constant of the dipole–dipole interaction is given by

$$C^{DD} = \left[\frac{\mu_0}{4\pi}\right]^2 30 \left[\frac{h}{2\pi}\right]^2 \gamma_S^2 \gamma_I^2 \tag{4.12}$$

and includes the effective electron magnetogyric ratio $\gamma_S = -g_e\beta$ where β is the bohr magneton and g_e is in the range >2.0023 depending on paramagnetic ion; $\gamma_S = [g_e/2.0023] \, 1.761 \, 10^{11}$ rad s^{-1} T^{-1}, γ_I, the nuclear magnetogyric ratio (proton $\gamma_I = 2.675 \, 10^8$ rad s^{-1} T^{-1}) and r_{IS}, the distance between the nuclear and the electron spin. For scalar interaction (SC), T_{-n}^1 is

$$T_{-n}^{1(SC)} = AS_{-n}^1 \tag{4.13}$$

with A as the scalar coupling constant. Equations (4.11) and (4.12) may be inserted in equation (4.10) to obtain the expressions defining the nuclear spin–lattice and the spin–spin relaxation rates in terms of spectral densities.

$$\frac{1}{T_{1M}} = -2\,Re\,\{K_1^{DD}\,(\omega_I) + K_1^{SC}\,(\omega_I) + 2\,K_1^{DD-SC}\,(\omega_I)\} \tag{4.14}$$

$$\frac{1}{T_{2M}} = K_0^{DD}\,(0) + K_0^{SC}\,(0) + 2\,K_0^{DD-SC}\,(0) + \frac{1}{2}\,\frac{1}{T_{1M}} \tag{4.15}$$

Here spectral densities appear due to SC–DD cross-correlation functions. We will not discuss these terms in this chapter but refer the interested reader to ref. [7] for a detailed presentation. Here we will almost exclusively concentrate on the dipole–dipole interaction with a time correlation function of the form

$$G_n^{DD}(\tau) = (-)^n C^{DD} \sum_{p \in (-1,0,1)} \sum_{q \in (-1,0,1)} \begin{pmatrix} 2 & 1 & 1 \\ n-p & p & -n \end{pmatrix} \begin{pmatrix} 2 & 1 & 1 \\ n-q & q & -n \end{pmatrix}$$

$$\mathrm{Tr}_L\left\{\frac{D_{0n-p}^2[\Omega_{LM}]^*}{r_{IS}^3}\,S_p^{1\dagger}\,e^{iL_L\tau}\,S_q^1\,\frac{D_{0n-q}^2[\Omega_{LM}]}{r_{IS}^3}\,\sigma_L^{eq}\right\} \tag{4.16}$$

The scalar interaction correlation function is

$$G_n^{SC}(\tau) = A^2\,\mathrm{tr}_L\{S_n^{1\dagger}\,e^{iL_L\tau}S_n^1\sigma_L^{eq}\} \tag{4.17}$$

Equations (4.14)–(4.17) are the main results of this section. In order to apply this theory we must be able to determine the spectral densities given in equations (4.14) and (4.15). This is feasible when we design a model for the dynamics of the lattice in terms of spin Hamiltonians and Markov operators.

2.2 THE LATTICE

In order to proceed we must focus on the dynamical processes of the lattice. The concept 'lattice' (L) refers to the rest of the system with respect to the nuclear spin degrees of freedom (I). Equations (4.14) and (4.15) expressed the nuclear spin relaxation rates in terms of lattice correlation functions or more precisely, in terms of its Fourier–Laplace transform evaluated at zero and at the nuclear spin resonance frequency. In order to determine the spectral densities we must specify the Liouville super-operator L_L ($L_L X \equiv [H_L, X]$) governing the dynamics of the lattice degrees of freedom. We wish to obtain a detailed description of the electron spin system and hence the environment of the paramagnetic species. Consequently, the description of the electron spin system has to be equally detailed in terms of electron spin Hamiltonians and Markov operators and not through phenomenological electron spin relaxation times.

The Liouville operator L_L governing the time evolution of the lattice correlation function $G_n(\tau)$ of equations (4.16) and (4.17) may be split into a sum of different contributions.

$$L_L = L_S^0 + L_{ST}(t) + \sum_k i\Gamma_k \qquad (4.18)$$

The first Liouville operator $L_S^0 = \omega_S[S_z, \ldots]$ is the super-operator generated by the *electron spin Zeeman Hamiltonian*, $H_S = \omega_S S_z$. By convention the sign of the resonance frequency ω_S is determined by the sign of the magnetogyric ratio γ_S. When describing the dynamics of the electron spin system (S) we sometimes use the term 'thermal-bath' (T) for the rest of the lattice (Figure 4.3), i.e. 'thermal-bath' is the 'lattice' with respect to the electron spin system. When the dynamics of the thermal-bath degrees of freedom are described by Markov operators Γ_k satisfying the equation of motion

$$\frac{\partial}{\partial t} P(\Pi, t \mid \Pi_0) = \Gamma_\Pi P(\Pi, t \mid \Pi_0) \qquad (4.19)$$

describing the time-dependence of the conditional probability function $P(\Pi t \mid \Pi_0)$ of the stochastic time dependent thermal bath variable Π. Constructing a model, these random variables are introduced through the spin Liouville operators $L_{ST}(t)$ and $L_{n_L}(t)$. The former refers to *electron spin-thermal bath* and the latter to a *nuclear spin-lattice* coupling. For a paramagnetic system with electron spin quantum number $S = 1/2$ the most important relaxation mechanisms are the chemical shift anisotropy, $(\mathbf{S} \cdot \mathbf{g} \cdot \mathbf{B})$ and the anisotropic hyperfine coupling, $(\mathbf{S} \cdot \mathbf{A} \cdot \mathbf{I})$. For paramagnetic ions with an electron spin quantum number $S \geqslant 1$ the most important relaxation mechanism is the (ZFS) Zero-Field Splitting interaction $(\mathbf{S} \cdot \mathbf{d} \cdot \mathbf{S})$. In the following and without loss of generality, we will confine to electron spin systems with spin quantum number $S \geqslant 1$. The electron spin-thermal bath coupling $H_{ST}(t)$ (in our case the ZFS Hamiltonian $(\mathbf{S} \cdot \mathbf{d} \cdot \mathbf{S})$) is expressed by a scalar contraction of irreducible spherical tensor operators

$$H_{ST}^{ZFS}(t) = \sum_{n \in (-2, -1, 0, 1, 2)} (-1)^n F_{-n}^2(t) S_n^2 \qquad (4.20)$$

Here $F_{-n}^2(t)$ is a stochastic time dependent second rank tensor function. S_n^2 is an irreducible spherical standard electron spin operator. In equation (4.20) we omit the higher order terms assuming that they give negligible contributions.

2.2.1 Constructing a complete set of lattice operators

In order to evaluate the Fourier–Laplace transform of the correlation functions of equations (4.16) and (4.17), a complete set of operators is formed as a direct product of eigenoperators/functions of the time independent Liouville/Markov operators of equation (4.18). The nuclear spin-lattice coupling, $T_n^1(t)$ is expanded in this set of basis operators and we obtain a matrix representation of the spectral density $K_n(\omega)$. In the following we will describe the general procedure to evaluate the spectral density $K_n(\omega)$. The details of this procedure for two particular models are given in Appendix C and D.

It is convenient to introduce a short notation for operators analogous to the ket and bra notation for spin state functions spanning the Hilbert space. An operator of rank Σ is written as a bra operator $(Q_{-n}^\Sigma|$ which is related to a ket operator by the convention $(-1)^n|Q_n^\Sigma)$. The metric of the Liouville space is written as: $(Q_n^\Sigma|Q_n^\Sigma)$ and defined by the trace; $\mathrm{tr}_L\{Q_n^{\Sigma\dagger}Q_n^\Sigma\}$. In this notation the correlation function of equation (4.10) takes the simple form $(1)^n(T_n^1(0)T_n^1(t))$.

2.2.1.1 A complete set of electron spin operators

A convenient set of normalized spherical electron spin tensor operators $\{Q_\sigma^\Sigma = |\Sigma, \sigma) \ (\Sigma = (0, 1, 2, ..., 2S-1, 2S)\}$ which span the electron spin-space of the Liouville space are chosen with defined angular momentum properties transforming as irreducible spherical tensor operators. The ket operator $|\Sigma, \sigma)$ has rank, Σ and projection number, σ. A vector ket operator proportional to S_z is written as $|1, 0)$ or $(|Q_0^1))$. Usually spin operators of spin Hamiltonians are written in terms of standard tensor operators, $\mathbf{S}$, which are not normalized. In Appendix A we give the relations between standard and normalized operators of rank 1 and 2. A complete set of electron spin operators of rank Σ and component $\sigma(-\Sigma, -\Sigma+1, \Sigma-1, \Sigma)$ are given in terms of ket–bra operators;

$$|\Sigma, \sigma) = \sum_m (-1)^{S-m-\sigma}\sqrt{2\Sigma+1} \begin{pmatrix} S & S & \Sigma \\ m+\sigma & -m & -\sigma \end{pmatrix} |S, m+\sigma\rangle\langle S, m| \quad (4.21)$$

where $|S, m\rangle$ is a Hilbert space eigenvector of S^2 and S_z.

2.2.1.2 A complete set of eigenfunctions of a Markov operator

The random fluctuation of the semiclassical spin Hamiltonian is modelled as Markov processes. For instance, isotropic reorientational diffusion is a Markovian process described by the Laplace operator on the surface of the unit sphere. In order to describe the dynamics of the lattice within our theoretical framework we must include the eigenfunctions $\{|\gamma)_i\}$ of these Markov operators $\{\Gamma_i\}$ into the direct product basis set.

Let us denote the complete orthonormal basis set of the lattice, $\mathbf{O} = \{\{O_h\}_1^\infty,$ $\{O_h\} = \{|\Sigma, \sigma)\}\otimes\{|\gamma_k)_1\}\otimes...,\}$ which is formed as a direct product of all eigenoperators/functions of L_L; that is, the electron spin system, and the classical degrees of freedom Λ. 1, 2..., etc., We have the norm,

$$(O_j|O_k) = \mathrm{tr}_L\{O_j^+ O_k\} = \int d\Lambda \, \mathrm{tr}_S\{O_j^+ O_k\} = \delta_{jk} \quad (4.22)$$

The vector operator, T_n^1, of $G_n(\tau)$ the time correlation function of equation (4.16) or (4.17), is expressed in terms of a complete set of lattice operators. In vector notation we write this expansion

$$T_n^1 = [c_1, \ldots, c_k \ldots] \begin{bmatrix} O_1 \\ \vdots \\ O_k \\ \vdots \end{bmatrix} \tag{4.23}$$

where $c_k = \text{tr}_L\{O_k^\dagger T_n^1\}$. Upon the introduction of this basis we obtain a matrix representation of the spectral density $K_n(\omega)$ of equation (4.8) and 4.9).

$$K_n(\omega) = \mathbf{C}^* \int_0^\infty e^{i\tau\Lambda} e^{-i\omega\tau E} \, d\tau \, \mathbf{C} = \mathbf{C}^* [i\mathbf{L} + i\omega E]^{-1} \mathbf{C} \tag{4.24}$$

where only a few $c_i \neq 0$ of the vector $\mathbf{C}$ which project out the relevant matrix elements of the inverted matrix $\mathbf{M}^{-1}$, where the original, and in principle infinite matrix $\mathbf{M}$ is equal to i $[\mathbf{L} + \omega E]$. E is a unit matrix. The problem of calculating the spectral density $K_n(\omega)$ has now been turned into a problem of determining only a few matrix elements of $\mathbf{M}^{-1}$.

2.3   THE DECOMPOSITION (DC) APPROXIMATION

A very important simplification of the nuclear spin–electron spin dipole–dipole correlation function (equation (4.16)) is a decomposition (DC) into a product of pure reorientational and electron spin correlation function.

$$\text{Tr}_L\left\{ \frac{D_{0n-p}^2[\Omega_{LM}]^* S_p^{1\dagger}}{r_{IS}^3} e^{iL_L\tau} \frac{S_q^1 D_{0n-q}^2[\Omega_{LM}]}{r_{IS}^3} \right\}$$

$$= {}^{DC}\delta_{pg} \langle S_p^1 | e^{iL_S\tau} S_q^1 \rangle \left\langle \frac{D_{0n-p}^2[\Omega_{LM}(0)] D_{0n-p}^2[\Omega_{LM}(\tau)]}{r_{IS}^3(0) \qquad r_{IS}^3(\tau)} \right\rangle \tag{4.25}$$

The DC approximation was first introduced by Bloembergen and Morgan [10]. In their paper the effect of the electron spin relaxation was introduced into an effective correlation time; $1/\tau_c = 1/\tau_R + 1/\tau_S$ of the Solomon–Bloembergen equations. The physical background behind the DC approximation in aqueous solution of paramagnetic ions is that on average the complex has an octahedral symmetry [1,10]. In a cubic symmetry the second rank ZFS tensor is zero by symmetry. However, owing to the collision with solvent the fluctuating ZFS interaction is expected to be almost an order of magnitude faster than the overall reorientation. Hence the latter will not depend on the rotational motion of the hydrated complex but only on the fluctuation of the deformation of the complex. Combining equations (4.25) and (4.16) we obtain the nuclear spin–electron spin dipole–dipole correlation function

$$G_n^{DD}(\tau) = (-1)^n C^{DD} \sum_p \sum_q \begin{pmatrix} 2 & 1 & 1 \\ n-p & p & -n \end{pmatrix} \begin{pmatrix} 2 & 1 & 1 \\ n-q & q & -n \end{pmatrix}$$

$$\delta_{pg} \left\langle \frac{D^2_{0n-p}[\Omega_R(0)]^* \, D^2_{0n-p}[\Omega_R(\tau)]}{r_{IS}^3(0)} \frac{r_{IS}^3(\tau)}{} \right\rangle (S_p^1 | e^{iL_s\tau} S_q^1) \quad (4.26)$$

where the term $(S_p^1 e[iL_s\tau]S_p^1)$ represents an electron spin correlation function. The prerequisite of DC and equation (4.26) is a time-scale separation. The spin bearing nuclei studied should not participate in the creation of the ligand field and thus indirectly in the creation of a static ZFS interaction. Formally, the DC approach is applicable when the complete lattice Liouville operator may be split $L_L \approx L_S + L_\Omega$, L_S do not contain any dependence of the degrees of freedom Ω present in the nuclear-electron spin dipole–dipole coupling. Clearly this approximation is valid for calculations of outer-sphere relaxation contributions.

2.3.1   Transition-metal ion complexes: $M(H_2O)_6^{n+}$

As we have already mentioned the DC approach has been applied successfully for hydrated transition-metal ions $M(H_2O)_6^{n+}$. However, it was recently questioned [16] for one system, namely the fast relaxing electron spin system of $Ni(H_2O)_6^{2+}$ $(S=1)$. For the majority of hexa-aquo complexes the nuclear spin relaxation rates $1/T_{1M}$ and $1/T_{2M}$ are given by equation (4.26). Ignoring internal rotation and assuming that the reorientation of the whole complex is well described by a single exponential correlation time we obtain (Appendix B)

$$G_1^{DD}(\tau) = (-1)^n C^{DD} \frac{e^{-\tau/\tau_R}}{5}$$

$$\left\{ \frac{1}{5}(S_{-1}^1 e^{-\tau iL_s} S_{-1}^1) + \frac{1}{10}(S_0^1 | e^{\tau iL_s} S_0^1) + \frac{1}{30}(S_1^1 | e^{\tau iL_s} S_1^1) \right\} \frac{1}{r_{IS}^6}$$

$$(4.27)$$

$$G_0^{DD}(\tau) = (-1)^n C^{DD} \frac{e^{-\tau/\tau_R}}{5}$$

$$\left\{ \frac{1}{10}(S_{-1}^1 e^{-\tau iL_s} S_{-1}^1) + \frac{2}{15}(S_0^1 | e^{\tau iL_s} S_0^1) + \frac{1}{10}(S_1^1 | e^{\tau iL_s} S_1^1) \right\} \frac{1}{r_{IS}^6}$$

The dipole–dipole correlation function is now written as a sum of electron spin correlation functions. Pure electron spin relaxation times may be introduced in the theoretical picture defined as the integral of the electron spin correlation functions of equation (4.27). An evaluation of the spectral density K_n, as described in Section 2.2 requires the Fourier–Laplace transform of the electron spin correlation function of equation (4.27), this turns into a problem of inverting a Liouville matrix **M** (Appendix D). This theoretical treatment solves the slow-motion problem for the electron spin system and takes into account multi-exponential relaxation of high spin systems.

2.3.2   The modified Solomon–Bloembergen and Morgan theory

We may now derive the SBM theory assuming that the electron spin is weakly coupled to the thermal bath and using the DC approximation. The RBW theory is applied in the Zeeman limit where $H_I > H_{ZFS}$. Then the electron spin correlation function of equation (4.27) is given by

$$\mathrm{tr}_S\{S_p^{1\dagger}\, e^{itL_S}\, S_p^1\} = \frac{S(S+1)}{3}\, e^{t[ip\omega - T_{1+|p|,s}^{-1}]} \tag{4.28}$$

and the modulation of the ZFS is fast $\omega_S \tau_v < 1$ giving extreme narrowing conditions which implies that $T_{1+|p|,s} = T_S$. The electron spin relaxation rate is given by [10]

$$T_S^{-1} = \frac{1}{5}\, \Delta^2\, \{4S(S+1) - 3\}\tau_v \tag{4.29}$$

The correlation time τ_v is the characteristic correlation time of the transient ZFS interaction, $\Delta = f_0^{2(P)}$ (assuming cylindric symmetric ZFS tensor) (rad s^{-1}). A generalization of equation (4.29) to non-extreme narrowing conditions has later been given by Rubenstein *et al.* [17]. Combining equations (4.27)–(4.29) gives the dipole–dipole contribution of the SBM theories and adding the scalar contribution equations (4.8) and (4.9) gives

$$\frac{1}{T_{1M}} = \frac{1}{225}\, S(S+1)\, \frac{C^{DD}}{r_{IS}^6}\, \{J_{-1}\,(\omega_S - \omega_I) + 3J_0(\omega_I) + 6J_1(\omega_S + \omega_I)\}$$

$$+ \frac{2}{3}\left(\frac{A}{\hbar}\right)^2 S(S+1)\, J_{-1}^{SC}\,(\omega_S - \omega_I) \tag{4.30}$$

$$\frac{1}{T_{2M}} = \frac{1}{450}\, S(S+1)\, \frac{C^{DD}}{r_{IS}^6}\, \{J_{-1}\,(\omega_S - \omega_I) + 3J_0(\omega_I) + 6J_1(\omega_S + \omega_I) + 6J_1(\omega_S)$$

$$+ 4J_0(0)\} + \frac{1}{3}\left(\frac{A}{\hbar}\right)^2 S(S+1)\, \{J_0^{SC}(0) + J_{-1}^{SC}\,(\omega_S - \omega_I)\} \tag{4.31}$$

$A/\hbar$ is the scalar interaction strength (in joules) and the reduced spectral density is defined as:

$$J_n(-n\omega_S + \omega_I) = \frac{\tau_{C,\,|n|+1}}{1 + (n\omega_S - \omega_I)^2\,\tau^2_{C,\,|n|+1}} \tag{4.32}$$

where the effective correlation times for the dipole–dipole interaction are

$$\tau^{-1}_{C,1} = \tau^{-1}_R + \tau^{-1}_M + T^{-1}_{1,S}$$
$$\tau^{-1}_{C,2} = \tau^{-1}_R + \tau^{-1}_M + T^{-1}_{2,S} \tag{4.33}$$

and for the scalar interaction

$$\tau^{-1}_{C,1} = \tau^{-1}_M + T^{-1}_{1,S}$$
$$\tau^{-1}_{C,2} = \tau^{-1}_M + T^{-1}_{2,S} \tag{4.34}$$

The parameter space of the SBM model of nuclear spin–electron spin dipole–dipole coupling comprise the nuclear spin-electron spin distance, r_{IS}, the reorientational correlation time τ_P, the transient ZFS interaction, Δ, and its characteristic correlation time τ_V and the static magnetic field B (see also Chapter 3, equations 3.129 to 3.137).

2.3.3   Outer sphere relaxation

The DC approach can also be applied to outer sphere relaxation contribution as mentioned. We will now present some expressions which describe this contribution in more detail. Nuclei of the second coordination shell may be considered to reorient independently of the reorientation of the paramagnetic complex. It is then reasonable to assume that the electron spin relaxation is not influenced by the molecules of the second coordination sphere. The relative translational motion of nuclear and electron spin modulates the classical correlation function $C^2_{n-p}(\tau)$ of equation (4.26):

$$\left\langle \frac{D^2_{0n-p}[\Omega_R(0)]^*\,D^2_{0n-p}[\Omega(\tau)]}{r^3_{IS}(0)\qquad\qquad r^3_{IS}(\tau)} \right\rangle = C^2_{n-p}(\tau) \tag{4.35}$$

The space-dependent time correlation function $C^2_{n-p}(\tau)$ has been discussed extensively by Hwang and Freed [18, 19] and Freed [20] and is expressed by

$$C^2_{n-p}(t) = \int d^3r \int d^3r_0 D^2_{0n-p}[\Omega_{LM}(t)]^*\,D^2_{0n-p}[\Omega_{LM}(t)]\,P(\mathbf{r}_0 \mid \mathbf{r}, t)\,g(r_0)/r^3_0\,r^3,$$

$$\tag{4.36}$$

A numerical finite difference method was developed to determine the correlation function of (4.36) also including the effect of the potential of averaged forces obtained from the radial distribution function $g(r)$. Analytical solutions were only obtained for the force-free model $[g(r)=0,\ r<d;$ $g(r)=1,\ r>d]$. The spectral density $J_{n-p}(\omega)$ which is the Fourier–Laplace transform of equation (4.41) for the cases of well defined *single* exponential electron spin relaxation times T_{1S} and T_{2S} [19–20], is

$$J_p(\omega_I) = \frac{8}{27}\frac{\eta_S}{dD}$$

$$Re\left[\frac{1+\dfrac{1}{4}(i\omega_I\tau+\tau/T_{pS})^{1/2}}{1+(i\omega_I\tau+\tau/T_{pS})^{1/2}+\dfrac{4}{9}(i\omega_I\tau+\tau/T_{pS})+(i\omega_I\tau+\tau/T_{pS})^{3/2}}\right] \quad (4.37)$$

where $\tau=d^2/D$, and $D=D_I+D_S$ is the sum of the diffusion coefficients of the molecule bearing the spin I and the paramagnetic complex bearing the electron spin S. Two limits are of particular interest. First, when $|\omega T_{iS}|\ll 1$ [19–20]

$$J_p(\omega_I) = \frac{8}{27}\frac{\eta_S}{dD}$$

$$Re\left[\frac{1+\dfrac{1}{4}(\tau/T_{pS})^{1/2}}{1+(\tau/T_{pS})^{1/2}+\dfrac{4}{9}(\tau/T_{pS})^{1/2}+\dfrac{1}{9}(\tau/T_{pS})^{3/2}}\right]\xrightarrow{\tau/T_{pS}\gg 1}\frac{2}{3}\frac{\eta_S}{d^3}T_{pS} \quad (4.38)$$

secondly, for slow relaxing electron spin systems $|\omega T_{iS}|\gg 1$ [19–20]

$$J_p(\omega_I) = \frac{8}{27}\frac{\eta_S}{dD}$$

$$\left[\frac{1+(5/4\sqrt{2})(\omega\tau)^{1/2}+\dfrac{1}{4}(\omega\tau)}{1+\sqrt{2}(\omega\tau)^{1/2}+(\omega\tau)+(2/3\sqrt{2})(\omega\tau)^{3/2}+(8/81\sqrt{2})(\omega\tau)^{5/2}+(\omega\tau)^3/81}\right] \quad (4.39)$$

$$\xrightarrow{\omega\tau\ll 1}\frac{8}{27}\frac{\eta_S}{dD}\left[1-\frac{3}{8}\left(\frac{2\omega d^2}{D}\right)^{1/2}\right] \quad (4.40)$$

$$\xrightarrow{\omega\tau\gg 1}\frac{6}{27}\eta_S D/(\omega d^2)^2 \quad (4.41)$$

With these expressions it is possible to obtain a rough estimation of the spectral densities of equations (4.26). However, a general treatment of outer-sphere relaxation, free or not, in the slow-motion regime of the electron spin is quite a complicated task.

2.4 A GENERAL EXPRESSION OF DIPOLAR CORRELATION FUNCTIONS FOR SLOW TUMBLING COMPLEXES

Several binding sites of paramagnetic ions may be present on macromolecules. A typical example is a hydrated paramagnetic ion which is bound to a protein or polyelectrolyte. We may distinguish between condensed ions with a partly deformed but intact hydration shell, free hexa-aquo ions and, finally, partly dehydrated ions which may form new first hydration shell complexes. In the theoretical description it is convenient to introduce a macromolecule fixed frame (M) at the binding site of the ion with the Z_M-axis defined by a local symmetry axis. The Euler angles $\Omega_{LM}(t)$ describe the relation between the laboratory frame (L) and the macromolecular fixed frame (M). We then rewrite the ZFS function of the Hamiltonian (equation (4.20))

$$F_{-n}^{2(L)}(t) = \sum_k \sum_m f_k^{2(P)} D_{km}^2[\Omega_{MP}(t)] D_{m-n}^2[\Omega_{LM}(t)] \qquad (4.42)$$

The time-dependence of the Euler angles $\Omega_{LM}(t)$ is due to local distortion of the ligand field symmetry (the first coordination shell) similar to a hexa-aquo complex. Two dynamic models of the ZFS will be discussed, namely the pseudo-rotation model (Section 3.2) and a vibrational model (Section 3.3). The time-dependence in $\Omega_{LM}(t)$ is due to the reorientation of the macromolecule and is expected to be slow compared to the relaxation processes of the electron spin system. The local fluctuation of the symmetry of a partially dehydrated paramagnetic complex is probably anisotropic owing to the perturbation from the molecular interface of the macromolecule. This means that the ZFS interaction averaged over the local fluctuation gives a non-zero value or a static ZFS interaction. Introducing a partial average $<D_{km}^2[\Omega_{MP}(t)]>_{\text{fast}} = S_{km}^2 \neq 0$ into equation (4.42) accounts for a deformation or the averaged low-symmetry of the environment of a metal ion. The index 'fast' refers to a partial average taken over the fast fluctuation of the symmetry of the complex. A well defined average means the presence of a *static* − ZFS interaction. With this in mind the ZFS Hamiltonian may be split into a *static*,

$$F_{-n}^{2(L)}(t) = \sum_k \sum_m f_k^{2(P)} S_{km}^2 D_{m-n}^2[\Omega_{LM}(t)] \qquad (4.43)$$

and a *transient* part

$$F_{-n}^{2(L)\,\text{transient}}(t) = \sum_k \sum_m f_k^{2(P)} \{D_{km}^2[\Omega_{PM}(t)] - S_{km}^2\} D_{m-n}^2[\Omega_{LM}(t)] \qquad (4.44)$$

Considering a paramagnetic ion *tightly bound* to a macromolecule we mean that a low symmetry complex of the type, $Me(H_2O)^{n+}{}_{6-k}(LM)_k$ is formed with atomic groups of the macromolecule as ligand (LM). When the reorientation of the r_{IS} vector is sufficiently slow compared with electron spin relaxation, it is possible to separate the dipole–dipole correlation function of equation (4.16) which is simplified to

$$G_n^{DD}(\tau, \beta) = (-1)^n C^{DD} \sum_p \begin{pmatrix} 2 & 1 & 1 \\ n-p & p & -n \end{pmatrix}^2 \frac{1}{r_{IS}^6} |d_{0n-p}^2(\beta)|^2 \, \text{tr}_S \{ S_p^{1\dagger} \, e^{iL_S(\beta)\tau} \, S_p^1 \}$$

(4.45)

where the Markov operator describing the reorientational diffusion of the whole complex has been removed from the lattice Liouville operator of equation (4.18), i.e. $L_S(\beta) = L_L - \Gamma_R$. It should be noted that this separation of the dipole–dipole correlation function differs from the DC approximation of equation (4.25) since the electron spin correlation function is governed by $L_S(\beta)$ which is dependent on the solid angles β_{LM}. The static and transient ZFS interaction Hamiltonians of equations (4.43) and (4.44) depend on the orientation, Ω_{LM}, of the macromolecule with respect to the laboratory frame.

The Fourier–Laplace transform of the electron spin correlation function of equation (4.45) using the basis set of Appendix D gives angle dependent coefficients of the **X** vector and also (correlation functions) matrix elements of **M**. If for instance the paramagnetic ion is bound with a low-symmetry coordination sphere we expect the existence of a static ZFS interaction fixed in the M-frame of the macromolecule.This operator introduces an angle dependent **M** matrix. In order to obtain the orientation independent relaxation rate we must integrate the orientation dependent nuclear spin relaxation rates over a distribution function $P(\Omega_{LM})$

$$K_n^{DD}(\omega_I) = \int P(\Omega) \int_0^\infty dt \, G_n^{DD}(t, \Omega) \, e^{-i\omega_I t} \, d\Omega = \int P(\Omega) \, K_n^{DD}(\omega_I, \Omega) \, d\Omega \quad (4.46)$$

where the angle dependent spectral density with the angle dependent coefficient **X** vector is given by:

$$K_n^{DD}(\Omega_I, \Omega) = \mathbf{X}(\Omega)^* [-i\mathbf{L}(\Omega) + i\omega_I \mathbf{E}]^{-1} \mathbf{X}(\Omega) \quad (4.47)$$

and $\mathbf{M} = [-i\mathbf{L}(\Omega) + i\omega_I \mathbf{E}]$ is inverted numerically.

2.5   THE LOW-FIELD APPROACH

A low-field formulation of equation (4.45) was first given by Lindner [13] in his analysis of proton spin lattice relaxation in solution of hexa-aquo Ni(II) complexes. Later this approach was used by the Florence group [14] for slowly rotating low symmetry Co(II) ($S = 3/2$) complexes. In the ZFS limit $H_{ZFS} > H_Z$ the electron spin is quantized in a molecular fixed frame M' rather in the laboratory frame (L) . In this regime the electron vector operators are expressed in the M' frame using the standard relation

$$S_q^{1(L)} = \sum_{m \in (-1, 0, 1)} S_m^{1(M')} D_{mq}^1 [\Omega_{LM'}] \quad (4.48)$$

where the Euler angles $\Omega_{LM'}$ specify the relative orientation betw ᴜ the laboratory frame and the molecular fixed frame M'. The lattice dipole–dipole operator may be rewritten with the electron spin vector operators expressed in the molecular fixed frame M'. The following new expression for the dipole–dipole correlation function is then obtained;

$$G_n^{DD}(\tau, \beta_{LM}, \beta_{LM'}) = (-1)^n C^{DD} \frac{1}{r_{IS}^6} \sum_p \begin{pmatrix} 2 & 1 & 1 \\ n-p & p & -n \end{pmatrix}^2 [d_{0n-p}^2(\beta_{LM})]^2$$

$$\sum_m [d_{mp}^1(\beta_{LM'})]^2 \times \mathrm{tr}\{S_m^{1(M')\dagger} e^{iL(\Omega)\tau} S_m^{1(M')}\} \quad (4.49)$$

where the Liouville operator governing the electron spin correlation function in the molecular fixed frame M' may be dependent on the Euler angles Ω_{LM} and $\Omega_{LM'}$ and is defined by the model Hamiltonians in equation (4.18). Equation (4.49) may not be investigated unless the electron spin correlation functions are determined. We can follow the recipe of Section 2.2 and determine the corresponding spectral density. Or an analytical expression may be found for the electron spin correlation function in the Redfield regime. Bertini et al. [14] investigated a case with electron spin quantum number $S = 3/2$ by introducing the following phenomenological electron spin correlation functions with one relaxation time τ_S

$$\mathrm{tr}_S\{S_m^{1(M')} e^{iL(\Omega_{LM'})\tau} S_{m'}^{1(M')}\} = \delta_{mm'} \frac{(S+1)S}{3} e^{-\tau/\tau_S}$$

$$\times \left[\delta_{m0} + \delta_{|m|1} \left[\frac{6 + 12 \cos[\omega\tau]}{5} \right] e^{im\omega_S x\tau} \right] \quad (4.50)$$

where ω is the transition frequency between zero-field split levels [$\omega = 2D$; $D = \sqrt{3}/2 f_0^{2(P)}$] and $x = \cos\beta_{LM'}$.

Another phenomenological approach has been published by Bayburt and Sharp [21] in their analysis of outer sphere proton–spin relaxation of solvent molecules in solutions of tris (acetylacetonato)Mn(III) in acetone. ($S = 2$). Sharp [15] later extended this approach to other electron spin quantum numbers and to the high-field limit, however, without removing the phenomenological description of the electron spin system. In the low-field description, the DC approximation is used by starting from equation (4.26). The electron spin operators are transformed to the principal frame of the ZFS interaction,

$$G_n^{DD}(\tau) = (-1)^n C^{DD} \sum_p \sum_q \begin{pmatrix} 2 & 1 & 1 \\ n-p & p & -n \end{pmatrix} \begin{pmatrix} 2 & 1 & 1 \\ n-q & q & -n \end{pmatrix}$$

$$\left\langle \frac{D_{0n-p}^2[\Omega_R(0)]^* D_{0n-q}^2[\Omega_R(\tau)]}{r_{IS}^3(0) \quad r_{IS}^3(\tau)} \right\rangle \sum_m \sum_{m'} \langle S_m^{1(M)} e^{iL\tau} S_{m'}^{1(M)} \rangle$$

$$\langle D_{mp}^1[\Omega_{LP}(t)] D_{m'q}^1[\Omega_{LP}(0)] \rangle \quad (4.51)$$

The translational diffusion correlation function of equation (4.51) was modelled using Freed's expression given in equation (4.37). The reorientation motion of the molecular fixed (P) frame relative to the laboratory frame is modelled as a independent free diffusion motion;

$$\langle D^1_{mp} [\Omega_{LP}(t) \, D^1_{m'q}[\Omega_{LP}]\rangle = \delta_{mm'} \, \delta_{pq} \, \frac{e^{-3D_R \tau}}{3} \tag{4.52}$$

The Liouville operator L_L which governs the electron spin correlation function in the molecular fixed frame(P) is generated by a cylindrical symmetrical ZFS interaction,

$$H^{ZFS}_{ST} = f^{2(P)}_0 \sqrt{\frac{3}{2}} \, ([S_Z]^2 - \frac{1}{3} \, S(S + 1)) \tag{4.53}$$

In this model the electron spin correlation functions are given by [21]

$$(S^1_0 | e^{iL_S \tau} S^1_0) = \frac{1}{3} \, S \, (S + 1) \, e^{-\tau/\tau_S} \tag{4.54}$$

$$(S^1_1 | e^{iL_S \tau} S^1_1) = \frac{1}{2} \, \frac{1}{2S + 1} \sum_m |C^+_{S, m}| \, e^{i\omega_m \tau} \, e^{-\tau/\tau_{S, m}} \tag{4.55}$$

The transition frequency can be written in terms of the axial ZFS parameter D

$$\omega_m = D(2m + 1); \quad f^{2(P)}_0 = \sqrt{\frac{2}{3}} \, D \tag{4.56}$$

and the coefficient is

$$c^+_{s,m} = \sqrt{\{(S - m)(S + m - 1)\}} \tag{4.57}$$

The spin relaxation times $\tau_{S,m}$ of the transitions at ω_m are all set equal to a phenomenological time constant τ_S. The Zeeman interaction generates a time dependent perturbation modulated by the reorientation of the molecule relative to the laboratory frame:

$$H_Z(t) = \omega_S \sum_m S^{1(P)}_m \, D^1_{m0}[\Omega_{LP}(t)] \tag{4.58}$$

thus causing relaxation or, in the slow reorientation limit, an angle-dependent transition frequency. A time-dependent ZFS interaction is probably the main relaxation mechanism modulated by ligand vibrations of the complex. However, if the effect of a time-dependent Hamiltonian in the molecular frame is explicitly taken into account, the phenomenological character of this approach would be removed.

3 Numerically Calculated NMRD Curves

We will now turn to some fundamental numerical results of the general theory. Two 'extreme' and one 'mixed' case is illustrated. The reorientation model, is one 'extreme' and refer to *rigid* low-symmetry complexes where no transient ZFS is present. The other 'extreme' is a model where only a transient ZFS interaction is present thus representing transition metal hexa-aquo complexes. Finally a model with *both* an active *transient* and a *static* ZFS. No phenomenological electron spin relaxation rate constants are introduced in the theory.

3.1 THE REORIENTATIONAL MODEL [6–8]

The reorientational model was first developed in order to describe *rigid* low-symmetry Ni(II) complexes [6–8]. It was later generalized to electron spin quantum number $S > 1$ [7]. The model is probably somewhat artificial since the only dynamic being present is the reorientation motion of the whole complex. Since the ZFS interaction and the dipole–dipole coupling are modulated by the same motion and when their principal frames coincide, the effects caused by the cross-correlation between the two interactions is at a maximum [6b]. A correction to the SBM theory of 60% was observed for the nuclear spin–spin relaxation rate ($S = 1$) in the perturbation regime as well as in the slow-motion regime [6b]. The cross-correlation effect was largest for $S = 1$ and became smaller, but still quite large, for higher spin quantum numbers. This model was also generalized to include anisotropic reorientation and non-coincident ZFS and DD tensors by Benetis and Kowalewski[8]. The three lattice Liouville operators of equation (4.18) for this model are,

$$L_L = L_S^0 + L_{ST}(t) + i\Gamma_R \qquad (4.18')$$

where the dynamics of the thermal bath (T) is described by a Markov diffusion operator, Γ_R.

$$\Gamma_R = D_r \nabla_\Omega^2 \qquad (4.59)$$

where ∇_Ω^2 refers to the Laplacian operator in spherical coordinates $\Omega_{LM} = (\alpha_{LM} \beta_{LM})$ describing the reorientation motion of the dipole–dipole interaction. D_r is the rotational diffusion constant. The reorientational degree of freedom is introduced in the electron spin-thermal bath coupling (ZFS) and in the nuclear spin–electron spin dipole–dipole coupling. Equation (4.20) becomes for a cylindrical symmetric ZFS Hamiltonian,

$$H_{ST}^{ZFS}(t) = \sum_{n \in (-2, -1, 0, 1, 2)} (-1)^n f_0^2 D_{0-n}^2 [\Omega_{LM}(t)] S_n^2 \qquad (4.20')$$

The parameter space of this lattice model includes only three different parameters except from the dipole–dipole coupling: the electron Zeeman frequency ω_S; a static ZFS interaction strength $f_0^{2(P)}$; the reorientational diffusion constant $6D_r = 1/\tau_R$.

Following the procedure of Section 2.2.1 given in detail in Appendix C, the conditional probability function which satisfies the rotation diffusion operator Γ_R is expanded in a set of normalized eigenfunctions $|LKM) = [(2L+1)/8\pi^2]^{1/2}D_{KM}^L KM[\Omega_R]$ of Γ_R. The complete set of basis operators is formed by the direct product of electron spin operator and normalized Wigner rotation matrix elements $\{|\Sigma, \sigma) \otimes |LKM)\}$ we obtain the matrix representation of the spectral density. The structure of **M** matrix is given in Appendix C. Some numerical results of this model are shown in Figures 4.4 and 4.5 where the dimensionless reduced spectral density $K_n(\omega_I)$

$$\mathbf{K}_n(\omega_I) = \frac{15K_n^{DD}(\omega_I)}{\tau_R C^{DD} S(S+1) r_{IS}^6} \tag{4.60}$$

is displayed as a function of field strength and ZFS interaction strength. The reduced spectral density varies in the range from zero to unity. The maximum value is unity when the effective correlation time τ_C of the dipole–dipole correlation function is equal to τ_R. When τ_C is smaller than τ_R for instance, because of fast electron spin relaxation, the reduced spectral density becomes smaller than unit. The reduced spectral density $\mathbf{K}_n(\omega_I)$ reduces to unity when $[\omega_S \tau_R] \ll 1$ and $[f_o \tau_R] \ll 1$. When $[f_o \tau_R] \ll 1$, $\mathbf{K}_1(\omega_I)$ is equal to the Solomon equation, $0.6J(\omega_S + \omega_I) + 0.3J(\omega_I) + 0.1J(\omega_S - \omega_I)$. The reduced spectral densities $\mathbf{K}_{DD}(\omega)$ and $\mathbf{K}_1^{SC}(\omega)$ are displayed in Figures 4.4 and 4.5 as a function of $\ln[\omega_S \tau_R]$ and $\ln[f_o \tau_R]$ for two different electron spin quantum numbers: (a) $S = 3/2$ and (b) $S = 5/2$.

When $\ln(f_o \tau_R) > 0$, slow-motion condition prevails. The characteristic low-field value of $\mathbf{K}_1^{DD}(\omega)$ of this model is unity and irrespective of the strength of the ZFS interaction. This is contrary to the SBM theory where $\mathbf{K}_1^{DD}(\omega) \ll 1$ in this regime, because of fast electron spin relaxation rates. The NMRD profiles of the reorientation model is similar to the SB profile even in the slow motion regime for the electron spin system. However, one discrepancy is found in the plateau structure which is due to multiexponential relaxation of the high spin systems $S > 1$ and a change in the position of the '$\omega_S \tau_R$' dispersion.

3.2 THE PSEUDO-ROTATION MODELS

The reorientation model was said to represent one extreme model. The other 'extreme' is denoted the pseudo-rotation (PR) model. The PR model represents a *generalized* SBM theory using the DC approximation which is valid in the low-field limit and under non-extreme narrowing condition and for strongly coupled electron spin systems. The physical idea behind the PR model may be traced to Rubenstein *et al.* [17] and Friedman [22–25]. The flickering model of Friedman [22, 23] attributes the fluctuation of the ZFS interaction to the orientation of water molecules of the first hydration shell. The physical picture

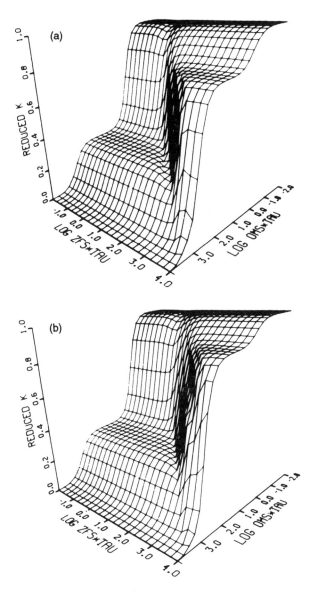

Figure 4.4 A chart of NMRD profiles are displayed for (a) $S = 3/2$ and (b) $S = 5/2$. The reduced spectral density $K_1^{DD}(\omega)$ is displayed on the z-axis as a function of the dimensionless quantities $\log(\omega_S \tau_R)$ and $\log(f_o \tau_R)$ (from ref. 7)

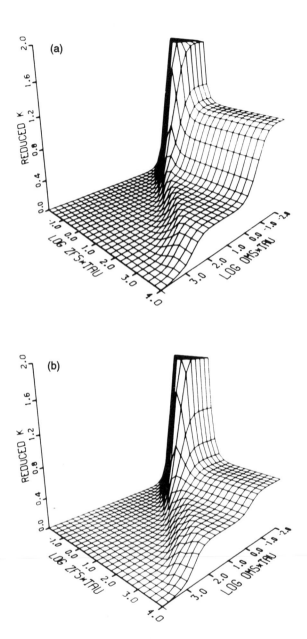

Figure 4.5   A chart of NMRD profiles are displayed for (a) $S = 3/2$ and (b) $S = 5/2$. The reduced spectral density $K_1^{SC}(\omega)$ is displayed on the z-axis as a function of the dimensionless quantities $\log(\omega_S \tau_R)$ and $\log(f_o \tau_R)$ (from Ref. 7)

behind the PR model is simple, following the fact that motions of the ligands of the first coordination shell are breaking the coordination symmetry is closely related to the modulation of the ZFS interaction. The hexa-aquo complex is nearly octahedral if one regards only the metal ion and the O oxygen atoms of water. For a perfect cubic symmetry the ZFS interaction equals zero by symmetry. In solution the flickering motions of the water ligands generates structures of lower symmetry. Hence a fluctuating and transient ZFS interaction is created with a characteristic correlation time τ_v, probably in the time range of $1-5$ ps. The magnitude of a transient ZFS interaction for the Ni(II) complex is expected to be about $1-3$ cm^{-1} [23]. In the PR model the time fluctuation in the ZFS is described as if the principal frame of the ZFS was diffusively reorienting relative to the laboratory frame [26]. The reorientation of the principal frame (P) of the ZFS interaction relative to the laboratory fixed frame (L) is modelled as an isotropic Brownian diffusion motion. For the PR model equation (4.18) becomes

$$L_L = L_S^0 + L_{ST}(t) + i\Gamma_{PR} \qquad (4.18'')$$

where the pseudo-rotational diffusion operator is

$$\Gamma_{PR} = D_{PR}\nabla_\Omega^2 \qquad (4.61)$$

and the ZFS interaction (4.20) has the form [26]

$$H_{ST}^{ZFS}(t) = \sum_{n \in (-2,-1,0,1,2)} (-1)^n f_0^2 D_{0-n}^2[\Omega_{LP}(t)]\, S_n^2 \qquad (4.20'')$$

assuming a fixed cylindrically symmetric ZFS tensor in the principal frame and the modulation owing to distortion of the first coordination sphere described by the Euler angles $\Omega_{LP}(t)$. An isotropic average over the fluctuating Wigner rotation matrix element is zero. $(\langle D_{k-n}^2[\Omega_{LP}(t)]\rangle = 0)$. In the extended pseudo-rotation model [16] we do not use the DC approximation and account for cross-correlation effects between the dipole–dipole interaction and the ZFS interaction. In this case the ZFS Hamiltonian takes the form

$$H_{ST}^{ZFS}(t) = \sum_{n \in (-2,-1,0,1,2)} (-1)^n f_0^2 D_{0m}^2[\Omega_{MP}(t)]\, D_{m-n}^2[\Omega_{LM}(t)]\, S_n^2 \qquad (4.20''')$$

where the Euler angles Ω_{LM} describe the relation between the laboratory frame (L) and the molecular frame (M), and Ω_{MP} the relative orientation between the M frame and the principal frame (P) of the ZFS interaction.

A complete set of lattice basis operators is introduced for the pseudo-rotation model [26] (equation (4.20″)) formed as the direct product $\{|\Sigma, \mu\rangle \otimes |lm\rangle\}$, where $|lm\rangle$ is the eigenfunction of the Markov operator Γ_{PR} of equation (4.61). The complete set of lattice basis operators for the extended pseudo-rotation model [16] equation (4.20‴)) is defined as $\{|\Sigma, \sigma\rangle \otimes |LKM\rangle \otimes |lm\rangle\}$.

The parameter space of both PR models is identical with that of the SBM theory. We have four parameters defining the models: the transient ZFS interaction strength, $f_o = \sqrt{2}/3D$, the characteristic correlation time of the transient ZFS interaction, τ_v, the reorientation correlation time of the whole complex, τ_R and the static magnetic field. The reduced spectral density $\mathbf{K}_1^{DD}(\omega)$ of the PR model depends on the dimensionless parameters: $\omega_S \tau_v$, $f_o \tau_v$ and the ratio τ_v/τ_R. In Figure 4.6 the spectral density chart of the reduced spectral density $\mathbf{K}_1^{DD}(\omega)$ is displayed.

In the front of Figure 4.6 we see a number of NMRD profiles for different values of the transient ZFS interaction with the characteristic shape of a frequency independent region at low-field and an increase of $\mathbf{K}_1^{DD}(\omega)$ at higher frequency. This is a shape which is often observed in experimentally determined NMRD profiles of a large variety of systems. At the back of the figure, that is, when $f_o \tau_v < 1$ the characteristic profile of the SBM equation is regained: $\qquad \mathbf{K}_1^{DD}(\omega) = 0.6J(\omega_S + \omega_I) + 0.3J(\omega_I) + 0.1J(\omega_S - \omega_I) \qquad$ with $J(\omega) = \tau_R/((\omega \tau_R)^2 + 1)$.

3.3   VIBRATION MODELS [26, 27]

Finally, we briefly present a third model which describes the effects of both a fluctuating low symmetry ligand field and a reorientationally modulated ZFS

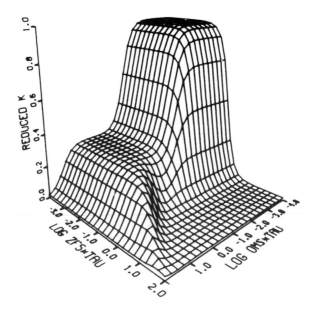

Figure 4.6   The reduced density $\mathbf{K}_1^{DD}(\omega)$ for the PR model and $S = 1$ is displayed as a function of $\log(\omega_S \tau_v)$ and $\log(f_o \tau_v)$. The distortion correlation time is $\tau_v 1 = 1.0$ ps and the reorientational correlation time $\tau_R = 30$ ps (from Ref. 26)

interaction. Instead of a rigid complex we have a deformable low-symmetry complex. The ZFS interaction induced by this deformation, is modelled by a one-dimensional Smoluchowsky diffusion equation. The Smoluchowsky diffusion operator is included instead of the pseudo-rotation operator and equation (4.18) now reads:

$$L_L = L_S^0 + L_{ST}(t) + i\Gamma_R + i\Gamma_S \tag{4.18'''}$$

The ZFS tensor component of equation (4.20') is described by a Taylor expansion in structure (normal) coordinates q_i [26, 26].

$$f_0^{2(P)}[q(t)] = f_0^{2(P)} + \sum_i \frac{\partial f_0^{2(P)}}{\partial q_i} + \frac{1}{2} \sum_{i,j} (\bar{q}_i - q_i^0)(\bar{q}_j - q_j^0) \frac{\partial^2 f_0^{2(P)}}{\partial q_i \partial q_j} + \cdots \tag{4.62}$$

The $\{q_i^0\}$ coordinates are chosen as the equilibrium configuration of the ligands which generate the ZFS interaction $f_0^{2(P)}$. The Smoluchowsky operator is defined as

$$\Gamma_S \phi = D_i \left\{ \frac{\partial}{\partial q_i} \left[\phi \frac{\partial (V(q_i/k_B T)}{\partial q_i} \right] + \frac{\partial^2}{\partial q_i^2} \phi \right\} \tag{4.63}$$

For a harmonic potential $V(q_i) = [1/2]K_i q_i^2$, the eigenfunctions of Γ_S (in our notation $|n\rangle$) are

$$\phi_n(q_i) = N_n \exp[-K_i q_i^2/2k_B T] H_n \left(\sqrt{\frac{K_i}{2k_B T}} q_i \right) \tag{4.64}$$

where $H_n(x)$ is a Hermite polynomial and k_B the Boltzmann constant. The corresponding eigenvalues are given by

$$\lambda_n = \frac{D_i K_i}{k_b T} n = \frac{n}{\tau_S} \tag{4.65}$$

The structure coordinates q may be chosen as normal vibrational modes, this implies that the lattice (L) has to be extended with appropriate Liouville super-operators describing the dynamics of an additional quantum mechanical vibration sub-system. Such a model has been presented, but not used in analysing an experimental NMRD profile [26, 27]. In Figure 4.7 the reduced spectral density $K_1^{DD}(\omega)$ is displayed for a number of different parameter sets. The parameter set includes a static and a transient ZFS interaction ($[f_0^{2(P)}, \partial f_0^{2(P)}/\partial q][\text{cm}^{-1}]$) together with their characteristic correlation times ($\{\tau_R \tau_v\}[\text{ps}]$). Figure 4.7 displays the drastic change of NMRD profiles after applying the DC approximation, in this case without justification. The low-field region of the spectral density may be decreased from unity by

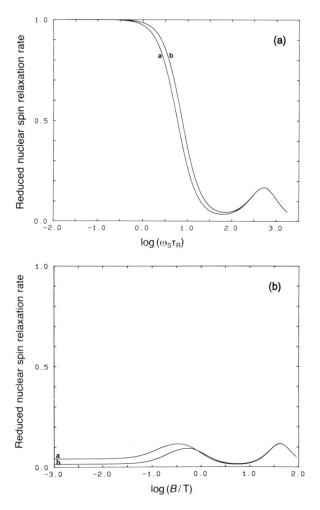

Figure 4.7 The reduced density $K_1^{DD}(\omega)$ is displayed for two sets of model parameters. The parameters in curly brackets are $\{\tau_R \tau_v\}$[ps] the static and transient ZFS interactions are given in $[f_0^{2(P)}, \partial f_0^{2(P)}/\partial q]$[cm^{-1}]. We have $\{100, 1\}$ and (a) [1.0, 4.0], (b) [2,0, 4.0]. In (A) the dipole–dipole correlation function of equation (4.16) is used and in (B) the DC approximation of equation (4.25) is applied (from Ref. 27)

tilting the principal frame of the ZFS interaction, by β degrees, relative to the principal frame of the reorientation diffusion tensor [8, 16]. Finally, in order to obtain a matrix representation of the spectral density, we must introduce a complete set of operators which span the Liouville space. For the case of one distortion mode, we have the direct product $\{|\Sigma, \sigma) \otimes |LKM) \otimes |n)\}$ [26, 27].

4 Experimental Studies

4.1 NMRD STUDIES OF PARAMAGNETIC SPECIES IN AN AQUEOUS SYSTEM

4.1.1 Application and validity of theoretical models

The pseudo-rotational model discussed in Section 3.3 represents one generaliz-ation of the SBM theory which is valid over the whole range of magnetic field strengths usually covered in NMRD experiments. However, numerical models are not as tempting to use as the closed form expressions of the SBM theory. A careful comparison between the prediction of our general theoretical approach and the SBM theory has been carried out only in the case of Ni(II) hexa-aquo complex [28]. The SBM expressions appeared to be very flexible and could, in fact, produce excellent fittings to the experimental data although the parameters provided fall outside the validity range of the theory. The parameter q/r_{IS}^3 of the dipole–dipole coupling which sets the height of the low-field level of the reduced spectral density is sensitive to theoretical models. Often one can find model parameters that reproduce the overall shape of the NMRD profile but where the field-independent level is at the wrong height. This problem may be hidden by scaling the dipole–dipole coupling with the coordination number or the electron-nuclear spin distance. In the analysis of the Ni(II) hexa-aquo complex [28] using the SBM expressions, the intermolecular spin-distance comes out quite small, $r_{IS} = 2.45$ Å, compared to the distance, $r_{IS} = 2.53$ Å obtained using the PR model. We do not yet know if these findings about the distance represent a general trend and furthermore if they are expected among other transition-metal ions and in microheterogeneous systems. It is evident that an excellent fit which conforms smoothly to the experimental NMRD curve, does not guarantee that reliable model parameters have been obtained.

An extensive experimental knowledge has emanated from the excess water proton spin-lattice NMRD profiles of transition metal aquo-ions published by Koenig and Brown [29–31] and others [32] which display relaxation data over a wide range of fields $\omega_I = 0.02$ MHz to ≈ 40 MHz.

In Figure 4.8 the water proton spin-lattice NMRD profiles of four transition metal ions are displayed $\{Gd^{2+}(S = 7/2), \; Mn^{2+}(S = 5/2), \; Fe^{2+}(S = 5/2), \; Cu^{2+}(S = 1/2)\}$ [29]. These hydrated metal ions are characterized by a relatively long electron spin relaxation time, thus representing quite effective relaxation agents for MRI studies. The T_1 NMRD profiles display a character-istic field-independent relaxivity at low field and $\omega_S \tau_R$ dispersion at about 0.2 T. One exception is found for the Mn(II) hexa-aquo complex, which has a low-field dispersion owing to a marked scalar contribution. In Table 4.1 we give the parameters describing the T_1 NMRD profile of the Mn(II) aqueous complex.

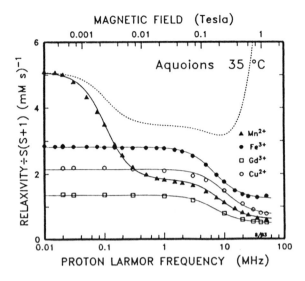

Figure 4.8 The $1/T_1$ NMRD profile of solvent in water solutions of four different transition metal ions at 35 °C. The relaxivity divided by the respective value of $S(S+1)$ for each type of ion [$Gd^{2+}(S=7/2)$, $Mn^{2+}(S=5/2)$, $Fe^{2+}(S=5/2)$, $Cu^{2+}(S=1/2)$]. The solid line through the data points result from a least-squares comparison of data with the SBM equations. For all but Mn^{2+}, the NMRD profiles for $1/T_2$ would be similar to those for $1/T_1$. For Mn^{2+} the expected NMRD profile for $1/T_2$ (derived from theory and supported by experiment) is indicated by the dotted line (from Ref. 28)

Table 4.1 Parameters of the water proton T_1^1-H NMRD profile of MN(II) hexa-aquo complex

Parameter	B (rad^2 s^{-2})	τ_v (ps)	τ_M (ns)	τ_R (ns)	q	r_{IS} (Å)
Mn(II) aqua-complex	0.1–0.074×10^{20}	2.0–3.3	760	0.03	6	2.74

Refs. [17, 33] and G. H. Reed *et al.*, *J. Chem. Phys.*, **55**, 3311 (1971).

The spin-lattice relaxation rate of water protons in solution of the Mn(II) hexa-aquo complex has also been reported by Kennedy and Bryan [32] in an extensive study of the effect of glycerol on the NMRD profile. A similar study has been carried out for Mn(II), Co(II), Ni(II), Cu(II) and Gd(III) salts in ethylene glycol by Banci *et al.* [33] in the temperature range -10 °C to $+40$ °C. Here we will focus on the results for the Mn(II) ion. Throughout the whole range of different glycerol concentrations, theory presupposes an averaged cubic Mn hexa-aquo complex. However, the fluctuation among different low symmetry structures which thus generates transient ZFS interactions will also be influenced by the presence of glycerol molecules. The DC approximation of SBM should be accurate. The electron spin relaxation rate is relatively slow

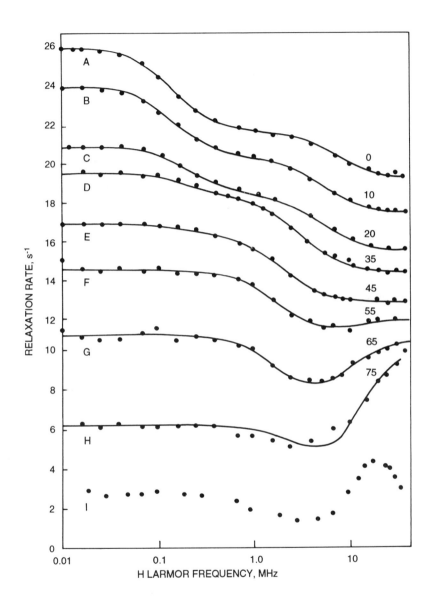

Figure 4.9 The $1/T_1$ NMRD profile of solvent protons in water solutions of 0.146 mM $MnCl_2$ at different concentrations of glycerol-d_5 at 286 K. The number to the right of each curve refers to the concentrations of glycerol-d_5 expressed as weight percent. The following amounts must be subtracted from each data set to convert for the vertical off-set used to display the data; (A) 18.0, (B) 16.0, (C) 14.0, (D) 12.0, (E) 10.0, (F) 8.0, (G) 5.0, (H) 2.0. To curve I 9.0 should be added (from Ref. 32)

and the Redfield treatment is thus sound. In Figure 4.9 the NMRD profiles are displayed [32]. The influence of an increasing amount of glycerol-d_5 on the NMRD profiles is manifest. The best fit parameters of the SBM theory corresponding to the NMRD profiles in Figure 4.9 are summarized in Table 4.2.

It is interesting to notice that for figures (A)–(D) the SBM theory is consistent with the physical picture of an hexa-aquo complex which reorients slowly combined with an unaffected transient ZFS but a slightly longer correlation time characterizing the deformation of the hydration shell. The physical picture is drastically changed for figures (E)–(H) where the best fit data yield coordination numbers from 5 to 4 which may indicate some sort of breakdown of the SBM theory. One problem is about the high-field SBM theory analysing data at low field [32]. A second problem concerns lowering the averaged symmetry of the hydrated ion. In our view the decrease in coordination number may in fact indicate the presence of a low-symmetry complex. An adequate PER theory could then provide a static ZFS parameter as a measure of the symmetry of the hydrated Mn(II) complex. The increase in the transient ZFS interaction for (G) and (H) may also indicate that some drastic change takes place in the first hydration shell and thus reflects a compensation of the SBM theory to take into account some defects of the model. However, these indications are purely speculative as long as the NMRD profiles have not been analysed within a general theoretical framework (cf. Section 2.3) valid in the whole region of magnetic field strength and taking into account both a transient and a static ZFS interaction.

4.1.2   The Ni(II) hexa-aquo complex (S = 1)

The Ni(II) hexa-aquo complex is probably the most difficult system among the hydrated transition metal ions to describe theoretically. The electron spin relaxation time is estimated to be about 1–2 ps which indicates that the symmetry breaking motions of the water molecules in the first hydration shell

Table 4.2   Fitting parameters of Figure 4.9 based on the SBM equations [32]

Curve	Weight % glycerol-d_5	Absolute viscosity (cP)	q	$10^{10}\tau_R$ (s)	$10^{12}\tau_v$ (s)	$10^7\tau_{ex}$ (s)	$10^{-19} B$ (rad^2 s^{-2})
A	0	1.20	6	38	3.6	0.6	2.8(?)
B	10	1.52	6	46	4.3	0.4–2.4	2.8
C	20	1.95	6	41	5.8	2.7	2.8
D	35	3.34	6	94	6.2	7.1	2.8
E	45	5.25	5	140	6.5	9.0	2.8
F	55	8.77	4.6	250	14	9.9	2.8
G	65	17.2	3.8	560	20	8.2	3.8
H	75	46.6	4.3	2700	25	11	7.8

Note: for all glycerol-d_5 concentrations: $r = 2.77$ Å; $A/h = 0.82$ MHz. The value of B is much larger than in Table 4.1.

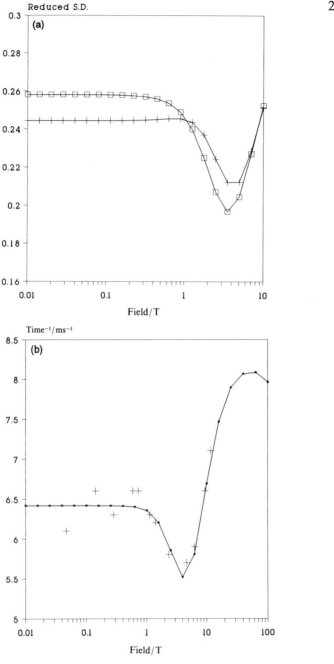

Figure 4.10 (a) The best fit using equation (4.20‴) to the experimental nuclear-spin lattice relaxation data of ref. [28]. (b) The unconstrained least-square fit using the PR model of equation 4.20″). the crosses are the corresponding dispersion curve using the ZFS Hamiltonian of equation (4.20‴) with the same model parameters (from Ref. 16)

probably take place in the same time scale as the electron spin relaxation. The slow-motion problem has to be solved for this electron spin system. A slow-motion theory was first presented by Friedman [23–25] using a memory function formalism. It was later used in an analysis of water proton spin lattice relaxation data of the Ni(II) aquo-ion [23]. The Ni(II) system was reinvestigated [28] using the PR model [26] of Section 3.2 and an extended data set by Hertz and Holz [34]. In Figure 4.10(a) the experimental proton T_1 NMRD profile is displayed together with the best fit curve using the extended PR model (PR2) where the adjusted parameters are summarized in Table 4.3.

The nuclear spin–electron spin distance comes out short and the correlation times of the dipole–dipole interaction, τ_R, and the ZFS interaction, τ_v, are similar at 51° and 71°C. The short distance 2.53 Å instead of a distance about 2.7 Å [35] may be due to underestimating the second sphere relaxation contribution.

In a recent paper [16] the DC approach was investigated in more detail. In Figure 4.10(b) the use of the DC approach is shown with the unconstrained least-square fit, using the pseudo-rotation model with (PR1) and without (PR2) the use of the DC-approach. The two NMRD profiles were calculated with the same parameter, and the discrepancy must be attributed solely to the cross-correlation effect between the nuclear-electron spin dipole–dipole and the ZFS interactions. Using a MD simulation of a divalent ion in aqueous solution, the dynamics of the first hydration shell was examined in greater detail. From the radial distribution function $g(r)$, the outer-sphere contribution was estimated to about 30% under the assumption that the contribution from the 25 ($=q_2$) water molecules of the second hydration shell only depends on the ratio of the coordination numbers q_2 and q_1 in the first and second sphere and on the intermolecular spin distances r_{IS} and d, the latter being the distance of closest approach ($[r_{IS}/d]^6(q_2/q_1) \approx (2.7/4.0)^6(25/6) \approx 30\%$) [16].

4.2 PARAMAGNETIC HYDRATED METAL IONS IN POLYELECTROLYTES AND BIOCHEMICAL SYSTEMS

4.2.1 *Ion–protein complexes*

In contrast to the NMRD curves of paramagnetic metal ions in homogeneous

Table 4.3 Best fit parameters of the SBM and pseudo-rotation models

	$\Delta/10^{12}$ (rad s^{-1})	τ_v (ps)	r_{IS} (Å)	τ_R (ps)	T (°C)
SBM	0.63	1.1	2.45	7.0	51
PR1	0.59	1.8	2.53	8.3	51
PR1	0.36	3.1	2.69	8.0	71
PR2	0.74	1.91	2.53	6.8	51

PR1 model from ref. [28] and PR2 from ref. [16]. For S = 1 we have $\Delta^2 = B$

aqueous solutions we now turn to a brief review of some characteristic NMRD profiles of ion–protein complexes and hydrated metal ions in microheterogeneous systems. Paramagnetic ions are useful probes for the investigation of active sites of metalloenzymes. Mn^{2+} may be substituted for $Mg^{2=}$ or Zn^{2+} as in carboxypeptidase A. The binding site may be characterized by structural parameters such as the number of exchangeable water molecules of the first coordination sphere ($q_w < 6$). In addition a quantitative PER analysis of solvent water proton spin relaxation rates also furnish a static ZFS interaction as a measure of the low-symmetry coordination shell. The transient ZFS interaction, Δ, and its correlation time τ_v rather characterize the flexibility of the first coordination shell. This is unlike a phenomenological theory which has much less information content and characterizes a binding site only with one electronic relaxation time τ_S.

Structural information about a binding site may be obtained more directly by measuring the PER effect on otherwise well resolved peaks of protein nuclei. Line-broadening and spin-lattice relaxation rates can be used to extract location and relative distances from the paramagnetic centre without too much theoretical difficulty. The investigation of the binding site of the Mn(II) hexaaquo complex in the channel opening of Gramicidin A is reported by Golovanov *et al.* [36a]. The peptide, which forms a transmembrane channel allowing only monovalent cations to cross the membrane, was incorporated

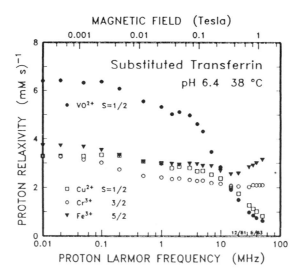

Figure 4.11 The $1/T_1$ NMRD profiles of solvent protons in water solutions of serum protein transferrin at 38 °C. The native protein was demetallized, and the two metal-binding sites per protein molecule were subsequently saturated with VO^{2+}, ($S = 1/2$), Cr^{2+} ($S = 3/2$), Fe^{2+} ($S = 5/2$) or Cu^{2+} ($S = 1/2$), which are of the same chemical environment. The relaxivities shown are per unit concentration of metal ion (not protein) (from Ref. 29)

into SDS micelles. The spin lattice relaxation rates of individual Gramicidin A protons were monitored. Distances between the divalent Mn ion and the nearest oxygen atom of Gramicidin A residues were determined [36]. In another rather typical study Lee and Nowak [36b] investigated the metal binding site of the enzyme yeast Enolase. The relaxation rates $1/T_{1,2}$ of protons and phosphorous nuclei were measured at two static magnetic fields as a function of Mn^{2+} concentration. The SBM theory was used.

The internal inconsistency of the SBM theory becomes obvious when we remember that it only provides a transient ZFS but may yield a coordination number q less than 6. The binding site of the ion must preserve a high (cubic) symmetry but may still replace ligand water molecules of the first coordination shell. The paramagnetic ion used (Mn^{2+}) may be insensitive to changes of its environment (ligand field) and does not respond by creating a static ZFS interaction. However, only a general (low-field) theory can provide information about the importance and magnitude of static and transient ZFS interactions and its influence on the NMRD profiles. The experimental ^1H NMRDs of Koenig and Brown [29] on different ions, VO^{2+} $(S = 1/2)$, Cr^{2+} $(S = 3/2)$, Fe^{2+} $(S = 5/2)$ and Cu^{12+} $(S = 1/2)$ bound to the same site of human transferrin are displayed in Figure 4.11.

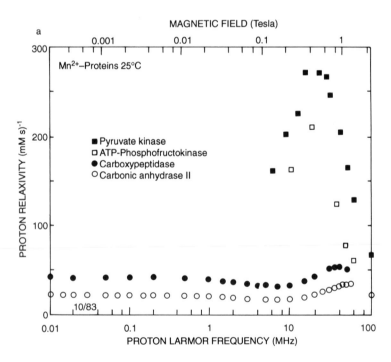

Figure 4.12 The NMRD profiles of $1/T_1$ of solvent protons in solutions of Mn^{2+} protein complexes (from Ref. 29)

In Figure 4.12 are shown the $1/T_1$ NMRD profiles of solvent protons in solutions of Mn^{2+} protein complexes: carbonic anhydrase, carboxypeptidase, ATP-phosphofructokinase, and pyruvate kinase complexes. The solvent water protons T_1 NMRD profile of Mn(II) carboxypeptidase A has been the subject of debate since the data of Koenig *et al.* [38] (0.01–100 MHz) and Navon [39] (6–100 MHz) were published in the early 1970s. Mn(II) carboxypeptidase A and Mn(II) bovine carbonic anhydrase B were first investigated by Navon in 1970]39] and later reinvestigated by Kushnir and Navon in 1984 [40]. Koenig *et al.* presented their original NMRD data of carboxypeptidase A in 1971 which was reinvestigated by Burton *et al.* in 1978 [41] using the SBM theory. Later also Mn(II)–Concanavalin A (Figure 4.13) has been investigated by Koenig and Brown [42]. The best fit data of Mn(II)–Concanavalin A was discussed by Bertini and Luchinat [14b] with a very strong dipole–dipole coupling corresponding to 8 water molecules! (Table 4.4).

In the analysis of water ^1H NMRD profiles of Mn(II) carboxypeptidase A as reported by Burton *et al.* [41] (Table 4.4), the use of the SBM equations illustrates some of the difficulties encountered in these types of system. Two inconsistent sets of parameters are obtained depending on the range of magnetic field included in the analysis. We may suspect that the paramagnetic

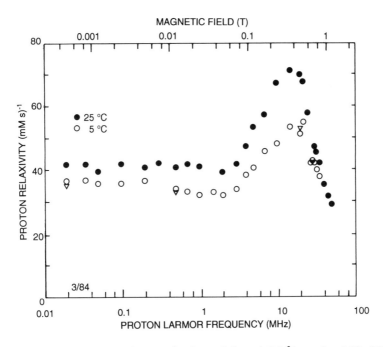

Figure 4.13 The paramagnetic contribution of bound Mn^{2+} to the $1/T_1$ NMRD profiles of solvent protons for a solution containing 0.39 mM apoCon-A protein; 0.1 M KAc, 0.9 M KCl buffer at pH 6.4; and 10% ^1H, 90% ^2H (from Ref. 42)

Table 4.4   Best fit parameters using the SBM theory from Burton *et al.* [41])

Parameter	B (rad^2 s^{-2})	τ_v (ps)	τ_M (ns)	q	r_{IS} (Å)
Data of Koenig et al. [38]					
Mn(II) Carboxypeptidase A	0.39×10^{20}	7.7	3.7	1	2.63
($v > 10$ MHz)					
(fit to all data)	1.3×10^{20}	16	190	1	2.02
Concanavatin A [14b]	0.151×10^{20}	39	160	8!	
Data of Navon [39]					
Carboxypeptidase	0.3×10^{20}	7.0	3	1	2.735
	0.53×10^{10}	7.6	300	2	2.735
Ref. [40]			100	1	
$B = \Delta^2\{4S(S+1)-3\}/25$					

complex is of low-symmetry type with both a static and transient ZFS interaction active. The relevant theoretical approach would correspond to equation (4.53)–(4.58) of Section 2.5.

The best fit parameters of the dispersion curves displayed in Figure 4.14 using the SBM theory varies substantially depending on the range of proton frequencies included. Two fits to the data of Mn(II) carboxypeptidase A are compared. One includes data over the whole frequency range and the other includes only frequencies >10 MHz. The coordination number is $q = 1$ and the transient ZFS interaction B = $[32/25]\Delta^2 = 3.9$–13×10^{19} rad^2 s^{-2}) [17, 33]. In a later paper Navon [39] obtained a coordination number $q = 1$, $\tau_{ex} \approx 50$–90×10^{-9} s, $\tau_c = 7.$–8×10^{-9} s. Furthermore, they interpreted the discrepancy found in their best fit parameters (giving a too long proton spin relaxation time) as evidence of an additional relaxation path.

We may summarize some facts critical to the use of the SBM expressions repeatedly mentioned in this chapter. A coordination number of water molecules in the first hydration shell, $q = 1$, is indicating a low symmetry complex. The problem is to know something about the implications of $q = 1$ in terms of a static ZFS interaction. Anyway, a static ZFS interaction should be included in a theoretical analysis but is completely ignored in the traditional use of the SBM approach. When we introduced the DC approach in Section 3.3, modelling low-symmetry Ni(II) complexes, we noticed that the NMRD profiles changed drastically. We may, however, not expect such a large effect for weakly coupled electron spin systems like Mn^{2+} complexes.

4.2.2   Counter–polyion interaction in polyelectrolyte solutions

Relaxation time measurements on water protons in polyelectrolyte solutions containing divalent paramagnetic counterions may reveal different types of

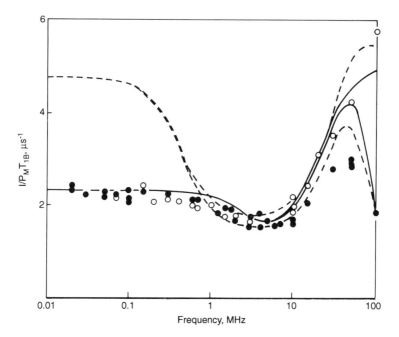

Figure 4.14 The normalized water proton relaxation rates in carboxypeptidase A–Mn(II) complex as a function of frequency according to Koenig *et al.* $q = 1, 2$. (----) fit to all data, (- - - -) fit to frequency >10 MHz (from Ref. 41)

counterions (see below). Polyphosphate (PP), polyacrylate (PA), polystyrene sulfonate (PSS) and Gantrez are the polysalts which were investigated by adding divalent paramagnetic ions and measuring longitudinal and transverse relaxation in the frequency range 4–90 MHz. Three spectroscopically distinguishable types of counter-ions are reported and two of them are among the Manning's fraction of condensed counterions [45]. Typical plots of $1/T_1$ and $1/T_2$ versus the concentration of Mn(II) ion are displayed in figure 4.15 for PSS and PA. In this figure the distinction between the two or three different types of ion is evident [45]. We also give their best fit parameters using the SBM theory for PSS and PA in Table 4.5.

4.2.3 Mn(II)–DNA interaction

Ion–DNA interactions are discussed by Kennedy and Bryant [46] using water proton T_1 NMRD data of a Mn(II)–DNA solution. The authors describe three different types of ion association. (i) The hydrated complex forms a new first coordination sphere complex. In terms of the theory presented in Section 3 this would mean the presence of a *static* as well as a *transient* ZFS interaction. The magnitude of the static ZFS interaction is difficult to estimate. However, the transient ZFS interaction, if partly generated by the motions of the hydrated

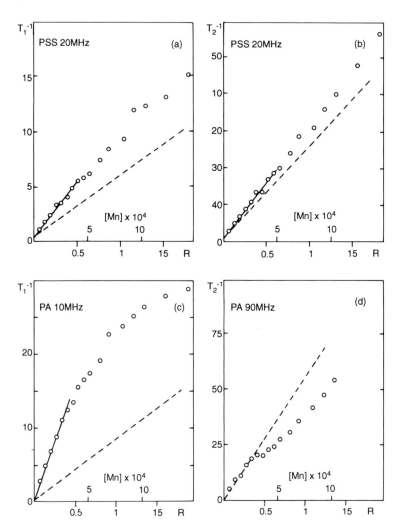

Figure 4.15 Relaxation rates $1/T_1$ and $1/T_2$ of water in the polyelectrolyte solution as a function of concentration of added Mn^{+2}. The R scale gives the corresponding Mn^{+2}/polymer ratio. The dotted line gives the relaxation rates in the absence of polyelectrolyte. (a) and (b)! with PSS ($1.6 \ 10^{-3} N$); (c) and (d): with PA ($1.5 \ 10^{-3} N$) (Ref. 45)

water molecules, would be similar in magnitude to that in the M-hexa-aquo complex. (ii) For metal ions that maintain a fully hydrated coordination sphere an outer sphere, association would slow down the fluctuations of the transient ZFS interaction compared to the hexa-aquo complex. A static ZFS interaction may be generated but would probably be smaller than in case (i). The third type (iii) is called '*territorially*' bound or *condensed* ion around the polyelectrolyte

Table 4.5 Best fit parameters

Parameter	B (rad^2 s^{-2})	τ_v (ps)	τ_M (ns)	τ_R (ns)	q	r_{IS} (Å)
Data of Karenzi et al. [45]						
Mn(II)–PA	0.08–0.1×10^{20}	10–13	25	1.8		
Mn(II)–PSS	0.082×10^{20}	8.5	18	0.057		
Data of Kennedy and Bryant [46]						
Mn(II)–DNA	0.27×10^{20}	14	760	170	5	2.74
Mn(II)–aquo-complex	0.1–0.074×10^{20}	2.1–3	760	0.03	6	2.74

owing to long-range electrostatic interactions. Perhaps we could characterize these ions as slightly perturbed hexa-aquo complexes. The NMRD profiles reported at different concentration of DNA are shown in Figure 4.16. Using the SBM theory the authors report the parameters which are summarized in Table 4.5, and compared to those of Karenzi *et al.* [45].

The reorientation correlation time ($\tau_R = 170$ ps) is slowed down considerably and it seems to characterize the reorientating DNA–Mn(II)–(H$_2$O)$_k$

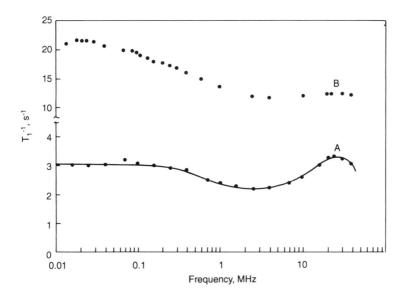

Figure 4.16 Nuclear spin relaxation rates of water protons obtained as a function of the applied magnetic field strength expressed as the proton Larmor frequency for (A) a 7.2 mM DNA phosphate solution containing 0.03 mM MnCl$_2$; (B) a 0.72-mM DNA phosphate solution containing 0;32-mM MnCl$_2$; Both data sets at 295 K. From ref. [46].

complex. The conclusion drawn by the authors, based on titration experiments, is that the assumption of a single bound state of Mn(II) on DNA is inaccurate.

4.2.4   Determination of counter-ion mobility in polyelectrolyte solutions

Leyte *et al.* [47] have shown how to determine the self-diffusion constant of monovalent ions in a polyelectrolyte solution by adding small amounts of Mn(II) to the solution. The method is based on the estimation of the outer-sphere translational diffusion relaxation contribution (cf. Section 2.3.3) to the spin-lattice relaxation rate of $^7Li^+$. In a polyelectrolyte solution the counter-ion concentration is not uniform throughout the solution owing to the strong electrostatic interaction of the polyelectrolyte. There is a high local concentration of divalent ions in the neighbourhood of the polyelectrolyte surface. We have already mentioned some classifications of hydrated metal ions such as tightly bound ions which form first coordination shell complexes or *condensed ions* which move 'freely' but are trapped by the polyion.

Leyte *et al.* [47] have demonstrated the use of NMRD measurements to determine the self-diffusion constant of Li^+ in polyacrylic acid (PAA) containing a small amount of $MnCl_2$. They compared data obtained from both the pulsed field gradient PGSE and NMRD methods applied in PAA solution. The diffusion constant measured, is an average over space. That is, if we consider the PGSE method, the diffusion constant of Li^+ has a uniform value averaged over space weighted by Li^+ concentrations. In NMRD measurements, on the other hand, the self-diffusion constant of Li^+ is dominated by the fraction of Li^+ in the high concentration region of Mn^{2+} and the average will be dominated by the axial diffusion along the polyelectrolyte. The authors noted that the Li^+ diffusion is slowed down close to the chain relative to the overall diffusion.

In NMRD measurements the intermolecular nuclear spin dipole–electron spin dipole coupling is the predominant relaxation mechanism for a $^7Li^+$ ion. The relaxation rate of $^7Li^+$ in a solution of LiCl and $MnCl_2$ was given by Abragam [48]

$$R_{1,2} = \frac{1}{3} \pi n_{Mn} \gamma_I^2 \gamma_S^2 \frac{1}{d^3} F_{1,2} [J(\omega)] \qquad (4.66)$$

$$F_1[J(\omega)] = 3J(\omega_{Li}) + 7J(\omega_S); \quad F_2[J(\omega)] = 2J(0) + \frac{3}{2} J(\omega_{Li}) + \frac{13}{2} J(\omega_S) \ (4.67)$$

where n_{Mn} is the Mn ion concentration, γ_I the gyromagnetic ratio of $^7Li^+$, γ_s the effective gyromagnetic ratio of the electron spin, and d the distance of closest approach of the centres of Li and Mn ions. $J(\omega)$ is the spectral density of translational diffusion at frequency ω.

The expression for the relaxation rate (R_1 or R_2) of lithium ions in a polyelectrolyte solution in the presence of manganese ions, taking into account the inhomogeneous distribution of ions, is given by

$$R_{1,2} = \frac{7\pi\gamma_I^2\gamma_S^2}{3n_{av}(R^2-a^2)} \int_a^R F_{1,2}\left[J(\omega)\right] n_{Li}(r)n_{Mn}(r)r\,dr \qquad (4.68)$$

where R is the cell radius of the cylindrical cell model, a is the chain radius, $n_{Li}(r)$, $n_{Mn}(r)$ are the Li^+, Mn^{2+} concentrations at a distance r from the chain, and n_{av} is the average concentration of Li^+.

The relaxation rate is an average over the Mn^{2+} and Li^+ ion concentration and the diffusion constant in $F_{1;2}$ is expected to be dependent on the distance to the chain. The spectral density $J(\omega)$ for relative translational diffusion is given in equation (4.38).

4.3 LYOTROPIC LIQUID CRYSTALLINE PHASES

4.3.1 The micellar phase

The use of paramagnetic ions in studies of lyotropic liquid crystals may give complementary information to ordinary ^{31}P and ^2H NMR investigations. One may analyse the $1/T_1$ NMRD profile of bulk-water protons or the broadening of resolved peaks of amphiphilic nuclei in the micelle aggregate. Gau, Wasylishen, Kwak and coworkers [49–53] suggested a method to determine the distribution coefficient of different molecules between the aqueous and the micellar phases. This method requires measurements of T_1 relaxation times of solubilizates nuclei in the presence and absence of paramagnetic relaxation agents. The fraction P of solute molecules in the micelle is then given by

$$P = 1 - \frac{R_1^P(\text{obs}) - R_1(\text{obs})}{R_{1,\text{aq}}^P - R_{1,\text{aq}}} \qquad (4.69)$$

where $R_1^P(\text{obs})$ and $R_1(\text{obs})$ are the observed spin-lattice relaxation rates of the solute molecule in the presence and absence of the paramagnetic ion, respectively. The paramagnetic species added to the solution should reside in the aqueous part and not influence the relaxation behaviour of the surfactant nor the solute protons in the micellar phase. A number of paramagnetic probes are discussed. The distribution coefficient of poly(ethylene oxide) PEO in sodium dodecyl sulfate(SDS) micelles was also determined [51,52]. The dynamics of solubilized hydrocarbons (benzene, naphthalene, cyclohexane, cyclododecane and others) in a series of micellar systems were investigated by relaxation measurements. ^1H-PER measurements were also used to determine the *mean distance* of hydrocarbons from the micellar surface where the paramagnetic $Mn(II)$ ion is located [53].

A detailed study of the location of a hydrated and condensed divalent ion at the surface of a charged molecular surface is presented by Chevalier and

Chachaty [54]. In this study the complexation of divalent ions by *micelles* of sodium monooctyl phosphate was investigated. Complex formation of paramagnetic ions may also be observed, perhaps more directly, by analysing the broadening of ESR lineshapes [55].

The interaction between a hydrated metal ion $M(H_2O)_{n-i}^{2k+}$ and an amphiphilic ligand molecule L may lead to complexes of the form $ML_i(H_2O)_{n-i}^{2+}$. Depending on the headgroup of the amphiphilic molecule the hydrated ion may be condensed, as for instance in the case of SDS micellar solutions [55], or a first hydration shell complex is forming as in sodium monooctyl $(C_8H_{17}OPO_3H^-, Na^+)$ phosphate micellar solutions [54]. When the ligand L is an anionic amphiphile, its complexing properties may drastically change in the monomeric state compared to the micellar phase. The cmc thus appears in a plot of the paramagnetic enhanced spin-lattice relaxation rate ^{31}P versus the amphiphile concentration. Chevalier and Chachaty report the concentration dependence of the paramagnetically induced T_{1p} of ^{31}P using both Mn^{2+} and Ni^{2+} as paramagnetic relaxation agents. T_{1p} depends linearly on the amphiphile concentration below the cmc and is drastically increased at cmc. It was also possible to notice that cmc is slightly smaller using Ni^{2+} ions compared to Mn^{2+} ions. Using the slowly relaxing Mn^{2+} ion it was also possible to observe an increase in T_{1p} at higher concentrations (0.6 M) which is probably due to growing micelles [54].

4.3.2 Determination of the lateral diffusion coefficient of the surfactant in the lamellar liquid crystalline phases

Structural and dynamical information about surfactant molecules in a lamellar phase may be obtained by analysing the time-modulation of intermolecular electron spin nuclear spin dipole–dipole correlation function [56]. The method is aimed at investigating the conformations of surfactant molecules in lyotropic liquid crystals. Chachaty and Korb [56] studied phosphorus and carbon-13 relaxation enhanced by Mn^{2+} VO^{2+} and Ni^{2+} ions exchanging among the surfactant polar heads. This method provide important complementary information to conventional NMR studies since it allows the determination of *the populations of the main conformers*, the reorientation correlation times and *lateral diffusion coefficients of the surfactant*. The lamellar phase of double-tailed sodium dibutyl phosphate (DBP)/water system and di-2-ethylhexyl phosphate (DEHP) with the presence of divalent paramagnetic ions has been investigated[56]. The paramagnetic ion is preferentially located at the water interface with the surfactant polar headgroups. Three paramagnetic probes were used: the slow relaxing ions Mn^{2+} and VO^{2+} and the fast relaxing Ni^{2+} ion. The electron spin relaxation rates of the former ions are expected to be slower than the translational diffusion motion of the ions at the interface. For Ni^{2+} ion, on the other hand, the result is reversed.

^{31}P and ^{13}C relaxation were measured on macroscopically oriented samples at three different concentrations of paramagnetic ions and at four different magnetic fields (2.15 T, 2.35 T, 7.05 T and 11.75 T). The paramagnetic enhanced spin-lattice relaxation rates were determined from the mean increment ratio $\Delta T_1^{-1}/\Delta p$, where p is the metal/surfactant molar ratio. The residence time τ_M of a paramagnetic divalent ion on a surfactant molecule was estimated to be about 5.10^{-15} s(303 K) so that fast exchange condition holds. Furthermore, τ_M is expected to be much larger than the effective correlation times governing the intra- and intermolecular nuclear spin–electron spin coupling. Intermolecular relaxation is described by means of the theory proposed by Korb et al.[58, 59] owing to dipolar interactions between unpaired electron spins and nuclear spins which diffuse laterally in parallel planes separated by a distance z. The paramagnetically enhanced intermolecular nuclear spin relaxation rate is obtained in operator form [58]

$$\frac{1}{T_{1,\text{inter}}} = \frac{8\pi}{3}\gamma_I^2\gamma_S^2\left[\frac{h}{2\pi}\right]^2 S(S+1)n_s Re\left\{\sum_{m=\pm2,\pm1,0} |d_{m1}^2 \leqslant (\beta)|^2 \alpha_m^2\right.$$

$$\left. \times\int_0^\infty \frac{d\rho}{\rho^2} E^m(z,\rho)\left[-D_L\left(\Gamma_\rho \frac{m^2}{\rho^2}\right) - i\omega_I\right]^{-1} E^m(z,\rho)\frac{g(\rho)}{\rho^3}\right\} \quad (4.70)$$

with the dimensionless dipolar interaction $E(z,\rho)$

$$E^0(z,\rho) = \frac{1-2(z/r)}{[1+(z/\rho)^2]^{5/2}}$$

$$E^{\pm 1}(z,\rho) = \frac{z/\rho}{[1+(z/\rho)^2]^{5/2}} \quad (4.71)$$

$$E^{\pm 1}(z,\rho) = \frac{1}{[1+(z/\rho)^2]^{5/2}}$$

where n_s is the number of paramagnetic species per unit area. $d_{m1}^2(\beta)$ represents a Wigner matrix element, β the angle between the lamellar director and the static magnetic field. The coefficients α_m, are $\sqrt{3}/8$, $3/2$ and $3/4$ for $m = 0$, ±1 and ±2, respectively. The relative lateral diffusion coefficient of spins I and S is given by D_L. Finally, Γ_ρ is a dynamical operator which appears from the radial part of the Smoluchowsky equation. The evaluation of this expression is accomplished using a finite difference technique and a diagonalization procedure.

The following parameters are needed in this treatment, namely, the distance of minimal approach, d, between two surfactant molecules and the radial distribution function $g(\rho)$ shown in the insert of Figure 4.17 [58]. The calculated variations of the intermolecular paramagnetically enhanced spin-lattice relaxation rate with the inter-planar z distance are displayed in

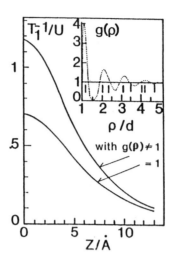

Figure 4.17 Calculated variation of the intermolecular paramagnetically enhanced longitudinal nuclear relaxation rate with the distance z between the planes where the paramagnetic ion and a ^{13}C atom diffuse. The calculated pair correlation function of head groups projected on the lamellar plane is given in the insert. The bold vertical marks indicate the positions of the successive shells of neighbours in a perfect hexagonal lattice of the same density (from Ref. 58)

Figure 4.17 for two different cases, namely uniform and non-uniform radial distribution.

4.4 POLYMER SOLUTIONS

The dynamics of $Mn(H_2O)_6^{2+}$ in an aqueous poly(ethyleneoxide) (PEO) solution were studied by Leyte *et al.* [60]. The paramagnetically enhanced water as well as polymer ^1H and ^{13}C nuclear spin relaxation rates are measured. Since PEO is uncharged we may expect that the $Mn(H_2O)_6^{2+}$ complex will remain fully hydrated. Thus the SBM theory is an adequate theory when high-field data are analysed. The bulk ^1H water relaxation rate for this slowly relaxing agent is mainly determined by the reorientational motion of the complex, which is about 30 ps in a homogeneous aqueous solution at 25 °C. The PER effects on polymer protons and carbons is determined by the translational diffusion correlation function equation (4.37) which is discussed in Subsection 2.3.3 concerning outer-sphere relaxation. ^1H-T_1 is measured at different PEO concentration. Then an increase in the reorientational correlation time of the Mn(II) hexa-aquo complex is observed. From ^1H and ^{13}C PEO relaxation rates versus the Mn(II) concentration, the diffusion constant of the Mn(II) hexa-aquo complex and the Mn^{2+} PEO distance of closest approach were estimated.

The experimental relaxation rate is given by equation (4.1), where the intermolecular proton–Mn(II) relaxation rate R_{1t} is estimated in detail. The expression for the nuclear relaxation rate due to translational diffusion of spherical particles was given by

$$\frac{1}{T_{1,0}} = \frac{1}{450} C^{DD} S(S+1) n_M \frac{1}{d^3} \{6J_t(\omega_I) + 14J_t(\omega_S)\}$$

$$\frac{1}{T_{2,0}} = \frac{1}{450} C^{DD} S(S+1) n_M \frac{1}{d^3} \{4J_t(0) + 3J_t(\omega_I) + 13J_t(\omega_S)\}$$

(4.72)

where n_M represents the number of Mn^{2+} ions per unit volume and the spectral density J_t is defined by [48]

$$J_t(\omega) = \frac{d^2}{D} \int_0^\infty [J_{3/2}(u)]^2 \frac{u \, du}{u^4 + x^4/4}, \quad x = \left[\frac{\omega d^2}{D}\right]^{1/2}$$

(4.73)

with $J_{3/2}(u)$ a spherical Bessel function of order $3/2$, D the average diffusion coefficient $D = 1/2[D_I + D_{Mn}]$, and d the distance of closest approach. The maximum value of the integral is $2/15$, which is obtained for $x \ll 1$.

5 Conclusion

There are many important applications of paramagnetic enhanced nuclear spin relaxation in homogeneous and microheterogeneous systems. However, a theoretical analysis requires a generalization of the traditional Solomon–Bloembergen–Morgan (SBM) theory. We have reviewed a general theoretical framework which is valid irrespectively of electron spin relaxation rates and the range of the static field. The main drawback of this theory is that all calculations must be performed numerically. However, NMRD profiles require model calculations of nuclear spin-lattice correlation functions which are easily performed on workstations.

Appendix A Standard spin operators of rank 2

Standard spin operators of rank 2 are constructed from vector operators as

$$S_\sigma^2 = \sum_m (-1)^\sigma \sqrt{5} \begin{pmatrix} 1 & 1 & 2 \\ m & \sigma - m & -\sigma \end{pmatrix} S_m^1 S_{\sigma-m}^1$$

(A1)

where the spherical standard vector operators are defined

$$S_1^1 = \frac{1}{\sqrt{2}} S_+; \quad S_{-1}^1 = \frac{1}{\sqrt{2}} S_-; \quad S_1^1 = S_z$$

$$S_\sigma^1 = [(S+2)(S+1)S]^{1/2} \sum_m (-1)^{m-S} \begin{pmatrix} S & S & 1 \\ m & -m-\sigma & \sigma \end{pmatrix} |m+\sigma\rangle\langle m|$$

(A2)

The normalized irreducible spherical tensor operator of rank Σ is denoted $|\Sigma, \sigma)$ and defined by the formula

$$|\Sigma, \sigma) = \sum_{M, M'} \sqrt{2\Sigma + 1} \begin{pmatrix} S' & S & \Sigma \\ M' & -M & -\sigma \end{pmatrix} (-1)^{S'-M'} |S' M'\rangle\langle SM| \quad (A3)$$

The relation between standard operators of rank 1 and rank 2 are:

$$|1, \sigma) = \sqrt{\frac{3}{(2S+1)(S+1)S}} S_\sigma^1;$$

$$|2, \sigma) = \sqrt{\frac{30}{(2S+3)(2S+1)S(2S-1)(S+1)}} S_\sigma^2 \quad (A4)$$

where the second rank spherical tensor operators are usually written as

$$S_2^1 = \frac{1}{\sqrt{6}} [3S_z^2 - S(S+1)]; \quad S_{\pm 1}^2 = \mp \frac{1}{2} [S_z S_\pm + S_\pm S_z]; \quad S_{\pm 2}^2 = \frac{1}{2} S_\pm S_\pm \quad (A5)$$

Appendix B   Correlation function in equation (4.16)

The nuclear spin–electron spin dipole–dipole correlation function of equation (4.16) defines a submatrix explicitly given by:

$$G_1^{DD}(\tau) = -C^{DD} \frac{S(S+1)}{5} \frac{1}{r_{IS}^6} \left[\frac{1}{\sqrt{5}}, \frac{-1}{\sqrt{10}}, \frac{1}{\sqrt{30}} \right]$$

$$\times \begin{bmatrix} (2, 2|1, -1|e^{iL_L\tau}1, -1)|2, 2) & (2, 2|(1 - 1|e^{iL_L\tau}1, 0)|2, 1) \\ * & (2, 1|(1, 0|e^{iL_L\tau}|1, 0)|2, 1) \\ * & * \end{bmatrix}$$

$$\begin{matrix} (2, 2|(1, -1|e^{iL_L\tau}1, 1)|2, 0) \\ (2, 1|(1, 0|e^{iL_L\tau}1, 1)|2, 0) \\ (2, 0|(1, 1|e^{iL_L\tau}1, 1)|2, 0) \end{matrix} \begin{bmatrix} \dfrac{1}{\sqrt{5}} \\ \dfrac{-1}{\sqrt{10}} \\ \dfrac{1}{\sqrt{30}} \end{bmatrix} \quad (B1)$$

and for the nuclear spin-spin relaxation the dipole–dipole correlation function matrix is $(n = 0)$;

$$G_0^{DD}(\tau) = -C^{DD} \frac{S(S+1)}{5} \frac{1}{r_{IS}^6} \left[\frac{-1}{\sqrt{10}}, \frac{\sqrt{2}}{\sqrt{15}}, \frac{-1}{\sqrt{10}} \right]$$

$$\times \begin{bmatrix} (2,1|1,-1|e^{iL_L\tau}1,-1)|2,1) & (2,1|(1-1|e^{iL_L\tau}1,0)|2,0) \\ * & (2,0|(1,0|e^{iL_L\tau}|1,0)|2,0) \\ * & * \end{bmatrix}$$

$$\begin{bmatrix} (2,1|(1,-1|e^{iL_L\tau}1,1)|2-1) \\ (2,0|(1,0|e^{iL_L\tau}1,1)|2-1) \\ (2,0|(1,1|e^{iL_L\tau}1,1)|2-1) \end{bmatrix} \begin{bmatrix} \frac{-1}{\sqrt{10}} \\ \frac{\sqrt{2}}{\sqrt{15}} \\ \frac{-1}{\sqrt{10}} \end{bmatrix} \quad \text{(B2)}$$

where the dipole–dipole coupling constant is

$$C^{DD} = \left[\frac{\mu_0}{4\pi} \right]^2 30 \left[\frac{h}{2\pi} \right]^2 \gamma_S^2 \gamma_I^2 \quad \text{(B3)}$$

Appendix C Liouville matrix for rigid low-symmetry complexes

The lattice Liouville superoperators are given by

$$L_L = L_0^Z + L_{ST}(t) - i\Gamma_\Omega \quad \text{(C1)}$$

Where the first Liouville operator is generated by the Zeeman interaction. The second Liouville operator is generated by the ZFS interaction. If we assume a cylindrically symmetric ZFS interaction which is parallel to the dipole–dipole vector and fixed in the molecular coordinate frame (M) we obtain

$$H_{ST}^{ZFS}(t) = f_0^{2(P)} \sum_m D_{0-m}^2 [\Omega_{LM}(t)] S_m^{2(L)} \quad \text{(C2)}$$

Here $f_0^{2(P)}$ is the zero field splitting parameter, the $D_{0-m}^2 [\Omega_{ML}]$ are second rank Wigner rotation matrix elements, Ω_{ML} are the Euler angles specifying the transformation from laboratory to a molecular fixed coordinate system (M), and $S_m^{2(L)}$ is the second rank standard electron spin tensor operator (A5).

For the isotropic molecular reorientation diffusion we have the Markov operator

$$\Gamma_\Omega = D \nabla_{ML}^2 \quad \text{(C3)}$$

The basis operators are formed by the eigenfunctions to the isotropic reorientation diffusion operators of (A3)

$$D_R \nabla_\Omega^2 | LM) = -D_R L(L+1) | LM) \tag{C4}$$

Expressions for the matrix elements of the different super-operators appearing in the Liouville matrix M are as follows:

For the identity operator

$$(L'M' | \Sigma' \sigma' | \mathbf{E} | \Sigma \sigma) LM) = \delta_{L'L} \delta_{M'M} \delta_{\Sigma'\Sigma} \delta_{\sigma'\sigma} \tag{C5}$$

the Markov operator

$$(L'M' | \Sigma' \sigma' | -\Gamma_\Omega | \Sigma \sigma) LM) = \delta_{L'L} \delta_{M'M} \delta_{\Sigma'\Sigma} \delta_{\sigma'\sigma} D(L+1)L \tag{C6}$$

the Zeeman interaction

$$(L'M' | \Sigma' \sigma' | iL_0 | \Sigma \sigma) LM) = \delta_{L'L} \delta_{M'M} \delta_{\Sigma'\Sigma} \delta_{\sigma'\sigma} i\omega_S \sigma \tag{C7}$$

and finally the ZFS interaction of equation (4.20′), Section 3.1.

$$(L'M' | \Sigma' \sigma' | L_{ZFS} | \Sigma \sigma) LM) = f_0 \delta_{\Sigma\Sigma'\pm1} \delta_{\sigma'+M'\sigma+M} R_{\Sigma'2\Sigma}$$

$$\sqrt{(2L'+1)(2L+1)(2\Sigma'+1)} (-1)^{\Sigma+\sigma-M'} \begin{pmatrix} L' & 2 & L \\ 0 & 0 & 0 \end{pmatrix}$$

$$\begin{pmatrix} L' & 2 & L \\ -M' & M'-M & M \end{pmatrix} \begin{pmatrix} \Sigma & 2 & \Sigma' \\ \sigma & M-M' & -\sigma' \end{pmatrix} \tag{C8}$$

The reduced matrix elements R introduced above are defined by equations (C9) and (C10) in analogy to the reduced matrix elements of the Wigner–Eckart theorem.

$$(\Sigma'\sigma' | S_m^N | \Sigma \sigma) = (-1)^{\Sigma-N+\sigma'} \sqrt{(2\Sigma'+1)} \begin{pmatrix} \Sigma & N & \Sigma' \\ \sigma & m & -\sigma' \end{pmatrix} R_{\Sigma'N\Sigma} \tag{C9}$$

and

$$(\Sigma'\sigma' | S_m^N | \Sigma \sigma) = tr_s \{ (\Sigma'\sigma' | S_m^N | \Sigma \sigma) - | \Sigma \sigma) S_m^N \} \tag{C10}$$

The structure of the matrix M is given below. We have marked the matrix elements in the inverted matrix M^{-1} representing the scalar (SC) contribution, the 3×3 submatrix representing the dipole–dipole (DD) contribution and the three elements of cross-correlation spectral densities SC/DD. We also give some of the diagonal matrix elements of M and where the ZFS interaction introduces off-diagonal elements. The Solomon equation is obtained in the limit of no ZFS interaction. Then the matrix M is diagonal and spectral density is given by three terms A, B and C; $K^{DD} = Re\{0.6 \ 1/A + 0.3 \ 1/B + 0.1 \ 1/C\}$. Its structure is given in Figure 4.18.

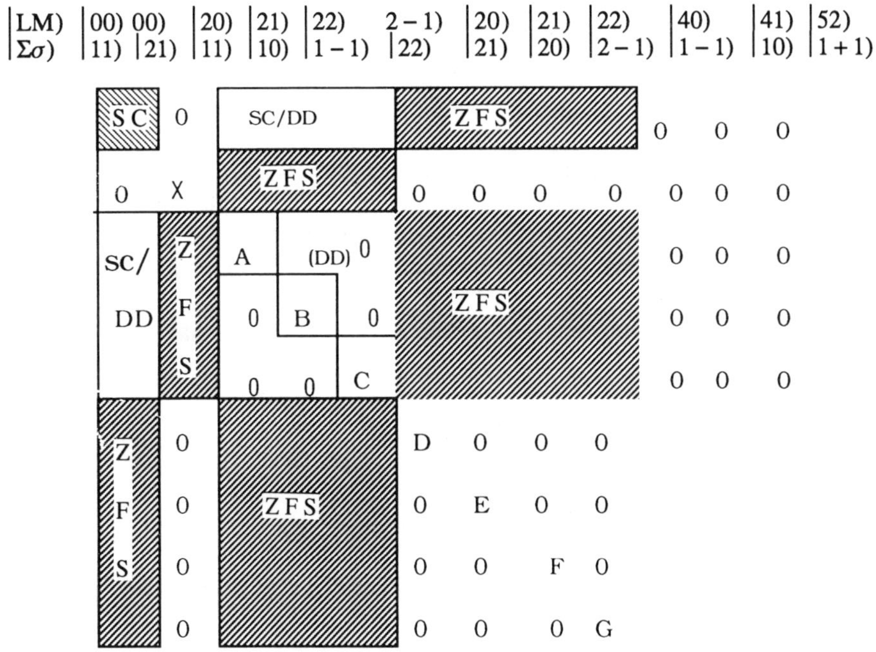

$$|LM) \quad |00)\,00) \quad |20) \quad |21) \quad |22) \quad 2-1) \quad |20) \quad |21) \quad |22) \quad |40) \quad |41) \quad |52)$$
$$|\Sigma\sigma) \quad |11)\,|21) \quad |11) \quad |10) \quad |1-1) \quad |22) \quad |21) \quad |20) \quad |2-1) \quad |1-1) \quad |10) \quad |1+1)$$

SC	0	SC/DD				ZFS			0	0	0
0	X	ZFS			0	0	0	0	0	0	0
SC/DD	Z F S	A	(DD)	0		ZFS			0	0	0
		0	B	0					0	0	0
		0	0	C					0	0	0
Z F S	0	ZFS			D	0	0	0			
	0				0	E	0	0			
	0				0	0	F	0			
	0				0	0	0	G			

Figure 4.18 The structure of the Liouville matrix **M** for rigid low-symmetry complexes

$SC = -i\omega_S$
$A = i(\omega_I - \omega_S) + 1/\tau_R$
$B = i\omega_I + 1/\tau_R$
$C = i(\omega_I + \omega_S) + 1/\tau_R$
$X = -i2\omega_S$
$D = i(\omega_I - 2\omega_S) + 1/\tau_R$
$E = i(\omega_I - \omega_S) + 1/\tau_R$
$F = i\omega_I + 1/\tau_R$
$G = i(\omega_I + \omega_S) + 1/\tau_R$

Appendix D Liouville matrix for the PR model of Section 3.2

The dynamic process active in modulating the transient ZFS interaction is modelled as an isotropic pseudo-rotation motion of the principal frame of the ZFS interaction relative to the laboratory frame. The lattice Liouville super-operators are given by

$$L_L = L_0^Z + L_{ST}(t) - i\Gamma_\Omega \tag{D1}$$

The first Liouville operator is generated by the Zeeman interaction. The second Liouville operator is generated by the transient ZFS interaction. If we assume a cylindrically symmetric ZFS interaction and with fixed amplitude in the principal coordinate frame (P), we obtain

$$H_{ST}^{ZFS}(t) = f_0^{2(P)} \sum_m D_{0-m}^2[\Omega_{PL}(t)]S_m^{2(L)} \tag{D2}$$

Here $f_0^{2(P)}$ is the zero field splitting parameter, the $D_{0-m}^2[\Omega_{PL}]$ are second rank Wigner rotation matrix elements, $\Omega_{PL}(t)$ are Euler angles specifying the transformation from laboratory to the principal frame of the ZFS interaction (P), and $S^{2(L)}$ is the same second rank standard electron spin tensor operator as in Appendix C.

For the isotropic molecular reorientation diffusion we have the Markov operator

$$\Gamma_{\Omega_{PL}} = D_{PR}\,\nabla_\Omega^2 \tag{D3}$$

The basis operators are formed by the eigenfunctions to the isotropic reorientation diffusion operators of (D3)

$$D_{PR}\,\nabla_\Omega^2|LM) = -D_{PR}(L+1)L|LM) \tag{D4}$$

where the pseudo-rotation diffusion coefficient D $(=1/6\tau_{PR})$ is closely related to the collective deformation of the symmetry of the paramagnetic complex owing to collisions with solvent molecules. Expressions for the matrix elements of the different super-operators appearing in $\mathbf{M}$ are as follows:

For the identity operator

$$(L'M'|\Sigma'\sigma'|\mathbf{E}|\Sigma\sigma)LM) = \delta_{L'L}\delta_{M'M}\delta_{\Sigma'\Sigma}\delta_{\sigma'\sigma} \tag{D5}$$

the Markov operator

$$(L'M'|\Sigma'\sigma'|-\Gamma_\Omega|\Sigma\sigma)LM) = \delta_{L'L}\delta_{M'M}\delta_{\Sigma'\Sigma}\delta_{\sigma'\sigma}D_{PR}(L+1)L \tag{D6}$$

the Zeeman interaction

$$(L'M'|\Sigma'\sigma'|iL_0|\Sigma\sigma)LM) = \delta_{L'L}\delta_{M'M}\delta_{\Sigma'\Sigma}\delta_{\sigma'\sigma}i\omega_S\sigma \tag{D7}$$

and finally the ZFS interaction

$$(L'M'|\Sigma'\sigma'|L_{ZFS}|\Sigma\sigma)LM) = f_0\delta_{\Sigma\Sigma'\pm1}\delta_{\sigma'+M',\sigma+M}R_{\Sigma'2\Sigma}$$

$$\sqrt{(2L'+1)(2L+1)(2\Sigma'+1)}\,(-1)^{\Sigma+\sigma-M'}\begin{pmatrix} L' & 2 & L \\ 0 & 0 & 0 \end{pmatrix}$$

$$\begin{pmatrix} L' & 2 & L \\ -M' & M'-M & M \end{pmatrix}\begin{pmatrix} \Sigma & 2 & \Sigma' \\ \sigma & M-M' & -\sigma' \end{pmatrix} \tag{D8}$$

Because of the DC approximation, the $\mathbf{M}$ matrix may be separated into three blocks, M_1, M_0 and M_{-1}, one for each electron spin correlation function $(S_m^1(t)\,S_m^1(0))$, with $m = 1$, 0 and -1, respectively. The M_1 block is shown in

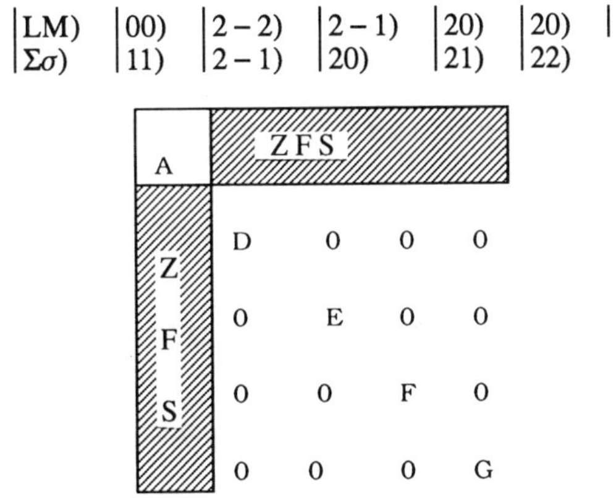

$$\begin{array}{cccccc}
|LM) & |00) & |2-2) & |2-1) & |20) & |20) \\
|\Sigma\sigma) & |11) & |2-1) & |20) & |21) & |22)
\end{array}$$

Figure 4.19 The structure of second order couplings in the Liouville matrix **M**. The M_1-block

Figure 4.19, where A, B and C refer to the corresponding diagonal terms of the DD matrix in Figure 4.18 and where $D = i(\omega_I + \omega_S) + 1/\tau_{PR}$, $E = i\omega_I + 1/\tau_{PR}$, $F = i(\omega_I - \omega_S) + 1/\tau_{PR}$, $G = i(\omega_I - 2\omega_S) + 1/\tau_{PR}$.

References

1. R. A. Dwek, *Nuclear Magnetic Resonance in Biochemistry, Applications to Enzyme Systems*, Clarendon Press, Oxford, 1973.
2. D. R. Burton, S. Forsén, G. Karlström and R. A. Dwek, *Prog. NMR Spec.*, **13**, 1 (1978).
3. J. Kowalewski, L. Nordenskiöld, N. Benetis and P.-O. Westlund, *Progress in NMR*, **17**, 141 (1985).
4. T. J. Swift and R. E. Connick, *J. Chem. Phys.*, **37**, 307 (1962).
5. Z. Luz and S. Meiboom, *J. Chem. Phys.*, **37**, 841 (1961).
6. N. Benetis, J. Kowalewski, L. Nordenskiöm, H. Wennerstróm and P.-O. Westlund, (a) *Molec. Phys.*, **48**, 329 (1983). ibid.; (b) *Molec. Phys.*, **50**, 515 (1983); ibid. (c) *J. Magn. Reson*, **58**, 261 (1984).
7. P.-O. Westlund, H. Wennerström, L. Nordenskiöld, J. Kowalewsk and N. Benetis, *J. Magn. Reson.*, **59**, 91 (1984).
8. N. Benetis and J. Kowalewski, *J. Magn. Reson.*, **65**, 13 (1985).
9. I. Solomon and N. Bloembergen, *J. Chem. Phys.*, **25**, 261 (1956).
10. N.Bloembergen and L. O. Morgan, *J. Chem. Phys.*, **34**, 842 (1961).
11. (a) A. G. Redfield, IBM *J. Res. Dev.*, **1**, 19 (1957). (b) R. K. Wagnsness and F.Bloch, *Phys. Rev.*, **89**, 728 (1953); (c) F. Bloch, *Phys. Rev.*, **105**, 1206 (1957); (d) C. P. Slichter, *Principles of Magnetic Resonance*. ch 5. 3rd edn. Springer-Verlag, 1990.
12. S. Koenig and R. D. Brown, *Prog. NMR Spec.*, **22**, 819, 1991.
13. (a) P.-O. Westlund, *SPR.N.M.R.* vols 20 and 22, ch. 14, 1991, 1993. (b) U. Lindner, *Ann. der Physik*, **7**, 320 (1965).
14. (a) I. Bertini, C. Luchinat, M. Mancini and G. Spina, *J. Magn. Reson.*, **59**, 213 (1984). (b) I. Bertini and C. Luchinat, *NMR of Paramagnetic Molecules in Biological Systems*, Benjamin/Cummings 1986.

15. R. Sharp, *J. Chem. Phys.*, **93**, 6921 (1990); *J. Chem. Phys.*, **98**, 6092 (1993); *J. Chem. Phys.*, **98**, 2507 (1993); *J. Chem. Phys.*, **98**, 912 (1993); *J. Magn. Reson.*, **100**, 491 (1992).
16. P.-O. Westlund, T. P. Larsson and O. Teleman, *Molec. Phys.*, **78**, 1365 (1993).
17. M. Rubenstein, A. Bararn and Z. Luz, *Molec. Phys.*, **20**, 67 (1971).
18. L.-P. Hwang and J. H. Freed, *J. Chem. Phys.*, **63**, 118 (1975).
19. L.-P. Hwang and J. H. Freed, *J. Chem. Phys.*, **63**, 4017 (1975).
20. J. H. Freed, *J. Chem. Phys.*, **68**, 4034 (1978).
21. T. Bayburt and R. R. Sharp, *J. Chem. Phys.*, **92**, 5892 (1990).
22. H. L. Friedman in P. Laszlo (ed.), (1978), *Protons and Ions in Fast Dynamic Phenomena*, Elsevier.
23. H. L. Friedman, M. Holz and H. G. Hertz, *J. Chem Phys.*, **70**, 3369 (1979).
24. C. F. Anderson, L. P. Hwang and H. L. Friedman, *J. Chem. Phys.*, **64**, 2806 (1976).
25. L. P. Hwang, C. F. Anderson and H. L. Friedman, *J. Chem. Phys.*, **62**, 2098 (1975).
26. P.-O. Westlund, H. Wennerström and N. Benetis, *Molec. Phys.*, **61**, 177 (1987).
27. P.-O. Westlund and T. P. Larsson, *Acta Chemica Scandinavlca*, **45**, 11 (1991).
28. J. Kowalewski, T. Larsson and P.-O. Westlund, *J. Magn. Reson.*, **77**, (1987).
29. S. H. Koenig and R. D. Brown III, *Magn. Reson. Med.*, **1**, 478 (1984).
30. S. H. Koenig, C. Baglin R. D. Brown III and C. F. Brewer, *Magn. Reson. Med.*, **1**, 496 (1984).
31. M. E. Fabry, S. H. Koenig and W. E. Schillinger, *J. Bio. Chem.*, **245**, 4256 (1970).
32. S. D. Kennedy and R. G. Bryant, *Magn. Reson. Med.*, **2**, 14 (1985).
33. L. Banci, I. Bertini and C. Luchinat, *Inorg. Chim. Acta*, **100**, 173 (1985).
34. H. G. Hertz and M. Holz, *J. Magn. Reson.*, **63**, 65 (1985).
35. G. W. Nellson, R. D. Broadbent, I. Howell and R. H. Tromp, *J. Chem. Soc. Faraday Trans.*, **89**, 2927 (1993).
36. (a) A. P. Golovanov, I. L. Barsukov, A. S. Arseniev, V. F. Bystrov, S. V. Sukhanov and L. I. Barsukov, *Biopolymers*, **31**, 425 (1991). (b) M. Eun Lee and T. Nowak, *Biochemistry*, **31**, 2171 (1992).
37. S. H. Koenig, C. Baglin, R. D. Brown III and C. F. Brewer, *Magn. Reson. Med.*, **1**, 496 (1984).
38. S. H. Koenig, R. D. Brown and J. Studebaker, *Cold Spring Harbor Symp. Quant. Biol.*, **36**, 551 (1971).
39. G. Navon, *Chem. Phys. Lett.*, **7**, 390 (1970).
40. T. Kushnir and G. Navon, *J. Magn. Reson.*, **56**, 373 (1984).
41. D. R. Burton, S. Forsen, G. Karlsröm and R. A. Dwek, *Progr. NMR Spectrosc.*, **13**, 1 (1979).
42. S. H. Koenig and R. D. Brown, *J. Magn. Reson.*, **61**, 426 (1985).
43. R. D. Brown, C. F. Brewer and S. H. Koenig, *Biochemistry*, **16**, 3883 (1977).
44. C. Coolbaugh Lester and R. G. Bryant, 'NMR of Paramagnetic Molecules', in L. J. Berliner and J. Reuben (eds), *Biological Magnetic Reson.*, **12**, Plenum Press (1993).
45. P. C. Karenzi, B. Meurer, P. Spegt and G. Weill, *Biophys. Chem.*, **8**, 181 (1979).
46. S. D. Kennedy and R. G. Bryant, *Biophys. J.*, **50**, 669, (1986).
47. R. H. Tromp, J. de Bleijser and J. C. Leyte, *J. Phys. Chem.*, **93**, 2626 (1989).
48. A. Abragam, *The Principles of Magnetism*, Oxford Univ. Press, 1961, Ch. VIII.
49. Z. Gau, R. E. Wasylishen and C. T Kwak, *J. Phys. Chem.*, **93**, 2190 (1989).
50. Z. Gau, R. E. Wasylishen and C. T. Kwak, *J. Chem. Soc. Faraday Trans.*, **87**, 947 (1991).

51. Z. Gau, R. E. Wasylishen, E. Roderick and C. T. Kwak, *J. Phys. Chem.*, **95**, 462 (1991).
52. Z. Gau, C. T. Kwak, R. Labonte, G. D. Marangoni and R. E. Wasylishen, *J. Colloid Interface Sci.*, **45**, 269 (1990).
53. R. E. Wasylishen, Z. Gau, C. T. Kwak, E. Verpoorte, J. B. MacDonald and R. M. Dickson, *Can. J. Chem.*, **69**, 822 (1991).
54. (a) Y. Chevalier and C. Chachaty, *J. Phys. Chem.*, **89**, 875 (1985). (b) Y. Chevalier and C. Chachaty, *J. Am. Chem. Soc.*, **107**, 1102 (1985).
55. B. Cabane, *J. Phys.* (Orsay, Fr), **42**, 847 (1981).
56. C. Chachaty and J.-P. Korb, *J. Phys. Chem.*, **92**, 2834 (1988).
57. Brület, P. and H. M. McConnell, *Proc. Natl. Acad. Sci. USA*, **72**, 1421 (1975).
58. J. P. Korb, M. Ahadi and H. M. McConnell, *J. Phys. Chem.*, **91**, 1255 (1987).
59. J. P. Korb, M. Ahadi, G. P. Zientara and J. H. Freed, *J. Chem. Phys.*, **86**, 1125 (1987); *J. Phys. Chem.*, **91**, 12551 (1987).
60. J. Breen, D. van Duijn, J. de Bleijser and J. C. Leyte, *J. Phys. Chem.*, **91**, 5354 (1987).

5 QUADRUPOLAR PROBES IN SOLUTION

J. Grandjean and P. Laszlo
University of Liège, Belgium and Ecole polytechnique, Paris, France

Dynamics of Solutions and Fluid Mixtures by NMR
Edited by J.-J. Delpuech   © 1995 John Wiley & Sons Ltd

1   Introduction

Quadrupolar nuclei have a quantum spin number greater than 1/2 and they are characterized by an electric quadrupole moment, together with a magnetic dipole. Their spin number may take integer or half-integer values between 1 and 7. Accordingly, they display wide diversity of their magnetic properties. Furthermore, theoretical descriptions may differ somewhat from one nucleus to another, depending on the spin number. Detection of quadrupolar nuclei in diamagnetic complexes of transition metals and that of lanthanide nuclei is often difficult. A large value of the electric quadrupole moment together with a non-symmetric environment induce extensive resonance broadening. This unfavourable situation, often conjugated with a low natural abundance, prevents their routine detection [1]. Therefore these nuclei are not easily amenable to dynamic studies and this chapter will concern itself chiefly with quadrupolar nuclei of the main group elements. Magnetic properties for the stable quadrupolar nuclei of this group are summarized in Table 5.1.

Complementary information on the NMR characteristics of the most studied nuclei can be found in standard references [2–4].

The structural environment of quadrupolar nuclei also presents a variety of types. There are two limiting cases. A quadrupolar nucleus may be covalently bonded to a molecular framework. Or it is ionic and may form only electrostatic bonds with appropriate ligands. With a nucleus of the first type, the origin of the quadrupolar interaction is intramolecular. It is intermolecular for a solvated ion nucleus. Theoretical descriptions of the NMR observables reflect also the diversity of these situations. In the next section, NMR equations for dynamical processes in isotropic solutions as they affect chemical shift and relaxation rates are summarized. In heterogeneous systems, local order may endure even in the liquid phase. Therefore, we shall review briefly theoretical equations for residual quadrupolar splitting and for relaxation rates in anisotropic media.

As indicated in Chapter 1, dynamic phenomena cover a variety of processes. Both ionic and covalently-bound quadrupolar nuclei can be involved. Examples reported in Section 3, indeed illustrate aggregation and complexation processes, interactions at a liquid–solid interface, molecular dynamics in proteins and polyelectrolytes, transport phenomena in model and biological membranes.

Previous reports on the use of quadrupolar nuclei to study dynamic processes are scattered in different reviews, dealing either with specific nuclei [4] such as halides [5], ^{23}Na[6], ^{27}Al[7], ^{51}V[8] or with particular applications such as

Table 5.1 Characteristics of quadrupolar nuclei from the main group. Adapted from *EPR/ENDOR Frequency Table*, Bruker Co. diary

Isotope	Natural abundance (%)	Spin	Electric quadrupole moment Q 10^{-24} cm^2	Receptivity [3] relative to ^{13}C^a
^{2}H	0.0148	1	0.002875	$8.2\ 10^{-3}$
^{6}Li	7.5	1	−0.000644	3.58
^{7}Li	92.5	3/2	−0.040	$1.54\ 10^3$
^{9}Be	100	3/2	0.053	78.8
^{10}B	19.8	3	0.08608	22.1
^{11}B	80.2	3/2	0.040	754
^{14}N	99.63	1	0.0193	5.69
^{17}O	0.038	5/2	−0.026	0.061
^{23}Na	100	3/2	0.108	525
^{25}Mg	10.00	5/2	0.22	1.54
^{27}Al	100	5/2	0.150	117
^{33}S	0.75	3/2	−0.064	$9.731\ 10^{-2}$
^{35}Cl	75.77	3/2	−0.08249	20.2
^{37}Cl	24.23	3/2	−0.06493	3.8
^{39}K	93.26	3/2	0.054	2.69
^{41}K	6.73	3/2	0.060	$3.28\ 10^{-2}$
^{43}Ca	0.135	7/2	<0.23	$5.27\ 10^{-2}$
^{69}Ga	60.1	3/2	0.168	237
^{71}Ga	39.9	3/2	0.106	319
^{73}Ge	7.8	9/2	−0.19	0.617
^{75}As	100	3/2	0.29	143
^{79}Br	50.69	3/2	0.293	226
^{81}Br	49.31	3/2	0.27	277
^{85}Rb	72.17	5/2	0.273	43
^{87}Rb	27.83	3/2	0.130	277
^{87}Sr	7.0	9/2	0.15	1.07
^{113}In	4.3	9/2	0.846	83.8
^{121}Sb	57.3	5/2	−0.33	520
^{123}Sb	42.7	7/2	−0.68	111
^{127}I	100	5/2	−0.789	530
^{133}Cs	100	7/2	−0.003	269
^{135}Ba	6.59	3/2	0.20	1.83
^{137}Ba	11.2	3/2	0.34	4.41
^{209}Bi	100	9/2	−0.46	777

^{a} Does not take into account the dynamic effects or different relaxation mechanisms that can dramatically affect the signal-to-noise ratio.

metalloproteins [9,10], complexation [10,11], membranes [12], biological systems [13], interfacial phenomena [14], etc. Here we focus on dynamic processes studied by quadrupolar nuclei through selective applications. Other examples can also be found in other chapters of this book.

2   Theory: The Main Results

As with dipolar nuclei, dynamic information on a chemical system may be obtained from the chemical shift, the relaxation rates or from the self-diffusion coefficient of a quadrupolar nucleus. In anisotropic medium, the quadrupolar splitting may also be influenced by dynamic processes.

2.1   CHEMICAL SHIFT AND CHEMICAL EXCHANGE

Nuclear shielding is a tensor. For simple molecules, its components are obtained by molecular orbital calculations using *ab initio* or semi-empirical methods [15]. Agreement of calculated values with experimental data is satisfactory. In anisotropic media, molecular dynamics may partially average the chemical shift tensor. Conversely, analysis of chemical shift data with an appropriate model may give access to details of the molecular motion. Unfortunately, with nuclei of concern here, the effect is usually overshadowed by the quadrupolar interaction. It prevents the determination of dynamical parameters.

In isotropic fluids, as a result of fast molecular motion, the time-averaged mean value of the nuclear shielding tensor is only available. Accordingly, information on molecular reorientation is lost.

When a quadrupolar nucleus exchanges between two or more sites at an appropriate rate, its chemical shift is affected. The corresponding behaviour for dipolar nuclei is covered in Chapter 3. Chemical exchange can be monitored by quadrupolar nuclei as well and with very similar descriptions. If the exchange rate is fast relative to the chemical shift difference of the observed nucleus in each i site, a single signal is detected. Its frequency is the weighted average of the values in each environment:

$$\delta_{obs} = \sum_i p_i \delta_i \qquad (5.1)$$

From the knowledge of populations p_i in each site, stability constants can be obtained, giving access to thermodynamic parameters from work at different temperatures. No exchange is detected in this manner when the exchange rate is much slower than the chemical shift difference: one signal is observed for each site and stability constants and thermodynamic parameters are then accessible from peak intensities. The stability constant may be expressed as the ratio between the forward and the reverse rate constant. Therefore, when one rate constant is obtained, the other can be calculated from the stability constant, using an extrapolated value at the appropriate temperature for instance.

Between these two limiting cases, at intermediate rates of exchange, when the exchange rate and the chemical shift difference between two sites are of the same order of magnitude, both chemical shift and linewidth are perturbed. Using line-shape analysis [16], non-stationary methods [17] or two-dimensional

NMR experiments (2D EXSY) [18], the kinetics of the dynamical process can be obtained.

2.2 QUADRUPOLAR SPLITTING

2.2.1 Molecules in an anisotropic medium

In anisotropic media, constraints on the molecular motion induce a line splitting of quadrupolar nuclei. It is due to the quadrupolar interaction between the electric field gradient at the nucleus site and the electric quadrupolar moment. For instance, water deuteron and oxygen-17 spectra exhibit a 1/1 doublet and a 5/8/9/8/5 quintet, respectively. To relate the splitting to the molecular orientation, the orientation of the electric field gradient tensor F has to be defined with respect to the direction of the laboratory frame L. Let us term 'the director' the axis along which the system tends to order itself (this name originates with studies in liquid crystals). This can be done through three changes of referentials: F–M between the electric field gradient tensor and the molecular coordinate systems, M–D which relates the molecular M and the director D frames and L–D, between the principal axes of the laboratory frame and the director. The principal axes characteristic of these second rank interactions (F) may differ from the molecular referential but, if the quadrupolar interaction is purely intramolecular, the angles between the M and F coordinate systems are time-independent. As the reorientation of the director relative to the laboratory frame of reference is usually (assumed to be) slow on the NMR timescale, the modulation of these interactions only occurs from the reorientational motion of the molecule.

The static part of the quadrupolar Hamiltonian may be written as [19]:

$$H_Q = \frac{eQ}{2I(2I-1)h} \sum_m (-1)^m A_{-m} V_m \qquad (5.2)$$

The A_m's are the standard components of the second rank spin tensor operator and the V_m's are the irreducible components of the second rank electric field gradient tensor (m runs from -2 to 2).

The NMR signal is split into $2I$ lines, separated by Δ [19]:

$$\Delta = \frac{3e^2qQ}{4I(2I-1)h} (3\cos^2\Theta_{LD} - 1)A \qquad (5.3)$$

with Θ_{LD}, the angle between the Z axes of the laboratory (L) and the director (D) coordinate systems and A, the residual (quadrupolar) anisotropy. The above equation postulates at least a threefold symmetry around the director. An extra geometric term must be taken into account for lower symmetry. Halle and Wennerström [19] have defined the two order parameters occurring in NMR theory, together with their relation to the elements of the Saupe's traceless tensors S_{ii} [20]. These are:

$$S_0 = \tfrac{1}{2}\langle(3\cos^2\Theta_{DM}-1)\rangle = S_{33} \qquad (5.4)$$

$$S_2 = \tfrac{1}{2}\sqrt{6}\langle\sin^2\Theta_{DM}\cos 2\Phi_M\rangle = 2(\sqrt{6})^{-1}(S_{11}-S_{22}) \qquad (5.5)$$

where the bracket $\langle\ \rangle$ stands for the time-average and Θ, Φ are the Eulerian angles.

Assuming purely intramolecular interactions, the residual anisotropies are [19]:

$$A(^2H)=\tfrac{1}{2}(3\cos^2\alpha - 1 + \eta\sin^2\alpha)S_0 - (1/2\sqrt{6})[3\sin^2\alpha + \eta(\cos^2\alpha + 1)]S_2 \qquad (5.6)$$

$$A(^{17}O) = -\tfrac{1}{2}(1-\eta)S_0 + (1/2\sqrt{6})(3+\eta)S_2 \qquad (5.7)$$

where α is half of the molecular angle and η is the asymmetry parameter defined as

$$\frac{V_{XX}-V_{YY}}{V_{ZZ}}$$

The components of the electric field gradient tensor by definition follow the sequence $V_{ZZ} \geqslant V_{YY} \geqslant V_{XX}$. The above equations are valid for planar systems. In cylindrical aggregates, the splitting is reduced by a factor 2 [19].

The above formulas (5.6) and (5.7) appropriate for a C_{2v} molecule such as the much studied water should be adapted to other molecular geometries. For instance, to obtain the residual quadrupolar anisotropies for acetonitrile-d_3, one should take into account its higher C_3 symmetry. This reduces to one the number of independent order parameters. The principal axes of the electric field gradient tensor for ^{14}N and ^2H, respectively identify with the C–N bond and the C–D bond. The residual anisotropies become:

$$A(^{14}N) = S_0 \qquad (5.8)$$

$$A(^2H) = \tfrac{1}{2}(3\cos^2\alpha - 1 + \eta\sin^2\alpha)S_0 \qquad (5.9)$$

For uniaxial anisotropic media, a five-term equation describes the residual anisotropy in the general case [21].

2.2.2 Ions in an anisotropic medium

When the anisotropic mesophase or the solid particles are charged, counterions are needed for electroneutrality. These counterion nuclei are often quadrupolar (^{23}Na, ^{35}Cl ...) and a residual splitting occurs in their NMR spectrum if the ion mobility is not fast enough to average out the quadrupolar interaction. The theory summarized above remains valid. The quadrupolar splitting is given by equation (5.3) replacing the residual anisotropy by the order parameter S defined as [21]:

$$S = \tfrac{1}{2}\langle(3\cos^2\Theta_{DM} - 1)\rangle + \eta\langle\sin^2\Theta_{DM}\cos^2\Phi_M\rangle \qquad (5.10)$$

Applicability of this equation is restrained to first-order effects excluding direct interaction between two quadrupolar nuclei.

The electric field gradient (efg) at the nucleus site originates from the ionic environment. Various quadrupolar relaxation theories have been proposed [22]. The most popular remains the molecular electrostatic approach of Hertz [23] where: (a) the electric field gradient is assumed to be generated by dipolar molecules and ions in the proximity of the nucleus (b) distortion of the spherical electronic cloud of the ion from the perturbing electric field of the dipoles causes a reinforcement of the field gradients. This effect is taken into account by the Sternheimer anti-shielding factor which is obtained from highly sophisticated atomic calculations. It is also assumed that the efg correlation function is a simple monoexponential. The senior author has always been unhappy with the Hertz model, because it is infallible: while it includes terms for all the factors expected to come into play, the postulate of their simple additivity has never been examined critically. Worse yet is the plethora of parameters this Hertz model suffers from: a judicious choice—even among the various experimental results from the literature—immunizes it against falsification!

This model has very recently been questioned, however. A simple picture of relaxation dynamics for such nuclei implies a bi-(multi)-exponential field gradient time correlation function, conversely to the usual assumption in electrostatic models [24]. The importance of solvent–solvent cross-correlations, particularly relevant for polar liquids, has also been stressed by these authors [24].

2.2.3 Chemical exchange

Molecules usually visit a number of different sites. If the exchange between these environments is fast compared with the difference in the quadrupolar splittings characterizing individual sites, the observed splitting is the weighted average:

$$\Delta = \sum_i p_i \Delta_i \qquad (5.11)$$

where p_i is the mole fraction in each site.

In the simplest case, molecules exchange fast between the bulk liquid phase and a 'bound' form. In the bulk, the quadrupolar splitting is zero-averaged as a result of fast reorientation. If the greatest fraction of the molecules are in the bulk, this can drastically reduce the observed splitting to a few Hz (see Section 3).

2.2.4 Amphiphilic molecules

Surfactants and phospholipids are typical examples in this group. Deuterium quadrupolar splittings have been used extensively to describe the chain mobility within the anisotropic phases formed by these molecules [12].

When the director is perpendicular to the magnetic induction field B_0, the geometric term vanishes in equation (5.3) and the quadrupolar splitting is:

$$\Delta = \frac{3e^2qQ}{4hI(2I-1)} S \qquad (5.12)$$

The order parameter S is expressed by equation (5.4), where Θ_{DM} is the angle between the bilayer normal (director) and the principal axis of the coordinate system associated with the electric field gradient tensor (the z-axis is taken as colinear with the C–D bond). Equation (5.12) remains valid for a rapid reorientation of the C–D vector. If the motion of the C–D vector is fast relative to the line separation, the quadrupolar interaction is partially averaged. This results in a decrease of the splitting and sometimes in a change in the lineshape. More details are provided in Chapter 8.

2.3 RELAXATION RATES

2.3.1 Isotropic medium

The fluctuating part of the Hamiltonian responsible for relaxation processes is expressed in the operator formalism [25] as:

$$H(t) = \sum_m (-1)^m F_m(t) A_m \qquad (5.13)$$

where the F_m's are random functions of time and the A_m's are operators acting on the spin variables of the system. For quadrupolar relaxation, the A_m's are second-rank spin tensor operators and the F_m's have the same meaning as the V_m's in equation (5.2). The F_m's describe reorientation of the spin, and thus of the quadrupole moment with respect to the electric field gradient. The Fourier transform of the appropriate correlation function of $F_m(t)$ gives the spectral densities [25]. In the simplest reorientational model, this correlation function is monoexponential and therefore characterized by a single correlation time. More sophisticated reorientational models are described in Chapter 2.

The longitudinal R_1 and transverse R_2 relaxation rates for a $I = 1$ nucleus are given by [25]:

$$R_1 = 6K[J(\omega_0) + 4J(2\omega_0)] \tag{5.14}$$

$$R_2 = 3K[3J(0) + 5J(\omega_0) + 2J(2\omega_0)] \tag{5.15}$$

with

$$K = \frac{\pi^2(e^2qQ)^2}{20h^2}(1 + \eta^2/3) \tag{5.16}$$

The meaning of η has been reported in Section 2.2.1. The quadrupolar coupling constant is e^2qQ/h and the spectral density has the form:

$$J(m\omega_0) = \frac{\tau_c}{1 + (m\omega_0)^2\tau_c^2} \tag{5.17}$$

In the extreme narrowing limit ($\omega_0\tau_c < 1$), the relaxation process is usually assumed to be monoexponential. In fact, it has been shown recently for nuclei with $I > 1$ that higher rank contributions may affect their relaxation and their lineshape [26–28]. Deviation from a monoexponential behaviour is more easily detected by triple quantum filtering techniques [29]. In the non-extreme narrowing condition ($\omega_0\tau_c \geqslant 1$) and for nuclei with $I > 1$, relaxation processes are no longer governed by a monoexponential law [30]. This multiexponential behaviour makes possible detection of multiquantum coherences [27, 31]. Too few applications have made use of this possibility (see Section 3). For half-integer spins, there are $I + 1/2$ decaying exponentials. Analytical expressions for longitudinal and transverse relaxations of a $3/2$ spin nucleus are [32]:

$$R_1 = 4K[0.2J(\omega_0) + 0.8J(2\omega_0)] \tag{5.18}$$

$$R_2 = 4K[0.3J(0) + 0.5J(\omega_0) + 0.2J(2\omega_0)] \tag{5.19}$$

Approximate analytical expressions have been also deduced for $5/2$ and $7/2$ spin nuclei when the effective spectral density is weakly frequency dependent ($\omega_0\tau_c \leqslant 1.5$) [33]. The above expressions (5.18 and 5.19) remain valid with:

$$K = \frac{3\pi^2(e^2qQ)^2(2I + 3)}{10h^2I^2(2I - 1)}(1 + n^2/3)$$

The increase of the transverse relaxation rate with the correlation time τ_c is of general concern. It results in line broadening which might prohibit signal detection. In contrast with the other transitions, the $+1/2 \rightarrow -1/2$ transition is such that the plot of the transverse relaxation rate versus the correlation time goes through a maximum [34, 35] (Figure 5.1). On the background of the extensive broadening of the signals from the other transitions, a rather narrow peak for the $+1/2 \rightarrow -1/2$ transition may be observed for slowly reorienting

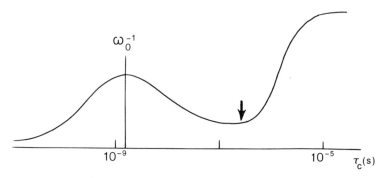

Figure 5.1 Schematic illustration of the qualitative dependence of the central $1/2 \rightarrow -1/2$ transition linewidth on the correlation time τ_c. Reprinted with permission from Butler and Eckert, *J. Am. Chem. Soc.*, **111**, 2802. Copyright (1981) American Chemical Society

molecules. An outstanding example is ^{51}V [35] and ^{27}Al [36] spectrum of transferrins, metalloproteins with a molecular weight of *ca.* 80 000. The estimated correlation time is indicated by an arrow (Figure 5.1).

Under non-extreme narrowing the NMR signal undergoes 'second-order dynamic shifts'. Their effect on the NMR lineshape of $I = 3/2$ [37], $I = 5/2$ and $I = 7/2$ nuclei have been investigated [38]. The NMR signal becomes asymmetric. The magnitude of the effect increases with the square of the quadrupolar coupling constant and is inversely proportional to the Larmor frequency.

For a covalently bound nucleus quadrupolar relaxation is intramolecular. The quadrupolar coupling constant can be determined from studies in the solid or in anisotropic phase. The above equations give directly access to the correlation time.

By contrast, the quadrupolar coupling constant of an ion cannot be measured experimentally. The electric field gradient at the nucleus originates in the ionic environment of the ion (see the previous section). As a result, in a non-exchanging system the above equations have two unknowns: the correlation time and the quadrupolar coupling constant.

If the extreme narrowing condition is obeyed, the problem is soluble via observation of only single quantum transitions, if two quadrupolar isotopes exist. Use of triple quantum filtering techniques may also solve this problem [29]. Under non-extreme narrowing conditions different values for transverse and longitudinal relaxation rates provide both the correlation time and the quadrupolar coupling constant. These values can also be obtained from the longitudinal relaxation rate at two different Larmor frequencies. Deconvolution of the superlorentzian bandshape or analysis of the multiexponential relaxation curve characterizing the non-extreme narrowing conditions also gives access to the correlation time. For a spin 3/2 nucleus it has been also proposed to analyse the spectral lineshape near the null point in inversion recovery experiments [39].

2.3.2 Anisotropic medium

So far an isotropic medium has been implicitly assumed. Fluctuation of the angle (Θ_{LD}) in equation (5.3) averages out the residual quadrupolar interaction. A single line is observed for each environment of the quadrupolar nucleus.

In anisotropic systems, the $2I$ transitions between the Zeeman levels perturbed by the quadrupolar interaction are nondegenerate and a line splitting is detected in the NMR spectrum. The relaxation rates differ from those in an isotropic medium and vary with the spin number of the quadrupolar nucleus. In the absence of cross-relaxation, these equations for a $I = 1$ nucleus are obeyed [40, 41]:

$$R_1 = 3K[J_1(\omega_0) + 4J_2(2\omega_0)] \tag{5.20}$$

$$R_2 = 1.5K[3J_0(0) + 3J_1(\omega_0) + 2J_2(2\omega_0)] \tag{5.21}$$

For nuclei with higher half-integer spin, $I + 1/2$ equations define the longitudinal or transverse relaxation processes. Appropriate equations for the transverse relaxation rate of a $3/2$ [28] or of a $5/2$ [19] spin nucleus are available. Information on dynamics is contained in the spectral densities $J(m\omega_0)$. The complexity of these systems prevents though a precise analysis from the relaxation times determination alone. Therefore, different NMR techniques to estimate the different spectral densities have been published for $I = 1$ [41] and $I = 3/2$ [42]. Responses of a $5/2$ spin nucleus to the 2D quadrupolar echo sequence [43, 44] and to the inversion recovery experiment [45] have been worked out.

2.3.3 Chemical exchange

When a molecule or an ion exchanges between different S environments, relaxation parameters are generally analysed in terms of a discrete exchange model. When the residence time of a molecule in a site is longer than the reorientational timescale and shorter than the macroscopic relaxation time (fast exchange), relaxation is a weighted average: for a two-site case:

$$1/T_{1,2} = R_{1,2} = (1 - p_B)R_{1,2}^A + p_B R_{1,2}^B \tag{5.22}$$

If the exchange and the relaxation rates are of the same order of magnitude, kinetic parameters can be obtained as well from the plot of the temperature dependence of the longitudinal relaxation rate. In the absence of a significant chemical shift difference for site A and B, a similar curve holds for the transverse relaxation rates (Figure 5.2).

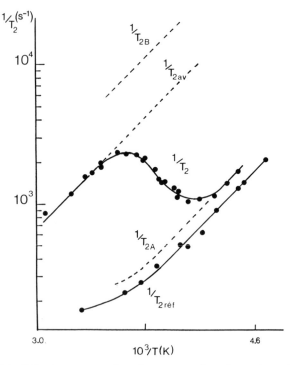

Figure 5.2 Plot of the transverse relaxation rate against the reciprocal temperature. Reprinted with permission from Schori *et al.*, *J. Am. Chem. Soc.*, **93**, 7133. Copyright (1971) American Chemical Society

The left and the right part of this curve describe the fast and the slow exchange limits, respectively. Intermediate exchange occurs in the middle temperature range. The lifetime in the site A is given by [46]:

$$1/\tau_A = (T_{1B}^{-1} - T_1^{-1})(T_1^{-1} - T_{1A}^{-1})p_B/(T_{1av}^{-1} - T_1^{-1}) \tag{5.23}$$

with

$$1/T_{1av} = p_A T_{1A}^{-1} + p_B T_{1B}^{-1}$$

In the presence of a chemical shift difference, lineshape analysis is mandatory for reliable determination of the exchange rate [47]. In complexation studies, this parameter may be obtained from measurements of T_1 and T_2 versus the (ion/ligand) molar ratio [48].

Chemical exchange in anisotropic media has been also considered for deuterium [49].

2.4   SELF-DIFFUSION COEFFICIENT

In an isotropic homogeneous system, the self-diffusion parameter D characterizes the radial distribution function of molecules with regard to their (arbitrary) original positions. Although the average displacement in all the three directions is zero, the mean-square displacement obeys the Einstein relationship:

$$\langle r^2 \rangle = 6Dt \qquad (5.24)$$

The most popular NMR method to determine D is the pulsed field gradient spin-echo experiment [50]. The quantitative relation between the attenuation A at the echo time 2τ and the duration δ of the pulsed field gradient G is:

$$A(2\tau) = A(0)\exp[-(2\tau/T_2) - (\gamma\delta G)^2 P(P - \delta/3)] \qquad (5.25)$$

where P represents the pulse interval, γ the gyromagnetic ratio of the observed nucleus and T_2 the transverse relaxation time. When the observed species exchanges fast between two or more sites, the measured parameter is population weighted-averaged like the other NMR parameters.

Improvements have been proposed to circumvent limitations of this basic experiment. Indeed, the investigated time domain can be drastically reduced for short relaxation time T_2. These technical aspects are discussed in two recent reviews [51, 52] and in Chapter 7. Since quadrupolar nuclei exhibit often broad resonances, the situation is unfavourable. Nuclei such as deuterium and lithium are suitable for self-diffusion studies, however (Section 3.2.3).

Obviously, the diffusion can be more complex than purely isotropic. Restricted diffusion or anisotropic diffusion are expected in heterogeneous systems. This is also discussed in the two above-quoted reviews [51, 52] and in Chapter 7.

3   Applications: Illustrative Examples

From its first applications to chemistry in the 1950s, nuclear magnetic resonance had an impact on chemical and molecular dynamics. Initially, internal rotation in molecules presented a problem that had been solved in part only by the existing techniques such as microwave and infrared spectroscopies. Thus, it was natural for investigators—Norman Sheppard was pre-eminent among them—to turn to the new technique for a solution. At the time, however, NMR was powerless when brought to bear on energy barriers below about 20 kJ mol^{-1}. In the intervening decades, dynamic NMR has been applied to the whole gamut of rate processes with half-lives ranging from seconds to picoseconds. These methodologies have been adequately and even profusely reviewed. Quadrupolar nuclei can also be profitably applied to the accurate determination of rate constants. Our approach will be through illustrative examples.

Generally speaking, NMR methods are used to study dynamic processes such as chemical exchange, molecular motion, chemical reaction or membrane processes. As seen in Section 2, when the probe nucleus is exchanged between two or more environments, NMR parameters are affected. Their variation gives access to structural, thermodynamic or kinetic information, depending on the exchange rate.

Numerous applications of quadrupolar nuclei to dynamic phenomena have been published. Typical results which illustrate the different aspects of the theoretical section are reported hereafter. Further examples can be found in other chapters of this book.

3.1 CHEMICAL EXCHANGE PROCESSES

3.1.1 Inorganic and organometallic compounds.

Nuclei such as ^{6}Li, ^{7}Li, ^{51}V (Table 5.1) have a small quadrupole moment. Their NMR spectra most often consist of rather narrow signals. NMR techniques used for dipolar nuclei (Chapter 3) are also appropriate for such quadrupolar nuclei.

Organolithium compounds play an important role in organic chemistry. Mechanistic studies demand knowledge of their structure and dynamics. Indeed, these lithiated substances have a high propensity to self-associate into higher aggregates. This oligomerization is solvent-dependent. Recently, ^{6}Li and to a lesser extent ^{7}Li NMR have emerged as new tools to study exchange processes in organolithiated compounds. ^{6}Li line-shape analysis has been performed on the addition-lithiation product of ^{6}Li enriched n-butyllithium and diphenylacetylene. Studies at several temperatures show two dynamical processes: dissociation of a dimer to a monomer as well as interchange of the lithium sites within the dimer [53]. Similar studies on lithium cyclopentadienide indicate a rapid exchange between the solvated lithium cation and a monomer bound lithium and a slower exchange between lithium sandwiched in a dimer and monomer-bound lithium [54]. Activation parameters for these dynamic processes were also reported [53, 54]. Neopentyllithium forms complexes with triamines such as pentaethyldiethylenetriamine. Two dynamic processes account for the temperature-dependent ^{6}Li NMR spectra: interaggregate exchange, or inversion of the N-dimethyl nitrogens [55]. 2D exchange spectroscopy (2D EXSY)[18] has been used also to display lithium exchange. Owing to the shorter relaxation time (broader signal) of ^{7}Li as compared to ^{6}Li, the former nucleus is less well adapted to such kinetic studies. Nevertheless both ^{6}Li and ^{7}Li 2D EXSY experiments have been reported in literature [53, 56]. The intra-aggregate ^{6}Li exchange in 3,4-dilithio-2,5-dimethyl-2,4-hexadiene $(CH_3)_2C=CLi—CLi=C(CH_3)_2$ in deuterated tetrahydrofuran is shown (Figure 5.3) [56].

Exchange between ^{7}Li has also been found for a dilithium salt of octaethylporphyrin^{2-} [57].

The ^{51}V nucleus has also a small quadrupolar moment and is well suited for rate studies. Oligomerization of vanadate has been investigated by the 2D EXSY method. Four exchanging vanadium species coexist and the major

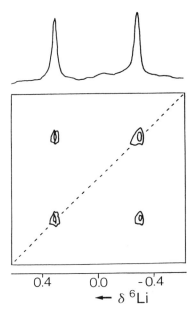

Figure 5.3 2D-^{6}Li, ^{6}Li exchange spectrum for the intra-aggregate ^{6}Li exchange in 3,4-dilithio-2,5-dimethyl-2,4-hexadine ($T = -83\,°C$). Reproduced by permission of VCH from Ref. 56

pathway for formation of each species can be inferred from the rate constants extracted from the data [58]. Vanadate esters have possible synthetic utility and biological activity. The 2D ^{51}V NMR spectroscopy has been used to study these reactions between vanadates and alcohols. From the intensities of the non-diagonal signals, microscopic exchange rates between various vanadate oligomers and vanadate esters have been obtained. The presence of imidazole in the reacting mixture affects both the equilibria and the kinetics of these reactions [59].

3.1.2 Ion–molecule interaction

Quadrupolar nuclei are particularly well suited to study ion–molecule interaction. Variable-temperature and variable-pressure studies of ligand substitution reactions have been monitored by ^{17}O NMR. These results, together with data obtained by observation of the ion nuclei, are described in the next chapter and are therefore not dealt with here. Preferential solvation of ions in binary liquid mixtures can also be monitored handsomely by NMR experiments using quadrupolar nuclei [60].

In the last two decades, much effort has been made to prepare cyclic polyethers (crown ethers), cryptands, podands or natural ionophores, all compounds able to complex strongly cations, alkali metal ions in particular.

The NMR of these nuclei provides an excellent means to describe cation complexation by such ligands. As fast exchange between the solvated and the complexed cation often prevails, the NMR spectrum of the cation nucleus shows but a single line (equation 5.1). The change in various observables, the chemical shift (equation 5.1), the linewidth or the longitudinal relaxation rates (equation 5.22), with an increase of the cation–ligand molar ratio, reflects the sequential displacement of solvent molecules by the complexing agent. The stoichiometry and the stability constant(s) of the complex(es) are deduced from the analysis of such curves. Using chiefly ^{23}Na and ^{133}Cs [61–63], Popov's group has performed a wide-ranging but rather descriptive survey of cation complexation by crown ethers and by cryptands. Typically, a 1/1 stoichiometry is obtained when the cation radius fits the cavity size of the crown ether. Larger cations form sandwich complexes in which the cation is located between two adjacent polyether rings. With cryptands, 'inclusive' and 'exclusive' complexes are elegantly distinguished by NMR. Let us remind the reader of the meaning of these terms. In the former, the cation sits in the middle of the cavity, it remains shielded from solvent molecules. When the cation is too large to enter the cavity, it remains in contact with solvent molecules ('exclusive' complex). Ion-pairing between the cation and the counteranion may then influence the formation constants of the complexes. Enthalpies and entropies for complex formation have been determined. More details are given in an excellent recent review [11].

In spiro-bis-crown ethers, two cavities are available for cation complexation. When both cavities are identical and fit the cation size, a strong negative cooperativity is intuitively expected and indeed observed between the two cationic sites [64]. Carboxylic ionophores are natural polyethers which form a pseudocyclic structure upon cation complexation. Formation constants are determined by a similar procedure [65, 66]. Synthetic polyethers also complex these cations as exemplified by studies on ^{7}Li [67] and ^{23}Na [68] nuclei.

Alkali metal NMR studies are not restricted to the observation of cations. Alkali metals in the presence of crown ether or cryptand solutions shows separate NMR signals for the complexed cation, a result of slow exchange between these species [69, 70].

When the rate constants of a kinetic process are of the same order of magnitude as the difference in the chemical shift of the cationic sites or as the value of the relaxation time, alkali metal NMR provides a choice technique to determine these parameters. Different situations can arise. The rate constant and the chemical shift difference between the two sites can be similar. The determination of the rate constant then requires a full lineshape analysis (Chapter 3). Typical data using ^{7}Li[71, 72], ^{23}Na [73–75] or ^{113}Cs [76, 77] are found in the literature. Sometimes a chemical shift difference between environments A and B is absent or not sufficient. Then the on rate constant can be obtained from a plot of the relaxation rate versus the reciprocal temperature (Figure 5.2 and equation 5.23) [46]. Several examples of this approach have been published, using ^{23}Na [46, 64, 78], ^{39}K [79, 80] or ^{87}Rb [79] NMR spectroscopy. Some ^{23}Na results have been analysed by a third method based upon T_1 and T_2 measurements, as described in the theoretical section [48, 81].

Knowing the stability constant of the complex, the determination of one rate constant, forward (on) or reverse (off) allows the calculation of the other rate constant. The formation rate of the complex is often diffusion-controlled, and the dissociation rate constant governs the selectivity of the cation formation.

In systems containing free (solvated) and ion-bound ionophore molecules, exchange may occur by a unimolecular dissociative process (equation (5.25)) or by a bimolecular associative mechanism (equation (5.27)) [46].

$$(M-L)^+ \underset{k_1}{\overset{k_{-1}}{\rightleftharpoons}} M^+ + L \qquad (5.26)$$

$$M^{*+} + (M-L)^+ \overset{k_2}{\rightleftharpoons} M^+ + (M^*-L)^+ \qquad (5.27)$$

Examples of both have been reported. Besides the relative sizes of the cation and the cavity of the complexant, solvent effects also influence these complexation mechanisms. No clearcut understanding of these has arisen yet [11]. This is an area in which there is room for further, incisive study.

3.1.3 Calciproteins

Calciproteins have an unique physiological role. Since calcium is a messenger ion, its binding on appropriate sites within proteins very often is a biological signal for transduction of information, from muscular contraction in vertebrates to neurotransmission in the brain. Furthermore, some calciproteins have a central importance to organisms.

Calmodulin (see below) is an ubiquitous molecule, well conserved through evolution, that functions as a universal modulator and activator of enzymatic activity in a very impressive lineage of organisms, from unicellular creatures to higher vertebrates. There are two ways in which quadrupolar NMR can be used to study calciproteins; directly through use of calcium-43 NMR, in which the Lund group (after cornering a sizeable chunk of the world supply of calcium-43!) excelled; or indirectly, through monitoring probe nuclei such as halide ions and sodium-23, since sodium ions and calcium ions share identical ionic and van der Waals' radii which make them isomorphous and able to substitute for one another.

In this section we deal with a class of homologous calcium binding proteins that includes parvalbumin (Par), troponin C (TnC), calmodulin (CaM) and the intestinal calcium binding protein (ICaBP). The crystal structure of these proteins shows a common feature of the calcium binding sites arranged pairwise. Par and ICaBP share two calcium binding sites of different affinities. Two helices surround every calcium binding loop. The coordinated groups involve three or four sidechain carboxyl groups, the remaining ones are neutral oxygens from either backbone carbonyls, sidechain serine or water. The unit formed by the two helices and the calcium binding loop is often referred as the 'EF hand'. In TnC and CaM four such EF hands are present, providing four calcium binding sites.

With physiological ions such as Na^+, K^+ or Cl^- fast exchange between solvated and protein-bound ion normally prevails. This prevents determination of the exchange rate constants and such results will not be further considered here. As the NMR sensitivity of calcium and magnesium nuclei is low these studies have been performed with ^{43}Ca or ^{25}Mg enriched proteins. In this way, both thermodynamic and kinetic data on cation binding by proteins have been obtained. Strategies used in ion-protein interaction do not differ from that reported in the former section. Practical details are available in the literature [47].

Calmodulin, an ubiquitous protein with four calcium binding sites, has been extensively studied by NMR methods. The binding of calcium to the protein is too strong for accurate determination of the formation constant ($K > 10^4$ M^{-1}). Temperature dependence of the ^{43}Ca linewidth between 0° and 70 °C covers the domain of the slow and the intermediate exchange (Figure 5.2). This curve has been used to determine the rate constant for the release of calcium ion in calmodulin and in its tryptic fragments. Different models for the calcium binding sites have been considered to fit this curve. A significant better fit was obtained by assuming two calcium ions exchanging with intermediate rates and two exchanging with slower rates but still fast enough to affect the lineshape within the high temperature range. The low affinity sites are characterized by a k_{off} value of 10^3 s^{-1} and the high affinity site by a value of 30 s^{-1} [47]. Tryptic digestion splits calmodulin in two peptide fragments with two calcium binding sites. The two fast exchanging sites have been located in the N-terminal half of the protein [82]. In the presence of 0.1 M KCl, the calculated k_{off} for the fast process is slowed down by a factor of two in the native protein but not in the N-terminal tryptic fragment [82]. The activation parameters for the dissociation of the calcium complexes agree with values obtained by other techniques. The presence of trifluoperazine, an antipsychotic drug, slows down the k_{off} exchange rate of the low affinity sites of calmodulin [83].

The influence of amino acid replacement on these rate constants has been also investigated for ICaBP [84]. The point mutations are restricted to the N-terminal Ca^{2+} binding site. The exchange rate for this N-terminal site varies then from 3 s^{-1} to 5000 s^{-1}, dependent on the amino acid substitution. Conversely, no effect on the rate constant has been detected for the C-terminal Ca^{2+} binding site [84].

^{27}Mg NMR has also been used to define the magnesium ion interaction with calmodulin. These results and others using alkali and alkaline-earth nuclei have been recently reviewed [10].

We will return in a later section to the isotropic reorientation of these globular macromolecules as approached quite accurately by quadrupolar methodologies.

These studies concern rather small proteins with molecular weight below 20 000. Analytical expressions of the relaxation times are thus still valid. With higher molecular weight proteins, extensive line broadening occurs. It results in a drastic decrease of the signal intensity. This does not necessarily preclude the observation of higher molecular weight metalloproteins from NMR of the metal nucleus (see the theoretical section). Indeed, the signal of the $1/2 \rightarrow -1/2$ transition remain rather narrow for a protein of higher molecular weight (Figure 5.1). Second order dynamic shifts may become important for such systems. No quantitative analysis of cation complexation by such macromolecules has appeared yet in the literature.

3.1.4 Exchange at a clay–liquid interface

In aqueous clay suspensions, fast exchange of water molecules occurs between the bulk phase and the clay–liquid interface. ^2H and ^{17}O quadrupolar splittings are observed in the NMR spectra of such suspensions [85]. As no splittings are detected in the bulk medium (see Section 2.2.3), this indicates a preferential orientation of water molecules near the solid surface.

Clays are minerals with a lamellar structure formed by layers of magnesio- or aluminosilicates. A tetrahedral layer consists of tetrahedras with a silicon ion at the centre and oxygen atoms at the corners. In octahedras occurring in octahedral layers, oxygens are seated on the corners and aluminium or magnesium ions at the centre. Clay platelets result from the association of these layers. In the investigated clays, platelets are formed by an octahedral layer sandwiched between two tetrahedral layers (phyllosilicates 2:1) (Figure 5.4).

When octahedras are centred on Al^{III} two thirds of octahedral sites are filled and one is dealing with a dioctahedral clay (Figure 5.4). When aluminium is replaced by magnesium the empty site is filled and this is a trioctahedral clay.

Clay platelets (lamellas) are negatively charged as a result of cation isomorphic substitution, either in the octahedral sheet or in the tetrahedral layer. Thus, Si^{IV} is partly replaced by Al^{III} with the attendant loss of positive charges. Similarly, Mg^{II} may substitute for Al^{III} or Li^{I} for Mg^{II} in the octahedral layers of dioctahedral and trioctahedral clays, respectively. Exchangeable cations are present in the interlamellar space to counterbalance these negative charges.

Thus interaction of water molecules at the surface of platelets causes a splitting in its ^2H and ^{17}O NMR spectrum. Obviously, fast water exchange between the clay surface and the bulk medium appears, thus reducing by three orders of magnitude the observed splitting (equation 5.11) [85]. The studied dioctahedral clays are montmorillonites, the first with cation isomorphic substitution in the octahedral layer only (MonO) and the second with cation

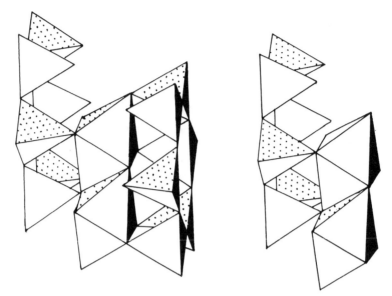

Figure 5.4 Structure of 2/1 clay minerals. Reproduced by permission of Academic Press from Ref. 145

replacement in both octahedral and tetrahedral sheets (MonOT). Two trioctahedral clays have been investigated: hectorite where Li^I substitutes for Mg^{II} in the octahedral layer and saponite where cations with lower oxidation state replace silicon ions in the tetrahedral layer. With aqueous suspensions of MonOT, a sign reversal of the water deuteron splitting is observed when starting with a sodium-exchanged clay, the (Ca^{2+}/Na^+) molar ratio in exchangeable cations is progressively increased [86] (Figure 5.5).

Fast exchange between two interfacial sites with opposite residual quadrupolar anisotropies (equation (5.3) and (5.11)) explains this observation [86, 87]. Averaged water orientation showing opposite anisotropies are schematized (Figure 5.6).

Suspensions of MonOT in water/CD_3CN, in water/CD_3OD, in water/$(CH_3)_2SO$ and in water/$(CH_3)_2CO$ binary systems have been studied in a similar manner [87]. Again, changes in the sign of ^2H splitting of the organic polar solvent occurs with an increase in the (Ca^{2+}/Na^+) molar ratio [87]. These observations can also be rationalized by fast exchange between two interfacial sites for organic cosolvent molecules [87].

For aqueous suspensions of each of the other three clays, no significant changes of the water deuteron splitting was observed with a similar change in the composition of exchangeable cations [88]. Isomorphic cation substitution in one layer, octahedral or tetrahedral, is the common feature of these three clays. Therefore, the presence of cation replacement in both octahedral and tetrahedral layers seems a prerequisite to interfacial sites with opposite

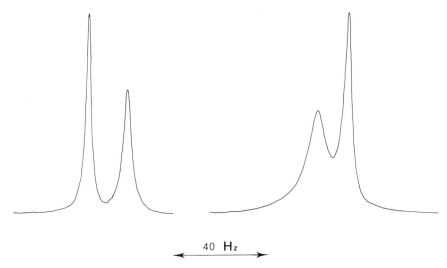

40 Hz

Figure 5.5 Typical ^{2}H NMR spectra of cation-exchanged montmorillonites in heavy water at 299 k (clay content: 23.8 mg/ml); Ratio of exchangeable cations: $Ca^{2+}/Na^+ = 0.02$ (left) and 0.13 (right). Reproduced by permission of Academic Press from Ref. 14

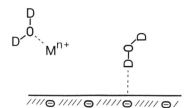

Figure 5.6 Appropriate orientations of heavy water at the clay interface

anisotropies [88]. Comparison of water deuteron splittings of aqueous suspensions of MonO-saponite or of hectorite-saponite with those of aqueous suspensions of these clays alone confirms the presence of two interfacial water sites with opposite residual anisotropies [89].

The differential broadening of the doublet component (Figure 5.5), allowing the detection of a sign reversal of the splitting is the cornerstone of such studies. It was explained by cross-correlation between quadrupolar and dipolar (paramagnetic) relaxations [90].

3.2 MOLECULAR DYNAMIC STUDIES

As seen in the theoretical section, knowledge of the relaxation rates of quadrupolar nuclei provides a handle onto molecular reorientation. For covalently bound nuclei this interaction is usually intramolecular. Relaxation times are related to molecular rotation as in the case of ^{13}C relaxation parameters. But for nuclei with a small quadrupole moment, relaxation processes are predominantly governed by the quadrupolar interaction. Data analysis is made easier as compared with ^{13}C or ^{1}H nuclei. Equations reported in Section 2.3.1 assume an isotropic diffusional rotation motion. Obviously, molecular reorientation can be more complex and adequate theoretical treatments have been provided (See chapter 2).

3.2.1 Water orientation

^{1}H, ^{2}H and ^{17}O nuclei are NMR probes into water dynamics. Proton relaxation is influenced by molecular motion, by proton exchange and by dipolar interaction with surrounding (macro) molecules. This complexity gave rise to a bewildering manifold of data analyses in the literature. To investigate molecular mobility and only it, proton NMR is definitely not the best available method. Observation of the two water quadrupolar nuclei (^{2}H and ^{17}O) provides much better information although deuteron–deuteron exchange may also affect ^{2}H relaxation. The analysis of ^{2}H and ^{17}O relaxation rates requires the knowledge of the quadrupolar coupling constants which can be measured or calculated rather accurately. The best value for water deuterons in the liquid phase is 255 kHz, as determined either experimentally [91] or by *ab initio* calculation [92]. Extrapolation from the solid or from the gas phase gives a value of about 8 MHz for the quadrupolar coupling constant of ^{17}O [93]. Asymmetry parameters of about 0.15 and 0.9 are often used for ^{2}H and ^{17}O, respectively [93].

Usually, water reorientation is assumed to be isotropic. To describe an anisotropic motion more than one correlation time or one diffusion constant is necessary. For an axially symmetric molecule such as water, two diffusion constants parallel and perpendicular to the molecular axis, are defined. A theoretical approach based upon $^{1}H-^{1}H$, $^{1}H-^{2}H$ and $^{1}H-^{17}O$ dipolar relaxation rates, together with ^{2}H and ^{17}O quadrupolar relaxation data has been used to define the best conditions to detect an anisotropy in water reorientation [94]. This approach has been applied to a study of water reorientation in fully neutralized poly(methacrylic acid) solutions. Small systematic differences between ^{2}H and ^{17}O water relaxation assuming an isotropic model argue for a small anisotropy in the water motion [95]. Similar studies have been performed in poly(acrylic acid) and in poly(styrene sulfonate) solutions [96]. A small anisotropy in water motion has been found in the high concentration limit. As compared with water in simple electrolyte solutions, water molecules in the proximity of the polyion are characterized by a reduced mobility which depends on the nature of counterions and on the side-chain of the polyion. This perturbation is restricted to the proximity of the surface of the polyelectrolyte [96]. This agrees with a rather general consensus on the state of water in heterogeneous systems: only one or two water layers close to the surface of the polyion, of the aggregate or of the solid are affected [14].

Numerous studies have dealt with water behaviour in protein solutions and in other colloidal systems. Most proton relaxation studies concern NMR imaging techniques, discussed in the last chapter. An extensive study of ^2H and ^{17}O relaxation times at several magnetic field strengths has been performed in colloidal systems [97–99]. The reduction in water mobility as compared with bulk water is rather small, less than one order of magnitude [97]. Differences between longitudinal R_1 and transverse R_2 relaxation rates and their frequency dependence point to the coexistence of fast and slow motions. R_1 is weakly affected by slow motions in the low frequency range only [97]. Discrete relaxation models based upon water exchange between two sites, usually the bulk medium and the macromolecule or solid surface (equation (5.22)) have been adapted, taking into account the anisotropic environment of water molecule near the surface [97]. A local rapid water reorientation accounts for the fast contribution to relaxation. The slow motion stems from translational mobility of hydration water, strongly reduced by comparison with bulk water [99]. The state of water in colloidal solutions and particularly in protein solutions has also been investigated by NMR field-cycling spectroscopy [100]. With appropriate equipment, one can readily measure proton or deuteron longitudinal relaxation rate as a function of magnetic field strength. The fast relaxation rates of ^7O presently preclude similar experiments for this nucleus. A multi-parameter non-linear regression analysis of the ^2H–NMR T_1 dispersion curve gives access to the protein rotational correlation time. Some contradictory results have been published, depending on data analysis [101, 102].

3.2.2   Reorientation of calciproteins

Alkali and alkaline–earth nuclei have been used to measure the rate of reorientation of calciproteins. Together with the homologous class of calciproteins (see Section 3.1.3), a few other macromolecules have also been considered. Native or cation-exchanged proteins (from the apoprotein) are prepared and metallic ion relaxation rates are then determined. From these values, one can extract a correlation time as described in Section 2.3.1. Some results are reported in Table 5.2.

A first conclusion emerges from these data. For these proteins the correlation time is not significantly affected by the probe nuclei, whether ^{43}Ca or ^{23}Na. These experimental results agree quite well with correlation times calculated with the classic Debye–Stokes–Einstein equation which characterizes the rotation of a spherical macromolecule in solution. As the overall shape of these globular proteins is well approximated by a sphere, this equation is indeed appropriate to determination of the correlation time. This indicates the cation binding site(s) reorient(s) at the rate of the whole macromolecule (strong binding). Such an observation is not fulfilled with haemocyanin [109] and

Table 5.2 Comparison of correlation times determined from relaxation rate experiments and calculated with the Debye–Stokes–Einstein equation

Protein or polypeptide fragment	MW(D)	τ_{DBS} (ns)	τ_c (ns)	Observed nucleus	Reference
Bovine Gla	5 700	2.6	2.5	^{23}Na	103
Parvalbumin	11 780	3.5	4.0	^{23}Na	104
			4.0	^{43}Ca	105
TR1(troponin C)	8 490	4.5	4.4	^{23}Na	106
Troponin C	17 965	9.0	9.0	^{23}Na	106
			11.0	^{43}Ca	105
Phospholipase A2	14 000	6.6	6.0	^{23}Na	107
TR1(calmodulin)	8 560	4.5	4.4	^{23}Na	108
			6.0	^{43}Ca	82
TR2(calmodulin)	8 114	4.5	4.5	^{23}Na	108
			5.0	^{43}Ca	82
Calmodulin	16 674	8.6	9.1	^{23}Na	108
			8.2	^{43}Ca	105
Lactalbumin	14 500	7.1	8.0	^{43}Ca	109
Crayfish MCBP	44 000	20.6	19.2	^{23}Na	107
Haemocyanin	450 000	250	14.0	^{43}Ca	110
			16.0	^{23}Na	110

internal motions, together with the overall motion, affect the cation binding sites of the protein.

3.2.3   Polyelectrolytes: counterion relaxation and diffusion studies

Polyelectrolyte dynamics have two main aspects: the motion of the organic moieties and the mobility of counterions near the polymer backbone. The first topic is covered in another chapter dedicated to polymers and biopolymers. Here, we summarize the main NMR results arising from observation of inorganic ion nuclei.

In polyelectrolyte solutions, the counterion relaxation rates R_1 and R_2 are larger than their values for simple electrolyte (salt) solutions. These NMR data have to be analysed by assuming fast exchange between solvated ions and counterions under the direct influence of the polyion (equation (5.22). For flexible polyelectrolytes, at low concentration, the extreme narrowing condition $R_1 = R_2$ fails to apply. This corresponds to slow motion of counterions. Different and conflicting models have been proposed to explain this observation. Ions can be tightly bound to the polymer [111] and they reorient at the rate of the segmental motion of the polymer.

Alternatively, escape of counterions from a polyion to the neighbouring macroion has also been considered as the origin of the slow process involved in quadrupolar relaxation [112]. Yet another mechanism has been suggested in ^{23}Na NMR studies of DNA: a local counterion motion correlated with hydration water of DNA (subnanosecond) and a slower motion associated with the mobility of ions in the vicinity of the macromolecule (in the timescale of several nanoseconds) [113]. A theoretical model supports these conclusions [114] and gives no support to the two other models [111, 112]. Furthermore, with flexible polyelectrolytes in the semi-dilute regime, the ion relaxation is correlated with the persistence length of the polyion [115]. Therefore, the mobility of condensed cations seems to be associated with the segmental motion of the charged polymer. A relaxation process involving an ion transfer from one polyion to another [112] appears to have been ruled out. Although it does not discard the site-binding model, the generality of the above conclusion for different polyelectrolytes does not favour it. Using ^7Li relaxation data together with neutron scattering results, it has been deduced that the lithium ion remains fully hydrated in polyacrylate solutions [116]. This result agrees with the concept of condensed ions. Site bindings for divalent cations on DNA appear to coexist with condensed ions as indicated by ^{25}Mg[117] and ^{43}Ca[118] NMR studies. In the above systems, the mole fraction p_B (equation (5.22)) of ions neutralizing the polyion are accounted for by the two-state condensation theory of Manning [119] (at least semi-quantitatively) and quantitatively by the Poisson–Boltzmann approach [120]. These two approaches, semi-empirical and theoretical, also reproduce the ^{23}Na relaxation rates for charged polysaccharides such as sodium heparinate [121] and sodium polyuronates [122].

Pulsed field gradient NMR techniques provide a means to measure self-diffusion coefficients of counterions. As a short relaxation time precludes its determination with the basic sequence (equation (5.25)), nuclei with a small quadrupolar moment have been mostly used. Similar self-diffusion coefficients have been found for ^7Li and ^{133}Cs cations in poly(acrylic acid) and poly-(methacrylic acid) at different degrees of neutralization [123]. This is consistent with non-specific electrostatic forces between inorganic ions and the polyion. ^7Li and ^9Be self-diffusion coefficients have been measured in lithium poly(styrenesulphonate) solutions with different contents of added beryllium sulphate [124]. Calculated values based upon two different models [125] do not agree with the experimental data and possible explanations have been put forward [124]. This technique yields as a diffusion constant a weighted average over space, including cation diffusion in the bulk liquid. Another diffusion coefficient can be obtained without pulsed field gradient accessories. Divalent cations interact more strongly with the polyion than alkali ions. Addition of a small amount of manganese(II) chloride to a lithium poly(acrylic acid) solution induces fluctuating dipolar interactions between these two cations. The average diffusion constant is mainly dominated by the fraction of monovalent cation near the polymer chain. Both methods have been used to characterize the lithium diffusion in poly(acrylic acid) solutions [126]. At the highest degrees of neutralization, diffusion of lithium ions near the chain can be 10 times slower than the overall diffusion. Immobilization on the chain is not observed, however [126].

In crosslinked poly(styrene sulphonate) systems, constraints on the polymer chain motion render the averaging of the counterion chain interaction less efficient [127]. The ^{7}Li and ^{23}Na relaxations are considerably enhanced and data at different magnetic fields reveal two processes outside of the extreme narrowing limit, contrary to the behaviour for non-crosslinked poly(styrene sulphonate). ^{7}Li and ^{23}Na self-diffusion coefficients allow calculation of the distance that an ion diffuses during the correlation time, thus characterizing these two processes (equation (5.24)). The smallest calculated distance is roughly equal to the length between two neighbouring crosslinks. The second distance extends over several crosslinks and reflects the crosslink concentration [127].

3.3   NMR AS A KINETIC METHOD: ION TRANSPORT IN MEMBRANES

NMR of quadrupolar nuclei has been extensively used to study cation transport across model or biological membranes. The observation of distinct cationic signals for internal and external compartments is a prerequisite. As usually intra and extracellular cation signals share similar chemical shifts, different approaches have been used. A membrane-impermeable paramagnetic agent added in one compartment interacts only with ions in this compartment. Depending on the added substance, it induces a shift of the relevant signal and/or a relaxation enhancement of the monitored ions. Magnetization transfer experiments or pulsed field gradient NMR techniques have been shown to be suitable for determining ion diffusion in biological systems. A last approach is based upon multiple quantum filtering techniques which eliminate signals characterized by relaxation processes within the extreme narrowing condition.

3.3.1   *Signal separation in the presence of paramagnetic compounds*

Inner and outer cationic resonances may be differentiated by paramagnetic addends [128, 129]. These membrane impermeable shift reagents are highly anionic. Salts of dysprosium triphosphate [128] are probably the most frequently used for this purpose. The presence of a small amount (a few mM) of the paramagnetic salt outside the cell or the vesicle induces an appropriate shift to separate the outer and the inner (unaffected) cation signals.

Ionophore-mediated cation transport in phospholipid vesicles has been studied by this NMR approach. The transport mechanism can be schematically described as a three-step process: cation complexation by the ionophore (natural polyether, see Section 3.1.2) at a membrane interface, diffusion of the formed complex through the membrane and dissociation of the complex at the other interface. Each of these steps is reversible. Addition of an ionophore to the lipidic suspension containing such a paramagnetic probe broadens both the inner and outer cationic signals, as a result of cation exchange between the two compartments. From the line-broadening, the cation translocation rate is obtained [130]. For the complexant studied, the overall rate is first order in ionophore concentration as determined by ^{23}Na or ^{39}K NMR [130–132]. Data analysis shows that the rate determining step is dissociation of the complex, which of course depends on the nature of the cation [130–132]. ^{23}Na and ^{39}K NMR spectra of cell suspensions, tissues or perfused organs in the presence of a paramagnetic shift reagent exhibit two well-separated lines. Variation of the line intensities gives access to the cationic transport rate. Numerous studies have been performed by this means and they have been adequately reviewed [13, 133, 134]. Together with the line intensity, the chemical shift of the ^{23}Na outer signal can be used to determine efflux of other cations, as it has been shown for erythrocytes suspensions [135].

3.3.2 Magnetization transfer and pulsed field gradient techniques

As pointed out in the theoretical section, non-stationary methods can be adopted to determine rate constant (Section 2.1). Magnetization transfer experiments have also been performed to study the ionophore-mediated cation transport through lipidic membranes [136, 137]. Selective excitation of the ^{7}Li [138] or ^{23}Na [139] external signal in the presence of an ionophore or in its absence allows the determination of the cation exchange rate. This method is useful when the relaxation rate is of the same order of magnitude as the chemical exchange rate.

Cation transport rate in such systems can also be determined by pulsed field gradient experiments. Such data have been reported recently [140]. A good agreement has been observed between these results and those obtained by magnetization transfer experiments [140].

3.3.3 Selection of the appropriate signal by NMR techniques

To avoid the presence of paramagnetic species multiple-quantum coherence filter techniques have been designed to select the internal ^{23}Na or ^{39}K signal. As the extreme narrowing condition is not fulfilled for these internal ions biexponential relaxation obtains. Use of a multiple-quantum coherence filter rejects the signal from the outer monoexponentially relaxing ions [26, 27, 141]. Quantitative analysis, however, requires that the filtered signal arises only from intracellular ions. In fact a fraction of the extracellular sodium ions, also shows a biexponential relaxation, probably due to interactions with cell surface and with plasma proteins [142, 143]. The addition of a paramagnetic anionic relaxation agent such as gadolinium triphosphate selectively eliminates the extracellular double-quantum coherence, providing a means for quantitative measurements of the intracellular signal [142]. Potassium ions are mainly present in the inside of cells and double quantum coherence techniques have been also employed to eliminate the narrower outer signal [144]. However, a decrease in the NMR sensitivity results from the use of this approach based upon multiple quantum coherence. Therefore its applicability to study dynamic processes seems to be limited to highly sensitive nuclei such as ^{23}Na.

4　Conclusions

Quadrupolar nuclei have not lagged behind in benefiting from some of the newer NMR techniques. Two-dimensional experiments and multiple quantum filtering have become useful auxiliaries, the latter for spectral simplification. Two-D methods have been useful to studies of chemical exchange within organolithium clusters or between vanadate oligomers. However, recording routine spectra is sufficient in many cases for reaping a wealth of information about dynamic processes, such as fluxional properties of organic and organometallic systems, conformational changes in proteins, transport through membranes, microdynamics at the interface of an heterogeneous catalyst, and so on.

The observed quadrupolar splitting, related to the mean molecular orientation, provides the investigator with a handy parameter to characterize ordered systems with. It reveals precise details of the ordering and of the averaging motions even for small populations, when a tiny fraction of the molecules only becomes oriented.

We do not want to give the impression that every single aspect, in dynamic applications of the NMR of quadrupolar nuclei, is easy. Analysis of relaxation data can be very complex for nuclei with spin quantum numbers greater than 3/2. Some 2D experiments have been proposed nevertheless for access to the individual spectral densities for each transition. In non-oriented macro-molecular samples, multiexponential relaxation prevents applicability of explicit analytical equations, except when $\omega\tau_c \leqslant 1.5$. Such a value is typically that for a 20-kD protein. Multiple quantum filtering will simplify the signal as mentioned. A rather fortunate circumstance also is the dependence of the $-1/2 \rightarrow +1/2$ transition on the correlation time (Figure 5.1). A flat minimum near $5 \ 10^{-6}$ results in metalloproteins with molecular weights in the 60–100 kD range displaying quite narrow resonances. Studies of the dynamics of such important biological objects can be safely predicted in the near future.

Technological improvements will also be brought to bear on use of self-diffusion coefficients for other nuclei than honorary dipolar nuclei graced with vanishingly small quadrupole moments.

Thus we can safely predict increasing use of quadrupolar nuclei to monitor with both accuracy and sophistication the local motions within a wide variety of chemical and biochemical systems.

References

1. R. G. Kidd, *Annu. Rep. NMR Spectrosc.*, **23**, 85–139 (1991).
2. R. K. Harris and B. E. Mann, *NMR and the Periodic Table*, Academic Press, London (1978).
3. C. Brevard and P. Granger, *Handbook of High Resolution Multinuclear NMR*, John Wiley & Sons, New York (1981).
4. P. Laszlo (ed.), *NMR of Newly Accessible Nuclei*, Vols 1 and 2, Academic Press, London (1983).
5. B. Lindman and S. Forsén, in P. Diehl, E. Fluck and R. Kosfeld (eds), *NMR Basic Principles and Progress*, Vol. 12, Springer-Verlag, Heidelberg (1976).
6. P. Laszlo, *Angew. Chem. Int. Ed. Engl.*, **17**, 254–266 (1978).
7. J. W. Akitt, *Progr. NMR Spectrosc.*, **21**, 1–149 (1989).
8. O. W. Howarth, *Progr. NMR Spectrosc.*, **22**, 453–485 (1990).
9. S. Forsén, T. Drakenberg and H. Wennerström, *Quarter. Rev. Biophys.*, **19**, 83–114 (1987).
10. C. Johansson and T. Drakenberg, *Annu. Rep. NMR Spectrosc.* **22**, 1–59 (1990).
11. C. Detellier, H. P. Graves and K. M. Brière, in E. Buncel and J. R. Jones (eds), *Isotopes in the Physical and Biomedical Science*, Vol. 2, 159–211, Elsevier Science Publishers, Amsterdam (1991).
12. J. Seelig and P. M. MacDonald, *Acc. Chem. Res.*, **20**, 221–228 (1987).
13. J. C. Veniero and R. K. Gupta, *Annu. Rep. NMR Spectrosc.* **24**, 219–265 (1992).
14. J. Grandjean, *Annu. Rep. NMR Spectrosc.*, **24**, 181–217 (1992).
15. J. A. Tossel, *Nuclear Magnetic Shieldings and Molecular Structure*, Kluwer, Dordrecht (1993).
16. G. Binsch, *Top. Stereochem.*, **3**, 97 (1968).
17. R. A. Hoffman and S. Forsén, *Prog. NMR Spectrosc.*, **1**, 15–204 (1966).
18. C. L. Perrin and T. J. Dwyer, *Chem. Rev.*, **90**, 935–967 (1990).
19. B. Halle and H. Wennerström, *J. Chem. Phys.*, **75**, 1928 (1981).
20. A. Saupe, *Z. Naturforsch.*, **19a**, 161 (1964).
21. M. H. Cohen and F. Reif, *Solid State Phys.*, **5**, 321 (1957).
22. M. Holz, *Progr. NMR Spectrosc.*, **18**, 327 (1986).
23. H. G. Hertz, *Ber. Bunsenges. Phys. Chem.*, **77**, 531 (1973).
24. J. E. Roberts and J. Schnitker, *J. Phys. Chem.*, **97**, 5410 (1993).
25. R. Abragam, *The Principles of Nuclear Magnetism*, Ch. VIII, Clarendon, Oxford (1970).
26. G. Jaccard, S. Wimperis and G. Bodenhausen, *J. Chem. Phys.*, **85**, 6282 (1986).
27. C. Chung and S. Wimperis, *Chem. Phys. Lett.*, **172**, 94 (1990).
28. J. R. C. van der Maarel, *J. Chem. Phys.*, **94**, 4765 (1991).
29. G. Kontaxis, H. Sterk and J. Kalcher, *J. Chem. Phys.*, **95**, 7854 (1991).
30. P. S. Hubbard, *J. Chem. Phys.*, **53**, 985 (1970).
31. W. D. Rooney, T. M. Barbara and C. S. Springer, Jr., *J. Am. Chem. Soc.*, **110**, 674 (1988).
32. T. E. Bull, *J. Magn. Reson.*, **8**, 344 (1972).
33. B. Halle and H. Wennerström, *J. Magn. Reson.*, **44**, 89 (1981).
34. T. E. Bull, S. Forsén and D. L. Turner, *J. Chem. Phys.*, **70**, 3106 (1979).
35. A. Butler and H. Eckert, *J. Am. Chem. Soc.*, **111**, 2802 (1989).
36. J. M. Aramini and H. J. Vogel, *J. Am. Chem. Soc.*, **115**, 245 (1993).
37. L. G. Werbelow and A. G. Marshall, *J. Magn. Reson.*, **43**, 443 (1981).
38. P.-O. Westlund and H. Wennerström, *J. Magn. Reson.*, **50**, 451 (1982).
39. W. S. Price, N.-H. Ge and L.-P. Hwang, *J. Magn. Reson.*, **98**, 134 (1992).
40. J. P. Jakobsen, H. K. Bildsoe and K. Schaumburg, *J. Magn. Reson.*, **23**, 153 (1976).

41. R. R. Vold and R. L. Vold, *J. Chem. Phys.*, **66**, 4018 (1977).
42. J. R. C. van der Maarel, *Chem. Phys. Lett.*, **155**, 288 (1989).
43. I. Furo, B. Halle and T. C. Wong, *J. Chem. Phys.*, **89**, 5382 (1988).
44. I. Furo and B. Halle, *J. Magn.. Reson.*, **98**, 388 (1992).
45. I. Furo and B. Halle, *J. Chem. Phys.*, **91**, 42 (1989).
46. E. Schori, J. Jagur-Grodzinski, Z. Luz and M. Shporer, *J. Am. Chem. Soc.*, **93**, 7133 (1971).
47. T. Drakenberg, S. Forsén and H. Lilja, *J. Magn. Reson.*, **53**, 412 (1983).
48. K. M. Brière and C. Detellier, *J. Phys. Chem.*, **91**, 6097 (1987).
49. K. Müller, R. Poupko and Z. Luz, *J. Magn. Reson.*, **90**, 19 (1990).
50. E. O. Stejskal and J. E. Tanner, *J. Chem. Phys.*, **42**, 288 (1965).
51. J. Kärger, H. Pfeifer and W. Heink, *Adv. Magn. Reson.*, **12**, 1 (1988).
52. P. Stilbs, *Prog. NMR Spectrosc.*, **19**, 1–45 (1987).
53. W. Bauer, M. Feigel, G. Müller and P. v. R. Schleyer, *J. Am. Chem. Soc.*, **110**, 6033 (1988).
54. L. A. Paquette, W. Bauer, M. R. Sivik, M. Bühl, M. Feigel and P. v. R. Schleyer, *J. Am. Chem. Soc.*, **112**, 8776 (1990).
55. G. Fraenkel, A. Chow and W. R. Winchester, *J. Am. Chem. Soc.*, **112**, 6190 (1990).
56. H. Günther, D. Moskau and D. Smak, *Angew. Chem. Int. Ed. Engl.*, **26**, 1212 (1987).
57. J. Arnold, *J. C. S. Chem. Commun.*, **976** (1990).
58. D. C. Crans, C. D Rithner and L. A. Theisen, *J. Am. Chem. Soc.*, **112**, 2901 (1990).
59. D. C. Crans, S. M. Schelble and L. A. Theisen, *J. Org. Chem.*, **56**, 1266 (1991).
60. C. Detellier and P. Laszlo, *Proc. Indian Acad. Sci. (Chem. Sci.)*, **94**, 291–336 (1985).
61. J. D. Lin and A. I. Popov, *J. Am. Chem. Soc.*, **103**, 3773 (1981).
62. E. Mei, J. L. Dye and A. I. Popov, *J. Am. Chem. Soc.*, **99**, 5308 (1977).
63. S. Khazaeli, J. L. Dye and A. I. Popov, *J. Phys. Chem.*, **87**, 1830 (1983).
64. J. Bouquant, A. Delville, J. Grandjean and P. Laszlo, *J. Am. Chem. Soc.*, **104**, 686 (1982).
65. J. Grandjean and P. Laszlo, *Angew. Chem. Int. Ed. Engl.*, **18**, 153 (1979).
66. J. Grandjean and P. Laszlo, *Tetrahedron Letters*, **24**, 3319 (1983).
67. U. Olsher, G. A. Elgavish and J. Jagur-Grodzinski, *J. Am. Chem. Soc.*, **102**, 3338 (1980).
68. J. Grandjean, P. Laszlo, W. Offermann and P. L. Rinaldi, *J. Am. Chem. Soc.*, **103**, 1380 (1981).
69. R. C. Phillips, S. Khaezaeli and J. L. Dye, *J. Phys. Chem.*, **89**, 606 (1985).
70. A. S. Ellaboudy, N. C. Pyper and P. P. Edwards, *J. Am. Chem. Soc.*, **110**, 1618 (1988).
71. Y. M. Cahen, J. L. Dye and A. I. Popov, *J. Phys. Chem.*, **79**, 1292 (1975).
72. M. Shamsipur and A. I. Popov, *J. Phys. Chem.*, **90**, 5997 (1986).
73. J. M. Cesaro, P. B. Smith, J. S. Landers and J. L. Dye, *J. Phys. Chem.*, **81**, 760 (1977).
74. P. Szczygiel, M. Shamsipur, K. Hallenga and A. I. Popov, *J. Phys. Chem.*, **91**, 1252 (1987).
75. B. O. Strasser, K. Hallenga and A. I. Popov, *J. Am. Chem. Soc.*, **107**, 789 (1985).
76. M. Shamsipur and A. I. Popov, *J. Phys. Chem.*, **92**, 147 (1988).
77. B. O. Strasser, M. Shamsipur and A. I. Popov, *J. Phys. Chem.*, **89**, 4822 (1985).
78. H. Degani, *Biophys. Chem.*, **6**, 345 (1977).
79. M. Shporer and Z. Luz, *J. Am. Chem. Soc.*, **97**, 665 (1975).

80. E. Schmidt and A. I. Popov, *J. Am. Chem. Soc.*, **105**, 1873 (1983).
81. A. Delville, H. D. H. Stöver and C. Detelier, *J. Am. Chem. Soc.*, **109**, 7293 (1987).
82. A. Teleman, T. Drakenberg and S. Forsén, *Biochim. Biophys. Acta*, **873**, 204 (1986).
83. H. J. Vogel, T. Andersson, W. M. Braunlin, T. Drakenberg and S. Forsén, *Biochem. Biophys. Res. Commun.*, **122**, 1350 (1984).
84. S. Linse, P. Brodin, T. Drakenberg, E. Thulin, P. Sellers, K. Elmden, T. Grundström and S. Forsén, *Biochem.*, **26**, 6723 (1987).
85. J. Grandjean and P. Laszlo, *J. Magn. Reson.*, **83**, 128 (1989).
86. J. Grandjean and P. Laszlo, *Clays Clay Minerals*, **37**, 403 (1989).
87. A. Delville, J. Grandjean and P. Laszlo, *J. Phys. Chem.*, **95**, 1393 (1991).
88. J. Grandjean and P. Laszlo, *Magn. Reson. Imag.*, **12**, 375, (1994).
89. J. Grandjean and P. Laszlo, *Clays Clay Minerals*, **42**, (1994).
90. D. Petit, J.-P. Korb, A. Delville, J. Grandjean and P. Laszlo, *J. Magn. Reson.*, **96**, 252 (1992).
91. R. P. W. J. Struis, J. De Bleijser and J. C. Leyte, *J. Phys. Chem.*, **91**, 1639 (1987).
92. R. Eggenberger, S. Gerber, H. Huber, D. Searles and M. Welker, *J. Chem. Phys.*, **97**, 5898 (1992).
93. D. Lankhorst, J. Schriever and J. C. Leyte, *Ber. Bunsenges. Phys. Chem.*, **86**, 215 (1982).
94. C. W. R. Mulder, J. Schriever and J. C. Leyte, *J. Phys. Chem.*, **87**, 2336 (1983).
95. C. W. R. Mulder, J. Schriever, W. J. Jesse and J. C. Leyte, *J. Phys. Chem.*, **87**, 2342 (1983).
96. J. R. C. van der Maarel, D. Lankhorst, J. de Bleijser and J. C. Leyte, *Macromolecules*, **20**, 2390 (1987).
97. L. Piculell, *J. Chem. Soc. Faradas Trans. I*, **82**, 387 (1986).
98. L. Piculell and B. Halle, *J. Chem. Soc. Faradav Trans. I*, **82**, 401 (1986).
99. B. Halle and L. Piculell, *J. Chem. Soc. Faraday Trans. I*, **82**, 415 (1986).
100. F. Noack, *Progr. NMR Spectrosc.*, **18**, 171–276 (1986).
101. G. Schauer, R. Kimmich and W. Nusser, *Biophys. J.*, **53**, 397 (1988).
102. L. T. Kakalis and T. F. Kumosinski, *Biophys. J.*, **43**, 39 (1992).
103. J. Grandjean and P. Laszlo, *C.R. Acad. Sc. Paris*, **294**, 1099 (1982).
104. J. Grandjean, P. Laszlo and C. Gerday, *FEBS Lett.*, **81**, 376 (1977).
105. T. Andersson, T. Drakenberg, S. Forsén, E. Thulin and M. Sward, *J. Am. Chem. Soc.*, **104**, 576 (1982).
106. A. Delville, J. Grandjean, P. Laszlo, C. Gerday, Z. Grabarek and W. Drabikowski, *Eur. J. Biochem.*, **105**, 289 (1980).
107. J. Grandjean and P. Laszlo, unpublished results.
108. A. Delville, J. Grandjean, P. Laszlo, C. Gerday, H. Brzeska and W. Drabikowski, *Eur. J. Biochem.*, **109**, 515 (1980).
109. J. M. Aramini, T. Drakenberg, T. Hiraoki, Y. Ke, K. Nitta and H. Vogel, *Biochem.*, **31**, 6761 (1992).
110. T. Andersson, E. Chiancone and S. Forsén, *Eur. J. Biochem.*, **125**, 103 (1982).
111. A. Delville, C. Detellier and P. Laszlo, *J. Magn. Reson.*, **34**, 301 (1979).
112. B. Halle, H. Wennerström and L. Piculell, *J. Phys. Chem.*, **88**, 2482 (1984).
113. L. Van Dijk, M. L. H. Gruwel, W. Jesse, J. de Bleijser and J. C. Leyte, *Biopolymers*, **26**, 261 (1987).
114. M. R. Reddy, P. J. Rossky and C. S. Murthy, *J. Phys. Chem.*, **91**, 4923 (1987).
115. C. J. M. van Rijn, A. J. Maat, J. de Bleijser and J. C. Leyte, *J. Phys. Chem.*, **93**, 5284 (1989).

116. J. R. C. van der Maarel, D. H. Powell, A. K. Jawahier, L. H. Leyte-Zuiderweg, G. N. Neilson and M. C. Bellissent-Funel, *J. Chem. Phys.*, **90**, 6709 (1989).
117. D. Murk Rose, C. F. Polnaszek and R. G. Bryant, *Biopolymers*, **21**, 653 (1982).
118. W. H. Braunlin, T. Drakenberg and L. Nordenskiöld, *Biopolymers*, **26**, 1047 (1987).
119. G. S. Manning, *Acc. Chem. Res.*, **12**, 443–449 (1979).
120. C. F. Anderson and M. T. Record Jr, *Ann. Rev. Biophys. Biochem.*, **19**, 423–465 (1990).
121. A. Delville and P. Laszlo, *Biophys. Chem.*, **17** 119 (1983).
122. H. Grasdalen and B. J. Kwam, *Macromolecules*, **19**, 1913 (1986).
123. R. Rymden and P. Stilbs, *J. Phys Chem.*, **89**, 2425 (1985).
124. L. G. Nilsson, L. Nordenskiöld and P. Stilbs, *J. Phys. Chem.*, **91**, 6210 (1987).
125. L. G. Nilsson, L. Nordenskiöld, P. Stilbs and W. H. Braunlin, *J. Phys. Chem.*, **89**, 3385 (1985).
126. R. H. Tromp, J. de Bleijser and J. C. Leyte, *J. Phys. Chem.*, **93**, 2626 (1989).
127. R. H. Tromp, J. R. C. van der Maarel, J. de Bleijser and J. C. Leyte, *Biophys. Chem.*, **41**, 81 (1991).
128. R. K. Gupta and P. Gupta, *J. Magn. Reson.*, **47**, 344 (1982).
129. J. A. Balschi, V. P. Cirillo and C. S. Springer Jr, *Biophys. J.*, **38**, 323 (1982).
130. F. G. Riddell and M. K. Hayer, *Biochim. Biophys. Acta*, **817**, 313 (1985).
131. F. G. Riddell, S. Arumugan, P. J. Brophy, B. G. Cox, M. C. H. Payne and T. E. Southon, *J. Am. Chem. Soc.*, **110**, 734 (1988).
132. F. G. Riddell and S. Arumugam, *Biochim. Biophys. Acta*, **984**, 6 (1989).
133. R. K. Gupta, P. Gupta and R. D. Moore, *Ann. Rev. Biophys. Bioeng.*, **13**, 221–246 (1984).
134. C. S. Springer Jr, *Ann. Rev. Biophys. Bioeng.*, **16**, 375–399 (1987).
135. E. Fernandez, J. Grandjean and P. Laszlo, *Eur. J. Biochem.*, **167**, 353 (1987).
136. G. A. Morris and R. Freeman, *J. Magn. Reson.*, **29**, 433 (1978).
137. J. Jeener, R. A. Meier, P. Bachmann and R. R. Ernst, *J. Chem. Phys.*, **71**, 4546 (1979).
138. F. G. Riddell and S. Arumugam, *Biochim. Biophys. Acta*, **945**, 65 (1988).
139. D. C. Shungu, D. C. Buster and R. W. Briggs, *J. Magn. Reson.*, **89**, 102 (1990).
140. A. R. Waldeck, A. J. Lennon, B. E. Chapman and P. W. Kuchel, *J. Chem. Soc. Faraday Trans.*, **89**, 2807 (1993).
141. J. Pekar, P. F. Renshaw and J. S. Leigh, *J. Magn. Reson.*, **72**, 159 (1987).
142. L. A. Jelicks and R. K. Gupta, J. Magn. Reson., **83**, 146 (1989).
143. R. B. Hutchinson, D. Malhotra, R. E. Hendrick, L. Chan and J. I. Shapiro, *J. Biol. Chem.*, **265**, 15506 (1990).
144. Y. Seo, M. Murakami, E. Suzuki, S. Kuki, K. Nagayama and H. Watari, *Biochemistry*, **29**, 599 (1990).
145. J. Thorez, X-ray Studies, in *Preparative Chemistry using Supported Reagents*, (ed. P. Laszlo) Academic Press, London, 1987, Chapter 11.

6 SOLVENT EXCHANGE ON METAL IONS: A VARIABLE PRESSURE NMR APPROACH

U. Frey, A. E. Merbach and D. H. Powell
University of Lausanne, Lausanne, Switzerland

Dynamics of Solutions and Fluid Mixtures by NMR
Edited by J.-J. Delpuech © 1995 John Wiley & Sons Ltd

1   Introduction

A general equation for substitution reactions on a metal ion M may be written

$$L_nMX + Y \rightarrow L_nMY + X \qquad (6.1)$$

In order to elucidate the mechanisms of such reactions, it is usual to study the dependence of the reaction rate on a number of parameters, both chemical (reactant concentrations, pH, ionic strength, solvent composition) and physical (normally temperature). The empirical rate law, the electronic and steric effects induced by variations of leaving X, entering Y, and non-reacting L ligands, the variable temperature activation parameters (activation enthalpy, $\Delta H^{\#}$ and activation entropy, $\Delta S^{\#}$) and any other experimental or theoretical information available on the system are then used to assign a reaction mechanism. Such information may, however, be insufficient to distinguish clearly between alternative reaction mechanisms. This is especially so for solvent exchange reactions ($X = Y$ in equation (6.1)), an important class of substitution reactions in inorganic chemistry, where one cannot vary the concentrations of the reactants (unless one works with an 'inert' diluent) so that the mechanistic assignment must often be made on the basis of activation parameters alone.

Temperature is, however, not the only physical variable which can be varied in kinetic studies. Pressure can also affect the rate of a substitution reaction in solution, and variable pressure measurements yield an activation volume, $\Delta V^{\#}$, which is now widely used as a supplementary (and often decisive) parameter for mechanistic assignment [1]. Although the first high pressure kinetic study in inorganic chemistry was made over thirty years ago [2], the rapid development of the field has occurred in the last twenty years, since the adaptation for variable pressure studies of most fast reaction techniques: stopped-flow [3], temperature-jump [4], pressure-jump [5] and NMR [6].

The pressure dependence of the rate of a chemical reaction is usually described by transition state theory. The activation volume, the difference between the partial molar volumes of the transition state and the reactants, at a temperature T, is defined by

$$(\partial \ln k/\partial P)_T = -\Delta V^{\#}/RT \qquad (6.2)$$

where k is the rate constant. The activation volume may itself depend on pressure, and it is useful to define the compressibility coefficient of activation, $\Delta \beta^{\#}$, as

$$\Delta \beta^{\#} = -(\partial \Delta V^{\#}/\partial P)_T \qquad (6.3)$$

The problem of finding an equation that describes correctly the pressure dependence of k has been dealt with by various researchers [7]. It is usual to assume that $(\partial^2 \Delta V^{\#}/\partial P^2)_T = 0$, so that $ln\ k$ has the quadratic form

$$\ln k = \ln(k)_0 - P\Delta V_0^{\#}/RT + P^2\Delta \beta^{\#}/2RT \qquad (6.4)$$

where $\Delta V_0^{\#}$ is the activation volume at zero pressure. It should be stressed, however, that there is no physical justification for such a quadratic form [8].

$\Delta V^{\#}$ may be positive or negative, bond stretching in a simple dissociative step giving rise to an increase in volume and bond formation in a simple associative process leading to a decrease in volume. For substitution reactions in general, the solvent electrostrictive effect due to the formation of ions or dipoles at the transition state also contributes to ΔV. The measured activation volume is usually considered to be the sum of an intrinsic contribution, $\Delta V_{int}^{\#}$, owing to changes in internuclear distances within the reactants on formation of the transition state and a solvent electrostrictive contribution, $\Delta V_{elec}^{\#}$. For substitution reactions involving charged species, $\Delta V_{int}^{\#}$ may be dominated by $\Delta V_{elec}^{\#}$, so that the sign of $\Delta V_{int}^{\#}$ may be different from $\Delta V_{int}^{\#}$. For solvent exchange reactions, however, there is no significant charge or dipole formation on going to the transition state, so that $\Delta V_{elec}^{\#}$ may be neglected. The sign of $\Delta V^{\#} \sim \Delta V_{int}^{\#}$ is then an immediate diagnostic of the activation step: dominant bond breaking results in positive $\Delta V_{int}^{\#}$ while dominant bond making gives negative $\Delta V_{int}^{\#}$.

The classification of substitution reaction mechanisms originally proposed by Langford and Gray [9], and widely used in coordination chemistry, is based on operational criteria. If kinetic tests show the presence of an intermediate of increased coordination number, an associative mechanism, A, is assigned. Conversely, if an intermediate of reduced coordination number is detected, a dissociative, D, mechanism is assigned. When no intermediate can be found, a concerted interchange, I, mechanism is assigned. This last category may be subdivided further, depending whether evidence is found for important incoming group influences (I_a) or not (I_d). Swaddle [10] has argued that this operational definition is too restrictive, and may be extended to include activation parameters as operational criteria. This is especially true for solvent exchange reactions, where few kinetic tests are applicable. The activation volume is of primary importance in this context. Microscopic reversibility for solvent exchange reactions means that the forward reaction path must be symmetrical to the reverse one Hence, for limiting A or D mechanisms, with a reactive intermediate, the two transition states along the reaction coordinate must have identical energies. Similarly for an interchange process, with no true minimum along the reaction coordinate, the bond lengths to the entering and leaving solvent molecules must be equivalent at the transition state. Thus for an I_a mechanism, the incoming and outgoing solvent molecules will both have relatively strong bonds, whereas for an I_d mechanism they will have relatively weak bonds. The difference between the two mechanisms will be manifested as a contraction or expansion at the transition state. If the bond distances between the central ion and non-exchanging solvent molecules are unaltered, $\Delta V^{\#}$ will be a direct measure of the degree of associativity or dissociativity at the transition state. A continuous spectrum of transition states for solvent exchange reactions can thus be envisaged (Figure 6.1), ranging from limiting associative,

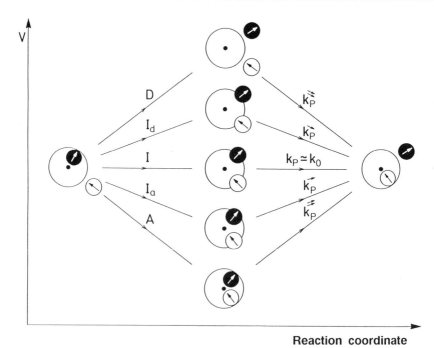

V

Reaction coordinate

Figure 6.1 Volume profiles for solvent exchange mechanisms

A, with $\Delta V^{\#} = -|\Delta V_{\text{lim}}^{\#}|$, via I_{a} and I_{d} to limiting dissociative, D, with $\Delta V^{\#} = +|\Delta V_{\text{lim}}^{\#}|$. For interchange mechanisms characterized by $\Delta V^{\#} \sim 0$, the activation mode cannot be specified, and no subscript is assigned to the I mechanism.

We have established the primary importance of $\Delta V^{\#}$, and hence of variable pressure kinetic measurements, for the classification of solvent exchange reaction mechanisms. For such measurements, the favoured technique is NMR, since the other commonly used kinetic techniques do not lend themselves to the study of symmetric reactions. In Section 2, therefore, we will discuss the different approaches used to perform NMR measurements on liquids at high pressures. The rates of solvent exchange mechanisms cover an enormous range of timescales: for example, water exchange rates at 298 K on metal cations vary from 10^{-8} s^{-1} for Rh^{3+} to almost 10^{10} s^{-1} (Figure 6.2). A large battery of NMR techniques must be called on, therefore, to study these reactions. In Section 3, we will describe these different techniques with the aid of examples drawn from our work in Lausanne. Finally, in Section 4, we will give a brief review of the work published in the field, classified according to the different types of solvated metal ion complexes, and concentrating on those areas where there has been recent progress. Readers interested in the use of variable pressure NMR in the study of chemical kinetics in general should refer to one of the more extensive reviews on this subject [11, 12].

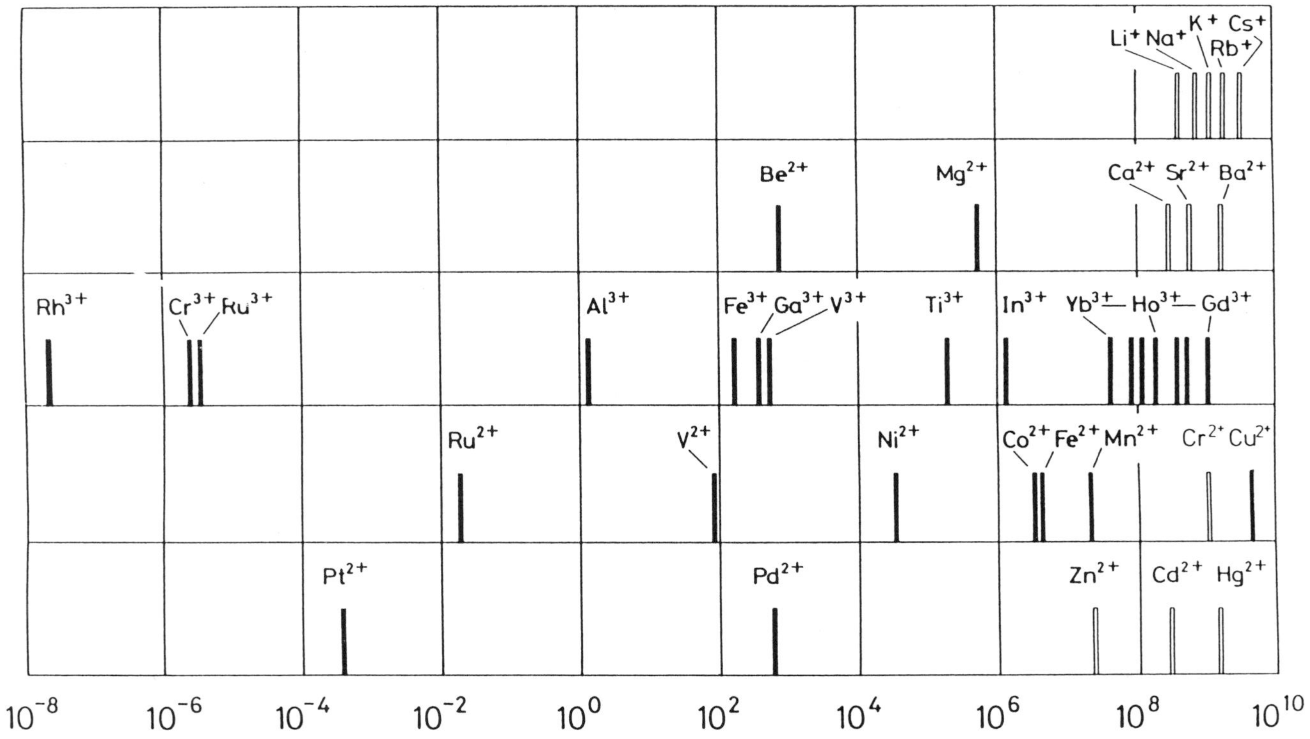

Figure 6.2 Rate constants at 298 K, k_{ex}^{298} for water exchange on metal cations measured by NMR (solid bars), except for Cr^{3+}, or derived from complex formation reactions data (open bars)

2 High-pressure NMR Instrumentation and Methodology

Two approaches have emerged for the measurement of NMR spectra at high pressure [6, 13]. The simplest approach is to design a pressurizable thick-walled glass sample tube which can be placed directly in a standard spectrometer probe. Such systems work well and sample spinning can be used, but there are certain disadvantages. If small pressure dependences (e.g. activation volumes) are to be measured with any precision, pressures of at least 1 kbar are required. This can be achieved with very thick-walled tubes where the sample is confined to a capillary, but the poor filling factor of these tubes results in low sensitivity. Furthermore, the temperature control of such tubes is not very good, and it cannot be emphasized too strongly that temperature has to be known as accurately as possible if the results of kinetic measurements are to be meaningful.

The alternative approach is to manufacture a dedicated high pressure probe-head which will replace the ambient pressure commercial probe-head. The probe is built around a high pressure chamber provided with electrical lead-through contacts so that both coil and sample are immersed in the high pressure transmitting fluid. The coil can be wound on a support placed closely around the sample tube and so have a high filling factor. The pressure fluid is also a good heat transfer medium and its heat capacity and that of the metallic bomb are such that temperature control can be more precise than is possible with the normal gas flow type unit. The sample cannot be made to spin, or at least there are practical difficulties which preclude this, but using field lock and small diameter tubes (5 mm), good resolution can be obtained from the static sample. Signal-to-noise ratio does not suffer too much from this approach because of the good filling factor and cryomagnets are capable of producing a remarkable magnetic field homogeneity over small volumes. In practice the static, specially built system has proved superior to the spinning pressure tube in all the factors outlined above, and especially for chemical kinetic studies.

For illustration we present a probe designed for a high field spectrometer (Bruker AM400) [14]. Figure 6.3 is a schematic drawing of a probe-head designed for a widebore (87 mm diameter) superconducting magnet (9.4 T). It is made of two aluminium supports. The lower one, Figure 6.3(a), contains the pressure bomb itself, which can be used up to 200 MPa and is made of a non-magnetic beryllium copper alloy. Three connectors are located at the bottom: one for the pressure transmitting liquid, and two for the inlet and outlet of the thermostating liquid. Figure 6.3(b) shows the upper aluminium support containing the HF cables, screwdrivers for the adjustment of the matching/tuning capacitors and the frequency adapter box. The adapter box contains a capacitive matching/tuning network specific for a small frequency range and which must therefore be changed to observe different nuclei. A

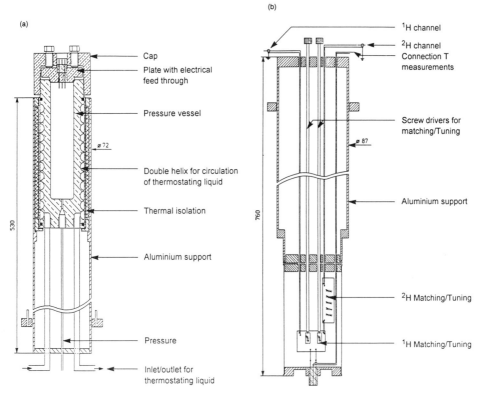

Figure 6.3 Schematic drawing of a wide-bore superconducting magnet multinuclear high-pressure bomb

double-tuned frequency network for ^1H (400 MHz) observation and ^2H (61.4 MHz) field lock is shown here.

The internal components of the bomb and the sample tube are shown in Figure 6.4. The radio-frequency coil is wound in a saddle-shaped form on 7 mm o.d. glass tubing to produce a radio-frequency field perpendicular to the vertical static magnetic field. The temperature is measured inside the bomb by a platinum resistor placed just above the sample tube. The temperature stability is better than ± 0.2 K and is not a limiting factor in the accuracy of measurements of activation volumes. The sample tube is a 5 mm o.d. NMR tube cut to the right length and closed with a movable seal, made of machinable glass, which allows pressure transmission to the sample. The sample tube was designed to minimize the volume (1 cm^3) to economize on expensive solvents such as ^{17}O enriched water. The surrounding pressure transmitting liquid is chosen so as not to contain the observed nucleus. The spectral resolution and magnetic field stability with this non-spinning high pressure probe is better than 0.5 Hz at 400 MHz and routinely better than 1 Hz.

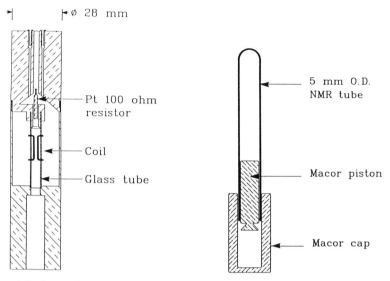

Figure 6.4 Internal component of the bomb shown in Figure 6.3 made of Vespel (left); and sample tube (right) [14]. Reproduced from *High Pressure Research*, 1990, Vol. 2, pp. 237–245, with permission of Gordon and Breach Science Publishers

3   NMR Techniques for the Study of Solvent Exchange

3.1   ISOTOPIC LABELLING: WATER AND ACETONITRILE EXCHANGE ON RUTHENIUM(II) AND RUTHENIUM(III)

Previous work on substitution reactions on Ru(II) and Ru(III) led to the conclusion that the exchange mechanisms were different on complexes of the two oxidation states; associative on Ru(III) and dissociative on Ru(II) [15]. It was important to test this conclusion using the effect of pressure on the exchange rate of two complexes with different oxidation state but identical coordination. The hexa-aqua complexes are available as salts of trifluoromethanesulphonic (triflic) acid (Ru(II)) or of *p*-toluenesulphonic (tosylic) acid (Ru(III)). The acetonitrile hexasolvate of Ru(II) is easily prepared and is much less labile than the aqua-ion due to metal to ligand back bonding. It was, therefore, of interest to compare its behaviour with that of the aqua complex. The exchange rates in these systems are sufficiently slow to be measured by the isotopic substitution technique, which has the advantage of requiring no assumptions to be made about individual relaxation rates.

Water exchange on the two hexa-aqua ions was followed using ^{17}O NMR spectroscopy [16], since the signals of coordinated water are resolved from the bulk water, even though the chemical shifts of bound water are very different in the two complexes: -196.3 ppm on Ru(II) and $+34.7$ ppm on Ru(III).

Typical spectra are shown in Figure 6.5. The relatively fast exchange on Ru(II) was followed by mixing ^{17}O enriched water with a solution of the complex in natural water. For the variable temperature measurements, mixing was done using the fast injection technique in a specially adapted NMR probe [17], the samples being kept in the probe throughout a run. Variable pressure measurements were made in a similar fashion, but the mixing was performed at ambient pressure on solutions cooled to 258 K, to minimize exchange prior to pressurization in a variable pressure NMR probe. The increase in height of the bound water signal with time (Figure 6.6) is proportional to the mole fraction x of ^{17}O labelled water coordinated to the metal, which is related to the water exchange rate, k_{ex} (from the bound to the free site, see equation (3.21)), by

$$x = x_\infty [1 - \exp(-k_{ex}t/(1 - x_\infty))] \qquad (6.5)$$

where x_∞ is the mole fraction of ^{17}O labelled water coordinated to the metal at exchange equilibrium. The exchange rate was found to be independent of acid concentration, so that possibility of a hydrolytic exchange pathway could be ruled out.

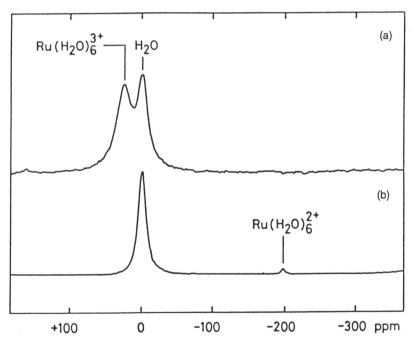

Figure 6.5   27.1 MHz ^{17}O NMR spectra of (a) ^{17}O enriched $Ru(H_2O)_6(tos)_3$ at 301.9 K, 200 min after mixing with normal water (b) isotopically equilibrated ^{17}O enriched aqueous solution of $Ru(H_2O)_6(trifl)_2$ at 297.2 K. Both solutions contain free acid. (Taken from Ref. 16, with permission)

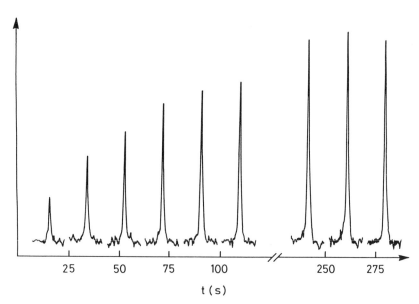

Figure 6.6 27.1 MHz ^{17}O NMR spectra, as a function of time at 297.7 K, of coordinated water of a solution of 0.25 molar $Ru(H_2O)_6(trifl)_2$ and 0.25 molar Htrif after mixing with ^{17}O enriched water

The signal of bound water in the Ru(III) system was not sufficiently separated from the bulk water signal (Figure 6.5) to allow the increase of the bound water signal on addition of the salt to ^{17}O enriched water to be followed. In this case the exchange was followed by the decrease of the bound water signal after the ^{17}O enriched complex was added to normal water. The exchange rate is much slower than for Ru(II), so that the mixing was done before transfer of the solution to either a commercial NMR probe or a variable pressure NMR probe. The hexaaquaruthenium(III)tosylate complex contains water of crystallization, so that the initial signal intensities of bound and bulk water were comparable. The evolution of both peaks was measured (Figure 6.7). The intensity of the bound water signal is proportional to the mole fraction, x, of ^{17}O labelled water in the bound water, which decreases according to

$$x = x_\infty + (x_0 - x_\infty)\exp[-k_{ex}t/(1 - x_\infty)] \qquad (6.6)$$

whereas the intensity of the bulk water signal increases with the mole fraction of ^{17}O labelled water in the bulk described by equation(6.5) (with x now referring to the mole fraction of ^{17}O labelled water in the bulk). The exchange rate was found to be dependent on acid concentration, and so was analysed to take account of a second, hydrolytic pathway for water exchange. The conjugated base $[Ru(H_2O)_5OH]^{2+}$ is in equilibrium with the hexa-aquaion

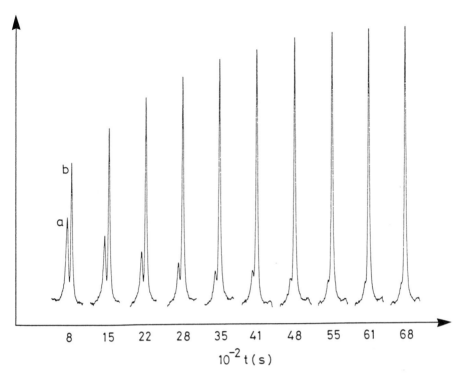

Figure 6.7   27.1 MHz ^{17}O NMR spectra, as a function of time at 315.3 K, of a 0.3 m $Ru(H_2O)_6(tos)_3$ (9% ^{17}O) and 0.5 m Htos aqueous solution obtained after mixing with normal water (for spectral assignment see Figure 6.5a) [16]. Reprinted with permission from Rappaport *et al.*, *Inorg. Chem.*, **27** 873. Copyright (1988) American Chemical Society

$[Ru(H_2O)_6]^{3+}$, so that the overall rate constant, k, is given by

$$k = k_{ex} + k_{OH}K_a/[H^+] \qquad (6.7)$$

where k_{ex} is the exchange rate on the hexa-aquaion, k_{OH} the exchange rate on the hydrolysed species and K_a the acidity equilibrium constant.

Acetonitrile exchange on Ru(II) was followed by observing the bound water ^{1}H NMR signal intensity after dissolution of $[Ru(CH_3CN)_6]^{2+}$ in CD_3CN [16]. The variable temperature measurements at high temperatures were made directly in the NMR probe, as for water exchange on Ru(III). At the lowest temperatures, where the exchange was very slow, the samples were kept in a constant temperature bath and transferred to the spectrometer at intervals for the measurements. For the variable pressure measurements, the samples were kept pressurized in the temperature bath. The pressure was released for the short time necessary for the measurements, so that a commercial variable temperature NMR probe could be used. The solvent exchange rate was determined from the decrease of the bound water signal using equation (6.6).

The results are shown in Table 6.1. The activation volumes for solvent exchange on the two Ru(II) solvates are essentially zero, indicating equal contributions from bond breaking and bond making during formation of the transition state, i.e. an interchange, I, mechanism. The much slower rate of acetonitrile exchange is due to the much higher enthalpy of activation and reflects the high thermodynamic stability of the complex. The activation volume for water exchange on $[Ru(H_2O)_6]^{3+}$ is markedly negative (as is the activation entropy), indicating an associative activation mode and an I_a mechanism. The loss of a proton to form $[Ru(H_2O)_5(OH)]^{2+}$ increases the exchange rate by two orders of magnitude. The much more positive activation volume indicates that the exchange is no longer associatively activated, due probably to the *trans* labilizing effect of the OH^- weakening the bonding to water and favouring a more dissociative process.

3.2 LINE-SHAPE ANALYSIS: TRIMETHYLPHOSPHATE
EXCHANGE ON ALUMINIUM(III) AND INDIUM(III)

Here we turn to proton NMR in order to observe the way trimethylphosphate (TMPA) exchanges on two related metal ions, both of which form octahedral solvates with this ligand: $M(TMPA)_6^{3+}$, with M = Al and In. The proton spectrum of TMPA is a doublet due to spin–spin coupling to the phosphorous atom and this doublet is maintained when the phosphate is complexed by a cation though it is shifted to low field (see Chapter 3, Section 2.3). The spectrum of a solution of $M(TMPA)_6^{3+}$ with excess TMPA, using deuterated nitromethane (CD_3NO_2) as a diluent, is thus two doublets, provided exchange is sufficiently slow. Exchange between free and bound solvent will cause line broadening and eventual collapse of the doublets. Calculated spectra were least-squares fitted to the experimental results (Figure 6.8), and exchange rates were extracted, using a lineshape analysis program derived from EXCGN[18].

The use of an inert diluent gives reduced viscosity leading to better resolution and definition of the line widths and also permits the free ligand concentration to be varied so that information can be obtained regarding the order of reaction. For TMPA exchange on Al(III) this is first order while for In(III) it is second order, the values of the rate constants and activation

Table 6.1 Rate constants and activation parameters for solvent exchange on various ruthenium solvates [16]

Solvate	k_{ex}^{298}/s^{-1}	$\Delta H^{\#}/kJ\,mol^{-1}$	$\Delta S^{\#}/J\,K^{-1}\,mol^{-1}$	$\Delta V^{\#}/cm^3\,mol^{-1}$
$Ru(MeCN)_6^{2+}$	$(8.9 \pm 2) \times 10^{-11}$	140.3 ± 2	$+33.3 \pm 6$	$+0.4 \pm 0.6$
$Ru(H_2O)_6^{2+}$	$(1.8 \pm 0.2) \times 10^{-2}$	87.8 ± 4	$+16.1 \pm 15$	-0.4 ± 0.7
$Ru(H_2O)_6^{3+}$	$(3.5 \pm 0.3) \times 10^{-6}$	89.8 ± 4	-48.3 ± 14	-8.3 ± 2.1
$Ru(H_2O)_5(OH)^{2+}$	5.9×10^{-4}	95.8	$+14.9$	$+0.9 \pm 2.0$

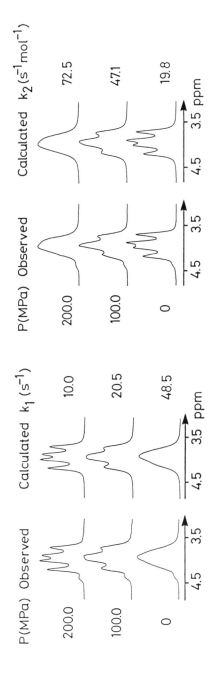

Figure 6.8 Observed and calculated 60 MHz ^1H NMR spectra of a solution of $M(TMPA)_6(ClO_4)_3$, dissolved in a mixture of TMPA and CD_3NO_2, recorded as a function of pressure. The spectra on the left are for M = Al and on the right for M = In. Reproduced by permission of IUPAC from Ref. 19

Table 6.2 Kinetic parameters for TMPA exchange on diamagnetic trivalent cations in CD_3NO_2 as diluent ($k_{ex} = k_1 + k_2$ [TMPA]) [19]

	r_i/pm	k_1^{298}/s^{-1}	k_2^{298}/s^{-1} mol^{-1}	$\Delta H^{\#}$/kJ mol^{-1}	$\Delta S^{\#}$/J K^{-1} mol^{-1}	$\Delta V^{\#}$/ cm^3 mol^{-1}	Mechanism
Al(TMPA)$_6^{3+}$	54	0.78		85.1	+38.2	+22.5	D
Ga(TMPA)$_6^{3+}$	62	6.4		76.5	+27.0	+20.7	D
Sc(TMPA)$_6^{3+a}$	75	736	85	34.1	−75.6	−23.8	A, I_a
Sc(TMPA)$_6^{3+}$			39	21.2	−143.5	−18.7	A, I_a
In(TMPA)$_6^{3+}$	80		7.6	32.8	−118	−21.4	A, I_a

^{a} In neat solvent.

parameters being shown in Table 6.2 [19]. The entropies of activation also have opposite signs and it is already clear from the ambient pressure data that the activation mode for TMPA exchange is dissociative on Al(III) and associative on In(III). The variable pressure kinetics, see Figure 6.8, provide a very good visual proof that the mechanisms are indeed different. The ^{1}H NMR spectrum for the TMPA exchange on In^{3+} at ambient pressure (Figure 6.8 right) shows a doublet at high field which is due to free TMPA and a doublet at low field due to the six TMPA molecules coordinated to In^{3+}. With increasing pressure, at constant temperature, one observes a coalescence of the signals. This means that the exchange rate is increasing and therefore the activation volume will be negative. For Al(TMPA)$_6^{3+}$ the change in spectra with pressure is the contrary: the deceleration of the ligand exchange with pressure indicates a positive activation volume. The difference in mechanistic behaviour can be explained in terms of ionic radii. The large In^{3+} cation (see Table 6.2) can easily accommodate, at least partially, a seventh solvent molecule in forming the activated complex. The small Al^{3+} ion, on the other hand, is rather crowded, and has to expel a solvent molecule from the first coordination sphere at the transition state, before a new molecule can be accepted.

3.3 RELAXATION RATE MEASUREMENTS I: DIMETHYLSULPHOXIDE EXCHANGE ON VANADIUM(III)

V(III) is one of four tripositive first row transition metal ions for which kinetic parameters, including activation volumes, for dimethylsulphoxide (DMSO) exchange, have been obtained [20]. The ^{1}H NMR spectrum of a solution of vanadium(III) triflate in DMSO consists of two resonances: a narrow, intense peak due to bulk DMSO and a smaller, wide peak, shifted to high frequencies, owing to the DMSO bound to the paramagnetic V^{3+} ion. The line width of the bound proton signal in neat DMSO is proportional to the transverse relaxation rate, $(1/T_2^b)_{neat}$, which can be written as a combination of paramagnetic broadening, $1/T_{2m}^b$, and exchange, k_{ex}, broadening terms

$$(1/T_2^b)_{neat} = 1/T_{2m}^b + k_{ex} \tag{6.8}$$

The temperature variation of $(1/T_2^b)_{neat}$ is shown as circles in Figure 6.9, and exhibits an exchange broadening region at high temperatures. The accuracy of the exchange rate determination is limited, however, by the temperature range over which the experiments can be performed. One way to extend the low temperature limit beyond the freezing point of DMSO is to make experiments in solutions containing an inert diluent. Such measurements also provide additional information on the rate law for the solvent exchange reaction, and hence its mechanism. Variable temperature studies were made in solutions containing CD_3NO_2 as the inert diluent. At very low temperatures, where the exchange is sufficiently slow, the linewidth measurements were supplemented by fast injection measurements similar to those described in Section 3.1. The

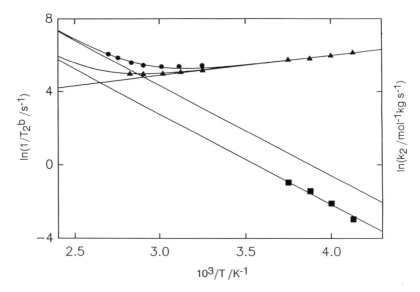

Figure 6.9 Temperature dependence of the transverse relaxation rate, $(1/T_2^b)_{\text{neat}}$, from the bound DMSO ^{1}H NMR signal of V(DMSO)$_6^{3+}$, in neat DMSO (●) and $(1/T_2^b)_{dil}$ in CD$_3$NO$_2$ (▲), including second-order exchange rates, k_2 (■), obtained from fast injection experiments in CD$_3$NO$_2$[20]. Reprinted with permission from Dellavia *et al.*, *Inorg. Chem.*, **31**, 792. Copyright (1992) American Chemical Society

rate constants, k_{obs} measured by fast injection were found to be linear in the free DMSO concentration, [DMSO], so that the second-order rate constant, k_2, could be calculated via

$$k_{obs} = k_2[\text{DMSO}] \tag{6.9}$$

The calculated second-order rate constants are shown as squares and the bound proton linewidth in solutions containing CD$_3$NO$_2$ diluent are shown as triangles in Figure 6.9. The bound proton relaxation rate in the diluent, $(1/T_2^b)_{dil}$, is given by

$$(1/T_2^b)_{dil} = (1/T_{2m}^b)_{dil} + k_{obs} = (1/T_{2m}^b)_{dil} + k_2[\text{DMSO}] \tag{6.10}$$

If one assumes that the second order behaviour may be extended to neat DMSO, the exchange rate, k_{ex} is related to k_2 via the neat DMSO concentration, [DMSO]$_M$ giving

$$(1/T_2^b)_{dil} = (1/T_{2m}^b)_{dil} + k_{ex}[\text{DMSO}]/[\text{DMSO}]_M \tag{6.11}$$

In fitting the data, however, [DMSO]$_M$ was treated as an adjustable parameter, as equation (6.9) probably does not hold to such high [DMSO]. The simultaneous fit of all the variable temperature in Figure 6.9 shows reassuring agreement between the linewidth and fast injection data. The two straight lines with negative slope represent k_{ex} and k_2, while the straight line with positive slope represents $1/T_{2m}^b = (1/T_{2m}^b)_{dil}$. The low temperature results in CD_3NO_2 diluent greatly facilitate the separation of the results in neat DMSO into exchange and paramagnetically broadened regions.

The pressure variation of the bound proton line-width in neat DMSO was measured in a variable pressure NMR probe. Measurements were made at four different temperatures in order to separate the pressure effects on paramagnetic relaxation and on chemical exchange. The data were fitted simultaneously using the assumption that $\ln(1/T_{2m}^b)$ and $\ln k_{ex}$ depend linearly on pressure with activation volumes $\Delta V_m^\#$ and $\Delta V^\#$, respectively.

The results of the variable temperature and pressure measurements are summarized in Table 6.3. The negative activation volume confirms an associative activation mode for DMSO exchange on V(III), as could be inferred from the second order rate law. Comparison of the activation volume with those for DMSO exchange on other tripositive first row transition metal ions (Table 6.4) shows that there is a changeover from associatively activated to dissociatively activated exchange on going from left to right across the series. Comparison with results for water exchange leads to the assignment of an interchange I_a mechanism for DMSO exchange on V(III).

3.4 RELAXATION RATE MEASUREMENTS II: N,N-DIMETHYLFORMAMIDE EXCHANGE ON TITANIUM(III)

We now turn to another study of solvent exchange on a paramagnetic trivalent first-row transition metal ion [21]. For solutions of Ti(III) in N,N-dimethylformamide (DMF), however, the bound DMF signal in both ^{1}H and ^{17}O NMR is too broad to be observed. The DMF exchange rate therefore had to be obtained from the observed line-width and chemical shift of the bulk DMF signals, which are influenced by the paramagnetic effect transmitted by exchange with the bound DMF. Variable temperature measurements were made using ^{17}O NMR at 9.4 T and ^{1}H NMR at 4.7 T and 9.4 T, for solutions of titanium(III) triflate in DMF and a pure DMF reference solution. Deuterated DMF-d_7 (25%) was used for the internal field lock and tetramethylsilane as the chemical shift reference in the solutions prepared for the ^{1}H NMR measurements. For the ^{17}O

Table 6.3 Derived NMR and kinetic parameters for the variable-temperature and variable-pressure study of DMSO exchange on vanadium(III) in neat solvent [20]

k_{ex}^{298}/s^{-1}	13.1 ± 1.5	$E_m^b/kJ \, mol^{-1}$	9.3 ± 0.5
$\Delta H^\#/kJ \, mol^{-1}$	38.5 ± 1.6	$[DMSO]_M/mol \, kg^{-1a}$	4.8 ± 1.0
$\Delta S^\#/J \, K^{-1} \, mol^{-1}$	-94.5 ± 4.6	$\Delta V^\#/cm^3 \, mol^{-1}$	-10.1 ± 0.6
$(1/T_{2m}^b)^{298}/s^{-1}$	$193.5 \pm 5.$	$\Delta V_m^\#/cm^3 \, mol^{-1}$	-2.1 ± 0.4

^{a} This leads to $k_2^{298}/mol^{-1} \, kg \, s^{-1} = k_{ex}^{298}/[DMSO]_M = 2.7$ in CD_3NO_2 diluent.

NMR measurements no field lock was necessary and no internal chemical shift reference was used. To avoid magnetic susceptibility corrections to the measured resonance frequency, the samples were sealed in spherical glass cells [22] that fit into 10 mm NMR tubes.

The effect of the paramagnetic ion on the bulk solvent linewidth and chemical shift can be expressed as the reduced transverse relaxation rate and reduced chemical shift, defined as

$$\frac{1}{T_{2r}} = \frac{1}{P_m}\left[\frac{1}{T_2} - \frac{1}{T_{2A}}\right] \qquad (6.12)$$

$$\Delta\omega_r = \frac{1}{P_m}(\omega - \omega_A) \qquad (6.13)$$

where T_2 and ω are the transverse relaxation rate and frequency of the bulk solvent signal in the solution containing the paramagnetic ion, T_{2A} and ω_A are the corresponding values in the pure solvent reference and P_m is the mole fraction of bound solvent molecules. Swift and Connick [23] derived expressions for these reduced quantities in terms of the transverse relaxation rate, $1/T_{2m}$, in bulk solvent, its chemical shift with respect to the pure solvent, $\Delta\omega_m$, and the solvent exchange rate, k_{ex}. These expressions have since been modified [24] to take account of the paramagnetic effect on 'outer sphere' solvent molecules giving

$$\frac{1}{T_{2r}} = k_{ex}\frac{T_{2m}^{-2} + k_{ex}/T_{2m} + \Delta\omega_m^2}{(k_{ex} + T_{2m}^{-1})^2 + \Delta\omega_m^2} + \frac{1}{T_{2os}} \qquad (6.14)$$

$$\Delta\omega_r = \frac{\Delta\omega_m}{(1 + 1/(k_{ex}T_{2m})^2 + \Delta\omega_m^2/k_{ex}^2} + \Delta\omega_{os} \qquad (6.15)$$

where $1/T_{2os}$ and $\Delta\omega_{os}$ take into account the relaxation rate and chemical shift enhancement of the 'outer sphere' solvent (see also Chapter 3, equations (3.66) and (3.67) where $\tau_M = 1/k_{ex}$).

The temperature variation of the reduced ^{17}O and formyl ^1H reduced relaxation rates and chemical shifts is shown in Figure 6.10. A changeover between several limiting regions of equations (6.14) and (6.15) can be seen. At high temperatures where the slope of $\ln(1/T_{2r})$ against inverse temperature is positive, the relaxation tends towards the fast exchange limit, where $k_{ex} \gg 1/T_{2m}$, $\Delta\omega_m$, giving (if outer sphere terms are neglected) $1/T_{2r} = 1/T_{2m} + \Delta\omega_m^2/k_{ex}$ and $\Delta\omega_r = \Delta\omega_m$. In the present system, $1/T_{2r} \sim \Delta\omega_m^2/k_{ex}$, so $1/T_{2m}$ was assumed to be zero in the analysis of the data. The effect of magnetic field on the ^1H $1/T_{2r}$ at high temperatures results from the $\Delta\omega_m^2$ term. At lower temperatures the slow exchange limit, $k_{ex} \ll 1/T_{2m} + \Delta\omega_m^2/k_{ex}$, where $1/T_{2r} = k_{ex} + 1/T_{2os}$, is reached. At moderate temperatures, where the slope of $\ln(1/T_{2r})$ against inverse temperature is negative, the exchange rate is dominant. Within this region $\Delta\omega_r$ decays to the value $\Delta\omega_{os}$. At low temperatures the

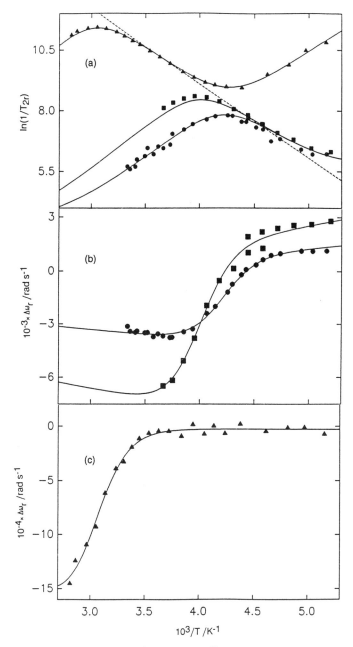

Figure 6.10 Variable-temperature ^{1}H NMR and ^{17}O NMR data for Ti(DMF)$_6$(trif)$_3$ in DMF, showing $\ln(1/T_{2r})$ (a) and $\Delta\omega_r$ (b) versus inverse temperature. The curves are the result of a simultaneous fit of all the data: ($\bullet$) ^{1}H NMR at 4.7 T; ($\blacksquare$) ^{1}H NMR at 9.4 T; ($\blacktriangle$) 54.2 MHz ^{17}O NMR at 9.4 T. The dashed line of Figure 6.10(a) shows the contribution of k_{ex} to $1/T_{2r}$

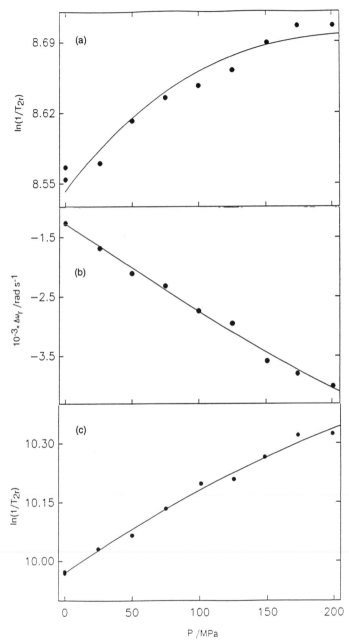

Figure 6.11 Variable-pressure 400 MHz ^{1}H NMR data (a and b) for Ti(DMF)$_6$(trif)$_3$ in DMF, showing $\ln(1/T_{2r})$ (a) and $\Delta\omega_r$ (b) versus pressure at 242.4 K. Variable-pressure 54.2 MHz ^{17}O NMR $\ln(1/T_{2r})$ data (c) for Ti(DMF)$_6$(trif)$_3$ in DMF at 270.3 K

Table 6.4 Rate constants and activation parameters for DMSO exchange on trivalent first-row octahedral transition-metal ions [a]

$[M(DMSO)_6]^{3+}$	r_i/pm	Electronic configuration	k_{ex}^{298}/s^{-1}	$\Delta H^{\#}$/kJ mol^{-1}	$\Delta S^{\#}$/J K^{-1} mol^{-1}	$\Delta V^{\#}$/cm^3 mol^{-1}	Reference
$[V(DMSO)_6]^{3+}$	64	t_{2g}^2	13.1	38.5	−94.5	−10.1	[20]
$[Cr(DMSO)_6]^{3+}$	61	t_{2g}^3	3.1×10^{-8}	96.7	−64.5	−11.3	[51]
$[Fe(DMSO)_6]^{3+}$	64	$t_{2g}^3 e_g^2$	9.3	62.5	−16.7	−3.1	[52]
$[Ga(DMSO)_6]^{3+b}$	62	$t_{2g}^6 e_g^4$	1.87	72.5	+3.5	+13.1	[48]

[a] By NMR analysis except for Cr^{3+} by isotopic labelling.
[b] In CD$_3$NO$_2$ as diluent.

Table 6.5 Rate constants and activation parameters for DMF exchange on trivalent first-row octahedral transition-metal ions [a]

$[M(DMF)_6]^{3+}$	r_i/pm	Electronic configuration	k_{ex}^{298}/s^{-1}	$\Delta H^{\#}$/kJ mol^{-1}	$\Delta S^{\#}$/J K^{-1} mol^{-1}	$\Delta V^{\#}$/cm^3 mol^{-1}	Reference
$[Ti(DMF)_6]^{3+}$	67	t_{2g}^1	6.6×10^{-4}	23.6	−73.6	−5.7	[21]
$[Cr(DMF)_6]^{3+}$	61	t_{2g}^3	3.3×10^{-7}	97.1	−43.5	−6.3	[51]
$[Fe(DMF)_6]^{3+}$	64	$t_{2g}^3 e_g^2$	61	42.3	−69.0	−0.9	[52]
$[Ga(DMF)_6]^{3+b}$	62	$t_{2g}^6 e_g^4$	1.72	85.1	+45.1	+7.9	[48]

[a] By NMR analysis except for Cr^{3+} by isotopic labelling.
[b] In CD$_3$NO$_2$ as diluent.

$1/T_{2os}$ term starts to influence $1/T_{2r}$ and becomes dominant in the case of ^{17}O NMR, the slope of $\ln(1/T_{2r})$ against inverse temperature becoming positive again. The variable temperature data were fitted using equations (6.14) and (6.15), with $1/T_{2os}$ represented by a simple exponential temperature dependence and $\Delta\omega_m$ and $\Delta\omega_{os}$ assumed to be inversely proportional to temperature. The values obtained for the parameters governing water exchange, as well as those describing $1/T_{2os}$, $\Delta\omega_m$ and $\Delta\omega_{os}$ were practically independent of whether the ^1H NMR and ^{17}O NMR data were fitted separately or simultaneously. This use of two nuclei, therefore, allows one to be very confident of the determination of k_{ex}, $\Delta H^{\#}$ and $\Delta S^{\#}$.

The ^1H linewidth and chemical shift for the titanium(III) solution were measured as a function of pressure at 242.4 K and 9.4 T in a variable pressure probe (Figure 6.11(a) and (b)). The temperature was chosen near the low temperature limit of the variable pressure system, and corresponds to the intermediate region between fast and slow exchange limits. Since $1/T_{2A}$ was less than 7% of $1/T_2$ it was assumed to be independent of pressure in the calculation of $1/T_{2r}$ (equation (6.12)). Similarly, ω_A was assumed to be pressure independent in the calculation of $\Delta\omega_r$ (equation (6.13)). Analysis of the data using equations (6.14) and (6.15) yielded $\Delta V^{\#} = -7.0 \pm 1.2 \text{ cm}^3 \text{mol}^{-1}$, independent of whether or not $\Delta\omega_m$ was assumed to depend on temperature.

The ^{17}O NMR linewidth for the titanium(III) solution was also measured as a function of pressure at 9.4 T in a variable pressure probe (Figure 6.11(c)). A higher temperature of 270.3 K was chosen, where $1/T_{2r}$ is in the slow exchange limit, and is practically identical to k_{ex}. Values of $1/T_{2A}$ for DMF, determined at variable pressure in a separate study [25], were used in the calculation of $1/T_{2r}$. Assuming the small (less than 4%) contributions from $1/T_{2os}$ and $\Delta\omega_m$ to be independent of pressure, a fit of the data using equation (6.14) yielded the activation volume $\Delta V^{\#} = -5.7 \pm 0.6 \text{ cm}^3 \text{mol}^{-1}$. This value is less negative than that obtained by ^1H NMR, although the error bounds of the two values overlap. The value obtained from ^{17}O NMR is to be preferred as it was determined at a temperature where $1/T_{2r}$ is practically equal to k_{ex}. Moreover, it was obtained nearer ambient temperature, and so can be more reasonably compared with values for related systems.

The parameters describing DMF exchange on trivalent first-row transition metal ions are given in Table 6.5. The results confirm a mechanistic changeover from associatively activated to dissociatively activated from left to right across the series, as observed for other solvents. An I_a mechanism is assigned to DMF exchange on titanium(III), the mechanism being significantly less associative than for water exchange on the same ion, due to the greater bulk of DMF.

3.5 RELAXATION RATE MEASUREMENTS III: WATER EXCHANGE ON LANTHANIDE(III)

We now turn to another type of paramagnetic system, where only the bulk water signal is observed as in Section 3.4, but where a slow exchange region in equation (6.14) is never reached. Water exchange on the lanthanide(III) aqua-ions is sufficiently rapid that the ^{17}O reduced transverse relaxation rate, $1/T_{2r}$, of aqueous paramagnetic lanthanide(III)perchlorate solutions is always in the fast exchange limit of equation (6.14), where

$$\frac{1}{T_{2r}} = \frac{1}{P_m}\left[\frac{1}{T_2} - \frac{1}{T_{2A}}\right] = \frac{1}{T_{2m}} + \frac{\Delta\omega^2}{k_{ex}} + \frac{1}{T_{2os}} \tag{6.16}$$

An analogous equation to equation (6.14) was derived by Zimmermann and Brittin [26] for the reduced longitudinal relaxation rate, $1/T_{1r}$ (see also Chapter 3, Section 4.2). In the fast exchange limit

$$\frac{1}{T_{1r}} = \frac{1}{P_m}\left[\frac{1}{T_1} - \frac{1}{T_{1A}}\right] = \frac{1}{T_{1m}} + \frac{1}{T_{1os}} \tag{6.17}$$

The paramagnetic lanthanide(III) ions Ce^{3+} to Tb^{3+} all (except Gd^{3+}) have low-lying electronic energy levels that result in very rapid electronic relaxation times (10^{-14} to 5×10^{-13} s [27]) This rapid electronic relaxation largely determines the relaxation of bound water molecules and results in a near extreme narrowing limit, $1/T_{2m} \sim 1T_{1m}$. Making the additional assumption that $1/T_{2os} \sim 1/T_{1os}$, one obtains from equations (6.16) and (6.17)

$$\frac{1}{P_m}\left[\frac{1}{T_2} - \frac{1}{T_1}\right] = \frac{\Delta\omega_m^2}{k_{ex}} \tag{6.18}$$

Variable temperature ^{17}O NMR measurements were made for aqueous solutions of lanthanide(III) perchlorates of Nd^{3+}, Eu^{3+}, Tb^{3+}, Dy^{3+}, Ho^{3+}, Er^{3+}, Tm^{3+} and Yb^{3+} sealed in glass spheres at 4.7 T and 9.4 T [28]. The temperature dependence of the transverse and longitudinal relaxation rates were measured by the Carr–Purcell–Meiboon–Gill spin echo [29] and inversion recovery [30] techniques, respectively. The spin-echo, rather than the linewidth method, was used to determine $1/T_2$ since the linewidths contain a significant inhomogeneity contribution owing to the magnetic susceptibilities of the paramagnetic samples. The chemical shifts were measured for the paramagnetic solutions and for a non-paramagnetic $La(ClO_4)_3$ reference. The relaxation rates for $Ho(ClO_4)_3$ are shown in Figure 6.12. The fact that the difference between $1/T_1$ and $1/T_2$ is proportional to the magnetic field squared confirms the approximations used in equation (6.18). The chemical shift was fitted with a quadratic equation in inverse temperature, and using the fact that in the fast exchange limit, ignoring outer sphere effects, $\Delta\omega_r = \Delta\omega_m$, a simultaneous fit of $\Delta\omega_r$ and the relaxation rate difference yielded the variable temperature activation parameters for water exchange. Reliable parameters were obtained for the series Tb^{3+} to Yb^{3+}. For the lighter lanthanides, Nd^{3+} and

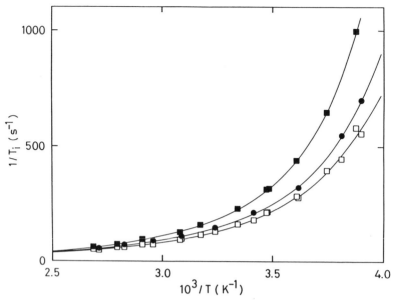

Figure 6.12 ¹⁷O NMR relaxation data obtained at variable temperature in a 0.3 molar Ho(ClO$_4$)$_3$/2 molar HClO$_4$ aqueous solution (□) $1/T_1$ at 4.7 T and 8.5 T; (●) $1/T_2$ at 4.7 T; (■) $1/T_2$ at 8.5 T [31]. Reprinted with permission from Cossy *et al.*, *Inorg. Chem.*, **27**, 1973. Copyright (1988) American Chemical Society

Eu^{3+}, the bound water chemical shift $\Delta\omega_m$ was too small for the relaxation rate difference in equation (6.18) to be measurable.

Variable pressure ¹⁷O relaxation rate measurements were made in a similar fashion in a high-pressure probe [31, 32]. The chemical shift was measured with respect to an internal reference, and was found not to vary significantly with pressure. The relaxation rate difference thus leads directly to the pressure variation of the water exchange rate via equation (6.18). The results for Ho(ClO$_4$)$_3$ are shown in Figure 6.13. The results for the series Tb^{3+} to Yb^{3+} were fitted using equation (6.5) with $\Delta\beta^{\#} = 0$ to obtain $\Delta V^{\#}$.

As mentioned above, Gd^{3+} is a case apart. The absence of low-lying electronic energy levels means that the electronic relaxation rates are relatively slow, so that the extreme narrowing limit for bound water relaxation does not apply. Indeed $1/T_{2r} \gg 1/T_{1r}$, and is dominated by relaxation owing to the scalar or contact interaction originating from the finite Gd^{3+} electron spin density at the ¹⁷O nucleus. This relaxation mechanism is so effective that the $\Delta\omega_m^2/k_{ex}$ term in equation (6.16) is negligible. In order to determine k_{ex} it is necessary to consider the microscopic origins of the scalar relaxation of the bound water. If the outer sphere term is also neglected one has, in an approximate form

$$\frac{1}{T_{2r}} = \frac{1}{T_{2m}} = \frac{S(S+1)}{3}\left(\frac{A}{\hbar}\right)^2\left[k_{ex} + \frac{1}{T_{1e}}\right]^{-1} \qquad (6.19)$$

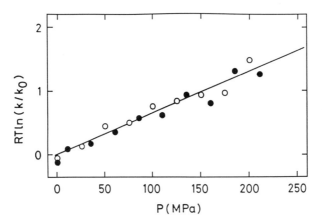

Figure 6.13 Pressure dependence at 269 K of water exchange rate constants on Ho^{3+} in 0.2 molar ($\bullet$) and 0.3 molar ($\bigcirc$) solutions

where $S = 7/2$ is the Gd^{3+} electron spin and $1/T_{1e}$ is the longitudinal electronic relaxation rate. The scalar coupling constant, $A/\hbar$ (in joules), is accessible from the chemical shift. Ignoring outer-sphere effects

$$\Delta\omega_r = \Delta\omega_m = \frac{g_L\mu_B S(S+1)B_o}{3k_B T}\frac{A}{\hbar} \qquad (6.20)$$

where g_L is the isotropic Landé g-factor ($g_L = 2.0$ for Gd^{3+}) and μ_B is the Bohr magneton (other notations as in Chapters 1 and 3, equations (3.68), (3.134) and (3.135)). Thus, a measurement of $1/T_{2r}$ and $\Delta\omega_r$ allows one to determine k_{ex}, provided that $1/T_{1e}$ is known. It is therefore necessary to combine ^{17}O NMR measurements with EPR measurements. The EPR linewidth is directly proportional to the *transverse* electronic relaxation rate, $1/T_{2e}$, which can be related to $1/T_{1e}$ using a theoretical model.

A combined ^{17}O NMR and EPR study of the Gd^{3+} aqua-ion was first made in 1980 [33], and has recently been improved by extension to a greater number of EPR [34] and NMR [35] magnetic fields. The temperature variation of the EPR linewidth, ΔH_{pp}, and the ^{17}O NMR reduced transverse relaxation rate, $1/T_{2r}$, and chemical shift, $\Delta\omega_r$, drawn from these recent studies are shown in Figure 6.14. The magnetic field dependence of $1/T_{2r}$ (Figure 6.14(b)) is due to the magnetic field dependence of $1/T_{1e}$ in equation (6.19). The values of ΔH_{pp} were analysed [34] using a model based on relaxation due to modulation of a zero field splitting (ZFS) of the degenerate electronic ground state. A least-squares fit (curves in Figure 6.14(a)) yielded a mean square ZFS energy, Δ^2, and a characteristic modulation time with value τ_v^{298} at 298 K and energy E_v. These parameters were used to calculate the magnetic field and temperature dependence of $1/T_{1e}$, so that the ^{17}O NMR results could be fitted with just three

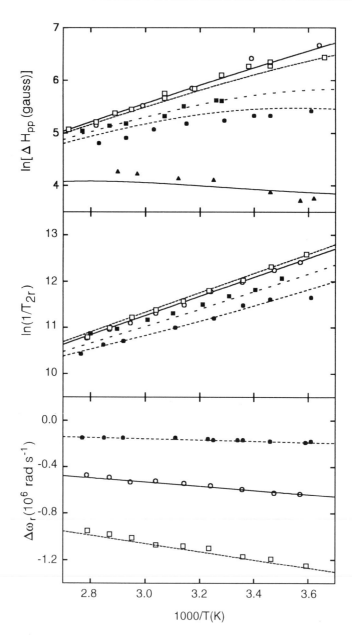

Figure 6.14 Temperature dependence of the peak-to-peak EPR linewidth, ΔH_{pp} (gauss), at 0.14 T (○), 0.34 T (□) 0.90 T(■) 1.2 T (●) and 5.0 T (▲) and the ^{17}O NMR reduced transverse relaxation rate, $1/T_{2r}$, and reduced chemical shift, $\Delta\omega_r$, at 1.4 T (●), 2.7 T (■), 4.7 T (○) and 9.4 T (□) for aqueous solutions of $Gd(ClO_4)_3$

parameters, $A/\hbar$, k_{ex}^{298} (or $\Delta S^{\#}$) and $\Delta H^{1\#}$ [35]. The quality of the least squares fit (curves in Figure 6.14(b) and (c)), especially to the field dependence of $1/T_{2r}$, shows the validity of the model used to describe the electronic relaxation. The calculated contribution of $1/T_{1e}$ to $1/T_{2r}$ decreases to less than 3% at 9.4 T, so that k_{ex} is determined well from these high field measurements.

The pressure dependence of $1/T_{2r}$ was measured at 9.4 T and 298.2 K in a high-pressure probe (Figure 6.15) [35]. The small contribution (<3%) from $1/T_{1e}$ can be neglected so that $1/T_{2r}$ is inversely proportional to k_{ex}. The decrease of $1/T_{2r}$ with pressure thus implies an increase of k_{ex} with pressure, and a least squares fit of the data (curve in Figure 6.15) with $\Delta\beta^{\#} = 0$ yields a negative activation volume.

The water exchange parameters for the lanthanide(III) aqua-ions Gd^{3+} to Yb^{3+} are shown in Figure 6.16. The negative $\Delta V^{\#}$ indicate an associatively activated exchange mechanism. The variation of k_{ex}^{298} across the series can be understood on terms of the known coordination number change from nine for the larger aqua-ions on the left of the series to eight for the smaller aqua-ions on the right of the series. The changeover occurs around Sm^{3+}, so that the increase of k_{ex}^{298} from Yb^{3+} to Gd^{3+} corresponds to greater stability of the nine-coordinate transition state.

3.6   MAGNETIZATION TRANSFER: DIMETHYLSULPHIDE EXCHANGE ON PLATINUM(II)

We consider the square-planar complex cis-PtR_2S_2, where R = phenyl (Ph) and S = dimethylsulphide (Me_2S), on which the exchange of Me_2S was studied in benzene. The ^1H NMR spectra of this complex show narrow signals for the bound and free solvent molecules over the accessible temperature domain. This

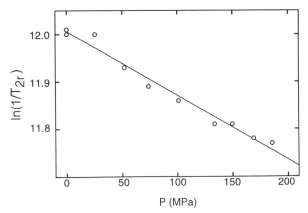

Figure 6.15   Pressure dependence of the reduced ^{17}O transverse relaxation rate, $1/T_{2r}$, of a 0.010 molar aqueous solution of $Gd(ClO_4)_3$, measured at 9.4 T and 298.2 K. Reproduced by permission of John Wiley & Sons from Ref. 35

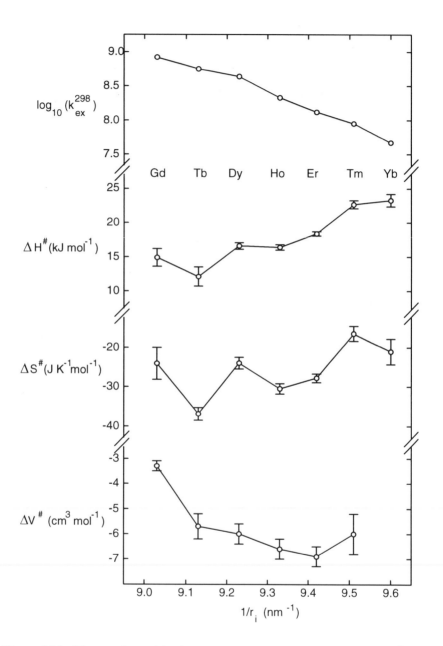

Figure 6.16   Water exchange kinetic parameters found for the heavy $Ln(H_2O)_8^{3+}$ ions, plotted against the inverse of the ionic radius, r_i. Reproduced by permission of John Wiley & Sons from Ref. 35

reaction is therefore too slow to be studied quantitatively by the NMR line-broadening technique. On the other hand, the exchange rate was found to be too high to use the isotopic labelling technique. However, magnetization transfer NMR is a useful technique to study exchange rates intermediate between those accessible by isotopic labelling and line-broadening and is also easily applied in a high-pressure probe.

The majority of square-planar complexes undergo solvent exchange by a pathway which involves associative behaviour. The search for factors promoting the conversion of the normal associative mode of activation into a dissociative process has attracted much attention. This work was undertaken to provide evidence for exchange mechanisms which are entirely dissociative [36].

The ^{1}H NMR spectrum of a solution of the complex in the presence of free solvent is a singlet for the solvent and a triplet-like structure for the solvent attached to platinum consisting of a singlet and the ^{195}Pt spin satellites, see Figure 6.17. The measurements were carried out by inverting selectively the latter resonance using the so-called '1, -3, 3, -1' pulse train [37] followed by a non-selective $90°$ pulse at a series of different time intervals, $22.5° - D_1 - 67.5° - D_1 - 67.5° - D_1 - 22.5° - t - 90° - D_2$, where $D_1 = 1/2\Delta\nu$, $\Delta\nu$ is the chemical shift difference in Hz, $D_2 = 5T_1$ and T_1 is the spin-lattice relaxation time. The variation of the intensity of the two resonances was

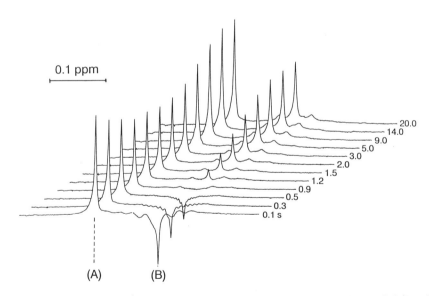

Figure 6.17 400 MHz ^{1}H NMR spectra of 0.1 molar $PtPh_2(Me_2S)_2$ and 0.2 molar Me_2S solution in benzene at 342 K and at 162 MPa as a function of the time interval between the inversion pulse train and the observation pulse (A) free Me_2S and (B) Me_2S bound to the platinum complex [36]. Reprinted with permission from Frey *et al.*, *J. Am. Chem. Soc.*, **111**, 8161. Copyright (1989) American Chemical Society

monitored as a function of time following the selective inversion. The evolution of the intensities is a function of T_1 of the two solvent signals and the exchange rate between them. For a quantitative inversion, the chemical shift difference between the two resonances must be significantly larger than the coupling to ^{195}Pt. This was assured by working at high-field (400 MHz) and by choosing diluents which gave adequate chemical shifts. Data were obtained at ambient pressure over a range of temperatures and then at variable pressure. A typical set of data is shown in Figure 6.17. The experimental magnetizations were obtained by direct measurement of the signal heights from the spectra and were fitted to the equations given by Led and Gesmar [38] . The contribution of the satellites to the height of the central line and the ratio between the signal linewidths were taken into account. An example of a fit is shown in Figure 6.18 and the derived kinetic parameters are given in Table 6.6.

The results show beyond doubt that for this and similar complexes (Table 6.6) the solvent exchange has a dissociative activation mode and it appears that the presence of the metal-carbon bond *trans* to the phenyl group causes the changeover in mechanism. It should also be noted that the activation entropy is very small and that it is therefore essential in this case to know the activation volume if mechanistic conclusions are to be made.

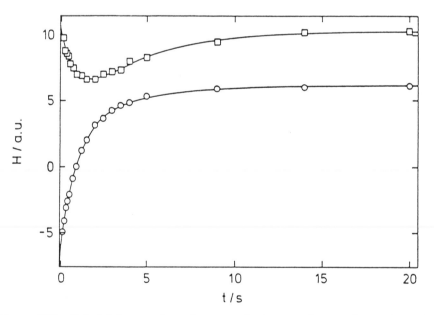

Figure 6.18 Calculated curve from the magnetization transfer experiment shown in Figure 6.16. (○) signal height of the central line from bound Me$_2$S and (□) signal height of the free Me$_2$S [36]. Reprinted with permission from Frey *et al.*, *J. Am. Chem. Soc.*, **111**, 8161. Copyright (1989) American Chemical Society

Table 6.6   Rate constants and activation parameters for ligand, L, exchange on *cis*-PtR₂L₂ [36]

R	L	k_{ex}^{298}/s^{-1}	$\Delta H^{\#}/kJ\ mol^{-1}$	$\Delta S^{\#}/J\ K^{-1}\ mol^{-1}$	$\Delta V^{\#}/cm^3\ mol^{-1}$
Ph	Me₂S	0.21	101	+42	+4.7
Ph	DMSO	1.24	64	−67	+5.5
Me	DMSO	1.12	84	−1	+4.9

3.7   2-D NMR: TRIMETHYLPHOSPHATE EXCHANGE ON TETRACHLOROBIS(TRIMETHYLPHOSPHATE)ZIRCONIUM (IV)

The cases discussed up to now have been relatively simple reactions with single reaction pathways, or if more complex, with straightforward means of separating the different reaction pathways such as testing the reaction order, or if there is a hydrolytic pathway, by following the effect of changes in pH. The system to be discussed here is much more complex. The two ligands, L, in $ZrCl_4 \cdot 2L$ can be either *cis* or *trans* with respect to one another and this means that there are three possible exchange reactions [39]:

$$\textit{cis-}ZrCl_4\cdot2L + {}^*L \overset{k_c}{\rightleftharpoons} \textit{cis-}ZrCl_4\cdot L^*L + L \qquad (6.21)$$

$$\textit{trans-}ZrCl_4\cdot2L + {}^*L \overset{k_t}{\rightleftharpoons} \textit{trans-}ZrCl_4\cdot L^*L + L \qquad (6.22)$$

$$\textit{cis-}ZrCl_4\cdot2L \overset{k_i}{\rightleftharpoons} \textit{trans-}ZrCl_4\cdot2L \qquad (6.23)$$

At 236 K, the ¹H NMR spectrum of a solution of $ZrCl_4 \cdot 2TMPA$ (TMPA = trimethylphosphate) in chloroform with excess TMPA consists of three doublets owing to the two forms of the complex and the free ligand with $^3J(^1H-^{31}P) = 14.0$ Hz for the latter and 11.7 and 11.8 Hz in the *cis*- and *trans*-complexes respectively. As the temperature is raised, the doublets of the *cis*- and *trans*-isomers broaden and eventually coalesce, indicating that the *cis*/*trans*-isomerization reaction occurs first. At slightly higher temperature, the doublet of the free-ligand also broadens and coalesces with the signal of the *cis*/*trans*-isomers. The source of this second broadening could be due to a ligand-exchange reaction between free ligand and the *cis*- and/or *trans*-isomer. In this case, NMR line-broadening simulation cannot easily be used to quantify the individual contributions of the three exchange reactions to the observed spectra.

It has now been shown that this problem can be overcome by using the very powerful 2-D NMR techniques. In the moderately slow exchange region where all the signals can be observed, a pulse sequence which is sensitive to exchange effects gives off-diagonal peaks where there is an exchange reaction. So, all the signals can be examined for exchange coupling and the individual exchange rate can be obtained. In addition, if a high-pressure probe is available for a suitable spectrometer, then it is possible to carry out a variable pressure 2-D experiment without too much difficulty.

The technique used was to obtain ^{1}H(400 MHz) EXCSY (or EXSY as in Chapter 3, Section 4.4.1) spectra on a Bruker AM-400 spectrometer equipped with a home built high-pressure probe. The pulse sequence used was: $90°(\Phi_1)-t_1-90°(\Phi_2)-\tau_m-90°(\Phi_3)-t_2$. The first pulse creates XY magnetization which evolves with the frequencies characteristic of the signals in the spectrum during the time interval t_1. The second pulse creates longitudinal magnetization which can undergo interconversion by chemical exchange during the mixing time, τ_m. Finally, the third pulse reconverts the longitudinal magnetization into observable transverse magnetization, and the resultant free induction decay is acquired during t_2. The phases $\Phi_1-\Phi_3$ were cycled so as to select populations during τ_m and to suppress axial peaks. The choice of τ_m was a compromise: it was chosen large enough to show up all cross-peaks, but small enough to fulfil the condition of initial rate $(\tau_m < 1/k_{max})$, where k_{max} is the largest exchange rate. Moreover, τ_m is at least by a factor 30 smaller than the spin-lattice relaxation time and, therefore, does not affect the intensities of both the autopeaks and the cross-peaks [40]. A double Fourier transformation with respect to t_1 and t_2 gives a rectangular stacked spectrum plot with the resonances of the normal spectrum displayed on a diagonal and any exchange is manifested by off-diagonal peaks at the intersection of the chemical shifts of the connected peaks in the two frequency directions (Figure 6.19). The relative intensities of the cross-peaks is related to the relative rates of exchange and so these can be measured for all pathways and changes in both relative and absolute rates with pressure can also be measured. The exchange rates given in Table 6.7 were obtained using the D2DNMR program [41].

The length of the diagonal in Figure 6.19 is 0.50 ppm and the doublet due to free ligand is at high field, to the right of the plot. The doublet at the extreme opposite end of the diagonal is due to the *trans* isomer and the doublet of the *cis* isomer can be discerned a little to high field. At ambient pressure there are strong off-diagonal peaks connecting the resonances of the *cis* and *trans* isomers as would be expected (see above). Weaker off-diagonal peaks connect the free ligand with the *trans* isomer and those connecting free ligand with the *cis* isomer are just discernible. An increase in the pressure produces a small increase in intensity of the off-diagonal peaks due to the *cis*–*trans* isomerization reaction. This isomerization, with a very small negative activation volume, is intramolecular and proceeds via a slightly contracted six-coordinated transition state. On the other hand, the intensities of the off-diagonal peaks owing to the intermolecular ligand exchange on both *cis*- and *trans*-isomers increase strongly with pressure, showing that the ligand exchange reactions are both associative. This is in marked contrast to the

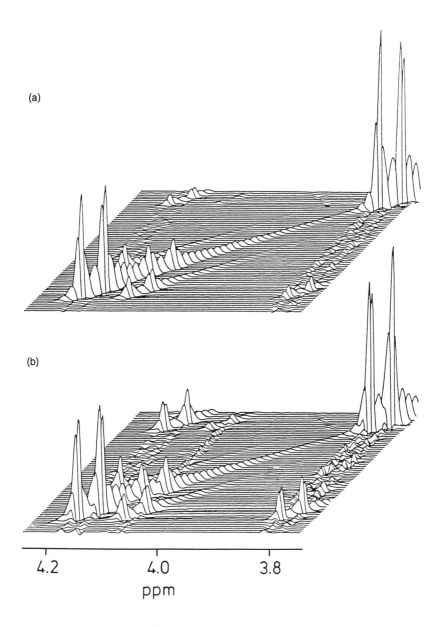

(a)

(b)

4.2   4.0   3.8

ppm

Figure 6.19 400 MHz 2-D ^{1}H EXCSY spectra of *cis*- and *trans*-ZrCl$_4$·2TMPA in the presence of excess TMPA, at two different pressures (a: 0.1 MPa and b: 198 MPa) demonstrating the effect of pressure on the intramolecular and intermolecular exchange pathways. Reproduced by permission of *Helvetica Chimica Acta* from Ref. 39

Table 6.7 Effect of pressure on the rate constants k_c and k_t for inter-molecular exchange of free TMPA on *cis-* and *-trans*[ZrCl$_4$ · 2TMPA], respectively, and on the rate constant k_i for *cis–trans* isomerization (CHCl$_3$/CDCl$_3$ solutions at 256 K) [39] [43]

Pressure/MPa	k_c/s^{-1}	k_t/s^{-1}	k_i/s^{-1}
0.1	0.1$_5$	0.5	5.2
198	0.4	1.7	7.1
$\Delta V^{\#}/$ cm^3 mol^{-1}		-11.1 ± 0.8	-1.6 ± 0.6

similar reactions occurring with the smaller Ti(IV) analogue ($\Delta V^{\#}_{cis} = +17.5$ cm^3 mol^{-1}). The results are summarized in Table 6.7. Given that the reaction of the *trans* complex predominates, it is then possible to obtain activation parameters in the normal way and show, by changing the free ligand concentration, that the ligand exchange is second order, as required for an associative process [42, 43].

4. Review of Work in the Field

4.1 OCTAHEDRAL COMPLEXES

4.1.1 Trivalent ions

The kinetic parameters for TMPA exchange on the diamagnetic ions Al^{3+}, Ga^{3+}, In^{3+} [19] and Sc^{3+} [44] are shown in Table 6.2 (Section 3.2). The results show a clear changeover of mechanism from dissociative (D) to associative (A or I$_a$) as the ionic radius increases. The larger Sc^{3+} and In^{3+} are able to accommodate a seventh water molecule in the transition state: the similarity of the activation volumes with the volume of reaction, ΔV^0, for addition of TMPA on Nd(TMPA)$_6$ [45] suggests a common A mechanism for both. Activation volumes measured for water [46, 47], DMSO and DMF [48] exchange on Al^{3+} and Ga^{3+} are all positive, also indicating dissociative activation modes for these ions. For DMSO and DMF exchange, the activation volumes together with the observed first order rate laws (working in CD$_3$NO$_2$ diluent) show conclusively that the mechanism is dissociative D. For water exchange the activation volume is less than the highest measured for water exchange on a hexa-aqua metal ion (+7.2 cm^3 mol^{-1} for Ni^{2+} [49]), so it can be concluded that an I_d mechanism operates.

The kinetic parameters available for DMSO and DMF exchange on the trivalent first-row transition metal cations are shown in Tables 6.4 and 6.5 [20, 21, 48, 50, 51, 52]. The activation volumes become decreasingly negative on crossing the series from left to right and are positive for Ga^{3+}. Similar results are obtained for water exchange [22, 47, 53, 55] and for TMPA exchange on Sc^{3+} [44] and Ga^{3+} [47]. There is thus clearly a changeover in mechanism from associative to dissociatively activated across the series.

For Ga^{3+} [47], Fe^{3+} [55], Cr^{3+} [54] and the low spin Ru^{3+} [16] and Rh^{3+} [56] in water, hydrolysis is kinetically important (see equation (6.7), Section 3.1 for analysis). The results for Ru^{3+} in Table 6.1 show that the exchange on the hydrolysed species $[Ru(H_2O)_5(OH)]^{2+}$ is two orders of magnitude faster than on the hexa-aqua ion $[Ru(H_2O)_6]^{3+}$, and takes place via a more dissociative mechanism (more positive activation volume) than the I_a mechanism for water exchange on the hexa-aqua ion. It is observed for these ions in general that water exchange on the hydrolysed species is two to four orders of magnitude faster, and more dissociative than on the hexa-aqua species. For Fe^{3+} in methanol [52] only the hydrolytic exchange pathway is observed.

4.1.2 Divalent ions

Activation volumes have been measured for solvent exchange on the high spin first-row transition metal ions in water [49, 57, 58, 59], methanol [60, 61, 62, 63], acetonitrile [64, 61, 65, 66, 67, 68], DMF [61, 69, 70, 71] and ammonia [72] and are shown in Table 6.8. As already discussed for the trivalent analogues, the results show a progressive change of activation mode from associative for the early elements to dissociative for the late elements. This change can be explained by the progressive filling of the d-orbitals and the decrease in ionic radii across the series, both of which disfavour bonding processes. The only solvent for which the mechanistic changeover is not clear cut is DMF, the activation volumes for DMF exchange being always positive, although very small for Mn^{2+}. This is presumably due to the greater bulk of DMF. The positive activation volume ($\Delta V^{\#} = +8.5 \pm 0.2$ cm^3 mol^{-1}) observed for DMF exchange on $[Mg(DMF)_6]^{2+}$, is consistent with a dissociative D mechanism [73]. For the Cu^{2+} ion, solvent exchange is very rapid owing to

Table 6.8 Volumes of activation (cm^3 mol^{-1}) for solvent S exchange on MS_6^{2+} of the first row transition metal series by NMR [11]

M^{2+}	V	Mn	Fe	Co	Ni	Cu
r_i/pm	79	83	78	74	69	$(73)^a$
	t_{2g}^3	$t_{2g}^3 e_g^2$	$t_{2g}^4 e_g^2$	$t_{2g}^5 e_g^2$	$t_{2g}^6 e_g^2$	$t_{2g}^6 e_g^3$
H_2O	-4.1	-5.4	$+3.8$	$+6.1$	$+7.2$	
MeOH		-5.0	$+0.4$	$+8.9$	$+11.4$	$+8.3$
MeCN		-7.0	$+3.0$	$+8.1$	$+8.5$	
DMF		$+2.4$	$+8.5$	$+9.2$	$+9.1$	
$NH_3{}^b$					$+5.9$	

^{a} Effective ionic radius.
^{b} In 15M aqueous NH_3

Jahn–Teller distortion. This was shown for the aqua-ion, where the water exchange rate is rapid ($k_{ex}^{298} = (4.4 \pm 0.1) \times 10^9$ s^{-1}) owing to labilizing on the axial water molecules in a Jahn–Teller distorted complex where the axis of distortion changes with a timescale $\tau_i^{298} = (5.1 \pm 0.6) \times 10^{-12}$ s [74]. No high-pressure measurements have been reported for water exchange on this ion. The exchange of the bidentate ethylenediamine (en) on [Ni(en)$_3$]$^{2+}$ has been measured in en solution [75]. The positive activation volume ($\Delta V^{\#} = +11.4 \pm 2.0$ cm^3 mol^{-1}) indicates a dissociatively activated exchange mechanism, as observed for monodentate solvent exchange on nickel(II).

There have been two theoretical studies of water exchange on the divalent first-row transition metal ions using gas phase *ab initio* calculations [76, 77]. The conclusions of these studies are contradictory. The most recent study concluded, in contrast to the above results, that, on the basis of a correlation between the binding energy of the sixth water molecule and the water exchange rate, the exchange is always dissociative. It should be noted, however, that the binding energy of a seventh water molecule was not calculated. The previous study concluded that the transition state is always seven coordinate, corresponding to an interchange reaction mechanism with the metal-oxygen distance of the entering and leaving water molecules determining whether the reaction is I_a or I_d. This is in better agreement with the experimental results, and indeed a correlation was observed between the calculated metal-oxygen distance of the entering and leaving water molecules and the measured activation volumes. Both these studies, however, suffer from the limitations of comparing gas-phase calculation with experimental results for solutions, i.e., they do not take into account the effect of second and further hydration shells.

4.2 TETRAHEDRAL COMPLEXES

The only tetrahedral species for which systematic solvent exchange kinetic studies have been carried out are the tetrasolvento beryllium(II) ions with water, DMSO, TMPA, DMF, tetramethyl urea (TMU) and dimethylpropylenurea (DMPU) [78]. The results are shown in Table 6.9. Use of inert diluents shows that for TMU and DMPU the exchange rate law is first order, and the positive activation entropy and volume are indicative of a D mechanism. For DMSO and TMPA, the rate law is second order and the activation entropy and volume are negative, the exchange being A or I_a. For DMF a first-order dissociative pathway and a second-order associatively activated pathway operate simultaneously. The activation volume, obtained in conditions where the second order mechanism dominates, is small but negative, consistent with an associatively activated mechanism. For water exchange, it can be seen that the small positive activation entropy is not very helpful in assigning a mechanism, whereas the large negative activation volume is unequivocal evidence for an associative activation mode. Indeed, the value is the most

Table 6.9 Rate constants and activtion parameters for solvent S exchanges on BeS_4^{2+} [a] [78]

	k^{298}	$\Delta H^{\#}$/kJ mol^{-1}	$\Delta S^{\#}$/J K^{-1} mol^{-1}	$\Delta V^{\#}$/ cm^3 mol^{-1}	Mechanism
H$_2$O	730[b,c]	59.2	+8.4	−13.6	A
DMSO	213[d]	35.0	−83.0	−2.5	A, I$_a$
TMPA	4.2[d]	43.5	−87.1	−4.1	A, I$_a$
DMF[e]	16[d]	52.0	−47.3	−3.1	A, I$_a$
	0.2[b]	74.9	−7.3	−	D
TMU	1.0[b]	79.6	+22.3	+10.5	D
DMPU	0.1[b]	92.6	+47.5	+10.3	D

[a] In CD$_3$NO$_2$ as diluent by ^{1}H NMR, except for H$_2$O in neat water by ^{17}O NMR.
[b] First-order rate constant (s^{-1}).
[c] Rate constant and $\Delta S^{\#}$ for exchange of a particular water molecule recalculated to second order units, i.e. $k_{ex}^{298}/55$: 13.1 m^{-1} s^{-1} and −24.9 J K^{-1} mol^{-1}.
[d] Second-order rate constant (m^{-1} s^{-1}).
[e] Two-term rate law.

negative value recorded for water exchange on a metallic aqua-ion, and is good evidence for an A mechanism.

4.3 LANTHANIDE IONS

Water exchange on the octa-aqua lanthanide ions was discussed in Section 3.5. DMF exchange has also been measured on the heavy lanthanide ions [Ln(DMF)$_8$]$^{3+}$ [79, 45, 80] and the results are shown in Table 6.10. The activation volumes show that the activation mode is always dissociative although the mechanism varies from I$_d$ for Tb^{3+} to D for Yb^{3+}.

Polyaminocarboxylate complexes with Gd^{3+} are used as contrast agents in medical magnetic resonance imaging. The contrast produced by these agents depends on their ability to induce NMR relaxation in the protons in surrounding water, which in turn depends on the water proton exchange rate. Until recently, it was assumed that the water exchange rate on these complexes was as rapid as for the octa-aqua Gd^{3+}. The water exchange rates have now been measured

Table 6.10 Kinetic parameters for DMF exchange on Ln(DMF)$_8^{3+}$ [80]

Ln^{3+}	$10^{-5}\ k_{ex}^{200}$/s^{-1}	$10^{-7}\ k_{ex}^{298}$/s^{-1}	$\Delta H^{\#}$/kJ mol^{-1}	$\Delta S^{\#}$/J K^{-1} mol^{-1}	$\Delta V^{\#}$/ cm^3 mol^{-1}
Tb	7.9 ± 0.2	1.9 ± 0.2	14.1 ± 0.4	−58 ± 2	+5.2 ± 0.2
Dy	2.8 ± 0.1	0.63 ± 0.03	13.8 ± 0.4	−69 ± 2	+6.1 ± 0.2
Ho	1.16 ± 0.02	0.36 ± 0.06	15.3 ± 0.8	−68. ± 4	+5.2 ± 0.5
Er	0.80 ± 0.03	1.3 ± 0.4	23.6 ± 1.8	−30 ± 9	+5.4 ± 0.3
Tm	0.29 ± 0.01	3.1 ± 0.3	32.2 ± 0.5	+10 ± 2	+7.4 ± 0.3
Yb	0.28 ± 0.01	9.9 ± 0.9	39.3 ± 0.6	+40 ± 3	+11.8 ± 0.4

Table 6.11 Water exchange kinetic parameters obtained for different Gd^{3+} complexes

	k_{ex}^{298}/s^{-1}	$\Delta H^{\#}/kJ\,mol^{-1}$	$\Delta S^{\#}/J\,K^{-1}\,mol^{-1}$	$\Delta V^{\#}/cm^3\,mol^{-1}$	Reference
$[Gd(H_2O)_8]^{3+}$	$(8.30 \pm 0.95) \times 10^8$	14.9 ± 1.3	-24.1 ± 4.1	-3.3 ± 0.2	[35]
$[Gd(PDTA)(H_2O)_2]^-$	$(1.02 \pm 0.10) \times 10^8$	11.0 ± 1.4	-54.6 ± 4.6	-1.5 ± 0.5	[35]
$[Gd(DTPA)H_2O]^{2-}$	$(4.1 \pm 0.3) \times 10^6$	52.0 ± 1.4	$+56.2 \pm 5$	$+12.5 \pm 0.2$	[81]
$[Gd(DTPA\text{-}BMA)H_2O]$	$(4.3 \pm 0.4) \times 10^5$	46.6 ± 1.3	$+18.9 \pm 4$	$+7.3 \pm 0.2$	[82]
$[Gd(DOTA)H_2O]^-$	$(4.8 \pm 0.4) \times 10^6$	48.8 ± 1.6	$+46.6 \pm 6$	$+10.5 \pm 0.2$	[81]

using ^{17}O NMR for the Gd^{3+} complexes with propylenediamine-tetraacetate ($PDTA^{4-}$) [35], diethylenetriaminepentaacetate ($DTPA^{5-}$) [81], tetraazacyclo-dodecane-tetraacetate ($DOTA^{4-}$) [81] and diethylenetriamine-pentaacetate-bismethylamide ($DTPA-BMA^{3-}$) [82]. The results are shown in Table 6.11. The exchange rates on the three contrast agents, $[Gd(DTPA)(H_2O)]^{2-}$, $[Gd(DOTA)(H_2O)]^-$ and $[Gd(DTPA-BMA)(H_2O)]$ are 200 to 2000 times lower than on the aqua-ion. This rate decrease is accompanied by a change of sign of the activation volume, the large positive values for the three contrast agents indicating dissociatively, rather than associatively, activated water exchange. The contrast agents can accommodate only one inner sphere water molecule, so that the incoming water molecule cannot participate in the water exchange, which will have a dissociative activation mode, and probably a limiting dissociative D mechanism. Without the participation of the incoming water molecule, more energy is required to break the bond between the outgoing water molecule and the highly charged Gd^{3+}, leading to the higher activation enthalpies and lower water exchange rates.

4.4   SQUARE PLANAR COMPLEXES

There have been several studies of Pt and Pd complexes with this stereochemi-stry. The results for homoleptic solvent exchange on square planar PtS_4^{2+} and PdS_4^{2+} complexes are summarized in Table 6.12 [83, 84, 85, 86, 87, 88]. For all these reported systems second order rate laws and negative values for both the entropies and volumes of activation are observed, indicating an associative mechanism, A or I_A, for the solvent exchange. This is consistent with the mechanism found previously for other substitution reactions on square planar complexes. From Table 6.12 it is clear that, with the exception of sulphur bound solvents, the activation volumes for the solvent exchange on Pt(II) and Pd(II) tetrasolvates are small and negative: those for Pt(II) are systematically more negative than those for Pd(II). The interpretation of these volumes in molecular terms is not trivial, as these solvates undergo large changes in coordination geometry on going from ground state to transition state. Tetrasolvate species may have their two axial sites occupied by loosely bound solvent molecules. The simultaneous loss of one of the axially bound solvent molecules with formation of a new bond with the second solvent molecule in going to a five coordinate transition state or intermediate may mean that through compensation, the overall volume change is small or negligible. The large magnitude of the activation volumes for the R_2S systems may be explained in terms of steric encumbrance. If the steric encumbrance of the solvents increases it becomes unlikely that two molecules of solvent will occupy sites above and below the plane defined by the complex. Formation of a pentacoordinate transition state in this case will lead to a significant decrease in volume. A similar observation has been noted for water exchange on $Pd(R_5dien)(H_2O)^{2+}$ complexes [89, 90]. The sterically unhindered complex

Table 6.12 Rate constants and activation parameters for solvent S exchange on PdS_4^{2+} and PtS_4^{2+} complexes in diluent CD_3NO_2

	k_2^{298} /mol dm³ s⁻¹	$\Delta H^{\#}$/kJ mol⁻¹	$\Delta S^{\#}$/J K⁻¹ mol⁻¹	$\Delta V^{\#}$/ cm³ mol⁻¹	Reference
$[Pd(Et_2S)_4]^{2+}$	5.0	50.4	−63	−11.6	[83]
$[Pd(H_2O)_4]^{2+}$ [a]	10	49.5	−60	−2.2	[84]
$[Pd(DMA)_4]^{2+}$	35	43.2	−76	−2.8	[83]
$[Pd(MeCN)_4]^{2+}$	49	45.4	−60	−0.1	[83]
$[Pd(DMF)_4]^{2+}$	153	41.9	−62	−0.2	[83]
$[Pd(Me_2S)_4]^{2+}$	2140	31.9	−74	−9.4	[83]
$[Pd(1,4\text{-dithiane})_2]^{2+}$	9800	22.9	−92	−9.8	[86]
$[Pd(MeNC)_4]^{2+}$ [b]	1.1×10^6	16.4	−75	−3.1	[83]
$[Pt(H_2O)_4]^{2+}$ [a]	7.1×10^{-6}	89.7	−43	−4.6	[87]
$[Pt(Me_2S)_4]^{2+}$	1.5	42.1	−100	−22.0	[86]
$[Pt(DMSO)_4]^{2+}$ [c]	2	47	−74	−5	[88]
$[Pt(1,4\text{-dithiane})_2]^{2+}$	29	32.9	−110	−12.6	[86]
$[Pt(DMSO)_4]^{2+}$ [d]	3200	32.8	−62	−2.5	[88]
$[Pt(MeNC)_4]^{2+}$ [b]	6.2×10^5	13.8	−88	−3.7	[83]

[a] In neat water, rate constant and $\Delta S^{\#}$ for the exchange of a particular water molecule recalculated to second-order units, i.e. $k_{ex}^{298}/55$.
[b] In diluent CD_3CN.
[c] S bonded.
[d] O bonded.

with R = H has an activation volume of $-2.8 \text{ cm}^3 \text{ mol}^{-1}$, but the value for the much more sterically encumbered complex R = Et is $-7.7 \text{ cm}^3 \text{ mol}^{-1}$.

The rather complex reactions on Pt(DMSO)_4^{2+} have also been given attention [88]. This complex contains two differently bonded pairs of ligands which are *cis* to each other, one pair bonded via oxygen and one pair via sulphur, and the two types of ligand exchange with free ligand at widely different rates, the oxygen bonded ligand exchanging the most rapidly (see Table 6.12). An interesting investigation has been carried out in the case of ligand exchange on *trans*-$\text{Pd(Me}_2\text{S)}_2\text{Cl}_2$ [91]. The activation volumes were obtained in several solvents with different electrostriction properties in order to test for any contribution from solvent electrostriction to the volume measured. There was a correlation with the value of the solvent electrostriction parameter q_p ($q_p = [3/(2\varepsilon + 1)^2]/(\delta\varepsilon/\delta P)_T$) and the intrinsic activation volume is estimated to be $-5.9 \text{ cm}^3 \text{ mol}^{-1}$. The ligand exchange on PtR_2L_2, which was found to go through a dissociative mode of activation, is described in Section 3.6.

4.5   OTHER RECENT STUDIES

Water and acetonitrile exchange on hexasolvento Ru(II) were discussed in Section 3.1. The near zero activation volumes show that interchange I mechanisms operate. Measurements have been made of acetonitrile exchange on two half-sandwich Ru(II) compounds, $[\text{Ru}(\eta^6\text{-C}_6\text{H}_6)(\text{CH}_3\text{CN})_3]^{2+}$ and $[\text{Ru}(\eta^5\text{-C}_5\text{H}_5)(\text{CH}_3\text{CN})_3]^+$ [92]. The activation volumes show that the exchange mechanism is I_d on the former and D on the latter compound. The exchange rate increases dramatically (by 5 orders of magnitude) at each step on going from $[\text{Ru(CH}_3\text{CN})_6]^{2+}$ to $[\text{Ru}(\eta^6\text{-C}_6\text{H}_6)(\text{CH}_3\text{CN})_3]^{2+}$ to $[\text{Ru}(\eta^5\text{-C}_5\text{H}_5)(\text{CH}_3\text{CN})_3]^+$. This is correlated with a stepwise change of 0.031 Å of the solid state Ru–N distances, and is due to the mutual *trans* influence of the ligands.

Studies have also been made of solvent exchange on oxo metal ions. High pressure studies have been made of DMSO exchange on oxotitanium(IV) [93]. Low temperature measurements in CD_3NO_2 diluent show signals corresponding to equatorial e-DMSO and axial a-DMSO in $[\text{TiO(e-DMSO)}_4(\text{a-DMSO})]$ as well as free DMSO. Two exchange processes are observed. For the first, corresponding to the fast exchange of the loosely bound a-DMSO with the bulk, a D mechanism is suggested. It is proposed that the second process takes place through migration of the a-DMSO molecule to the equatorial plane and concerted loss of an e-DMSO molecule. This leads to an expanded transition state in which the breaking of the Ti(IV)-e-DMSO bond dominates, as suggested by the positive activation volume. The exchange of HMPA (hexamethylphosphoramide) and TMPA has been measured on dioxouranium(VI) [94]. The rate law for HMPA exchange on $[\text{UO}_2(\text{HMPA})_4]^{2+}$ is second order, indicating an associative A mechanism. The negative activation volume ($\Delta V^{\#} = -11.3 \pm 1.4$ cm^3 mol^{-1}) is consistent with this assignment, but is only a fraction of the partial molar volume of HMPA calculated from the density of the pure solvent: $V_s^0 = 175.0$ cm^3 mol^{-1}. This suggests that as the entering HMPA forms the fifth equatorial bond the other four bound water molecules move outwards, substantially counterbalancing the expected decrease in volume. The rate law for TMPA exchange on [UO$_2$(TMPA)$_5$]$^{2+}$ is first-order, indicating a D mechanism. The activation volume is indeed positive, but it is very small ($\Delta V^{\#} = +2.1 \pm 1.5$ cm^3 mol^{-1}). The primary characteristic of a D process transition state is the elongation of the U–O bond of the leaving TMPA to the point at which there is no possibility of return. Simultaneously with this the other four U–O bond distances will decrease, leading to a small activation volume, which also represents a very small fraction of the partial molar volume of TMPA: $V_s^0 = 117.95$ cm^3 mol^{-1}.

The exchange of water and acetonitrile have been studied on the bimetallic Rh(II) complexes [Rh$_2$(H$_2$O)$_{10}$]$^{4+}$ and [Rh$_2$(CH$_3$CN)$_{10}$]$^{4+}$ [95]. High-pressure measurements have been made for the CH$_3$CN complex. The NMR spectra and crystal structure indicate eight equatorial e-CH$_3$CN and two axial a-CH$_3$CN per Rh$_2^{4+}$ unit. The a-CH$_3$CN exchanges very rapidly (too rapidly to be measured) with the bulk, most probably through a dissociatively activated mechanism. The e-CH$_3$CN is slower ($k_{ex}^{298} = (3.1 \pm 0.4) \times 10^{-5}$ s^{-1}), and a reaction pathway involving concerted migration of the a-CH$_3$CN and release of an e-CH$_3$CN is suggested by similarity with the case of [TiO(DMSO)$_5$]$^{2+}$ described above, although direct intermolecular exchange cannot be excluded. The negative activation volume ($\Delta V^{\#} = -4.9 \pm 0.2$ cm^3 mol^{-1}) indicates a contracted transition state.

5. Conclusions

We hope that we have been able to demonstrate how high-pressure NMR can be applied to the study of chemical reaction kinetics, and to show the enormous impact such studies have made on our understanding of a class of reaction mechanisms that is fundamental in inorganic chemistry, i.e. solvent exchange reactions on metal ions. In particular, we have tried to show that the determination of the sign of an activation volume, in the absence of electrostrictive changes, allows unequivocal assignment of at the very least the activation mode and that this is often sufficient to elucidate the mechanism of the reaction. It should not be forgotten that the methodology presented here can be applied to a wide variety of systems. The unequivocal nature of the activation volume indicates that variable pressure studies will be rewarding in a great number of fields. It will also be interesting to examine the basic concept of the activation volume more deeply in terms of real molecular interactions and so obtain a better physical understanding of the way molecules can approach and react with one another within a solvent matrix.

Acknowledgements

We wish to thank the large number of people who have contributed over the years to the work performed at Lausanne, as well as all those groups elsewhere with whom we have had fruitful collaborations. We also wish to thank the Swiss National Science Foundation for continued and generous support of our work.

References

1. R. van Eldik (ed.), *Inorganic High Pressure Chemistry: Kinetics and Mechanisms*, Elsevier, Amsterdam 1986.
2. H. R. Hunt and H. Taube, *J. Am. Chem. Soc.*, **80**, 2642 (1958).
3. K. Heremans, J. Snauwaert and J. Rijkenberg, in K. D. Timmerhaus and M. S. Barber (eds), *High Pressure Science and Technology*, Vol. 1, Plenum, New York, 1979.
4. E. F. Caldin, M. W. Grant and B. B. Hasinoff, *J. Chem. Soc., Faraday I*, **68**, 2247 (1972).
5. K. R. Brower, *J. Am. Chem. Soc.*, **90**, 5401 (1968).
6. H. Vanni, W. L. Earl and A. E. Merbach, *J. Magn. Reson.*, **29**, 11 (1978).
7. H. Kelm (ed.), *High Pressure Chemistry*, Reidel, Dordrecht, 1978.
8. D. A. Palmer and H. Kelm, *Coord. Chem. Rev.*, **36**, 89 (1981).
9. C. H. Langford and H. B. Gray, *Ligand Substitution Processes*, Benjamin, New York, 1965.
10. T. W. Swaddle, *Adv. Inorg. Bioinorg. Mechanisms*, A. G. Sykes (ed.), **2**, 95 (1983).
11. A. E. Merbach and J. W. Akitt, *NMR Basic Principles and Progress*, **24**, 189 (1990).
12. R. van Eldik and A. E. Merbach, *Comments Inorg. Chem.*, **12**, 341 (1992).
13. W. L. Earl, H. Vanni and A. E. Merbach, *J. Magn. Reson.*, **30**, 571 (1978).
14. U. Frey, L. Helm and A. E. Merbach, *High Press. Res.*, **2**, 237 (1990).
15. H. Taube, *Comments Inorg. Chem.*, **1**, 17 (1981) and refs. therein.
16. I. Rappaport, L. Helm, A. E. Merbach, P. Bernhard and A. Ludi, *Inorg. Chem.*, **27**, 873 (1988).
17. P. Bernhard, L. Helm, A. Ludi and A. E. Merbach, *J. Am. Chem. Soc.*, **107**, 312 (1985).
18. J. J. Delpuech, J. Ducom and V. Michon, *Bull. Soc. Chim. Fr.*, 1848 (1971).
19. A. E. Merbach, *Pure & Appl. Chem.*, **54**, 1479 (1982).
20. I. Dellavia, P.-Y. Sauvageat, L. Helm, Y. Ducommun and A. E. Merbach, *Inorg Chem.*, **31**, 792 (1992).
21. I. Dellavia, L. Helm and A. E. Merbach, *Inorg. Chem.*, **31**, 2230 (1992).
22. A. D. Hugi, L. Helm and A. E. Merbach, *Inorg. Chem.*, **26**, 1763 (1987).
23. T. J. Swift and R. E. Connick, *J. Chem. Phys.*, **37**, 307 (1962).
24. K. E. Newman, F. K. Meyer and A. E. Merbach, *J. Am. Chem. Soc.*, **101**, 1470 (1979).
25. P. Furrer, Ph. D. Thesis, Univ. Lausanne (1993).
26. R. Zimmermann and W. E. Brittin, *J. Phys. Chem.*, **61**, 1328 (1957).
27. B. M. Alsaadi, F. J. C. Rossotti and R. J. P. Williams, *J. Chem. Soc., Dalton Trans.*, 2147 (1980).
28. C. Cossy, L. Helm and A. E. Merbach, *Inorg. Chim. Acta*, **139**, 147 (1987).
29. S. Meiboom and D. Gill, *Rev. Sci. Instrum.*, **29**, 688 (1958).
30. R. V. Vold, J. S. Waugh, M. P. Klein and D. E. Phelps, *J. Chem. Phys.*, **48**, 3831 (1968).

31. C. Cossy, L. Helm and A. E. Merbach, *Inorg Chem.*, **27**, 1973 (1988).
32. C. Cossy, L. Helm and A. E. Merbach, *Inorg. Chem.*, **28**, 2699 (1989).
33. R. V. Southwood-Jones, W. L. Earl, K. E. Newman and A. E. Merbach, *J. Chem. Phys.* **73**, 5909 (1980).
34. D. H. Powell, A. E. Merbach, G. Gonźalez, E. Brücher, K. Micskei, M. F. Ottaviani, K. Köhler, A. von Zelewsky, O. Ya. Grinberg and Ya. S. Lebedev, *Helv. Chim. Acta*, **76**, 2129 (1993).
35. K. Micskei, D. H. Powell, L. Helm, E. Brücher and A. E. Merbach, *Magn. Res. Chem.*, **31**, 1011 (1993).
36. U. Frey, L. Helm, A. E. Merbach and R. Romeo, *J. Am. Chem. Soc.*, **111**, 8161 (1989).
37. P. J. Hore, *J. Magn. Reson.*, **55**, 283 (1983).
38. J. J. Led and H. Gesmar, *J. Magn. Reson.*, **49**, 444 (1982).
39. U. Frey, L. Helm and A. E. Merbach, *Helv. Chim. Acta*, **73**, 199 (1990).
40. R. Willem, *Progr. Nucl. Magn. Reson. Spectrosc.*, **20**, 1 (1987).
41. E. W. Abel, T. P. J. Coston, K. G. Orrel, S. Sick and D. Stephenson, *J. Magn. Reson.*, **69**, 92 (1986).
42. M. Turin-Rossier, D. Hugi-Cleary and A. E. Merbach, *Inorg. Chim. Acta*, **167**, 245 (1990).
43. M. Turin-Rossier, D. Hugi-Cleary, U. Frey and A. E. Merbach, *Inorg. Chem.*, **29**, 1374 (1990).
44. L. Helm, C. Ammann and A. E. Merbach, *Z. Phys. Chem. Neue Folge*, **155**, 145 (1987).
45. D. L. Pisanello, P. J. Nichols, Y. Ducommun and A. E. Merbach, *Helv. Chim. Acta*, **65**, 1025 (1982).
46. D. Hugi-Cleary, L. Helm and A. E. Merbach, *Helv. Chim. Acta*, **68**, 545 (1985).
47. D. Hugi-Cleary, L. Helm and A. E. Merbach, *J. Am. Chem. Soc.*, **109**, 4444 (1987).
48. C. Ammann, P. Moore, A. E. Merbach and C. H. McAteer, *Helv. Chim. Acta*, **63**, 268 (1980).
49. Y. Ducommun, W. L. Earl and A. E. Merbach, *Inorg. Chem.*, **18**, 2754 (1979).
50. D. L. Carle and T. W. Swaddle, *Inorg. Chem.*, **20**, 4212 (1981).
51. S. T. D. Lo and T. W. Swaddle, *Inorg. Chem.*, **14**, 1878 (1975).
52. F. K. Meyer, A. R. Monnerat, K. E. Newman and A. E. Merbach, *Inorg. Chem.*, **21**, 774 (1982).
53. A. D. Hugi, L. Helm and A. E. Merbach, *Helv. Chim. Acta*, **68**, 508 (1985).
54. S. Xu, H. R. Krouse and T. W. Swaddle, *Inorg. Chem.*, **24**, 267 (1985).
55. T. W. Swaddle and A. E. Merbach, *Inorg. Chem.*, **20**, 4212 (1981).
56. G. Laurenczy, I. Rapaport, D. Zbinden and A. E. Merbach, *Magn. Reson. Chem.*, **29**, 45 (1991).
57. Y. Ducommun, D. Zbinden and A. E. Merbach, *Helv. Chim. Acta*, **65**, 1385 (1982).
58. Y. Ducommun, K. E. Newman and A. E. Merbach, *Helv. Chim. Acta*, **62**, 2511 (1979).
59. Y. Ducommun, K. E. Newman and A. E. Merbach, *J. Am. Chem. Soc.*, **101**, 5588 (1979).
60. W. L. Earl, F. K. Meyer and A. E. Merbach, *Inorg. Chim. Acta*, **25**, 91 (1977).
61. F. K. Meyer, K. E. Newman and A. E. Merbach, *Inorg. Chem.*, **18**, 2142 (1979).
62. F. K. Meyer, K. E. Newman and A. E. Merbach, *J. Am. Chem. Soc.*, **101**, 5588 (1979).
63. L. Helm, S. F. Lincoln, A. E. Merbach and D. Zbinden, *Inorg. Chem.*, **25**, 2550 (1986).
64. K. E. Newman, F. K. Meyer and A. E. Merbach, *J. Am. Chem. Soc.*, **101**, 1470 (1979).

65. Y. Yano, M. T. Fairhurst and T. W. Swaddle, *Inorg. Chem.*, **19**, 3267 (1980).
66. M. J. Sisley, Y. Yano and T. W. Swaddle, *Inorg. Chem.*, **21**, 1141 (1982).
67. A. R. Monnerat, P. Moore, K. E. Newman and A. E. Merbach, *Inorg. Chim. Acta*, **47**, 139 (1981).
68. M. Ishii, S. Funahashi, K. Ishihara and M. Tanaka, *Bull. Chem. Soc. Japan*, **62**, 1852 (1989).
69. C. Cossy, L. Helm and A. E. Merbach, *Helv. Chim. Acta*, **70**, 1516 (1987).
70. P. Moore and L. Fielding, *J. Chem. Soc., Chem. Comm.*, 49 (1988).
71. M. Ishii, S. Funahashi and M. Tanaka, *Chem. Letts.*, 871 (1987).
72. R. L. Batstone Cunningham, D. W Dodgen and J. P. Hunt, *Inorg. Chem.*, **21**, 3831 (1982).
73. P. Furrer, U. Frey, L. Helm and A. E. Merbach, *High Pressure Research*, **7**, 144 (1991).
74. D. H. Powell, L. Helm and A. E. Merbach, *J. Chem. Phys.*, **95**(12), 9258 (1991).
75. S. Soyama, M. Ishii, S. Funahashi and M. Tanaka, *Inorg. Chem.*, **31**, 536 (1992).
76. R. Akesson, L. G. M. Petterson, M. Sandström, P. E. M. Siegbahn and U. Wahlgren, *J. Phys. Chem.*, **97**, 3765 (1993).
77. S.-K. Kang, B. Lam, T. A. Albright and J. F. O'Brien, *New J. Chem.*, **15**, 757 (1991).
78. P.-A. Pittet, G. Elbaze, L. Helm and A. E. Merbach, *Inorg. Chem.*, **29**, 1936 (1990).
79. C. Cossy and A. E. Merbach, *Pure & Appl. Chem.*, **60**, 1785 (1988).
80. D. L. Pisaniello, L. Helm, P. Meier and A. E. Merbach, *J. Am. Chem. Soc.*, **105**, 4528 (1983).
81. K. Micskei, L. Helm, E. Brücher and A. E. Merbach, *Inorg. Chem.*, **32**, 3844 (1993).
82. G. González, D. H. Powell, V. Tissières and A. E. Merbach, *J. Phys. Chem.*, **98**, in press.
83. N. Hallinan, V. Besançon, M. Forster, G. Elbaze, Y. Ducommun and A. E. Merbach, *Inorg. Chem.*, **30**, 1112 (1991).
84. L. Helm, L. I. Elding and A. E. Merbach, *Helv. Chim. Acta*, **67**, 1453 (1984).
85. B. Moullet, C. Zwahlen, U. Frey, G. Gervasio and A. E. Merbach, *to be submitted*.
86. U. Frey, S. Elmroth, B. Moullet, L. I. Elding and A. E. Merbach, *Inorg. Chem.*, **30**, 5033 (1991).
87. L. Helm, L. I. Elding and A. E. Merbach, *Inorg. Chem.*, **24**, 1719 (1985).
88. Y. Ducommun, L. Helm, A. E. Merbach, B. Hellquist and L. I. Elding, *Inorg. Chem.*, **28**, 377 (1989).
89. J. Berger, M. Kotowski, R. van Eldik, U. Frey, L. Helm and A. E. Merbach, *Inorg. Chem.*, **28**, 3759 (1989).
90. L. Helm, A. E. Merbach, M. Kotowski and R. van Eldik, *High Pressure Res.*, **2**, 49 (1989).
91. M . Tubino and A. E. Merbach, *Inorg. Chim. Acta*, **71**, 149 (1983).
92. W. Luginbühl, P. Zbinden, P.-A. Pittet, T. Armbruster, H.-B. Bürgi, A. E. Merbach and A. Ludi, *Inorg. Chem.*, **30**, 2350 (1991).
93. I. Dellavia, L. Hehm and A. E. Merbach, *Inorg. Chem.*, **31**, 4151 (1992).
94. A. Abou-Hamdan, N. Burki, S. F. Lincoln, A. E. Merbach and S. J. F. Vincent, *Inorg. Chim. Acta*, **207**, 27 (1993).
95. P.-A. Pittet, L. Dadci, P. Zbinden, A. Abou-Hamdan and A. E. Merbach, *Inorg. Chim. Acta*, **206**, 135 (1993).

7 APPLICATIONS OF FIELD GRADIENTS IN NMR

D. Canet

Université Henri Poincaré, Nancy I, France

and

M. Décorps

Université Joseph Fourier, Grenoble I, France

Dynamics of Solutions and Fluid Mixtures by NMR
Edited by J.-J. Delpuech © 1995 John Wiley & Sons Ltd

1 Introduction

Most NMR experiments require the best possible homogeneity for both magnetic fields: the static magnetic field B_0 and the alternative magnetic field B_1. Concerning B_0, this commitment is dictated by the necessity of obtaining narrow lines whereas an homogeneous B_1 field permits to irradiate all parts of the sample in an identical manner and to deal with well defined pulse lengths and uniform receptivity. However, in some instances, inhomogeneous fields may be advantageous for determining dynamical or structural parameters. As a matter of fact, it was recognized very early [1] that the presence of a static field gradient could alter significantly the results of spin echo experiments, because of translational diffusion of molecules within the sample. In turn, this experiment, if conveniently performed, can lead to the determination of the self-diffusion coefficient. Likewise, at the same period it was noticed that when saturated spins are replaced by flowing unsaturated spins, a simple NMR experiment becomes sensitive to this coherent motion [2].

The interest in using magnetic field gradients has tremendously increased for the past twenty years. As everyone knows, the most spectacular application is NMR imaging [3] (most of the time coined MRI for Magnetic Resonance Imaging). This important topic of NMR will be covered briefly here within the context of dynamics of fluids; however, other important fields of applications relevant to this context are either well established (determination of translational diffusion parameters, measurements of flow velocities etc.) or have been the subject of recent developments (localized spectroscopy, selection of coherence pathways, solvent suppression, etc.).

In a general way, it can be stated that the classical relaxation parameters lend themselves to rotational dynamics whereas information about translational motions is better derived from spatial labeling techniques. In NMR, field gradients constitute the easiest way to reach this goal.

2 Gradients

The definition of a field gradient stems from its *spatial* variation. Generally, we shall be concerned with *uniform* field gradients; this means that, over the sample or the object under investigation, the derivative of B with respect to a given spatial variable, say X, is constant. As an example, $g_{0X} = \partial B_0/\partial X$ denotes a uniform B_0 gradient in the X direction. We shall in the following distinguish the laboratory frame (X, Y, Z), with Z coinciding with the direction of B_0, from the rotating frame (x, y, Z) where spin dynamics is better handled because this frame is defined in such a way that the radio-frequency field appears stationary along x, y or any other direction in the xy plane according to its phase.

The first question to address is to understand how a field gradient acts on a spin system. This can be outlined by two keywords: *spatial labeling* and *defocusing* (or *dephasing*). The former means that a spectroscopic property will become spatially dependent (or spatially encoded) if a gradient has been used in the course of the NMR experiment. As explained below, this will be the *precession* frequency (rotation around Z) when B_0 gradients are used, or the *nutation* frequency (rotation around the rotating frame axis coinciding with B_1) in the case of B_1 gradients (Figure 7.1). The second keyword means that, if the whole sample is concerned, nuclear magnetization is scattered or dispersed in a plane perpendicular to the magnetic field applied in the form of a gradient. If the gradient is applied for a time long enough, this process leads to the cancellation of magnetization (on an average); it must nevertheless be noticed that this process is reversible since refocusing occurs if the gradient is reversed.

In order to make these statements more clear, let us consider the simple experiment which consists in flipping the magnetization from its equilibrium axis (the Z axis) to the xy plane by means of a 90° radio-frequency pulse (rf pulse). In the presence of a homogeneous B_0 field, precession occurs at a single frequency, ν_0, the same for any location within the sample, given by:

$$\nu_0 = (\gamma B_0 / 2\pi)(1 - \sigma) \tag{7.1}$$

where σ is the shielding constant which accounts for the chemical shift phenomenon and which will be omitted almost systematically in the following because we shall consider most of the time a single chemical species. In fact,

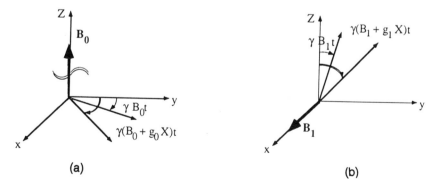

Figure 7.1 (a) Precession with respect to a static magnetic field involving a gradient: the magnetization rotates in the xy plane by an angle which depends on the spatial position. (b) nutation with respect to a radio-frequency field gradient along the x direction of the rotating frame; again the magnetization rotates according to the location of the molecule to which it belongs

because a NMR spectrometer operates in the rotating frame (for receive operations) the actual precession frequency is:

$$f_0 = \nu_0 - \nu_r = (\gamma B_0 / 2\pi) - \nu_r \tag{7.2}$$

where ν_r is the so-called carrier frequency (frequency of the transmitter and generally the reference frequency of the receiver). As a consequence, f_0 lies in the domain of audio-frequencies whereas ν_0 lies in the domain of radio-frequencies. Now, let us suppose that B_0 is not homogeneous but varies linearly according to one spatial coordinate, say X:

$$B_0(X) = B_0 + g_X^0 X \tag{7.3}$$

The precession frequency becomes spatially dependent:

$$f(X) = f_0 + (\gamma/2\pi) g_X^0 X \tag{7.4}$$

In other words, precession frequencies are distributed over a range depending on the gradient magnitude (g_X^0) and of the space occupied by the sample (object) under investigation (Figure 7.2). The NMR signal is weighted by the spin density (or by the local concentration of the considered species) at location X. If a frequency analysis is performed, for instance through Fourier transformation, the spectrum (in the frequency domain) can be displayed as a function of X and its amplitude, at each point, shows up the local spin density. This is the principle of NMR imaging (Section 4). Now, if instead of sampling the precession signal as a function of time in view of performing a Fourier analysis, we look at the global magnetization components, we find a net result which is essentially zero, provided that the gradient is sufficiently strong to scatter the whole magnetization in the xy plane and to average it out. The dispersed magnetization could, however, be *refocused* by the application of the same gradient pulse if a rf inverting pulse is inserted between the two gradient

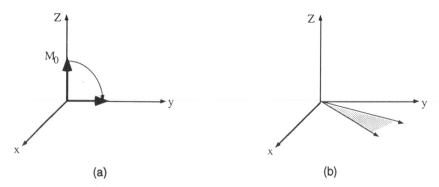

(a)   (b)

Figure 7.2 A 90° rf pulse (B_1 is supposed to coincide with the x axis of the rotating frame) takes the equilibrium magnetization to the y axis. (b) If precession occurs in the presence of a gradient, magnetization spreads in the xy plane according to the gradient amplitude and to the object shape

pulses or by inverting the polarity of the second gradient pulse. This feature is exploited for the study of motions, coherent if one is dealing with flow, or incoherent in the case of translational diffusion. Sensitivity to translational motion is usually increased by inserting a time interval between the defocusing and refocusing gradient pulses (Figure 7.3). During this time interval, motions manifest themselves leading to an attenuation or a dephasing of the refocused magnetization. This is the basis of all experiments yielding a quantitative evaluation of motional parameters (Section 5).

At this point, it may be worth discussing merits (and possibly drawbacks) of B_1 gradients versus the more widely used B_0 gradients. This discussion will also take into consideration the instrumental aspect of both types of gradient. Formally, B_1 gradient or B_0 gradient pulses, if associated with proper homogeneous rf pulses, should produce the same effects. Let us recall that a B_0 gradient acts on precession, or, in other words, spatially labels the angular frequency in the xy plane of the rotating frame. On the other hand, a B_1 gradient acts on nutation, that is spatially labels the angular frequency in a plane of the rotating frame, perpendicular to the B_1 direction. Exchanging these two planes (switching from one to the other) is easily performed by means of $\pi/2$ homogeneous rf pulses. Several possible arrangements are shown in Figure 7.4.

Despite this formal equivalence, things are quite different in practice. Concerning B_0 gradients, it is easy to create gradients in the three directions X, Y, and Z without paying attention to the transmit–receive system of the spectrometer. Proper coil geometries can fulfil this objective. Furthermore, B_0 gradients are active irrespective of the Larmor frequency. In other words, they could be viewed as universal. There exists, however, some drawbacks: (i) B_0 gradient pulses involve non-negligible rise and fall times since the relevant coils are not part of a resonant circuit; (ii) the problem of eddy currents is especially acute; these arise from the response of the probe and magnet materials which may alter the NMR signal; active shielding methodology is capable of partly overcoming this problem [4]; (iii) the field-frequency stabilization system (lock system) is invariably perturbed by any B_0 gradient pulse, leading to severe instabilities.

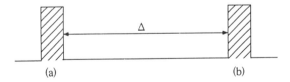

Figure 7.3 (a) and (b): gradient defocusing and refocusing pulses, respectively. If motion occurs during the time interval Δ, refocusing is incomplete and the NMR signal will be attenuated according to the relevant motion. The way in which gradient pulses may be constructed for defocusing or refocusing purposes will be indicated later by reference to the actual experiments and to the type of motion to be investigated

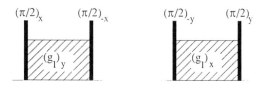

Figure 7.4 Possible schemes including B_1 gradients (g_1) and homogeneous rf pulses, equivalent to a single static field gradient pulse (g_0)

The production of B_0 gradients requires specific coil geometries, which may be of the Maxwell type for the Z gradient or a Golay arrangement (saddle-shaped coils) for the X and Y gradients. The former is schematized in Figure 7.5(a) and consists of two coils, whose plane is perpendicular to Z, with current in opposite directions so that magnetic fields produced by the two coils tend to subtract in the region comprised between them: the largest field toward positive Z is near the left coil whereas the largest field toward negative Z is in the vicinity of the right coil. Similarly, as shown in Figure 7.5(b) transverse gradients are generated by saddle-shaped coils.

At the present stage of development of B_1 gradients, the experimental arrangement is rather basic. The most efficient coil design consists simply in a single turn coil positioned in such a way that B_1 varies linearly across the object under investigation [5]. This is shown in Figure 7.6 and is suitable for creating a gradient along a transverse direction. Of course, the coils used for generating B_1 gradients are part of a resonant circuit tuned at the measurement frequency and eventually impedance matched. For this reason, the switching of B_1 gradients is nearly instantaneous, without any perturbation of the spectrometer and without causing eddy currents. Further advantages of B_1 gradients come from their immunity to magnetic susceptibility variations across the sample or to static field gradients created by heterogeneities of the object under investigation. Unfortunately, they suffer from two drawbacks: (i) large gradient

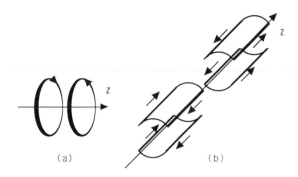

Figure 7.5 Coils for generating B_0 gradients (a) a Maxwell pair and (b) a saddle coil. Current directions are indicated by arrows

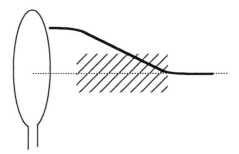

Figure 7.6 Schematic representation of the B_1 field produced by a simple turn coil which should be properly located with respect to the object under investigation so as to produce an uniform gradient in the region of interest (dashed area)

values are not presently available; (ii) because of the necessity to prevent any coupling between the various rf coils inside the probe, it is very difficult (or even impossible) to create B_1 gradients acting independently along the three spatial directions.

Finally, because the use of gradients, in view of determining dynamical parameters, is frequently associated with a spin echo experiment, it may be useful to recall here the basic features of the spin echo experiment, although it has been already described in Chapter 2 in the context of transverse relaxation time determination. The sequence of rf pulses is as follows

$$\left(\frac{\pi}{2}\right)_x - \tau - (\pi)_y - \tau \tag{7.5}$$

where x and y denotes axes of the rotating frame. In that case refocusing occurs along y. The main interest of this pulse sequence is to refocus any effect of precession owing to either chemical shift or B_0 inhomogeneities. This is schematized in Figure 7.7 for a magnetization associated with a single chemical shift evolving in an inhomogeneous static field. Clearly, after the second time interval τ, all elementary magnetizations (at different locations within the sample), which spread out due to B_0 inhomogeneities, are taken back together along the y axis. In other words, they have been *refocused* (any precession phenomenon is compensated for) and lead to the formation of an echo. In any case, the echo amplitude is attenuated according to $\exp(-2\tau/T_2)$. It can be mentioned that the experiment works equally well with the two pulses of identical phase (i.e. $(\pi/2)_x - \tau - (\pi)_x - \tau$). It is a simple matter to show that if both pulses are of identical phases, refocusing occurs along $-y$. However, any imperfection of the refocusing pulse introduces amplitude and phase distortions. This can be circumvented by phase cycling the refocusing pulse over the four transmitter channel phases x, y, $-x$, $-y$. The signal resulting from the cycle is obtained by averaging the signals according to an add–subtract process, giving rise to the well known EXORCYCLE phase

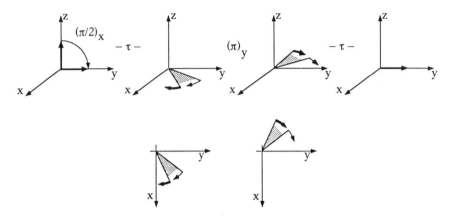

Figure 7.7 The fate of magnetization processing in an inhomogeneous B_0 field under the sequence $(\pi/2)_x - \tau - (\pi)_y - \tau$. The dashed area corresponds to the frequency spread due to variations of B_0 across the sample. The overall precession arises from the difference between the resonance frequency and the carrier frequency

cycling [6]. Another solution is to apply a gradient pulse to ensure a complete dephasing of spins before application of the refocusing pulse. Rephasing is then produced by applying, after the π pulse, a second gradient pulse. Such a method eliminates the need for EXORCYCLE phase cycling.

3   Applications to Water Suppression and Selection of Coherence Pathways

3.1   WATER SUPPRESSION

An interesting application of the gradient defocusing properties in the field of dynamics of fluids concerns the selective suppression of the huge solvent signal. As far as the determination of relaxation parameters by pure spectroscopic methods is concerned (T_1, T_2, $T_{1\rho}$, NOE factors), this is, in fact, the sole instance where gradients can be useful. There exists actually a wide range of methods for solvent suppression including more or less elaborated rf pulse sequences aiming at the elimination of a frequency band around the solvent resonance [7, 8]. Methods for solvent suppression which operate before data acquisition fall into two classes. The first class comprises methods which destroy the solvent longitudinal magnetization (saturation techniques) and the second class those which leave it unchanged (selective excitation techniques). Both techniques can benefit from the use of magnetic field gradients.

The saturation method implies the continuous application for a duration of the order of relaxation times, of a rf field of small amplitude at the solvent frequency, leading to its saturation, with the unavoidable saturation of resonances involving exchange with the line primarily saturated; this might become unacceptable when dealing with protons exchanging with the solvent (for example, the amide protons of a peptide in aqueous solution). The time required for saturating the solvent resonance may be strongly reduced (avoiding most difficulties caused by spin exchange or chemical exchange) if this single long pulse is replaced by a frequency selective (at the solvent frequency) 90° pulse followed by a gradient pulse which spatially randomizes the transverse solvent magnetization [9]. Solvent suppression may be improved further through repetitive application of the selective pulse followed by a gradient pulse [8] (Figure 7.8). However, in contrast with the standard saturation technique, saturation by gradients ($\langle \mathbf{M} \rangle = 0$) is a coherent and thus reversible process.

Gradient pulses can also be useful in the second group of techniques involving frequency selective excitation. For instance, Figure 7.9 shows the [90°, τ, 90°], $T_E/2$. [90°, 2τ, 90°], $T_E/2$. Acq. pulse sequence [10, 11], where two strong spoiler gradients are applied around the [90°, 2τ, −90°] refocusing

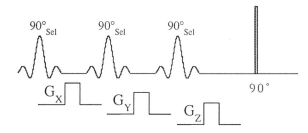

Figure 7.8 Water presaturation using selective pulses at the water frequency and transverse magnetization spoiling

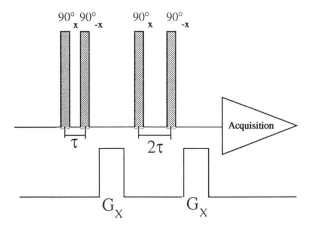

Figure 7.9 Water suppression using frequency selective excitation and gradients around the refocusing pulse

pulse. B_1 gradients can also be used for reaching the same goal [5] (Figure 7.10).

3.2   SELECTION OF COHERENCE TRANSFER PATHWAY

A general description of the use of the gradients to select for specific coherence pathways would be beyond the scope of this chapter. However, the invest-

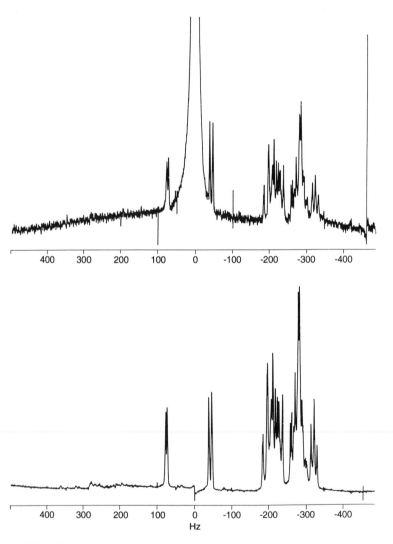

Figure 7.10 Elimination (bottom spectrum) of the huge water resonance in a 1 M solution of glucose

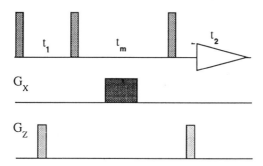

Figure 7.11 A pulse sequence for gradient-enhanced exchange spectroscopy

igation of dynamic processes such as Nuclear Overhauser Effect, spin diffusion, chemical exchange, etc., involves the selection of the proper coherence pathways in 1D and 2D experiments. The simplest sequence employed in 2D exchange spectroscopy is shown in Figure 7.11. In the mixing period all pathways involving $p \neq 0$ must be suppressed. Since a p quantum coherence gets a phase proportional to p when evolving in the presence of gradient, the application, during the mixing time, of a gradient pulse (a homospoil pulse) of sufficient length and strength destroys [12] all coherence orders with $p \neq 0$. Similarly, a signal arising from magnetization which is longitudinal during the preparation period, may be efficiently suppressed by applying gradient pulses of the same strength and length following the first and third pulse, selecting the coherence pathway $p = 1 \rightarrow 0 \rightarrow -1$ [13]. Gradient selection of coherence pathways eliminates the need for phase cycling. However, gradient techniques frequently result in a loss of a factor of two in signal-to-noise ratio.

4 Basics of Localization and Imaging Experiments

4.1 LOCALIZED SPECTROSCOPY

In this section we shall present an overview of techniques that provide spectroscopic information from a *limited region of a whole sample*. Ideally, the size and position of the volume of interest (VOI) would be easily modifiable, the spectrum would be free of contaminations from outside the VOI and the techniques would acquire information from an image-selected volume element. All the techniques require spatial encoding which may be ensured with either B_1 or B_0 gradients.

A high resolution spectrometer exploits the oldest localization technique since it retains signal only from that part of the NMR tube placed into the coil. This crude localization is based on the use of B_1 gradients. *Surface coils* (for a review see Ref. 14) work on the same principle since the spatial selectivity relies on the strongly inhomogeneous field of a flat coil. It is the method of choice for many applications where the accuracy of localization is not of prime importance. The selected volume is a function of both the coil geometry and the pulse sequence. However, the volume selected using a B_1 gradient method with a single surface coil technique is not well defined in shape and difficult to vary in position. Furthermore, surface coil spectroscopy is limited to the study of superficial regions.

When using B_0 gradients, volume selection may rely upon the use of a highly homogeneous static field in the region of interest and an inhomogeneous field outside this region. More often, volume selection involves applying slice selective rf pulses and the selected volume is formed at the intersection of the selected planes (Figure 7.12). Slice selective pulses can be seen as the transposition of one of the frequency selective procedures employed in NMR spectroscopy. These procedures make use of pulses which are selective with respect to resonance frequencies: they may be square-shaped pulses of low amplitude or, in order to avoid spurious oscillations in the selectivity profile, shaped according to a gaussian or a sinc function. Here, the spectral discrimination is achieved by applying a gradient in the Z direction so that the frequency dimension is now associated with a spatial dimension (Z). The principle of the slice selection method is schematized in Figure 7.13. The low-amplitude shaped pulse is devised so as to tip to the transverse plane those magnetizations whose resonance frequencies are close to the carrier frequency. These resonance frequencies are defined by the applied gradient and are associated with a slice perpendicular to the Z direction, whose thickness depends obviously on the gradient magnitude and also on the efficiency of the selective pulse. Magnetization pertaining to other regions will remain stored along the Z axis (corresponding to thermal equilibrium). In fact, the selected magnetization is frequently defocused in the transverse plan. For sensitivity reasons, its refocusing is mandatory; this is usually achieved by inverting the gradient polarity for an appropriate duration. Of course, at this stage, the

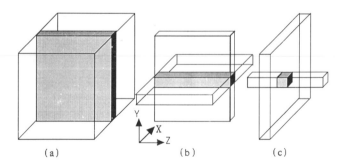

Figure 7.12 The successive stages for selecting a cubic sensitive volume. (a), (b), (c): slice selection in the X, Y and Z directions, respectively

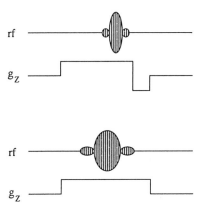

Figure 7.13 Possible slice selection schemes using a radio-frequency selective pulse together with a gradient applied along the direction perpendicular to the slice of interest. Top, for tipping the selected magnetization into the transverse plane, with a reversion gradient. Bottom, for inverting the selected magnetizations

gradient acts only on the magnetization in the transverse plane, in other words, on the selected magnetization.

Spectroscopic information from the selected cube is obtained after the normal one-dimensional Fourier transform and without further signal processing. The size of the region of interest can be modified by varying the pulsed gradient strength, and its position can be moved within the object by changing on the frequency of the selective pulses. The size and location of the volume of interest (VOI) are defined using a standard NMR image. Single voxel techniques can be divided into three groups:

(i)   *Difference techniques.* In this group of experiments volume selection results from an add–subtract process. The basic scheme is ISIS [15] which has become one of the most successful techniques to achieve spatial localization. It consists of applying a combination of selective inversion pulses which cause inversion of the longitudinal magnetization in the selected planes.

$$(\pi)_{on/off} \quad (\pi)_{on/off} \quad (\pi)_{on/off} \text{—Magnet stabilization delay—Read-out pulse}$$
$$\scriptstyle g_x \qquad\quad g_y \qquad\quad g_z$$

A three-dimensional volume selection requires eight experiments which are averaged with a sign $(-1)^N$ depending on the number N of pulses used in the experiment. This produces a constructive averaging of signals from inside the VOI, whereas signals from outside the VOI cancel. ISIS is not T_2 weighted; this may be a major advantage for spectroscopy of short T_2 species.

(ii)   *Destruction of the magnetization from outside the volume of interest.*
        Three pulses, pulse sandwiches or pulse-trains are successively applied
        in the presence of field gradients to retain longitudinal magnetization
        only in the region of interest. Techniques such as VSE [16] and SPARS
        [17] were among the first 3D techniques available and they aroused
        much interest. They have now been superseded by echo techniques.

(iii)   *Direct excitation of the spins from inside the volume of interest.* Two
        types of techniques are currently used in *in vivo* spectroscopy to directly
        excite the spins inside the VOI and to achieve 3D selection. One of them
        exploits stimulated echoes (STEAM [18]), the other exploits Carr and
        Purcell echoes (PRESS [19]), which are, however, inappropriate for
        short T_2 species. An important advantage of these methods is that a
        localized spectrum may be obtained in a single scan.

As discussed in Section 4.5 spatially resolved spectroscopic information may
be also obtained with spectroscopic imaging techniques.

4.2   TWO-DIMENSIONAL FOURIER IMAGING

NMR imaging has proved to be an invaluable tool among many medical
imaging techniques. This is the result of the wealth of information available
from the NMR signal which can be measured by a variety of experimental
techniques. The rapid development of medical imaging has triggered new
applications of NMR in material sciences and fluid dynamics. A wide variety
of methods allows accurate measurements of many physical properties.

We shall be concerned here with heterogeneous samples for which the
objective is to determine the transverse magnetization distribution or, to be
more specific, its variations inside the object under investigation. As explained
in the introduction, if only one dimension is to be investigated, this would
require a single gradient pulse acting in that direction, which would serve to
frequency encode the nuclear spins as a function of their location. A three-
dimensional image would thus need to perform a sufficient number of
experiments with gradients acting along the three cartesian directions, and
varied in an appropriate manner. In practice, for keeping the measuring time
within reasonable limits, most NMR imaging experiments reduce to a two-
dimensional examination of a selected slice perpendicular to a given direction
(say Z). Hence, one begins usually by a slice selection procedure as explained
in the previous section. Once a particular slice perpendicular to Z has been
selected, two methods are available for imaging the X, Y plane (we recall that
capital letters are reserved for the laboratory frame while small letters are
associated with the rotating frame).

Fourier imaging [20] is the most widely used method, not only for medical
applications, but also in material sciences and in studies involving the mobile
parts (fluids) of heterogeneous objects. Most of the time, it is the 'spin-warp'

version [21] of the experiment which is preferred. The method implies obviously two gradients in the Y and X directions. The first one is applied after the slice selection procedure for a fixed duration τ but with an amplitude incremented from one experiment to the next. This incrementation is similar to the time incrementation (t_1) of a classical two-dimensional NMR experiment. It is generally schematized by a series of horizontal bars going from negative to positive values of the gradient (see Figure 7.14). This gradient is dubbed phase-encode gradient because it produces a dephasing (precession) of transverse magnetization, by an angle which can be expressed as $2\pi k_Y Y$ with

$$k_Y = (2\pi)^{-1}\gamma g_Y \tau \qquad (7.6)$$

The direction X is examined during acquisition; this is performed under the application of gradient of constant amplitude g_X (frequency encoding). This amounts to sampling a line in the k plane (along k_X, for a given k_Y). An echo sequence is usually employed and the global imaging sequence is displayed in Figure 7.14.

Denoting by k_X the variable associated with acquisition $k_X = (2\pi)^{-1}\gamma g_X t_2$, the final signal can be expressed as:

$$S(k_X, k_Y) = \iint\limits_{-\infty}^{+\infty} \rho(X, Y)\exp[i2\pi(k_X X + k_Y Y)] \, dX \, dY \qquad (7.7)$$

where $\rho(X, Y)$ represents, in fact, the transverse magnetization at coordinates (X, Y) in the selected slice. $\rho(X, Y)$ is of course proportional to the spin density at (X, Y) but can be modified by various dynamical parameters (T_1, T_2, $T_{1\rho}$, diffusion, flow). It is precisely this quantity that we want to determine; it

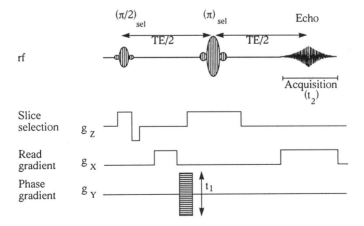

Figure 7.14   The conventional spin warp imaging sequence

will constitute the two-dimensional image and stems from the inverse Fourier transform of (7.7) with respect to k_X and k_Y:

$$\rho(X, Y) = \int_{-\infty}^{\infty} \int_{-\infty}^{\infty} S(k_X, k_Y) \exp[-i2\pi(k_X X + k_Y Y)] \, dk_X \, dk_Y \qquad (7.8)$$

4.3   TWO-DIMENSIONAL PROJECTION RECONSTRUCTION [22]

It may be, in some instances, more judicious to sample the k plane by means of polar coordinates, that is the angle φ (with respect to the k_X axis) and the distance k with respect to the origin of the (k_X, k_Y) frame. This comparison is shown in Figure 7.15. This is achieved by rotating the gradient according to the angle φ. The experimental procedure consists in applying simultaneously the g_X and g_Y gradients, which is actually equivalent to a gradient g of constant amplitude $(g = \sqrt{g_X^2 + g_Y^2})$ along the direction defined by the angle φ ($\tan \varphi = g_Y/g_X$). g remains constant and is rotated by varying φ from 0 to 2π. Acquisition is performed during gradient application, and no echo procedure is required. The whole experiment is schematized in Figure 7.16. After a slice selection procedure involving a gradient in the Z direction which tips magnetization to the transverse plane, acquisition takes place in the presence of gradients in the X and Y directions (read gradients). Here, the line in the k plane, making the angle φ with k_X, is sampled.

The obvious advantage of the method comes from the fact that it is less prone to signal alteration by T_2 (transverse relaxation) effects than the conventional Fourier spin echo technique. It is therefore well suited for

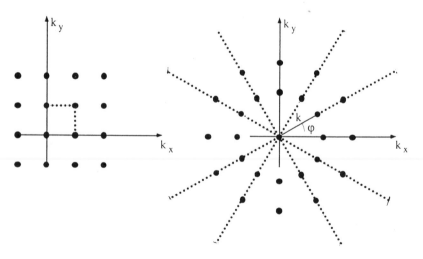

Figure 7.15 Comparison of the cartesian and polar coordinates (left and right respectively) used for scanning the k_x, k_y plane. Dots indicate the actual sampling procedure

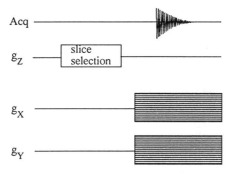

Figure 7.16 Principle of imaging by the projection-reconstruction method

systems involving short transverse relaxation times, as found for example, in heterogeneous systems where the mobile part strongly interacts with a rigid matrix. Data treatment is not as straightforward as for two-dimensional Fourier imaging. The collected signal has the general form:

$$S(k_X, k_Y) = \int_{-\infty}^{+\infty} \int_{-\infty}^{+\infty} \rho(X, Y) \exp[i2\pi \mathbf{kr}]\, dX\, dY \tag{7.9}$$

The problem is to retrieve $\rho(X, Y)$ from (7.9) by an inverse transformation which must necessarily involve the two variables which are at hand, namely $k = (2\pi)^{-1}\gamma gt$ and φ. In a first step, we consider the usual inverse Fourier transform of (7.9)

$$\rho(X, Y) = \int_{-\infty}^{+\infty} \int_{-\infty}^{+\infty} S(k, \varphi) \exp[-i2\pi(k_X X + k_Y Y)]\, dk_X\, dk_Y$$

where we have to substitute k_X and k_Y by φ and k. This is easily performed by replacing the 'cartesian coordinates' k_X and k_Y by the 'polar coordinates' k and φ: $k_X = k \cos \varphi$, $k_Y = k \sin \varphi$, with $dk_X\, dk_Y = k\, dk\, d\varphi$. Hence

$$\rho(X, Y) = \int_0^{\pi} \left[\int_0^{\infty} S(k, \varphi) \exp[-i2\pi(k_X X + k_Y Y)]k\, dk \right] d\varphi \tag{7.10}$$

where $k_X X + k_Y Y$ has still to be expressed according to the variables k and φ, owing to the fact that X and Y are fixed for the calculation of (7.10).

Since the g components are $g_X = g \cos \varphi$ and $g_Y = g \sin \varphi$, this can be expressed as

$$k_X X + k_Y Y = k(X \cos \varphi + Y \sin \varphi) \tag{7.11}$$

The integral between brackets is simply the Fourier transform of the signal $S(k, \varphi)$ multiplied by k (usually called the filtered signal because the signal S is multiplied by k). This quantity can be expressed as

$$\rho = \int_0^{\infty} kS(k, \varphi) \exp[-i2\pi k(X \cos \varphi + Y \sin \varphi)]\, dk \tag{7.12}$$

Because we are looking at $\rho(X, Y)$, the calculation of (7.10) is performed for each required set of X and Y values inside the region to be investigated. Integral (7.12) is evaluated by considering the whole range of available k values, deduced from the applied gradients ($k = (2\pi)^{-1}\gamma\sqrt{g_X^2 + g_Y^2}\,t$), the angle φ involved in (7.30) resulting from the ratio $g_Y/g_X = \tan\varphi$. The calculation of (7.11) ends up with the numerical integration over φ. This algorithm is known under the acronym of FBP (Filtered Back Projection) [23].

4.4 IMAGING BY B_1 GRADIENTS

NMR imaging can be performed by resorting only to B_1 gradients (rf gradients). The method is a sort of transposition of procedures using B_0 gradients. Advantages as well as pitfalls of B_1 gradients (listed above) support the conclusion that, at the present time, B_1 gradients can better be used for NMR microscopy with applications essentially in the field of materials science [24]. This basic scheme of 2D imaging by B_1 gradients [25] involves at the onset a slice selection procedure. The latter is carried out by the rf gradient (along the Z direction) of a purposely designed saddle coil, as the one which is used in a conventional NMR probe for transmit–receive operations. The rf read-gradient (acting along a transverse direction) is generated by a single turn coil which is orthogonal to the saddle coil. The XY plane is examined by means of projections along the gradient axis, collected for different orientations of the object with respect to the Z axis (the NMR signal is acquired by means of the saddle coil in short windows during which the read-gradient is switched off). To this end, the experiment, including the initial slice selection, is repeated for as many sample rotations as necessary, performed in a stepped manner around Z. The NMR signal pertaining to *one* projection (defined by φ which specifies the angle by which the object has been rotated with respect to its initial position) relies on the nutation angle produced by the rf read-gradient and can be expressed as

$$S(k, \varphi) = \iint \rho(X, Y)\cos[2\pi k(X\cos\varphi + Y\sin\varphi + D)]\,dX\,dY \qquad (7.13)$$

where X and Y are the coordinates of the considered elementary magnetization, D is the distance between the rotating axis and the virtual point where B_1 is zero whereas k is related to the time t during which the gradient has been applied ($k = (2\pi)^{-1}\gamma g_1 t$). The whole set of data (projections) can be treated with the help of the filtered back projection algorithm mentioned in the previous section.

4.5 SPECTROSCOPIC IMAGING

One of the more promising localized spectroscopy techniques combines spectroscopy and imaging and is known under the name spectroscopic imaging

(SI) or chemical shift imaging. This technique, which maps out the distribution of different substances in an object, produces an image where the pixel intensity $I(\mathbf{r}, \delta)$ is a function of a space variable (multidimensional, in general) and of the chemical shift δ. Several methods have been proposed [26–31]. One group of experiments is based on chemical-shift-selective imaging of predefined spectral components such as 'oil' and 'water'. Chemical shift selectivity may be obtained for instance by selective excitation of the resonance of interest which is then imaged. However, most of the techniques belong to the Fourier SI group where spatial and chemical shift information is obtained by sampling the signal in the reciprocal space k_x, k_y, k_σ and Fourier transforming the data set. We shall restrict this discussion to the description of the basic technique where the spatial information is phase encoded prior to acquisition of a free induction decay or echo, in the absence of gradient. The principle will be presented in the case of a one-dimensional object but the extension to two or three dimensions is straightforward. Successive acquisitions are made using different amplitudes of a phase encoding static field gradient. After processing of the data array resulting from N spatially encoded experiments, spectroscopic information can be obtained from N voxels simultaneously. It is, in fact, an imaging pulse sequence without a read-out gradient. A one-dimensional (one spatially resolved dimension) spectroscopic imaging spin echo sequence is depicted in Figure 7.17. The major advantage of spectroscopic imaging over single voxel techniques is that spectra can be obtained simultaneously from several slices (1D) or bars (2D) or cubes (3D) inside the object.

SI techniques were used to image sandstones [32–35], allowing fluids of different chemical composition (typically oil and water) in porous materials to be separated and to follow the injection of water and the displacement of oil.

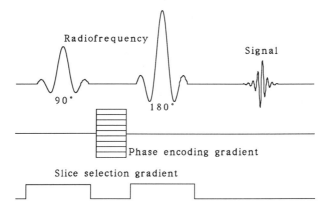

Figure 7.17 Principle of the spectroscopic imaging method. Two different phase encoding gradients (corresponding to two spatial directions) should be shown, leading to an actual three-dimensional experiment

Oil displacement in dolomite was recently observed in a similar way, using a spectroscopic imaging technique [36]. However, this technique has some limitations such as susceptibility broadening which increases with the field intensity. A SI experiment may be, however, useful simply for imaging purposes in the case of large linewidth. Sandstone cores may contain significant concentration of iron increasing the linewidth to several kHz. This linewidth increases when B_0 increases. Thus the gradient strength required to achieve a reasonable spatial resolution in the read-out direction may be one order of magnitude greater than the available gradient strength. This difficulty can be circumvented by using phase encoding instead of frequency encoding. However, diffusion in the internal field gradients still limits the spatial resolution.

4.6   CONTRAST IN NMR IMAGING

A unique feature of NMR imaging is the variety of physical properties which can be imaged. It should be borne in mind that the physical variable which is imaged is the transverse magnetization during acquisition. Of course, the amplitude of the acquired magnetization is primarily dependent on the spin density. However, depending on the preparation period and of the evolution of the magnetization prior acquisition, the signal intensity may be affected by a number of physical phenomena. For instance, the use of an inversion-recovery preparation period introduces T_1 weighting but also possibly bulk flow weighting, a spin echo delay introduces T_2 and diffusion weighting, etc. The enormous potential of NMR imaging for studying dynamical processes is based on the sensitivity of NMR to the physical environment of the nucleus. This potential is still largely unexplored.

5   Basics of Motion-sensitive Experiments

5.1   FLOW-SENSITIVE TECHNIQUES

Early in its development, it was recognized that NMR could be used to detect the macroscopic motion of spins [1, 2]. There is now a long history of NMR applied to the measurement of flow. A number of imaging or non-imaging techniques have been developed and applied to various fields of investigations and reviews of techniques imaging or measuring flow by NMR are available [37, 38]. A flow-sensitive experiment generally comprises a preparation period which prepares the magnetization away from equilibrium, a waiting delay and finally a read-out period during which changes caused by the flow are determined. For a fluid moving extremely slowly and with a discernible boundary, simple imaging experiments may be repeated and compared. However measurement of flow may be obtained directly, through the changes in the NMR signal intensity or phase due to flow occurring during the NMR experiment. The various methods may be classified into three groups, inflow/outflow, time-of-flight and phase sensitive methods.

A very simple flow-sensitive experiment exploits the spin-lattice relaxation time T_1: if a part of the flow experiment is outside the magnet, the transit time limits the buiding-up of longitudinal magnetization. This simple experiment exploits the effect of magnetic field gradients. Another crude experiment exploits B_1 gradients: consider a fluid flowing through a coil; the replacement of partly saturated spins in the coil by unsaturated spins produces an increase of the NMR signal with increasing velocity. These two experiments exploiting either B_1 or B_0 gradients (or both), which are based on the effects arising from inflow of fresh spins and outflow of excited spins, are the basis of inflow/outflow methods which may be spatially resolved (imaging) or not.

The second class of experiments, the time-of-flight methods, comprises imaging techniques which involve selectively exciting a particular group of spins and imaging of the displacement of these labelled spins during the pulse sequence. An example of a time-of-flight technique is depicted in Figure 7.18. A slice selective 90° pulse in the presence of X-gradient, excites transverse magnetizations in a plane YZ parallel to the flow direction (Y). Then a selective 180° pulse refocuses only the spins in a plane (ZX) orthogonal to the flow direction at $Y = Y_0$. The result is a tagging of the spins of the YZ plane at the coordinate Y_0. Finally a second 180° refocuses the magnetizations into the YZ plane. Frequency encoding the spin positions with a Y read-out gradient reveal the position of spins during this detection period, allowing measurement of the displacements of the tagged spins.

The last class of experiments, the phase sensitive techniques, exploit the phase shift of the transverse magnetization component occurring in the

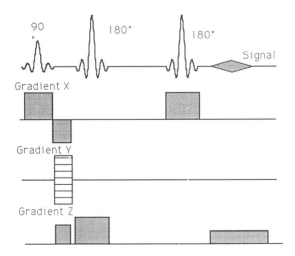

Figure 7.18 A sequence for imaging spin displacement (time-of-flight method)

presence of a magnetic field gradient in the direction of flow. The presence of a phase shift induced by a magnetic field gradient to a magnetization flowing in a convection current has been discovered by Carr and Purcell [39]. Hahn was the first to take advantage of such a phase shift to measure coherent flow [1], followed by Stejskal [40], who proposed the use of pulsed gradients, as opposed to continuous gradients, in combination with a spin-echo sequence. A number of imaging and non-imaging techniques now exist which exploit the phase shift of the transverse magnetization from the moving spins relative to the static ones. However, most of the phase-sensitive methods are based on the well known pulsed gradient spin echo (PGSE) sequence [41], which gives access to coherent or incoherent flow (diffusion) and which is described hereafter in more detail.

5.2   THE PULSED GRADIENT SPIN ECHO (PGSE) EXPERIMENT [41]

Let us suppose that two identical static field gradient pulses (of duration δ) are inserted in each half of a spin echo sequence and let us denote their separation by Δ (Figure 7.19). As far as defocusing or refocusing processes are concerned, nothing is changed with respect to the original Hahn experiment; this is just because an equal 'amount of B_0 inhomogeneity' (producing defocusing) has been placed on each side of the central π pulse whose purpose is to refocalize magnetization provided that the latter spreads in the same way in each half of the sequence. To this extent, the location of each of the gradient pulses (or their time separation Δ) is irrelevant. This is true as long as translational motions can be disregarded. If this is not the case, it is conceivable that something will happen. Qualitatively, let us consider a molecule at location X when the first gradient pulse is applied. It may be at location $(X + \Delta X)$ at the time of the second gradient pulse. It will therefore be subjected to another B_0 field and, for that reason, some modification of the echo is

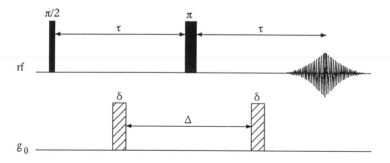

Figure 7.19   The PGSE experiment: static field gradient pulses (g_0) of equal duration are applied in each half of the sequence

expected in relation with molecular translational motions within the sample. They may be incoherent (diffusion processes related to brownian motions) or coherent (flow processes). Both types of motions will now be considered separately in a more quantitative fashion.

5.2.1 Diffusional processes.

This incoherent process stems from the so-called diffusion equation, valid for any molecular property ψ:

$$D \frac{\partial^2 \psi}{\partial X^2} = \frac{\partial \psi}{\partial t} \tag{7.14}$$

We shall first verify that the solution to this equation is a gaussian function whose standard deviation is proportional to $\sqrt{t}$. To this end, let us look for a solution of the type:

$$\psi = \frac{1}{\sigma\sqrt{2\pi}} \exp\left(-\frac{X^2}{2\sigma^2}\right) \tag{7.15}$$

with σ^2 assumed to be of the form $\sigma^2 = kt$ and where k has to be determined with respect to D, as far as the function (7.15) satisfies the diffusion equation (7.14). After some straightforward calculations, we obtain from (7.15) and from the assumed expression of σ^2:

$$\frac{\partial \psi}{\partial t} = \frac{1}{2t} \psi + \frac{X^2}{2kt^2} \psi$$

and

$$\frac{\partial^2 \psi}{\partial X^2} = \frac{X^2}{2kt^2} \psi - \frac{1}{kt} \psi,$$

which are seen to fit equation (7.14) provided that k is identified with $2D$. As a consequence $\sigma = \sqrt{2Dt}$ and the solution of equation (7.14) for a diffusion time Δ (as this will be the case in the following) can be written as:

$$\psi = \frac{1}{\sqrt{4\pi D\Delta}} \exp\left(-\frac{X^2}{2D\Delta}\right) \tag{7.16}$$

We return now to the PGSE experiment where the first gradient pulse, in addition to precession (see Figure 7.19), produces a dephasing which, for a location defined by X, can be expressed by:

$$\theta(X) = \gamma g \, \delta X = 2\pi q X \tag{7.17}$$

where the variable $q = (2\pi)^{-1} \gamma g \delta$ is in accord with the usual notation [3]. This dephasing is entirely refocused by the central π pulse and by the second gradient pulse unless molecules have moved because of the diffusional process.

When diffusion is taken into consideration, the precession angle produced by the second gradient pulse has to be changed from $-\theta$ to $-(\theta + \varphi)$, the additional phase angle φ accounting for translational motion. To be more specific, and following the approach of Karczmar *et al.* [42], let us consider the two transverse components of the magnetization located at X. We shall disregard precession due to chemical shift since it is anyhow refocused by the central π pulse. After the first gradient pulse, one has for the transverse components of magnetization located at X (and scaling the magnetization amplitude to unity):

$$\begin{cases} M_x = \sin\theta \\ M_y = \cos\theta \end{cases}$$

which becomes after the $(\pi)_y$ pulse:

$$\begin{cases} M_x = -\sin\theta \\ M_y = \cos\theta \end{cases}$$

and after the second gradient pulse:

$$\begin{cases} M_x = -\sin\theta\,\cos(\theta + \varphi) + \cos\theta\,\sin(\theta + \varphi) \\ M_y = \sin\theta\,\sin(\theta + \varphi) + \cos\theta\,\cos(\theta + \varphi) \end{cases} \tag{7.18}$$

After expansion of the right hand sides of the two above equations, an average over all the locations within the sample must be considered; this will concern the phase angle θ and will be denoted with brackets. Conversely, the sine and cosine functions involving the additional phase angle φ have to be evaluated according to the distribution given by equation (7.16). The relevant averages will be wriitten as $\overline{\sin\omega}$ and $\overline{\cos\varphi}$, they are assumed to be independent of the space average. This means that, for systems considered in this section, diffusion processes are assumed to be location independent.

As a consequence, the variables θ and φ can be separated; thus, equations (7.18), for the whole sample, can be written as:

$$\begin{cases} \langle M_x \rangle = \overline{\sin\varphi} \cdot (\langle \sin^2\theta \rangle + \langle \cos^2\theta \rangle) \\ \langle M_y \rangle = \overline{\cos\varphi} \cdot (\langle \sin^2\theta \rangle + \langle \cos^2\theta \rangle) \end{cases} \tag{7.19}$$

If the gradient is sufficiently strong, it can be assumed that all values of θ occur with the same probability, so that:

$$\langle \sin^2\theta \rangle = \langle \cos^2\theta \rangle = \tfrac{1}{2} \tag{7.20}$$

$\langle M_x \rangle$ and $\langle M_y \rangle$ reduce to:

$$\begin{cases} \langle M_x \rangle = \overline{\sin \varphi} \\ \langle M_y \rangle = \overline{\cos \varphi} \end{cases} \tag{7.21}$$

Because a self-diffusion process implies that molecules can go in either direction with the same chance, it is obvious that the two values φ and $-\varphi$ correspond to the same probability. Therefore, $\overline{\sin \varphi} = 0$ and $\langle M_x \rangle = 0$; we are thus left with the calculation of $\overline{\cos \varphi}$ for evaluating the echo decay as a function of translational diffusion. Going back to equation (7.16) with the q notation ($q = (2\pi)^{-1}\gamma g \delta$ and $\varphi = 2\pi q r$), where r represents the displacement of a diffusing molecule during the time interval Δ, we can write:

$$\overline{\cos \varphi} = \frac{1}{\sqrt{4\pi D\Delta}} \int_0^\infty \cos(2\pi q r)\exp\left(-\frac{r^2}{4D\Delta}\right) dr = \exp(-4\pi^2 q^2 D\Delta) \tag{7.22}$$

Finally, inserting the damping factor arising from transverse relaxation and denoting by M_0 the magnetization of the whole sample at thermal equilibrium, we arrive at the actual echo amplitude in the PGSE experiment:

$$M(\tau, \Delta) = M_0 \exp\left(-\frac{2\tau}{T_2}\right) \exp\left(-\gamma^2 g^2 D\Delta\delta^2\right) \tag{7.23}$$

In order to measure the self-diffusion coefficient, the most straightforward procedure consists in performing a series of experiments, with fixed values of Δ and τ ($\Delta \leqslant 2\tau$) and varying δ. The slope of the logarithm of $S(\delta)$ versus δ^2 should directly yield D:

$$\ln[S(\delta)] = K - (\gamma^2 g^2 D\Delta).\delta^2 \tag{7.24}$$

It must be noticed that diffusion has been neglected for the duration δ of gradient pulses because they are generally much shorter than the interval Δ. If this is not the case, it can be shown [41] that Δ must be replaced by $(\Delta - \delta/3)$.

5.2.2   Flow processes

So far, we have considered incoherent motions due to free diffusion, which are accounted for by an echo attenuation (equations (7.23) and (7.24)). The question we address now is that of a *coherent* motion. To simplify, let us suppose that the fluid under investigation is flowing with a constant and uniform velocity v *along the gradient direction X*. The angle φ in equation (7.18) can then be expressed as $\varphi = \gamma g \delta v \Delta$; it is the same for any location within the sample and from equations (7.21), one has:

$$\begin{cases} \langle M_x \rangle = \sin(\gamma g \delta v \Delta) \\ \langle M_y \rangle = \cos(\gamma g \delta v \Delta) \end{cases} \tag{7.25}$$

This amounts to *no echo attenuation* but rather to *a phase shift* depending directly on the velocity v. In fact, magnetization does not refocus only along the y axis but along an axis of the xy plane whose position depends on flow velocity. Measuring the phase modulation of the echo consequently yields the velocity v (see Section 5.1).

5.3 THE EQUIVALENT OF THE PGSE EXPERIMENT WITH B_1 GRADIENTS

The interest of B_1 gradient pulses comes from the simple idea that they are capable of acting on nuclear spins (as any rf pulse) and, at the same time, they can be used for labeling purposes by means of nutation angles directly related to spatial location. Because there exists a formal analogy between B_0 and B_1 gradients, it should be possible to devise analogous experiments (to the one involving B_0 gradients) in a much simpler way. This is indeed the case as far as the PGSE experiment is concerned: two B_1 gradient pulses, each of duration δ, separated by a time interval Δ, lead to the same type of molecular information with the additional advantage of being weighted by longitudinal relaxation rather than by transverse relaxation. As explained below, a slight inconvenience may be an intrinsic loss of sensitivity by a factor two, owing to the inherent defocusing/refocusing features of B_1 gradient pulses. Discarding the original pulse sequence with B_1 gradients aimed at measuring self-diffusion coefficients [43], we shall turn to the simplest scheme adapted from recent approaches [42–44] whose phase cycle is designed to probe only longitudinal magnetization. To this end, let us consider the sequence schematized in Figure 7.20 which includes two B_1 gradient pulses and a read-pulse (which does not need to be set with great accuracy as far as it is the same for all experiments carried out for different values of δ; as a matter of fact, it does not really require an homogeneous B_1 field and can possibly be delivered by the coil used for generating the rf gradient).

Straightforward calculations show that shifting the phase of the second gradient pulse by 180° every other scan (without altering the phases of the

Figure 7.20 A sequence involving two B_1 gradient pulses (g_0) of identical duration δ for monitoring molecular motion during the time interval Δ. The simple phase cycle applied to the second gradient pulse makes it possible to probe only longitudinal magnetization

read-pulse and of the acquisition) insures the cancellation of any other contribution than longitudinal magnetization prior to the read-pulse. This precludes the formation of stimulated echoes that eventually would interfere with the free induction following the $\pi/2$ read-pulse. Following arguments similar to those of the previous section, we arrive, *after the two scans* implied by the phase cycling, to equations analogous to equations (7.21):

$$\begin{cases} \langle M_x \rangle = 0 \\ \langle M_y \rangle = \overline{\sin \varphi} \\ \langle M_z \rangle = \overline{\cos \varphi} \end{cases} \tag{7.26}$$

which reduce to $\langle M_z \rangle$ in the case of diffusion processes with, however, an additional damping term expressed as $\exp(-\Delta/T_1)$ (contrary to the PGSE experiment where transverse relaxation is involved). It can be further noticed that this result is obtained with two scans whereas equation (7.21) describes the situation pertaining to only one scan. The measured signal is of the form (after two scans):

$$S(\delta, \Delta) = M_0 \exp\left(-\frac{\Delta}{T_1}\right) \exp\left(-\gamma^2 g_1^2 D\Delta\delta^2\right) \tag{7.27}$$

Concerning coherent motions (flow), they lead to a phase shift analogous to the one previously described for the PGSE experiment.

5.4 DISPLACEMENT IMAGING [45]

Returning to the PGSE experiment, it can be noticed that, in a general way and assuming narrow gradient pulse conditions, the echo amplitude can be written as

$$E_\Delta(\mathbf{q}) = \int \rho(\mathbf{r}) \int P(\mathbf{r}' \mid \mathbf{r}, \Delta) \exp\left[(i2\pi\mathbf{q}(\mathbf{r}' - \mathbf{r}))\right] d\mathbf{r}' \, d\mathbf{r} \tag{7.28}$$

where the q notation ($q = (2\pi)^{-1}\gamma g\delta$) has been used (see equation (7.22) and quadrature detection assumed. $P(\mathbf{r}' \mid \mathbf{r}, \Delta)$ is the conditional probability that a spin starting at position $\mathbf{r}$ will migrate to $\mathbf{r}'$ during the diffusion time Δ ($\mathbf{q}$, $\mathbf{r}$ and $\mathbf{r}'$ are vectors so as to keep (7.28) in its more general form). Defining an ensemble averaged probability $\bar{P}(\mathbf{R}, \Delta)$, where $\mathbf{R} = \mathbf{r}' - \mathbf{r}$, enables one to rewrite (7.28) as

$$E_\Delta(\mathbf{q}) = \int \bar{P}(\mathbf{R}, \Delta) \exp(i2\pi\mathbf{q}\mathbf{R}) \, d\mathbf{R} \tag{7.29}$$

which is seen to represent a Fourier transform relationship between E and $P(\mathbf{R}, \Delta)$. Thus, the Fourier transform of the echo amplitude with respect to $\mathbf{q}$ provides the average probability as a function of $\mathbf{R}$. This is an 'image' of the displacement distribution obtained by scanning the q-space, whereas an image of the density distribution is obtained by scanning the k-space. The accessible resolution in **R** is expected to suffer from less limitations than in k-space imaging [46].

6 Applications

Obtaining information on polyphasic flows, diffusion and restricted diffusion in porous media, without adding contrast agents, may contribute to the knowledge of their transport properties. The investigation of fluids in porous media is important in a number of applications ranging from the diffusion of solvents into plastic materials to the uptake of water into building materials. This also contributes to the renewed interest in the application of NMR to oil exploration and recovery. A number of NMR parameters such as the relaxation times T_1, $T_{1\rho}$, T_2, D, etc. are sensitive to molecular mobility. Maps based on these parameters may be produced which visualize fluids and the interactions of these fluids with the surface of the pores. We shall limit this overview to the presentation of some results concerning direct measurements of net fluid transport and molecular self-diffusion.

6.1 FLOW STUDIES

The recent introduction of MRI has given a new impetus to use NMR for the determination of flow characteristics. MRI now allows non-invasive measurement of localized fluid flow velocity to be performed inside porous media. Very slow movements may be monitored using standard MRI techniques whereas more rapid fluids movement requires the use of flow-sensitive techniques. For example, MRI was recently used for obtaining local values of the porosity and fluid flow velocity in sandstone, during water injection [47]. MRI was also used [48] to map the distribution of water in natural soils, the drainage or the wetting of soils by water. A high-speed imaging technique was used to image the velocity distribution resulting from the injection of water through a sandstone sample [49]. Flow-weighted imaging may be also useful to examine the connectivity of pores in a limestone and to visualize the various flow paths [50]. A particularly interesting application of flow imaging concerns the evaluation of water transport in plants. Another field of application is natural convection induced by thermal gradients. MRI now allows detailed information regarding fluid velocities during natural convection to be obtained [51, 52].

6.2 DIFFUSION

6.2.1 Free diffusion

Free diffusion means that molecules in the course of their translational motion are not subjected to any sort of physical barrier which would limit the distance they could in principle travel during the interval Δ (see the basic PGSE experiment). As a consequence, this definition is to be understood according to the actual experiment set-up, since restrictions to apparent free motions depend strongly on the timescale of the diffusion interval Δ with respect to the mean displacement of molecules within the considered fluid. Manifestations of restricted diffusion will be considered in the next section. If any manifestation of this kind is not apparent, a first approach to interpretation of self-diffusion coefficients rests on the familiar Stokes–Einstein relation

$$D = k_B T / 3 \pi \eta d \qquad (7.30)$$

This formula is strictly valid for a molecular entity of spherical shape (d denoting its diameter), when molecules do not encounter large particles in the course of their translational motion, or in other words when they are solely subjected to frictional effects from the solvent (or medium) of viscosity η. The factor $k_B T$ (k_B: Boltzmann constant, T: absolute temperature) accounts for brownian motions. The important point in formula (7.30) is that the self-diffusion coefficient is directly related to the size of the molecular entity via its apparent diameter d. In other words, for systems without local or global organization or not involving more or less large particles, the self-diffusion coefficient allows one to probe any kind of association which would manifest itself by an increase of the apparent diameter d. In liquid systems, typical D range from 10^{-9} m^2 s^{-1} (10^{-5} cm^2 s^{-1}) to 10^{-12} m^2 s^{-1} (10^{-8} cm^2 s^{-1}). Reference data obtained by NMR, published by Holz and Weingärtner [53], are given in Table 7.1. Unless otherwise specified, they concern pure liquids at 25 °C. It can be mentioned that formula (7.30) is valid in the case of 'stick boundary' conditions according to which the molecule species under investigation perturbs other molecules in its vicinity (tends to stick to them). Conversely, when no perturbation is assumed (slip boundary conditions), the factor 3 in the denominator of (7.30) must be replaced by a factor 2 [54]. Since viscosity is pressure and temperature dependent, the Stokes–Einstein formula lends itself to PVT studies. Self-diffusion coefficients of the three simple alcohols, methanol, ethanol and 1-propanol, have been investigated that way [55]. The product $D\eta$ was shown to correlate with the reduced density $\rho^* = \rho d^3$, where ρ is the number density. Self-diffusion data can also be interpreted according to the rough-hard-sphere model.

Another interesting use of self-diffusion coefficients concerns the investigation of attractive or repulsive interactions between two species i and j; such interactions can be characterized by the parameter A_{ij} [56].

$$A_{ij} = (1/T_1)_{\text{inter}} \bar{D} / r \qquad (7.31)$$

where $(1/T_1)_{\text{inter}}$ is the relaxation rate pertaining to the intermolecular dipole–dipole interaction of two spins 1/2 nuclei resting on the molecular entities i and j, respectively; $\bar{D}$ is the mean self-diffusion coefficient of species i and j, and ρ the number density defined above. The difficulty of the method

Table 7.1 Typical self diffusion coefficients deduced from NMR measurements performed at 25 °C on pure liquids (unless otherwise specified). (From Ref. [53])

Species	Observed nucleus	10^9 D$(m^2 s^{-1})$
Acetonitrile	^{1}H	4.37
n-Hexane	^{1}H	4.26
Cyclopentane	^{1}H	3.10
Methanol	^{1}H	2.415
Fluorobenzene	^{1}H	2.395
Water	^{1}H	2.30
Benzene	^{1}H	2.207
N,N-dimethylformamide	^{1}H	1.63
Cyclohexane	^{1}H	1.43
Ethanol	^{1}H	1.075
Dimethyl sulfoxide	^{1}H	0.73
Tetradecane	^{1}H	0.56
Cyclooctane	^{1}H	0.55
Deuterium oxide in H_2O (10 mol %)	^{2}H	2.23
Perdeutero benzene	^{2}H	2.09
Deuterium oxide	^{2}H	1.87
Lithium chloride (4M in H_2O)	^{7}Li	0.70
Perfluoro benzene	^{19}F	1.46
Potassium fluoride (3M in H_2O)	^{19}F	1.135
Sodium chloride (2M in H_2O)	^{23}Na	1.135
Tribenzylphosphite (3M in C_6D_6)	^{31}P	0.365
Caesium chloride (2M in H_2O)	^{133}Cs	1.895

lies in the determination of $(1/T_1)_{inter}$, that is the extraction of the intermolecular contribution from the total relaxation rate. Anyway, once A_{ij} has been determined from experimental data, it can be interpreted in terms of the closest distance approach a and of the atom pair correlation function $g_{ij}(r)$:

$$A_{ij} = \frac{K}{a^4} \int_a^\infty (a/r)^6 g_{ij}(r) r^2 \, dr \qquad (7.32)$$

From this latter equation (in which K is a constant depending on the isotope used for the NMR measurements) it has been shown [57] that A_{ij} could be viewed as being proportional to C_{loc}/C, where C_{loc} is the local concentration of species j around one molecule belonging to species i, whereas C is the mean concentration of species j in the solution. Concentration-dependent measurements of A_{ij} makes it possible to detect association, since, in that case, when lowering the amount of species j, C_{loc} must remain higher than C, and A_{ij} must increase. The method has proved to represent a powerful tool for the detection of solute–solvent or solute–solute interactions in electrolyte solutions as well as in non-electrolyte mixtures [58].

A rather pragmatic use of differential self-diffusion coefficients is for solvent suppression [59]. For instance, when dealing with a large molecule in aqueous solution, high degree of water elimination can be obtained by applying an appropriate PGSE sequence prior to the NMR measurement, so as to irreversibly cancel water magnetization through diffusion processes. Conversely, magnetization of large molecules is little modified, because the self-diffusion coefficient of large molecules (or large molecular entities) is generally several orders of magnitude smaller than the one of water.

Aggregation processes are conveniently studied by self-diffusion measurements, even in multi-component systems since the PGSE method affords in principle the possibility to separately determine the self-diffusion coefficient of each component [60]. Since the whole subject is covered in Chapter 8, just one aspect will be considered here, which deals with water self-diffusion in a micellar solution. The measured quantity, D_{obs} can be written as

$$D_{obs} = A(1 - P_b)D_W + P_b D_{mic} \qquad (7.33)$$

where A is the so-called obstruction factor [61] to be discussed below, P_b the proportion of water bound to the micelle, D_W the diffusion coefficient of pure water and D_{mic} the diffusion coefficient of the micelle. Equation (7.33) implies a rapid exchange between bound and free states. Since $D_{mic} \ll D_W$, equation (7.33) reduces to

$$D_{obs} = A(1 - P_b)D_W \qquad (7.34)$$

D_W is of course perfectly known, so that $A(1 - P_b)$ can be determined as a function of concentration. Now, for a spherical micelle at low concentration of the species constituting the micelle, A is very close to 1, so that P_b can be determined unambiguously and consequently the number of water molecules bound to each individual species. Conversely, assuming this number is independent of concentration makes it possible to evaluate A at higher concentrations. It turns out that the evolution of A with concentration constitutes a clear fingerprint of the micelle shape [61]. The method has been applied successfully for discriminating between spherical, prolate or oblate aggregates.

6.2.2 Restricted diffusion

Let us suppose that, during the diffusion interval Δ of the PGSE experiment, the diffusing molecules may reach the limits of the space allowed to their translational displacements. This space could be for instance a spherical cavity whose radius R is of the same order of (or smaller than) the mean displacement of molecules during Δ. To be more specific, and owing to the usual meaning of a diffusion coefficient ($D = \langle r^2 \rangle / 6\Delta$), where r^2 is the mean-square displacement in time Δ), this situation would occur if R is smaller than $\sqrt{6D\Delta}$. In such a situation, we shall say that diffusion is restricted. Concerning the approach of Section 5.2, this implies that in equation (7.22) integration can no longer go to infinity but can account for the shape and size of the cavity in which the molecules are embedded. Experimentally, restricted diffusion manifests itself by a non-exponential magnetization decay as a function of δ^2 (δ being the duration of the gradient pulses in the PGSE experiment). Conversely, this non-exponentiality should provide additional information. As a matter of fact, analytical expressions of magnetization decay have been derived for confinement in a sphere [62] and between two infinite planes [63]. The sphere case is especially important in view of determining the droplet size, R, in emulsion systems [64–66], since the magnetization decay depends on R as shown by the expression given below which substitutes to equation (7.23)

$$M(\tau, \Delta) = M_0 \exp(-\tau/T_2) \exp(-8\pi^2 q^2 R^2) \sum_{k=1}^{\infty} \frac{1}{a_k^2(a_k^2 - 2)} \left[\frac{2}{a_k^2(D\delta/R^2)} - \right.$$

$$\left. \frac{2 + \exp[-a_k^2 D(\Delta - D)/R^2] - 2\exp(-a_k^2 D\delta/R^2) - 2\exp(-a_k^2 D\Delta/R^2) + \exp[-a_k^2 D(\Delta + \delta)/R^2]}{a_k^2(D\delta/R^2)^2} \right]$$

$$\text{(7.35)}$$

(The summation in (7.35) is generally limited to some of the first terms because, beyond a certain stage, subsequent terms become negligibly small; the coefficients a_k are available in the literature [67]: $a_1 = 2.081576$, $a_2 = 5.940370$, $a_3 = 9.205839$, $a_4 = 12.40444$, $a_5 = 15.57923$, $a_6 = 18.74265$, $a_7 = 21.88279$, $a_k = k\pi$ for $k > 7$.) As can be seen from equation (7.35) a fit of experimental data yields in principle D (the diffusion coefficient within the spherical cavity) and R (the radius).

The characterization of porous media seems to be a very promising field of application of restricted diffusion as probed by the PGSE sequence. The problem here is more complicated by the fact that not only diffusion inside a pore, but also hopping between pores, must be accounted for. Using the displacement imaging method (Section 5.4) and appropriate models, Callaghan *et al.* [41, 68, 69] have been able to derive the structural parameters of porous media (pore diameter and pore spacing). Another approach [70] makes use of the dependence of the apparent diffusion coefficient as a function of the diffusion time Δ to extract the pore surface to volume ratio (at short times) and a parameter characterizing the tortuosity of the pore space (at long times).

7   Conclusion

Owing to their spatial labeling properties, field gradients used in conjunction with NMR experiments yield unique information, not easily attainable by other spectroscopic techniques. The key point is the possibility to derive the information *in situ* and in a non-destructive manner, in contrast with most other techniques which probe only surfaces. Pieces of information about fluids or fluid mixtures embedded in more or less rigid materials range from their local distribution (imaging) to their motional properties (translational diffusion or flow). Of course, NMR techniques devoted to these latter parameters apply as well to homogeneous liquid samples. Technical improvements of the so-called NMR microscopy techniques should lead in the near future to localized spectroscopic parameters (chemical shifts, coupling constants, relaxation times, self-diffusion coefficients, etc.), obtained with a standard accuracy and a spatial resolution of the order of 10 μm or better.

References

1. E. L. Hahn, *Phys. Rev.*, **80**, 580 (1950).
2. G. Suryan, *Proc. Indian Acad. Sci.* **A33**, 107 (1951).
3. P. T. Callaghan, *Principles of Nuclear Magnetic Resonance Microscopy*, Clarendon Press, Oxford, 1991.
4. P. Mansfield and B. Chapman, *J. Phys. E*, **19**, 540 (1986).
5. D. Canet, B. Diter, A. Belmajdoub, J. Brondeau, J. C. Boubel and K. Elbayed, *J. Magn. Reson.*, **81**, 1 (1989).
6. G. Bodenhausen, R. Freeman and D. L. Turner, *J. Magn. Reson.*, **27**, 511 (1977).
7. M. Guéron, P. Plateau and M. Décorps, *Prog. NMR Spectrosc.*, **23**, 135 (1991).
8. P. C. M. van Zijl and C. T. W. Moonen, *NMR Basic Principles and Progress*, **26**, 67 (1992).
9. A. Haase, J. Frahm, W. Hänicke and D. Mattei, *Phys. Med. Biol.*, **30**, 341 (1985).
10. V. Sklenar and A. Bax, *J. Magn. Reson.*, **74**, 469 (1987).
11. M. von Kienlin, M. Décorps, J. P. Albrand, M. F. Foray and P. Blondet, *J. Magn. Reson.*, **76**, 169 (1988).
12. J. Jeener, B. H. Meier, P. Bachman and R. R. Ernst, *J. Chem. Phys.*, **71**, 4546 (1979).
13. C. T. W. Moonen, P. van Gelderen, G. W. Vuister and P. C. M. van Zijl, *J. Magn. Reson.*, **97**, 419 (1992).
14. C. S. Bosch and J. J. Ackerman, *NMR Basic Principles and Progress*, **27**, 3 (1992).
15. R. J. Ordidge, A. Connely and J. A. B Lohman, *J. Magn. Reson.*, **66**, 283 (1986).
16. W. P. Aue, S. Müller, T. A. Cross and J. Seelig, *J. Magn. Reson.*, **56**, 350 (1984).
17. P. R. Luyten and J. A. den Hollander, *Magn. Reson. Imag.*, **4**, 237 (1986).
18. J. Frahm, H. Bruhn, M. L. Gyngell, K. D. Merboldt, W. Hänicke and R. Sauter, *Magn. Reson. Med.*, **9**, 79 (1989).
19. P. A. Bottomley, *Ann. N. Y. Acad. Sci.*, **508**, 333 (1987).
20. A. Kumar, D. Welti and R. R. Ernst, *J. Magn. Reson.*, **18**, 69 (1965).
21. W. A. Edelstein, J. M. S. Hutchinson, G. Johnson and T. Redpath, *Med. Biol.*, **25**, 748 (1980).
22. P. C. Lauterbur, *Pure Appl. Chem.*, **40**, 149 (1974).
23. A. L. Robinson, *Science*, **140**, 542 (1975).
24. P. Maffei, L. Kiéné and D. Canet, *Macromolecules*, **25**, 514 (1992).
25. P. Maffei, P. Mutzenhardt, A. Retournard, B. Diter, R. Raulet, J. Brondeau and D. Canet, *J. Magn. Reson.*, **A107**, 40 (1994).
26. T. R. Brown, B. M. Kincaid and K. Ugurbil, *Proc. Natl. Acad. Sci. USA*, **79**, 3523 (1982).
27. L. D. Hall and S. Sukumar, *J. Magn. Reson.*, **50**, 161 (1982).
28. A. A. Maudsley, S. K. Hilal, W. H. Perman and H. E. Simon, *J. Magn. Reson.*, **51**, 147 (1983).
29. A. Haase and D. Matthaei, *J. Magn. Reson.*, **71**, 550 (1987).
30. H. W. Park, Y. H. Kim and Z. H. Cho, *Magn. Reson. Med.*, **7**, 340 (1988).
31. D. G. Norris and W. Dreher, *Magn. Reson. Med.*, **30**, 641 (1993).

32. L. D. Hall, V. Rajanayagam and C. Hall, *J. Magn. Reson.*, **68**, 185 (1986).
33. L. D. Hall and V. Rajanayagam, *J. Magn. Reson.*, **74**, 139 (1987).
34. J. M. Dereppe, C. Moreaux and K. Schenker, *Magn. Reson. Imag.*, **9**, 809 (1991).
35. J. L. A. Williams, D. G. Taylor, G. Maddinelli, P. Enwere and J. S. Archer, *Magn. Reson. Imag.*, **9**, 767 (1991).
36. P. D. Majors, J. L. Smith, F. S. Kovarik and E. Fukushima, *J. Magn. Reson.*, **89**, 470 (1990).
37. A. Caprihan and F. Fukushima, *Phys. Rep.*, **4**, 195 (1990).
38. H. van As and T. J. Schaafsma, in *An Introduction to Biomedical Resonance*, G. Thieme, Verlag, Berlin, 1985, pp. 68–96.
39. H. L. Carr and E. M. Purcell, *Phys. Rev.*, **94**, 630 (1954).
40. E. O. Stejskal, *J. Chem. Phys.*, **43**, 3597 (1965).
41. E. O. Stejskal and J. E. Tanner, *J. Chem. Phys.*, **42**, 288 (1965).
42. G. S. Karczmar, D. B. Twieg, T. J. Lawry, G. B. Matson and M. W. Weiner, *Magn. Reson. Med.*, **7**, 111 (1988).
43. D. Canet, B. Diter, A. Belmajdoub, J. Brondeau, J. C. Boubel and K. Elbayed, *J. Magn. Reson.*, **81**, 1 (1988).
44. R. Dupeyre, Ph. Devoulon, D. Bourgeois and M. Décorps, *J. Magn. Reson.*, **95**, 589 (1991).
45. P. T. Callaghan, D. MacGowan, K. J. Packer and F. O. Zelaya, *J. Magn. Reson.*, **90**, 177 (1990).
46. P. T. Callaghan and Y. Xia, *J. Magn. Reson.*, **91**, 326 (1991).
47. M. R. Merrill and Z. Jin, *Magn. Reson. Imag.*, **12**, 345 (1994).
48. M. H. G. Amin, L. D. Hall, R. J. Chorley, T. A. Carpenter, K. S. Richards and B. W. Bache, *Magn. Reson. Imag.*, **12**, 319 (1994).
49. D. N. Guilfolte and P. Mansfield, *Magn. Reson. Imag.*, **9**, 775 (1991).
50. J. W. Gleeson and D. E. Woessner, *Magn. Reson. Imag.*, **9**, 879 (1991).
51. D. Bourgeois and M. Décorps, *J. Magn. Reson.*, **94**, 20 (1991).
52. S. J. Gibbs, T. A. Carpenter and L. D. Hall, *J. Magn. Reson., Series A*, **105**, 209 (1993).
53. M. Holz and H. Weingärtner, *J. Magn. Reson.*, **92**, 115 (1991).
54. H. J. V. Tyrrel and K. R. Harris, *Diffusion in Liquids*, Butterworths, London, 1984.
55. S. Meckl and M.D. Zeidler, *Molec. Phys.*, **63**, 85 (1988).
56. M. Holz, *Prog. NMR Spectrosc.*, **18**, 327 (1986).
57. K. J. Müller and H. G. Hertz, *Chem. Scr.*, **29**, 277 (1989).
58. M. Holz, R. Grunder, A. Sacco and A. Memameo, *J. Chem. Soc. Faraday Trans.*, **89**, 1215 (1993).
59. P. C. M. van Zijl and C. T. W. Moonen, *J. Magn. Reson.*, **87**, 18 (1990).
60. See for instance: B. Lindman and P. Stilbs, in K. L. Mittal and B. Lindman (eds), *Surfactants and solutions*, Plenum, New York, 1984, vol. 3, pp. 1651–1662.
61. B. Jönsson, H. Wennerström, P. Linse and F. G. Nilsson, *Colloid Polym. Sci.*, **264**, 77 (1986).
62. J. S. Murday and R. M. Cotts, *J. Chem. Phys.*, **48**, 4938 (1968).
63. J. E. Tanner and E. O. Stejskal, *J. Chem. Phys.*, **49**, 1768 (1968).
64. K. Packer and C. Rees, *J. Colloid Interface Sci.*, **40**, 206 (1971).
65. P. T. Callaghan, K. W. Jolley and R. Humphrey, *J. Colloid Interface Sci.*, **93**, 521 (1991).
66. I. Lönnqvist, A. Khan and O. Söderman, *J. Colloid Interface Sci.*, **144**, 401 (1991).
67. R. M. Kleinberg and M. A. Horsfield, *J. Magn. Reson.*, **88**, 9 (1990).
68. P. T. Callaghan, A. Coy, D. MacGowan, K. J. Packer and F. O. Zelaya, *Nature*, **351**, 467 (1991).

69. P. T. Callaghan, A. Coy, T. P. J. Halpin, D. MacGowan, K. J. Packer and F. O. Zelaya, *J. Chem. Phys.*, **97**, 651 (1992).
70. L. L. Latour, P. P. Mitra, R. L. Kleinberg and C. H. Sotak, *J. Magn. Reson.*, **A101**, 342 (1993).

8 SURFACTANT SOLUTIONS: AGGREGATION PHENOMENA AND MICROHETEROGENEITY

B. Lindman, U. Olsson and O. Söderman
University of Lund, Sweden

Dynamics of Solutions and Fluid Mixtures by NMR
Edited by J.-J. Delpuech   © 1995 John Wiley & Sons Ltd

1   Introduction

Amphiphilic molecules, such as surfactants and lipids, self-assemble in water—and some other polar solvents—to form aggregates of many different shapes and sizes. For general reviews see Refs [1–5]. These aggregates are the building blocks of the large number of phases which may form. The macroscopic properties of the various phases vary enormously. One important example of practical significance is the consistency and rheology. Simple surfactant–water systems may vary from being low viscous, viscosity of neat water or below, to being 'thick' solutions, gels and rather hard solids. The differences in the conditions for forming the different states may be quite subtle, and minor changes in concentration, temperature or surfactant chemical structure may induce transitions between different states, quite often in an unexpected direction. The relation between molecular interactions and dynamics and the macroscopic properties is far from trivial and, for example, a solid-like appearance does not generally reflect any quenched dynamics of individual molecules. Instead, it has been demonstrated that molecular interactions and local dynamics do not differ appreciably between different phases, even for extreme changes in macroscopic appearance. The understanding of these effects has to a major extent resulted from NMR relaxation and self-diffusion studies and dynamic NMR has been the main approach which has allowed a quantitative characterization of a wide range of basic dynamic processes in surfactant solutions, not only on the molecular level but also on the aggregate level.

A large part of this chapter will be concerned with the characterization of the dynamics of surfactant solutions. We will present in Sections 3 and 4 a general account on this topic which includes:

- a description of the special models developed for the interpretation of relaxation data of surfactant systems and an account of how information on molecular and aggregate dynamics is derived,
- a detailed account of deduced molecular dynamics for a representative system of spherical micelles

- a brief overview of different systems investigated, elucidating the effect on molecular dynamics of such factors as surfactant chemical structure and aggregate and phase structure.

It should be remarked that often the information one wants to extract from the NMR parameters is *static* in nature. Examples are micelle size and/or shape, degree of micellar association of both surfactants and counter-ions and microstructure of microemulsion solutions. On the other hand, the NMR observables are often dynamic quantities, such as relaxation rates and self-diffusion coefficients. One of the key issues involved in NMR studies of surfactant solutions is thus to go from dynamic observables to static properties of the system under study. This distinction is not always clearly pointed out, but should always be kept in mind.

While NMR is certainly the most powerful tool for quantifying the molecular dynamics of surfactant systems, the significance of dynamic NMR in the surfactant field goes far beyond that. Thus dynamic NMR has been most significant for resolving central issues in the field during the last two decades. In many cases the approaches used are non-trivial and it is not immediately realized how dynamic NMR is related to many of the problems. In Section 5 we will selectively consider a number of problem areas in the field of surfactants and discuss how NMR can provide information and what limitations there are. We will also briefly mention alternative experimental techniques and the advantages and disadvantages of dynamic NMR compared to those.

The developments in the surfactant field in general has been quite rapid recently and several novel phenomena have come to our attention. For those readers who are less familiar with these developments we will provide in Section 2 an introductory account of surfactant self-assembly in general, and on recent progress in particular.

The field of NMR studies of surfactant systems has been systematically reviewed by a number of workers in the past [6, 7]. For this reason the present account will focus on recent work and especially on new types of applications of dynamic NMR in the surfactant field.

2 Surfactant Self-assembly, a Brief Account of Recent Developments

Surfactant and lipid molecules are amphiphilic, i.e. they possess parts which are distinctively lipophilic and hydrophilic. They therefore combine the properties of polar solutes, like electrolytes, with those of hydrocarbons (typically). In aqueous systems this ambivalence leads to two main effects:

- adsorption at interfaces (water/air, water/hydrocarbon, water/solid and water/macromolecule),
- self-assembly (alone or with low molecular weight or macromolecular cosolutes).

Recent findings have emphasized a close connection between the two types of phenomena, in particular that surfactant adsorption can also be best considered as a self-assembly process. The driving force of the two types of phenomena is the same, i.e. the hydrophobic interaction, based on the free-energy penalty of introducing nonpolar species, like hydrocarbon groups or molecules, in water with its strong intermolecular association. We will review below different types of surfactant systems following generally a line of increasing complexity of the system.

2.1 TYPES OF SURFACTANTS

Surfactants are best classified after the chemical structure and we generally use a classification based on the character of the polar head-group. Generally, surfactants are divided into non-ionic, zwitterionic, anionic, cationic and catanionic. The nonpolar hydrophobic part may also vary in nature, there may be one, two or three alkyl chains, there may be branched chains, aromatic groups may occur, there may be a fluorocarbon group (which has some lipophobic character as well), etc. Several developments in recent years as regards new surfactants may be noted like bolaforms (α, ω surfactants), divalent surfactants but, in particular, macromolecular surfactants. Surface-active polymers are generally of two types, block copolymers of polar and nonpolar segments or graft copolymers, the latter being mainly of the type where alkyl chains have been attached to a hydrophilic backbone.

2.2 MICELLIZATION

Surfactant self-association is typically strongly cooperative and often leads, as for single alkyl chain surfactants with strongly polar head-groups, in the first step to closely spherical micelles as depicted in Figure 8.1. It is characteristic of micellization, and surfactant self-assembly in general, that it is governed by two opposing forces, one driving force which is the hydrophobic interaction and one opposing force given by hydrophilic interactions, i.e. head-group repulsions, in general mediated by an interaction with the solvent.

The onset of micelle formation occurs at a quite well-defined concentration, termed the critical micelle concentration (CMC), above which the surfactant unimer concentration and the activity stay closely constant. It is an excellent first approximation to consider micelle formation as analogous to a phase separation.

CMC values are about two orders of magnitude higher for ionic than for non-ionic surfactants because of the entropy penalty paid by confining the counterions in the vicinity of the highly charged micellar surfaces; this negative entropy term becomes less significant in the presence of electrolyte—thus salt depresses the CMC—and at higher surfactant concentrations.

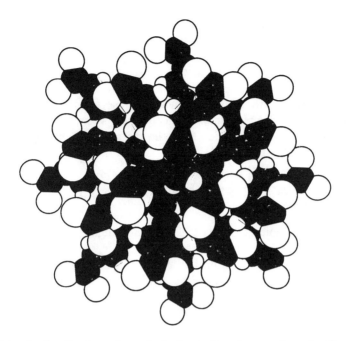

Figure 8.1 A visualization of a spherical micelle of approximately 60 surfactant molecules. Reproduced by permission of Bengt Jönsson

The CMC value decreases strongly on lengthening the surfactant alkyl chain as expected from the hydrophobic interaction being the driving force; the CMC is reduced by roughly a factor of 3 for non-ionics and a factor of 2 for ionics on adding one methylene group to the alkyl chain.

Micelle size also shows a simple relation to the length of the alkyl chains in that the radius of a spherical micelle is close to the length of an extended (all-trans) surfactant molecule. While basic features of micelle formation, like the thermodynamics of micelle formation and micelle structure and dynamics, were already established in the 1930s, much refinement and confirmation are due to work during the last two decades. Dynamic NMR has been the main technique behind this development and the picture of micelles it has provided will be described below.

2.3   BEYOND THE SIMPLE SPHERICAL MICELLE

The simple spherical micelle is not the only type of aggregate formed in surfactant–water solutions and is not necessarily the first in the self-assembly. For surfactants with relatively weak polar groups, and thus a weak opposing force, the first-formed aggregates may be rod- or thread-type or there may be bilayer structures, including disc-like micelles and vesicles; the latter are generally not the equilibrium state but recently support for thermodynamically stable vesicles has accumulated for the class of catanionic surfactants in the presence of electrolyte.

The most interesting developments recently have concerned *bicontinuous* systems, i.e. where the aggregates extend over microscopic distances in all three dimensions. It will in the following be useful to distinguish between systems based on discrete units (micelles or vesicles) and nondiscrete structures characterized by continuity of the surfactant aggregates, in one, two or 3 three dimensions (Figure 8.2).

2.4   SURFACTANT SELF-ASSEMBLIES IN PHASES OF LONG-RANGE ORDER

In addition to isotropic solutions lacking long-range correlations, surfactant self-assemblies form the basis of several liquid crystalline phases; these are characterized by long-range order but are as will be further illustrated below characterized by short-range disorder. The most studied liquid crystalline phases are the lamellar, the (normal and reversed) hexagonal and cubic phases; especially for the latter there are many different types and in particular we can

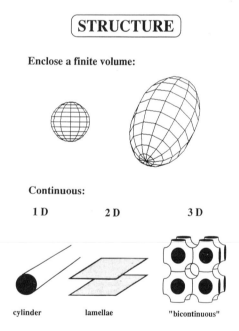

Figure 8.2   A surfactant film may divide space in many different ways and in particular it is important to distinguish between surfactant self-assemblies which are of the discrete micellar type, with different possible shapes of the micelles, and those which extend over macroscopic distances in one, two or three dimensions

distinguish between discrete and bicontinuous type structures. Many additional phases, termed intermediate, deformed etc., have been observed in recent years and have become more or less well established and characterized.

2.5   MORE COMPLEX SURFACTANT SYSTEMS

Both from a fundamental, a biological and a technical point of view, interest is generally focused on more complex systems, i.e. those containing three or more components. Common additional components are an electrolyte, an oil like a hydrocarbon, a weak amphiphile or a second surfactant. The same structures and phases as mentioned above form for the more complex systems, but because of the larger freedom of variation, phase diagrams may become very complex, with quite a large number of distinct phases (as well as concomitant multi-phase regions). A complete characterization of the phase equilibria may then become a formidable task, not the least since a macroscopic separation of different phases may be extremely difficult owing to high viscosities and small differences in density.

Two further denotations of surfactant self-assemblies should be mentioned here, i.e. reversed micelles and microemulsions. A surfactant dissolved in a nonpolar solvent shows typically an essentially noncooperative association into small or moderately sized oligomers. However, in the presence of water, even in quite minute amounts, there may be a very significant organization into self-assemblies of discrete small domains of water surrounded by a surfactant monolayer.

The term microemulsion is generally used to denote isotropic solutions in systems having as main components one surfactant, oil (like a hydrocarbon) and water. A microemulsion is distinct from an emulsion, which is a thermodynamically unstable dispersion of two or more phases, and the term, used for historical reasons, is, therefore, rather unfortunate. Microstructure of microemulsions has since the late 1970s attracted an enormous interest. We now know that it can vary strongly depending on conditions (surfactant, temperature, electrolyte, etc.) and that it can encompass normal micellar and reversed micellar structures as well as bicontinuous structures. This matter, to which dynamic NMR has provided the main insight, will be discussed in some detail below.

2.6   SYSTEMS OF POLYMERIC SURFACTANTS AND POLYMER–SURFACTANT MIXTURES

High molecular weight surfactants give a self-assembly pattern similar to that of the low molecular counterparts, but self-assembly may become much stronger with very low CMCs and phase transitions produced by subtle changes in, for example, temperature and concentration. Systems of both a surfactant and a (homo- or co-) polymer have in recent years attracted very great interest in general. Dynamic NMR should here have a very great potential and studies start to emerge. Different types of polymer–surfactant systems should be mentioned. Adding a surfactant to a polymer solution may lead to micellization at the polymer chain. The lowered CMC reflects that micelle formation is favoured at the polymer surface compared to the bulk and this may be due to hydrophobic interactions for a polymer with nonpolar parts, but typically the driving force is a modification of the electrostatic interactions. For this reason, ionic surfactants show a much larger tendency to associate with polymers than nonionics. Other polymer–surfactant systems that we will consider are polymer molecules incorporated in liquid crystalline phases or in microemulsions.

2.7   DISPERSE SYSTEMS AND MACROSCOPIC INTERFACES

These are areas where the applications of NMR is still in its infancy but where we can see a great potential. Progress has mainly been retarded because of experimental difficulties but significant studies of emulsions as well as of adsorbed surfactant layers start to emerge. In both cases questions related to the clustering of surfactant molecules at the interfaces and the role of surfactant molecule dynamics are central. Studies of vesicles is another area where progress should become pronounced in the near future.

3   NMR Relaxation Studies

3.1   INTRODUCTION

NMR relaxation is a useful experimental technique to study surfactant aggregation in liquid solutions and liquid crystals [6, 8, 9]. The experiment yields information on the local dynamics and the conformational state of the surfactant hydrocarbon chain and has, for example, demonstrated the liquid-like interior of surfactant micelles. However, the aim of NMR relaxation studies of isotropic micellar solutions and microemulsions is often to study properties such as the surfactant aggregate size.

The reorientation dynamics of aggregated surfactant molecules, which we will discuss in more detail below, is characterized by a locally preferred orientation, in that they essentially form a monomolecular film of oriented molecules. Locally, surfactant molecules undergo rapid internal motions, such as *trans–gauche* isomerizations in the hydrocarbon chain. These motions are, due to the preferred orientation, slightly anisotropic. The situation can be pictured as the polar headgroups being anchored at the polar/apolar interface leaving room for dangling motion of the hydrocarbon chains, with certain restrictions on the number of conformations due to packing constraints in the film. The residual interaction or anisotropy is in isotropic solutions further averaged by the thermal tumbling of the aggregates in combination with the diffusion of surfactant molecules within the curved surfactant film.

NMR spectra from surfactants in micellar solutions and microemulsions are generally in the motional narrowing regime, and the spin dynamics are characterized by well defined relaxation rates. Exceptions can be found in solutions of very long ($>10^3$ Å) wormlike cylindrical micelles where the large size in connection with a steric overlap of the micelles give rise to a very slow ($\approx 10^{-4}$ s) aggregate reorientation [10, 11]. On the other hand, the dynamics are typically outside the extreme narrowing regime, and one often observes $R_2 \gg R_1$, where R_1 and R_2 are the longitudinal and transverse relaxation rates, respectively. Micellar aggregates reorient typically on the timescale of 10^{-9} s or slower. Thus, at conventional magnetic field strengths, the essential information on the surfactant aggregates is stored in the transverse relaxation rate, R_2.

Hydrocarbon, typically an alkyl chain of 12–16 carbons, constitutes the major part of surfactant molecules, offering ^{1}H and ^{13}C nuclei for NMR studies. Being $I = 1/2$ nuclei the relaxation is mainly due to dipole–dipole interactions (although chemical shift anisotropy and scalar interaction can sometimes be important in the case of ^{13}C). The easy access and the relatively high sensitivity (particularly important in early continuous wave NMR studies) has made ^{1}H NMR (in particular bandwidth measurements) studies particularly popular. Micellar growth can easily be studied in a qualitative manner by monitoring the bandwidth of aliphatic ($-CH_2-$) protons in the spectrum. A quantitative analysis is, however, complicated by the coupling of the various protons, having locally different motional characteristics, along the alkyl chain. In an alkyl chain of say 16 carbons, the protons constitute a coupled spin system of approximately 2^{16} spin states, which can have a broad spectrum of transition rates, resulting in significant deviations from a Lorentzian band shape of the CH_2 resonance [10, 12]. This complexity can, however, be turned into an advantage in that it allows for a continuum description of the transition rate spectrum. The ^{1}H band shape problem was analysed by Wennerström et al. [10, 12], within the two-step formalism (see below), and quantitative ^{1}H band shape analyses have been performed in systems of large, slowly reorienting micelles [10, 11]. In contrast to the non-Lorentzian band shape, one often observes an exponential longitudinal relaxation due to rapid internal equilibration of the spin system (spin diffusion), and the frequency dependence of R_1 has been measured by field cycling techniques [13, 14].

In an alkyl chain, the ^{13}C nuclei are relaxed mainly by the fluctuating dipole–dipole coupling to the directly bound protons. The various carbons along the chain are often resolved, allowing for individual characterizations of the different segments of the hydrocarbon chain. Usually one applies broadband proton decoupling, resulting in exponential longitudinal relaxation of the individual ^{13}C magnetizations, and it is possible to determine R_1 and the nuclear Overhauser enhancement (NOE) of the different methylene segments. On the other hand, it is difficult to measure R_2 of ^{13}C. Therefore ^{13}C is mainly useful for studying local chain dynamics and segmental order, since the aggregate dynamics occur on time-scales which normally are significantly longer than the inverse ^{13}C resonance frequency.

Certain surfactants carry phosphate or ammonium polar head groups, offering ^{31}P and ^{14}N nuclei, respectively, for relaxation studies. The phosphate ^{31}P $(I = 1/2)$ has a relatively large chemical shift anisotropy which is strongly dependent on the pH (degree of protonation), complicating the analysis of relaxation data. ^{14}N is a $I = 1$ nucleus and its relaxation is dominated by the strong quadrupolar interaction. Except for the case of quaternary ammonium groups, the magnitude and the asymmetry of the electric field gradient tensor is also pH dependent, and special care has to be taken.

The most suitable nucleus for relaxation studies of surfactant systems is ^{2}H, which can be synthetically incorporated into the hydrocarbon chain of the surfactant molecule. Normally one selectively labels one single methylene segment, typically adjacent to the polar head group, in order to avoid resonance overlap in the spectrum. ^{2}H is a $I = 1$ nucleus, and the dominating interaction governing the relaxation is the strong quadrupolar interaction. For aliphatic (C–^{2}H) deuterons, the analysis is further simplified by the fact that the electric field gradient tensor has essentially cylindrical symmetry around the C–^{2}H bond direction, resulting in a vanishing asymmetry parameter. The ease with which both R_1 and R_2 can be accurately measured, the relatively simple interaction Hamiltonian, and the sufficiently high sensitivity to allow measurements at low resonance frequencies (down to 2 MHz using variable field electromagnets), gives ^{2}H NMR relaxation a particularly important role among the different experimental tools available to study surfactant systems.

To make the significance of the NMR technique as an experimental tool in surfactant science more apparent, it is important to compare the strengths and the weaknesses of the NMR relaxation technique, in relation to other experimental techniques. In comparison with other experimental techniques to study for example micellar size, the NMR relaxation technique has two major advantages, both of which are associated with the fact that it is reorientational motions which are measured. One is that the relaxation, i.e. R_2, is sensitive to small variations in micellar size. For example in the case of a sphere, the rotational correlation time is proportional to the cube of the radius. This can be compared with the translational self-diffusion coefficient which varies linearly with the radius. The second, and perhaps the most important advantage is the fact that the rotational diffusion of particles in solution is essentially independent of inter-particle interactions (electrostatic and hydrodynamic). This is in contrast to most other techniques available to study surfactant systems or colloidal systems in general, such as viscosity, collective and self-diffusion, scattered light intensity, etc. In this latter respect, NMR relaxation experiments are comparable with form factor determinations with small angle X-ray or neutron scattering experiments. A weakness of the NMR relaxation approach to aggregate size determinations, compared with form factor determinations, would be the difficulties of absolute calibration, since the transformation from information on dynamics to information on structure must be performed by means of a motional model.

3.2   TIMESCALE SEPARATION: THE TWO-STEP MODEL

In this section we will discuss a motional model [15–17] of aggregated surfactant molecules, which involves a timescale separation of fast local and slow global motions. This model has been extensively applied to, and been shown to rationalize, NMR relaxation data from aggregated surfactant systems. We will discuss it in connection with ^{2}H relaxation experiments, since the most extensive experimental studies have been performed on ^{2}H labelled surfactants. We begin by recalling that the longitudinal (R_1) and transverse (R_2) relaxation rates of a $I = 1$ nucleus, due to a quadrupolar interaction in an isotropic solution, is given by: [18]

$$R_1 = \frac{3\pi^2}{40} \chi^2 [2j(\omega_0) + 8j(2\omega_0)] \qquad (8.1a)$$

$$R_2 = \frac{3\pi^2}{40} \chi^2 [3j(0) + 5j(\omega_0) + 2j(2\omega_0)] \qquad (8.1b)$$

Here, χ is the quadrupolar coupling constant and $j(\omega_0)$ the reduced spectral density function evaluated at the Larmor frequency, ω_0.

Based on existing knowledge of surfactant aggregate structures, Wenner-ström et al. [15–17] suggested that the reorientational motion of surfactant molecules can be divided into (i) fast local motions (equilibration of internal modes), which are slightly anisotropic owing to the preferred orientation of the surfactant molecules, and (ii) slow isotropic motions associated with the aggregate tumbling and the lateral diffusion of surfactant molecules within the micellar interface. If these two motions occur on significantly different time scales, the fast motions generate a well defined residual anisotropy to be averaged by the slow motions, and the spectral density can be written as a weighted sum of the fast and the slow components, respectively:

$$j(\omega_0) = (1 - S^2)j_f(\omega_0) + S^2 j_s(\omega_0) \qquad (8.2)$$

Here, $j_f(\omega_0)$ and $j_s(\omega_0)$ represent the reduced spectral density functions describing the fast and slow motions respectively. The parameter S is often referred to as an order parameter and is given by the average:

$$S = \tfrac{1}{2} \langle 3 \cos^2 \theta_{MD} - 1 \rangle_f \qquad (8.3)$$

where θ_{MD} is the angle between the axis of the maximum component of the electric field gradient tensor and the director axis.

The situation is illustrated in Figure 8.3 for the case of a ^{2}H labelled methylene segment of a surfactant molecule residing in a micelle. The director

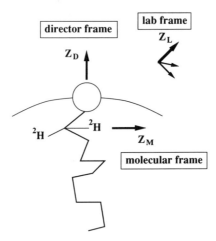

Figure 8.3 A schematic illustration of the various coordinate frames considered within the two-step model, for the case of a specifically deuterium labelled methylene segment in the surfactant hydrocarbon chain. The lab frame (L) is set by the direction of the external magnetic field, where Z_L is the field direction. In this frame the nuclear quadrupolar moment tensor is diagonal. The director frame (D) is associated with the micellar aggregate, where Z_D specifies the micellar surface normal. It is assumed that the fast local dynamics occur with an essentially cylindrical symmetry around Z_D. The molecular frame (M) corresponds to the principal axis of the electric field gradient tensor. For the case of a methylene segment, Z_M specifies the maximum component of the field gradient tensor, which is furthermore cylindrically symmetric around Z_M

axis, Z_D corresponds to the micellar surface normal. The field gradient tensor is cylindrically symmetric around the C–D bond axis, with the maximum component in the bond axis direction, Z_M.

In equation (8.2) it is also assumed that the fast motions have an effective threefold or higher symmetry around the director. For most situations one may expect an effectively cylindrical symmetry. For lower than threefold symmetry, equation (8.2) has to be extended to include a second order parameter, describing the residual order in the plane perpendicular to the director. A similar extension is also necessary when the electric field gradient has a non-vanishing asymmetry parameter.

3.3   INTERPRETING THE SPECTRAL DENSITIES

The fast motions, described by the spectral density function $j_f(\omega_0)$, are expected to occur on a similar timescale ($\tau_f \approx 10^{-11}$ s) as in the corresponding liquid alkane. Thus this motion is in extreme narrowing ($\tau_f \ll 1/\omega_0$) in the accessible frequency range, and we may assume: $j_2(2\omega_0) \approx j_f(\omega_0) \approx j_f(0) = 2\tau_f$, where τ_f should be considered as an *effective* correlation time, representing the integral over a non-exponential correlation function.

The slow motion is a combination of micelle tumbling and the lateral diffusion of surfactant molecules within the surfactant film. For spherical aggregates, this motion is described by a Lorentzian spectral density function:

$$j_s(\omega_0) = \frac{2\tau_s}{1 + (\omega_0\tau_s)^2}$$ (8.4)

where τ_s is the correlation time. The tumbling and the lateral diffusion are expected to be statistically independent, in which case the correlation function can be factorized:

$$G_s = G_t G_d = e^{-t/\tau_t} e^{-t/\tau_d} = e^{-t/\tau_s}$$ (8.5)

where subscripts t and d refer to tumbling and lateral diffusion, respectively, and the slow correlation time can be written as:

$$\frac{1}{\tau_s} = \frac{1}{\tau_t} + \frac{1}{\tau_d}$$ (8.6)

The correlation time associated with the tumbling of a sphere of radius R is given by:

$$\tau_t = \frac{4\pi\eta R^3}{3k_B T}$$ (8.7)

where η represents the solvent viscosity and $k_B T$ the thermal energy. The effect of lateral diffusion can be described in terms of diffusion on the surface of a sphere of radius R. In this case, the correlation time can be written as:

$$\tau_d = \frac{R^2}{6D_{lat}}$$ (8.8)

where D_{lat} is the lateral diffusion coefficient.

With the assumptions of (i) extreme narrowing conditions for the fast motions and (ii) a Lorentzian spectral density function of the slow motion, the expressions for the relaxation rates become with equation (8.2):

$$R_1 = \frac{3\pi^2}{40} \chi^2 \left[(1 - S^2)20\tau_f + S^2 \left(\frac{4\tau_s}{1 + (\omega_0\tau_s)^2} + \frac{16\tau_s}{1 + (2\omega_0\tau_s)^2} \right) \right]$$ (8.9a)

$$R_2 = \frac{3\pi^2}{40} \chi^2 \left[(1 - S^2)20\tau_f + S^2 \left(6\tau_s + \frac{10\tau_s}{1 + (\omega_0\tau_s)^2} + \frac{4\tau_s}{1 + (2\omega_0\tau_s)^2} \right) \right]$$ (8.9b)

These equations (8.9) have been applied in a large number of ^{2}H NMR relaxation studies of micellar systems. It turns out that frequency dependent relaxation studies are particularly useful to study small spherical micelles. These have a typical radius of about 20 Å. Using the water viscosity we obtain with equation (8.7) $\tau_t \approx 7$ ns at room temperature. Including the effect of lateral diffusion, a typical value of the slow correlation time is $\tau_s \approx 5$ ns. Thus, the relaxation dispersion from the slow motion occurs around $\nu_0 \approx (2\pi\tau_s)^{-1} \approx$ 30 MHz which is well within the accessible Larmor frequency range for ^{2}H.

4 Self-diffusion Studies

4.1 GENERAL CONSIDERATIONS

Measurements of self-diffusion coefficients by means of PGSE techniques have evolved to become one of the most important tools in the characterization of surfactant systems. In particular this is true for those surfactant systems which are isotropic liquid solutions such as micellar systems and micro-emulsions. There are essentially two reasons for this state of affairs, one of which has to do with the measuring technique as such. Thus the PGSE approach offers a convenient method of determining self-diffusion coefficients that really has very few competitors with the same versatility and ease of determination. The technique as such will be outlined in another contribution in this volume (see Chapter 7) and has also been described in a number of review articles [19, 20]. An account of the most recent developments of the method can be found in [21]. Thus we shall not dwell on the technical aspects here, but merely note that the technique requires no isotopic labelling (avoiding possible disturbances due to addition of probes); furthermore, it gives component resolved diffusion coefficients with great precision in a minimum of measuring time.

The second reason for its success rests on the fact that the method monitors transport over macroscopic distances (typically in the micrometer regime). It is very important to differentiate between restricted diffusion and free diffusion. The former situation holds for systems where the molecules experience some sort of boundary for their diffusion on this length scale. An example would be molecules confined to the interior of an emulsion droplet. With the latter situation we mean molecules that undergo Gaussian diffusion. In fact, one of the main powers of the PGSE technique is the ability to differentiate between these two situations. We will return to the case of restricted diffusion below, and first concentrate on molecules undergoing free diffusion. For such systems the determined diffusion coefficients reflect aggregate sizes, obstruction effects and possible aggregate confinements, yielding information which is easily interpretable in terms of microstructure of surfactant solutions as well as conveying information on such essential properties as degree of micellization and counter-ion binding. The fact that the information is obtained without the need to invoke complicated models, as is the case for the NMR relaxation approach (see above), is particularly important. In this context it should be stressed that with the PGSE approach one obtains the self-diffusion rather than the collective diffusion coefficient.

The foundation for the use of NMR to monitor diffusion was laid in 1950 in the seminal paper by Hahn [22]. The next important development occurred around the middle of the 1960s, when Stejskal and Tanner demonstrated [23], following a suggestion by Douglass, McCall and Anderson [24], the use of pulsed gradients. Finally towards the end of the 1960s the FT approach was introduced [25], and thereby the method became component resolved in that individual diffusion coefficients of different components of multicomponent systems could be measured simultaneously as long as they have resolved NMR signals.

In its simplest version, the PGSE experiment consists of two equal and rectangular gradient pulses of magnitude g and length δ, sandwiched on either side of the 180°, rf-pulse in a simple Hahn echo experiment. For molecules undergoing free diffusion characterized by a diffusion coefficient of magnitude D, the echo attenuation due to diffusion is given by [23, 26]:

$$E(\Delta, \delta, g) = E_0 \exp[-\gamma^2 g^2 \delta^2 (\Delta - \delta/3)D] \qquad (8.10)$$

where Δ represents the distance between the leading edges of the two gradient pulses, γ is the magnetogyric ratio of the monitored spin and E_0 denotes the echo intensity in the absence of any field gradient. By varying either g, δ or Δ (while at the same time keeping the distance between the two rf-pulses constant) D can be backed out by fitting equation (8.10) to the observed intensities.

The range of diffusion coefficients that can be measured with the PGSE method covers the range from fast diffusion of small molecules in solutions with D-values typically around $10^{-9} \, \mathrm{m^2 \, s^{-1}}$ to very slow diffusion of, for instance, polymers in the semi-dilute concentration regime, where D-values down to $10^{-16} \, \mathrm{m^2 \, s^{-1}}$ can be measured [27]. Measurements of such very slow diffusion require gradients of extreme magnitudes and place severe demands on the actual experimental set-up [21]. Perhaps the most important experimental problem is gradient pulse matching, which, if not performed in a proper way, induces uncorrectable phase errors in the spectra. Other problems are connected with the inevitable vibrations of the samples after the application of such large gradients (note that the rms movement of a molecule diffusing with a diffusion coefficient of $10^{-16} \, \mathrm{m^2 \, s^{-1}}$ is only of the order of 100 Å during a time span of a second!).

What often limits the lowest value of D that can be measured is the value of T_2. As a general rule, slow diffusion is often found in systems which also show rapid transverse relaxation. As a consequence the echo intensity gets severely damped by T_2-relaxation in such systems.

4.2 RESTRICTED DIFFUSION

As mentioned above, one of the key properties of the PGSE diffusion experiment is the fact that the transport of molecules is measured over a time which we are free to choose at our own will in the range from around a few ms to several seconds. If the molecules experience some sort of boundary with regard to their diffusion during this time, the outcome of the experiment becomes drastically changed [28–30] and equation (8.10) above is no longer a valid description of the echo intensity. In fact, the exact equations governing the echo amplitude as a function of the relevant parameters are not known for any other case than for free diffusion and free diffusion superposed on flow. As a consequence one resorts to various levels of approximations. In one such approximation scheme, one considers gradient pulses which are so narrow that no transport during the pulse takes place. Thus we are in effect describing the gradient pulses as delta functions. This has, for obvious reasons, been termed the Short Gradient Pulse (SGP) (or Narrow Gradient Pulse, NGP) limit and leads to a very useful formalism whereby the echo attenuation can be written as:

$$E(\delta, \Delta, g) = \iint \rho(\mathbf{r}_0) P(\mathbf{r}_0 | \mathbf{r}, \Delta) \exp[i\gamma g\delta \cdot (\mathbf{r} - \mathbf{r}_0)] \, d\mathbf{r} \, d\mathbf{r}_0 \qquad (8.11)$$

where $P(\mathbf{r}_0 | \mathbf{r}, \Delta)$ is the propagator which gives the probability of finding a spin at position $\mathbf{r}$ after a time Δ if it was originally at position $\mathbf{r}_0$. For free diffusion $P(\mathbf{r}_0 | \mathbf{r}, \Delta)$ is a Gaussian function and if this form is inserted in equation (8.11), equation (8.10) above with the term $(\Delta - \delta/3)$ replaced with Δ is obtained. This is the SGP result for free diffusion. For cases other than free diffusion, other expressions for $P(\mathbf{r}_0 | \mathbf{r}, \Delta)$ have to be used. Stejskal and Tanner solved the problem of reflecting planar boundaries, while the case of molecules confined to a spherical cavity, was recently solved by Balinov et al. [31]. The result is:

$$E(\delta, \Delta, g) = \frac{9[\gamma g\delta R \cos(\gamma g\delta R) - \sin(\gamma g\delta R)]^2}{(\gamma g\delta R)^6} + 6(\gamma g\delta R)^2$$

$$\times \sum_{n=0}^{\infty} [j_n'(\gamma g\delta R)]^2 \sum_m \frac{(2n+1)\alpha_{nm}^2}{\alpha_{nm}^2 - n^2 - n} \exp\left(-\frac{\alpha_{nm}^2 D\Delta}{R^2}\right) \frac{1}{[\alpha_{nm}^2 - (\gamma g\delta R)^2]^2} \qquad (8.12)$$

where $j_n(x)$ is the spherical Bessel function of the first kind and α_{nm} is the mth root of the equation $j_n'(\alpha) = 0$ or equivalently, expressed in terms of Bessel functions of the first kind, $nJ_{n+1/2}(\alpha) - \alpha J_{n+3/2}(\alpha) = 0$. D is the bulk diffusion coefficient of the entrapped liquid, and the rest of the quantities are defined above. The main point to notice about equation (8.12) is that the echo decay does depend on the radius, and thus droplet radii can be obtained from the echo decay for molecules confined to the sphere as would be the case for the dispersed phase in an emulsion, provided that the conditions underlying the SGP approximation are met.

The second approximation scheme used is the so called Gaussian phase distribution, which was originally introduced by Douglass and McCall [32]. The approach rests on the assumption that the phases accumulated by the spins on account of the action of the field gradients are Gaussian distributed. Under these conditions, and for the case of a steady field gradient, Neuman [33] derived the echo attenuation for molecules confined within a sphere, within a cylinder and between planes. For spherical geometry Murday and Cotts [34] derived the equation for pulsed gradients and obtained:

$$\ln[E(\delta, \Delta, g)] = -\frac{2\gamma^2 g^2}{D} \sum_{m=1}^{\infty} \frac{\alpha_m^{-4}}{\alpha_m^2 R^2 - 2}$$

$$\times \left\{ 2\delta - \frac{2 + \exp[-\alpha_m^2 D(\Delta - \delta)] - 2\exp(-\alpha_m^2 D\delta) - 2\exp(-\alpha_m^2 D\Delta) + \exp[-\alpha_m^2 D(\Delta + \delta)]}{\alpha_m^2 D} \right\}$$

$$(8.13)$$

where α_m is the mth root of the Bessel equation $1/(\alpha R)J_{3/2}(\alpha R) = J_{5/2}(\alpha R)$ or equivalently, $(\alpha R)J'_{3/2}(\alpha R) - \frac{1}{2}J_{3/2}(\alpha R) = 0$. Again, D is the bulk diffusion coefficient of the entrapped liquid. Equation (8.13) has been used to obtain sizes of emulsion droplets. The validity of the Gaussian phase approximation has recently been tested by very accurate computer simulations and it was found that it is indeed a sufficiently accurate approximation for a wide range of parameter values [31].

The NMR sizing method, which was apparently first suggested by Tanner in [26] has been applied to a number of different emulsions ranging from cheese to crude oil emulsions [35–40].

A very important application of PGSE NMR as applied to restricted diffusion is concerned with the studies of fluid saturated porous media. For such systems it is of importance to characterize the transport of fluids, as well as structure and morphology of the media, and both these properties may be obtained from PGSE work. An example of such systems within the field of surfactants is constituted by highly concentrated emulsions, which may have up to as much as 99% by volume of dispersed phase. Due to this property, these systems are used in a variety of different applications, ranging from aviation fuels to cosmetics.

Callaghan and coworkers have developed a pore-hopping formalism which may be used to interpret PGSE data from such porous systems [29]. The analysis is performed within the SGP limit mentioned above. An alternative approach has been developed by Söderman and Linse, who used Brownian dynamic simulations to interpret echo decays. The advantage of the latter approach is that it does not rely on the narrow pulse approximation, but can be used for any length of the gradient pulses.

4.3 SOLVENT DIFFUSION IN PARTICLE SUSPENSIONS. OBSTRUCTION EFFECTS

The diffusion properties of a small entity such as a water molecule, through a solution of large colloidal particles may convey information concerning the shape and size of the colloidal particles. This follows since the degree of obstruction owing to the presence of particles, depends on the particle geometry and size. The relevant theory for this kind of diffusion has been worked out by Jönsson and coworkers [41]. In short, the presence of spherical and rod-shaped micelles gives rise to minor obstruction effects, while oblate or disk-shaped particles cause a larger effect. Given in Figure 8.4 is the obstruction effect, defined as the ratio of the diffusion coefficient in the presence and absence of obstructing particles. The dependence of the obstruction effect for water molecules on the volume fraction of the obstructing particles provides a convenient way of qualitatively judging the geometry of the colloidal particles. A quantitative analysis is often hampered by the fact that the colloidal particles are always hydrated to some extent. Since the water of hydration has other dynamical properties than ordinary water, hydration also reduces the observed diffusion coefficient as compared to bulk water.

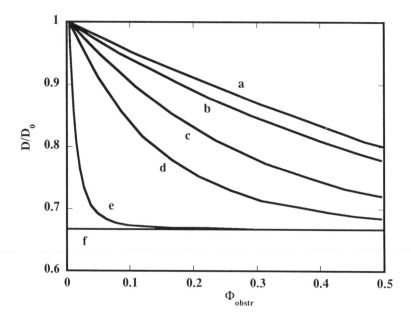

Figure 8.4 The variation of the reduced self-diffusion coefficient D/D_0 of small solvent molecules in solutions as a function of the volume fraction of obstructing particles of different geometries. Curve (a) denotes spheres, (b) long prolates, (c), (d) and (e) oblates of axial ratios 1:5, 1:10 and 1:100, respectively, and (f) denotes large oblates

4.4   SELF-DIFFUSION OF COLLOIDAL PARTICLES

As noted above one way to judge the size/shape of colloidal particles such as micelles is to investigate the diffusion of small particles in the presence of the colloid particles. An alternative approach is to monitor the self-diffusion of the colloidal particle as such. The self-diffusion coefficient is in the absence of interactions (i.e. at infinite dilution) given by the Stokes–Einstein relation:

$$D_0 = \frac{k_B T}{6\pi\eta R_H} \tag{8.14}$$

where k_B represents the Boltzmann constant, T the absolute temperature, η is the solvent viscosity and R_H is the hydrodynamic radius. At higher concentrations, interactions become important and the self-diffusion coefficient decreases with increasing concentration [42]. The concentration dependence is often expressed in terms of an expansion

$$D = D_0(1 - k_s\Phi + \cdots) \tag{8.15}$$

where Φ is the particle volume fraction and k_s a dimensionless constant which depends on the aggregate geometry and interactions. For hard sphere systems [42, 43], $k_s \approx 2.5$.

Strictly speaking, equation (8.14) above is valid for spheres. However, it can also be taken as the definition of the hydrodynamic radius for particles of arbitrary geometry, and thus it can be used to characterize also particles of unknown shape.

An example of this approach is shown in Figure 8.5 in which the hydrodynamic radii for micelles formed by non-ionic surfactants of the oligo(ethylene oxide) type as a function of the temperature are displayed. Questions pertaining to the aggregate growth as a function of temperature and concentration in these systems have been the subject of rather lively discussions over the years [44]. Clearly, data such as those presented in Figure 8.5 are important in this context [45, 46].

Although the method outlined above is certainly useful, one should be aware of the limitations inherent in the approach. First, a direct application of equation (8.14) assumes infinite dilution. As micelles are self-assembled units, one must be aware of the fact that dilution may change their properties. At higher concentrations interactions become important (cf equation (8.15)), and these must be taken into account [47, 48]. In fact, the obstruction effects can be used to an advantage as information pertaining to the interparticle interactions can be derived from these.

4.5   DIFFUSION IN BICONTINUOUS MICROSTRUCTURES

Anderson and Wennerström [49] have calculated the geometrical obstruction factor of the self-diffusion of surfactant and solvent molecules in bicontinuous

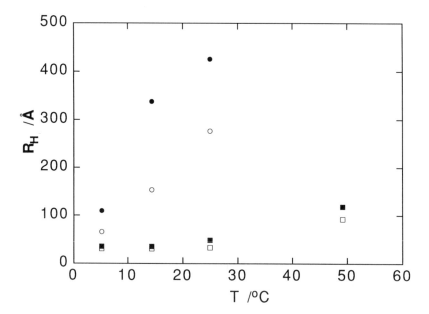

Figure 8.5 Micelle hydrodynamic radii for two concentrations of $C_{12}H_{25}(OCH_2CH_2)_5OH$ (● corresponds to 2.73 wt% and ○ to 0.92 wt% surfactant, respectively) and $C_{12}H_{25}(OCH_2CH_2)_8OH$ (■ corresponds to 3.6 wt% and □ corresponds to 1.02 wt%, respectively). Refs [45, 46]

model-cubic microstructures. The geometrical obstruction factor is defined as the relative diffusion coefficient D/D^0, where D is the diffusion coefficient in the structured surfactant system and D^0 is the diffusion coefficient in the pure solvent. As model structures, they considered mainly those of constant mean curvature surfaces, but they also obtained results for the interconnected cylinder model. Although considering the case of triply periodic microstructures, their results apply also to the disordered bicontinuous microemulsions and L_3 (sponge) phases. In a bicontinuous microemulsion the geometrical obstruction factor depends on the water-to-oil ratio. An expansion around the balanced (equal volumes of water and oil) state gives to leading order:

$$D_w/D_w^0 = 0.66 - \beta(\Phi_o - 1/2) \tag{8.16}$$

$$D_{oil}/D_{oil}^0 = 0.66 + \beta(\Phi_o - 1/2) \tag{8.17}$$

where Φ_o is the oil volume fraction. The expansion coefficient, β, depends on the coordination number of the structure, i.e. on the family of constant mean curvature surfaces. For the P-family (coordination number of 6) $\beta = 0.78$, while for the D-family (coordination number of 4) $\beta = 0.54$. The surfactant diffusion coefficient is maximum for the balanced state and the corresponding expansion gives to leading order

$$D_s/D_s^0 = 2/3 - \beta'(\Phi_o - 1/2)^2 \qquad (8.18)$$

Here D_s^0 denotes the (two-dimensional) lateral diffusion coefficient of the surfactant within the surfactant film. The expansion coefficient β' has the value 1.8 for the P-family of surfaces.

In the bilayer continuous structures, occurring in the bicontinuous cubic and L_3 phases, the diffusion can be described by essentially the same equations. Here, the solvent in each of the two sub-volumes makes up a volume fraction of $(1 - \Phi_s)/2$, where Φ_s is the surfactant, or more generally the bilayer volume fraction. Thus the relative diffusion coefficient of the solvent is given by equation (8.16) when replacing $(\Phi_o - 1/2)$ by $\Phi_s/2$:

$$D_{solvent}/D_{solvent}^0 = 0.66 - \beta\Phi_s/2 \qquad (8.19)$$

In the case of the surfactant, which experiences a surface diffusion, the actual interface where the diffusion is evaluated has to be specified since the metric of a surface depends on the mean and Gaussian curvatures. If this surface is chosen as the bilayer–solvent interface, the relative surfactant diffusion coefficient is given by equation (8.18) when replacing Φ_o by $(1 - \Phi_s)/2$:

$$D_s/D_s^0 = 2/3 - \beta'\Phi_s^2/4 \qquad (8.20)$$

The equations above have been applied to the analysis of self-diffusion data from a number of bicontinuous microemulsions, L_3 phases and bicontinuous cubic phases [50–52]. One should note that the equations above describe only the effect of geometrical obstruction. In many cases, solvation effects (solvent molecules have a lower mobility in the vicinity of the surfactant film) may be of similar or even higher magnitude. The solvation effects increase with the surfactant concentration and care has to be taken to separate solvation from obstruction effects. It is often possible, however, to obtain information on solvation from other experiments, like nuclear spin-relaxation experiments on solvent molecules. By combining self-diffusion data and spin-relaxation results, Balinov et al. were able to estimate the coefficient β in the L_3 phase of the AOT–brine system and concluded that the microstructure has a low average coordination number [51, 52].

5 Some Central Problems of Surfactant Systems as Studied by NMR

5.1 MOLECULAR TRANSPORT

Studies of molecular transport in simple and complex surfactant systems can be made from two starting-points. Information on the rate of migration can have a direct implication on biological processes and the functioning of a surfactant formulation as in pharmaceutical delivery systems and detergents, or diffusion coefficients can be used for monitoring association processes or microstructure. Through the spin echo experiment in the presence of (normally pulsed) magnetic field gradients, molecular transport over macroscopic distances (10^{-7} to 10^{-4} m) can be monitored accurately and conveniently. For complex systems, Fourier transformation is essential and thereby a very detailed picture of the molecular mobility in the system can be obtained. Although the technique is rather generally applicable to studies of molecular displacements, and can be applied to study rates of macroscopic flow as well, the interest for surfactant systems is focused on the possibility to monitor self-diffusion processes. Alternative ways of studying self-diffusion are generally based on using (most frequently radioactively) labelled compounds, such as in the capillary tube technique. These techniques may be more sensitive, and sometimes more accurate, but NMR approaches have many advantages, such as generally no need for labelling, speed and the possibility to study many compounds simultaneously in complex mixtures.

While self-diffusion coefficients of isotropic phases can be obtained accurately and rapidly for complex surfactant systems, the case of anisotropic phases is experimentally demanding, the most important reason being the signal broadening owing to static effects. However, these broadening effects can be eliminated by macroscopical orientation and rotating the sample so as to have the phase director oriented at the magic angle. Lindblom and co-workers [53–55] have been able to monitor the self-diffusion along the surfactant bilayer in a large number of lamellar phases oriented in thin layers between glass slides.

Self-diffusion studies of non-oriented liquid crystals require special considerations, but, as discussed above, novel pulse techniques have opened this field recently. For systems showing small residual interactions, as would be the case for ^2H_2O in many liquid-crystalline phases, Callaghan et al. [56] have shown that the experiment is feasible. Since the molecular diffusion in such systems is different in different directions relative to the phase symmetry axis and since the quadrupolar powder pattern is a superposition of splittings whose values depend on this direction, this opens up the possibility of investigating in detail the anisotropic water diffusion in such systems. For the case that diffusion in one direction is strongly hindered, like that perpendicular to the bilayers in a lamellar phase, self-diffusion may be determined by structural defects rather than by the intrinsic molecular translation. Therefore, observed self-diffusion coefficients can be dependent on microcrystallite size.

5.2   PHASE DIAGRAMS

Strikingly, NMR has become a standard technique to establish phase diagrams of surfactant systems, with some very distinct advantages compared to alternative methods. Surfactant phase diagrams generally include both solution phases and liquid-crystalline phases [3, 4, 57–59]. In a three- or four-component system there are usually two isotropic solution phases, one rich in water, termed L_1, and one poor in water, termed L_2. In certain systems, these may be connected so that there is a single extensive solution region.

Methods to elucidate phase behaviour based on microscopic observation in polarized light, calorimetry, X-ray diffraction, and so on, have been elaborated to great perfection, but various NMR observations are important complements and it can in fact often be practical to use NMR to determine entire phase diagrams. A big advantage of the NMR method is that it is not necessary to achieve a macroscopic phase separation, and two- or three-phase character can be detected with single-phase domains on the micrometer scale.

Phase diagram studies by NMR are generally based on the fact that the rapid molecular dynamics cause an elimination of any spin interactions to an extent that is directly related to the degree of anisotropy of the structure. The most important use of NMR is to distinguish between phases giving residual dipolar or quadrupolar interactions and phases that do not; in general, this also marks the difference between optically anisotropic and optically isotropic phases. Whether residual couplings are seen depends on the extension of the anisotropic domains (microcrystallite sizes) in combination with the rate of molecular diffusion.

The presence of even very minute amounts of isotropic phase is generally visible in ^1H and ^{13}C spectra, whereas because of large signal width due to static dipolar couplings, the signals from anisotropic liquid crystals are much more difficult to observe. While ^1H and ^{13}C NMR can give some initial hints, they do not provide sufficiently detailed and unequivocal information on phase diagrams. It is much more useful and general to base an analysis on quadrupole splittings or chemical shift anisotropies.

For quadrupolar nuclei the simplest case is with $I = 1$ rather than $I = 3/2$, $5/2$, or $7/2$ since any central signal peak can be referred to the presence of isotropic phase, while a doublet signifies an anisotropic phase; for $I = 3/2, 5/2$, or $7/2$ nuclei, on the other hand, there is always a central peak, but otherwise the same considerations apply. Furthermore, the magnitude of the splitting depends on the degree of anisotropy of the phase, so that, under identical local molecular conditions and composition, a lamellar phase should give twice as large a splitting as a hexagonal phase. For nuclei with sizeable shift anisotropies, the isotropic and anisotropic phases are again easily identified. Both the sign and the magnitude of the shift anisotropy is different for lamellar and hexagonal phases, so in this case, the distinction is even more straightforward.

The ^2H quadrupole splitting method (generally using heavy water) has been used to establish entire phase diagrams for a number of two- and three component systems [60, 61]. We exemplify with a recent investigation of a mixture of a cationic and an anionic surfactant (Figure 8.6) [62]. In spite of the enormous complexity, the different phase regions could be well established. The NMR approach is particularly powerful in delineating multi-phase regions

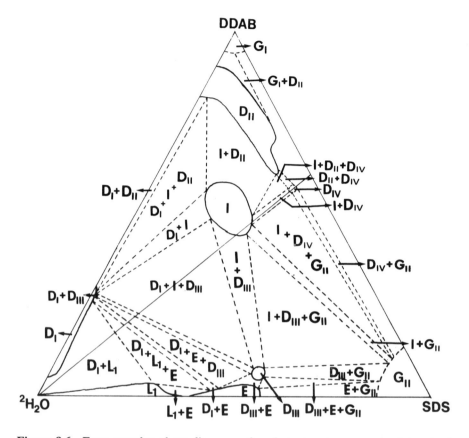

Figure 8.6 Even complex phase diagrams of surfactant systems may be determined using the ^{2}H NMR technique. The figure shows single-, two- and three-phase regions for the system of didodecyldimethylammonium bromide, sodium dodecyl sulfate and water and is taken from [62], where phase notation is explained. Reprinted with permission from Marques *et al.*, *J. Phys. Chem.*, **97**, 4729. Copyright (1994) American Chemical Society

involving two or more liquid crystalline phases, often extremely difficult to separate macroscopically, and was the first to prove the coexistence of different lamellar phases.

5.3 PHASE STRUCTURE

Basic structural questions of self-assembled surfactant systems are firstly whether there are discrete structural units (micelles) or whether they are continuous in one, two or three dimensions. Further structural aspects concern the shapes of discrete or nondiscrete structural units. Regarding the first problem, diffusion NMR provides direct insight and has become widely used.

The second type of problem often requires more detailed considerations but NMR relaxation approaches have been constantly developed to meet with these problems. The problem of phase structure is, of course, intimately linked to that of phase behaviour, and some of the basic principles of phase structure studies by NMR have already been touched on.

An initial aim is normally to obtain information on whether a phase is isotropic or anisotropic and on whether, in the latter case, it has a uniaxial or biaxial structure, as well as on the degree of anisotropy of an anisotropic phase. If a phase is isotropic on the relevant NMR timescale, static dipolar, quadrupolar, and shift anisotropy interactions are averaged to zero by molecular motion. If a phase is anisotropic, the spectrum should contain static interaction effects (e.g. quadrupole splittings). If the local order parameter can be predicted theoretically or if there is adequate reference data, one may from the magnitude of the splitting decide whether the phase is built up of rod aggregates with rapid diffusion around the rods or if it is built up of repeating planar aggregates. In a hexagonal liquid crystalline phase the rods are cylindrically symmetric and the phase uniaxial, but structures based on rods which are biaxial have also been identified. The shape of the NMR signal is different for uniaxial and biaxial structures, as illustrated in Ref.[63].

If we turn to the isotropic phases, several possibilities exist both as regards liquid solution phases and cubic liquid-crystalline phases. The latter, which have a long-range order, can be distinguished from the former in giving rise to low-angle X-ray diffraction patterns. A number of different cubic phase structures are possible, but the X-ray low-angle diffractograms normally do not give a sufficient number of reflections to permit a complete structural assignment [64]. However, here NMR self-diffusion studies [65–67] may very clearly distinguish between possible alternatives, as illustrated for the dodecyltrimethylammonium chloride–water system [65]. For the micellar phase the surfactant self-diffusion coefficient decreases strongly with increasing concentration; the micelles stay closely spherical and the decrease at higher concentrations can be referred mainly to intermicellar interactions (repulsive electrostatic interactions). As one passes over into the first cubic liquid-crystalline phase there is a regular change in the D value and the self-diffusion results require a structured based on discrete repulsive micellar-like units in a water continuum; later work has shown that the micelles are short rods rather than spheres [68–70], and spheres are also inconsistent with the relatively rapid NMR relaxation. The cubic phase occurring at higher surfactant contents is characterized by a very much higher surfactant self-diffusion coefficient, and this phase cannot be discontinuous for the surfactant molecules. Instead these data are consistent with a bicontinuous structure with rodlike surfactant structures forming extended networks. Lindblom *et al.* [66] have gone a step further and used a quantitative comparison between self-diffusion data of cubic phases with self-diffusion along a surfactant bilayer (obtained from experiments on macroscopically oriented lamellar phases) to provide a more detailed structural characterization. This has provided evidence for a bicontinuous cubic phase based on bilayer units. In recent years detailed interpretation of self-diffusion data for several cubic phases has been successfully carried out in terms of minimal surface structures [50, 51].

For microemulsions, early structural descriptions were mainly in terms of discrete structures, i.e. oil-in-water (o/w) or water-in-oil (w/o) droplet models, but subsequent work revealed a more complex situation. Here self-diffusion studies, especially of the FT NMR type, provide one of few general experimental approaches and have become extensively used [71]. The principles of interpretation are simple. For a microemulsion constituent confined to droplets, the macroscopic self-diffusion will be that of the droplets and thus very low. On the other hand, for a constituent occurring only in the continuous medium, the macroscopic self-diffusion coefficient will essentially be that of the neat liquid and thus high. There are minor effects due to solvation and obstruction on molecular diffusion, as well as to interdroplet interactions on droplet diffusion which have been worked out theoretically.

The interpretation of molecular self-diffusion coefficients is thus relatively straightforward and it is possible from a structural model of microemulsions to predict its self-diffusion characteristics. It is useful to start from the following three clear-cut limiting cases:

(1) *Water-in-oil droplet structures.* $D_{water} \ll D_{oil}$ and $D_{surfactant} \sim D_{droplet}$. D_{oil} will be of the same order of magnitude as D of neat oil or oil solution.

(2) *Oil-in-water droplet structure.* $D_{water} \gg D_{oil}$ and $D_{surfactant} \sim D_{oil} \sim D_{droplet}$. D_{water} will be of the same order of magnitude as D of neat water or a relevant aqueous solution.

(3) *Bicontinuous structure:* D_{water} and D_{oil} are both high while $D_{surfactant}$ is low. $D_{surfactant}$ is low because surfactant molecules occur in large aggregates, but is predicted to be higher than for the droplet structures because diffusion is unrestricted (although hindered). In recent years there has been an important development as regards the quantitative interpretation which has provided strong support for a structure of multiply connected surfactant monolayers with a low mean curvature separating oil and water domains. The methods used in interpreting diffusion data from both disordered and ordered bicontinuous phases are described above under Section 4.5.

5.4 SURFACTANT SELF-ASSOCIATION

Several NMR parameters are markedly different in the unimeric and micellized states and are therefore possible candidates for a characterization of surfactant self-association processes. Although NMR can be a convenient alternative for the determination of critical micellar concentrations (CMCs), an analysis of variable concentration NMR data generally provides a much richer characterization of self-association than that. We will here only describe how self-diffusion can be used to measure the free unimer concentration in simple and complex surfactant systems. In practice it is an alternative to surfactant-selective electrodes to measure surfactant activities. With a two-state assumption, we have under the normal rapid exchange conditions

$$D = p_M D_M + p_{free} D_{free} \qquad (8.21)$$

which with the phase-separation model gives

$$C < \text{CMC}: D = D_{free}$$

$$C > \text{CMC}: D = D_M - \frac{\text{CMC}}{C} (D_M - D_{free})$$

The predicted behaviour of two straight-line segments intersecting at the CMC is in rough agreement with observations and becomes a better approximation the longer the alkyl of a surfactant, as expected.

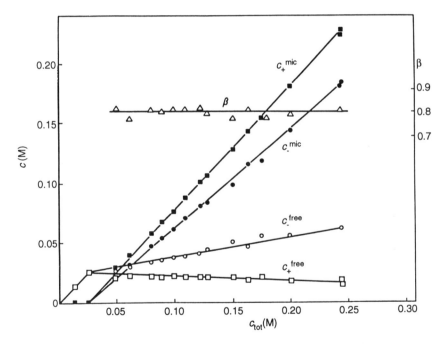

Figure 8.7 Typical evolution with total surfactant concentration of the concentrations of different entities for an ionic surfactant. The concentrations were deduced from FT NMR self-diffusion data. The example shown is decylammonium dichloroacetate. Superscripts mic and free refer to micellar and umimeric states, respectively, while subscripts + and − refer to surfactant ion and counter-ion, respectively. β is the degree of counter-ion binding. Reprinted with permission from Ref. 6. Copyright (1987) Marcel Dekker, Inc.

Surfactant self-diffusion coefficients can be analyzed directly assuming a two-state model,

$$D = \frac{C_M D_M + C_{free} D_{free}}{C} \qquad (8.22)$$

to give the concentrations of free and micellized surfactant since D_{free} and D_M can be separately measured or estimated to a good approximation. An example of results is given in Figure 8.7. The marked decrease in the monomeric surfactant concentration above the CMC is a general finding for a large number of ionic surfactants and can be referred to electrostatic effects [72]. The observations are quantitatively rationalized by a theory of surfactant self-association where the electrostatic effects are treated using the Poisson–Boltzmann equation [73, 74]. For non-ionic and zwitterionic surfactants, no corresponding decrease in the surfactant monomer concentration is found [75].

The variation of the surfactant monomer concentration with the total surfactant concentration has also been studied using ^{2}H NMR relaxation for sodium dodecylsulphate (SDS) in formamide [76]. Also, here a quantitative agreement with the Poisson-Boltzmann equation was obtained.

5.5   MICELLE SIZE AND SHAPE

Properties pertaining to micelle size/shape can conveniently be obtained from NMR relaxation data. The problem inherent in such a process is that one has to resort to a modelling of the relaxation parameters, as the information is not directly obtainable from the raw relaxation data. Needless to say, the models used must be as realistic as possible. One model that has proven to be useful in this regard is the so-called two-step model, described above.

To investigate the applicability of the two-step model model of relaxation, Söderman et al. studied aqueous micellar systems of alkyltrimethylammonium chloride surfactants of two different alkyl chain lengths, dodecyl and hexadecyl chains, respectively [69, 77]. The alkyl chains were ^{2}H labelled in the α-methylene segment, adjacent to the ammonium polar head group, and the relaxation rates, R_1 and R_2, were measured in the Larmor frequency range 2–55 MHz, to cover the dispersion of the slow motion. The lower frequencies were obtained on a variable field electromagnet. In Figure 8.8 the results [77] are shown for the hexadecyltrimethylammonium chloride micelles (the concentration was 13 wt. %). The solid lines are best fits of the equations (8.9) to the whole data set, using τ_f, τ_s and S as adjustable parameters. As is seen in Figure 8.8 the data are well described by a Lorentzian spectral density function superimposed on a constant offset in the investigated frequency range, supporting the assumption of timescale separation. The fit yielded $\tau_f = 42 \pm 1$ ps, $\tau_s = 7.6 \pm 0.4$ ns and $S = 0.186 \pm 0.003$, where the uncertainties correspond to an approximately 80% confidence interval taking only random

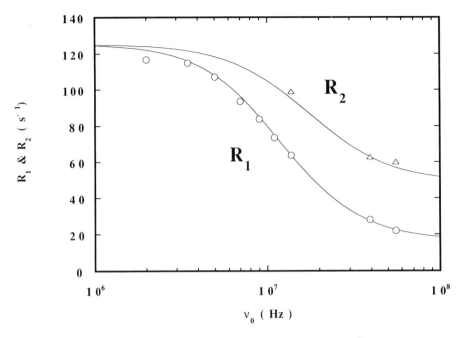

Figure 8.8 Variation of R_1 and R_2 with the Larmor frequency for ^{2}H-nuclei bound to the α-methylene segment of the surfactant hexadecyltrimethylammonium chloride. The sample is an aqueous micellar solution containing 13 wt.% of the surfactant. The experiments were performed at 27 °C. Data taken from Ref. 77

errors into account, and the value $\chi = 167$ kHz was used for the quadrupolar coupling constant.

The slow motion correlation time was interpreted in terms of equations (8.6)–(8.8). They contain two unknown parameters, R and D_{lat}. However, D_{lat} can be measured in phases of extended surfactant films, such as bicontinuous cubic phases, and R is expected to be close to the overall length of the surfactant molecule. Using $D_{\text{lat}} \approx 1 \times 10^{-11}$ m^2 s^{-1} as measured in a bicontinuous cubic phase of the similar surfactant hexadecyltrimethylammonium fluoride and the radius $R = 23.7$ Å, a slow motion correlation time of approximately 10 ns is calculated from equations (8.6)–(8.8), which is in reasonable agreement with the experimentally determined value.

The applicability of the two-step model receives further support when comparing the S value obtained from the fit to the relaxation dispersion with that measured from quadrupolar splittings in liquid crystalline phases forming at higher surfactant concentrations. In the liquid crystalline phases, extended surfactant aggregates are 'locked' in a crystalline array. Two very common structures are the lamellar phase where surfactant bilayers are stacked with one dimensional order and the hexagonal phase where cylindrical aggregates are packed into a two-dimensional hexagonal lattice. In these phases one has effectively frozen the slow motion and the nuclei experience a static quadrupolar Hamiltonian, $\langle H_Q \rangle_f$, in addition to the Zeeman term, where the average is taken over the fast local motions. This results in a quadrupolar splitting of magnitude [15]:

$$\Delta_{lam} = \tfrac{3}{4} \chi |S| \tag{8.23}$$

in the lamellar phase with planar films. In the case of the hexagonal phase, the rapid surfactant diffusion around the cylinder axis results in an additional averaging of the interaction by a factor of two and the splitting is given by:

$$\Delta_{hex} = \tfrac{3}{8} \chi |S| \tag{8.24}$$

One may expect that the local motions, and hence the S values should be similar in the liquid crystalline surfactant aggregates and the micelles in the dilute solution phase. This is also found to be the case. As mentioned above, the S-value obtained from the fit to the relaxation dispersion (Figure 8.8) was found to be $S = 0.186$. A very similar value, 0.192, was measured in the hexagonal phase at higher concentrations [77].

The small spherical micelles ($R \approx 20$ Å) can be seen as a limiting case and in general surfactant aggregates can be of much larger extensions. Increasing the size of the aggregates, τ_s increases, and the relevant relaxation dispersion moves out of the accessible frequency window. In some extreme cases, for example with long, entangled wormlike micelles, the dynamics may even move out of motional narrowing, showing slow motion effects in the NMR spectrum [11]. However, for most micellar and microemulsion systems, the motional narrowing condition still applies. In these cases, valuable information may also be obtained by measuring R_1 and R_2 for ^{2}H labelled surfactants. For larger aggregates, the relevant information on the aggregate size is contained in the zero frequency spectral density only. Assuming that the fast motions are in extreme narrowing, in which case they contribute to a constant offset to both R_1 and R_2, it is useful to consider the difference $\Delta R = R_2 - R_1$, which then only depends on the slow motions. For the case of very slow motions, we have also

$$j_s(0) \gg j_s(\omega_0) \approx j_s(2\omega_0)$$

and ΔR can be approximated from:

$$\Delta R = \frac{9\pi^2}{40} (\chi S)^2 j_s(0) \tag{8.25}$$

This expression has been used to study shape variations of surfactant aggregates in different microemulsion systems [78, 79]. Here, the micellar aggregates are swollen by the addition of a third component (normal micelles in aqueous solution are swollen by oil, and reverse micelles in oil are swollen by water), where the spherical aggregates can have radii of the order of 100 Å. Quantitative applications of equations (8.25) require accurate determinations of S from adjacent liquid crystalline phases. Furthermore, they require assumptions of the aggregate shape, and a knowledge of D_{lat} to determine, for example, an aggregate radius of spherical aggregates. So, obviously, there are some uncertainties involved concerning the 'absolute calibration' of the method.

The situation can, however, be optimized by noting that the aggregates are formed under the constraint of a constant average interfacial area to enclosed volume ratio, set by the concentration ratio of surfactant to the internal component. With this constraint, the minimum aggregate size corresponds to a spherical shape. Due to translational entropy, spherical aggregates, representing the minimum aggregate size, will always be stable at high dilutions, where the experiment can be calibrated. In many cases, D_{lat} is also known rather accurately from the various bicontinuous phases of the different systems.

Deviations from spherical shape can be modelled as a growth into prolate or oblate shapes. The area to enclosed volume constraint implies a constraint in the possible values of the two semi-axes describing the particle size. Halle [80] has calculated the correlation functions for the combined particle tumbling and surface diffusion of prolate and oblate particles. His results have, for example, been applied to microemulsion systems [78, 79] focusing on the ratio $j_{s,pr}(0)/j_{s,sph}(0)$, where subscripts pr and sph refer to prolate and spherical particles, respectively. For a given interfacial area to enclosed volume ratio, which specifies the radius, R, of the sphere, $j_{s,pr}(0)/j_{s,sph}(0)$ is a function of the prolate axial ratio, D_{lat} and R. Knowing D_{lat} and R from other experiments it is possible to determine the aggregate axial ratio from the relaxation experiment.

As mentioned above, the strength of the NMR relaxation technique in comparison to other (say scattering) techniques for studying micelles and microemulsions, is related to the fact that rotational dynamics are monitored. This has the advantage that (i) the dynamics are sensitive to small size variations and (ii) the dynamics are essentially independent of interparticle interactions. In an extensive study of a three-component microemulsion system composed of the non-ionic surfactant pentaethyleneoxide dodecylether, water and decane, oil swollen micelles dispersed in water were investigated over a large concentration range [78].

With non-ionic surfactants, spherical aggregates are formed at lower temperatures, while the aggregates grow in size with increasing temperature [44]. In Figure 8.9 we show the variation of ΔR with temperature for two compositions, $\Phi = 0.12$ and $\Phi = 0.23$ in the microemulsion phase [78]. Here, $\Phi = \Phi_s + \Phi_o$ is the total volume fraction of surfactant (Φ_s) and oil (Φ_o), and the ratio $\Phi_s/\Phi_o = 0.815$ was kept constant.

ΔR decreases when the temperature is decreased in the microemulsion phase and levels off at lower temperatures at a minimum value, ΔR_{min}, when the aggregates have reached the limiting spherical shape, corresponding to the minimum size. Note that the minimum ΔR value is the same for the two concentrations, as expected since the radius, dictated by Φ_s/Φ_o, should be the

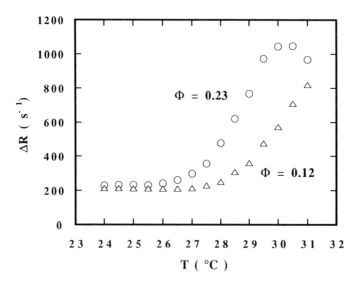

Figure 8.9 The variation of ^{2}H $\Delta R = R_2 - R_1$ with temperature of the ^{2}H labelled non-ionic surfactant pentaethyleneoxide dodecylether $(CH_3(CH_2)_{10}C^2H_2(OCH_2CH_2)_5OH)$ in a ternary system with water and decane. The ratio Φ_s/Φ_o is 0.815, where the Φ_s and Φ_o are the surfactant and oil volume fractions respectively. The concentrations are $\Phi = \Phi_s + \Phi_o = 0.12$ and 0.23, respectively. Data taken from Ref. 78

same. By considering the reduced ΔR value, $\Delta R/\Delta R_{min} = j_{s,pr}(0)/j_{s,sph}(0)$, it was possible to calculate the increase in the axial ratio of the aggregates with increasing temperature, assuming a growth into prolate aggregates. The variation obtained of the axial ratio is shown in Figure 8.10.

Very large micelles may also form in binary surfactant systems. These are long worm-like micelles which become entangled at higher concentrations, giving rise to rheological properties similar to those in polymer solutions. Such systems have been examined by ^{1}H band shape analysis [10, 11]. The protons of the surfactant hydrocarbon chain form a very large dipolar coupled spin system, with essentially a continuum distribution of transverse relaxation rates. The distribution of relaxation rates is related to the distribution of order parameters [12], which can be obtained from analysing the band shape in the liquid crystalline (lamellar or hexagonal) phases, where the band shape is associated with a continuum distribution of residual dipolar splittings [81].

Slow motions have also been studied by measuring the differential line broadening (DLB) of proton J-coupled ^{13}C nuclei [82]. Here the difference in bandwidth of the different resonances in a ^{13}C multiplet can be attributed to the interplay between the dipole-dipole and chemical shift anisotropy relaxation mechanisms, and by analysing the difference, information on the slow motion may be obtained.

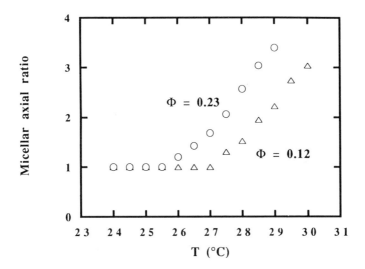

Figure 8.10 Variation of the micellar axial ratio with temperature, assuming a monodisperse prolate growth, as deduced from the relaxation data of Figure 8.9, by using the correlation function due to Halle [80] for the combined motion of aggregate tumbling and surface diffusion

5.6 COUNTER-ION BINDING

There are typically distinct differences in properties between ionic and non-ionic surfactant systems, which can be referred to the much less favourable self-assembly and phase separation for ionic. This can be referred essentially to the entropy in the counter-ion distribution. In an ionic surfactant system there is an inhomogeneous distribution of counter-ions, with a higher concentration close to the charged microscopic interfaces than at large distances from these. A number of NMR parameters give insight into the counter-ion distribution as well as into specific ionic interactions. The following is a partial list of possible approaches:

(1) The translational mobility of an ion is different by one to two orders of magnitude between moving freely in a solution and diffusing with an entity of colloidal dimensions, like a micelle. This allows counterion self-diffusion coefficients to be used for the characterization of counter-ion distribution, for example in terms of a counter-ion association degree.

(2) Quadrupole relaxation reflects both the distribution of counterions and the dynamics of counterions associated with an aggregate. A full interpretation is typically difficult since information on the quadrupole interaction of the associated state is not available and is difficult to obtain independently since the relaxation dispersion is typically not observed.

(3) For anisotropic liquid crystalline systems, both quadrupole relaxation and quadrupole splittings provide direct information into counter-ion interactions.

The two-site model leads to the definition of the degree of counter-ion binding, β, as the ratio of counter-ions to surfactant ions in a surfactant self-assembly. This is a useful but incomplete characterization of the counter-ion distribution. The value of β is directly obtainable from self-diffusion data since the self-diffusion coefficients of free ions are easily obtainable. For the free counter-ion diffusion a correction for the obstruction effect is made. The micellar D value is obtained as described above or estimated; as $D_M \ll D_{free}$ an exact D_M value is not critical.

An example of a study of counter-ion binding by FT NMR self-diffusion is given in Refs [83, 84] and illustrated in Figure 8.7. The method is generally applicable to organic counter-ions, whereas only a few inorganic ions, such as Li^+ (see Ref. [72]), F^-, Cs^+, and Be^{2+}, can be studied. A general observation for a large number of surfactant systems is a roughly concentration-independent counter-ion binding. This so-called ion condensation behaviour has also been predicted theoretically in Poisson–Boltzmann calculations. A more satisfactory approach than using the two-site model would be to make use of the full theoretically deduced counter-ion distribution around micelles to calculate counter-ion self-diffusion coefficients. This is discussed in Ref. [85]. The FT NMR self-diffusion method is very convenient for studying ion competition effects, as illustrated nicely in a study of the interaction of substituted acetate ions with cationic micelles [86].

The constancy of β to changes in micelle concentration is also directly borne out in studies of counter-ion quadrupole relaxation rates for a large number of systems using ^{23}Na, ^{35}Cl, ^{85}Rb, ^{133}Cs and ^{81}Br NMR (reviewed in Ref. [87]). These studies also demonstrate that the intrinsic relaxation rates of micellar counter-ions are nearly independent of micelle concentration and of the alkyl chain length of the surfactant.

5.7 HYDRATION

Surfactant self-assembly is a delicate balance between hydrophobic and hydrophilic interactions and the interactions between the head-groups and the solvent are decisive both for the onset of self-assembly and for the curvature of the surfactant films and thus aggregate shape and phase behaviour. Both ^1H, ^2H, and ^{17}O NMR have been successfully used to study the hydration of surfactant aggregates. The three approaches by far most used are ^1H (or ^2H) self-diffusion, ^{17}O quadrupole relaxation, and ^2H quadrupole splittings. We stress at the outset that a division into free and bound water molecules on which the concepts of hydration and hydration number are based is far from unambiguous and, furthermore, this division is dependent on the physicochemical parameter monitored.

Water self-diffusion can be investigated using either ^1H or ^2H NMR. In studies of water diffusion as a function of surfactant concentration, one observes a changed behaviour after the CMC. Thus the rate of decrease of D with surfactant concentration is relatively high below the CMC, while above the CMC it is much lower. Using a model with free water molecules and water molecules diffusing with monomeric and micellized surfactant molecules, one can directly conclude that the hydrocarbon–water contact is almost completely eliminated on micellization. For a number of ionic surfactants this type of investigation has yielded micellar hydration numbers in the range of 10 to 15 water molecules per surfactant molecule diffusing with the micelles [72, 88]. Accounting for head group and bound counterion hydration, this suggests very little hydration of the alkyl chains. For non-ionic surfactants of the ethylene oxide (EO) variety, self-diffusion work has demonstrated a decreased hydration with increasing temperature, of significance in discussions of the aggregation and phase behaviour of these surfactants [89]. A striking observation is that at a given temperature, hydration is dependent only on the concentration of EO groups, not on whether the EO groups occur in spherical micelles, in rod micelles, or in poly(ethylene oxide).

Quadrupole effects in the NMR of water nuclei can be used to study hydration phenomena in exactly the same way as quadrupole effects of ions can be used to study counter-ion binding. ^2H and ^{17}O are rather unique in providing a detailed molecular picture of the hydration in these types of systems. The water exchange between different sites is (in all the cases investigated so far) rapid, and often a simple two-site model is appropriate. The fact that the field gradients (being mainly of an intramolecular origin) are little influenced by changes in the system and that the quadrupole coupling constants are known to a good approximation (0.222 MHz for ^2H and 6.7 MHz for ^{17}O) facilitates the analysis considerably.

Water NMR relaxation studies would be expected to be the richest source of information on hydration in colloidal and macromolecular systems in general. ^1H and ^2H NMR have been widely used during a long period of time, but unfortunately major problems of interpretation make many studies questionable [17]. The problems are associated with a complex relaxation mechanism of water protons and the very strong (often dominant) contributions from exchange with macromolecular ^1H or ^2H nuclei. These difficulties led Halle [90] to investigate in detail the possibilities of employing ^{17}O relaxation. Halle demonstrated that in interpreting multifield relaxation data using the two-step model of relaxation, it was, indeed, possible to map in detail the hydration of a large number of systems and that problems in previous work were often related to the neglect of the fast motions. Some general conclusions from water quadrupole relaxation and splitting studies are:

(1) Bound water molecules reorient quite rapidly, in fact less than a factor of 10 slower than in bulk water, and the degree of anisotropy of this motion is low.

(2) One or two layers of water molecules are perturbed appreciably and there are no indications of long-range hydration structures.
(3) Fewer than two methylene groups in the alkyl chain are significantly exposed to water in surfactant micelles.
(4) For lamellar phases there is often an 'ideal swelling' behaviour where water molecules in the vicinity of the surfaces have their orientation probability distribution almost constant on increasing the water content and where water addition only increases the amount of free water. There is no appreciable long-range ordering effect on the water molecules.
(5) For liquid crystalline phases in general even the water molecules in direct contact with the surfaces are very mobile and have a quite low anisotropy in their motion; generally, S_b is on the order of 5×10^{-2}, which is a factor of 10 below the order parameter of the polar head group.

5.8   SOLUBILIZATION

Solubilization, or incorporation of otherwise water-insoluble compounds into surfactant aggregates, is one of the most important aspects of surfactant self-assembly. The most basic aspect of solubilization, besides that of phase diagrams, concerns the thermodynamic parameters of the solubilization equilibria, such as

$$A_n + S_{aq} \rightleftharpoons A_n S$$

that is, the equilibrium between the aqueous and the micellar states of a solubilized molecule. This seemingly simple aspect of micellization has been confused by the use of inapplicable approaches (such as solubility studies) and the lack of reliable methods. Solubilizate self-diffusion studies, however, provide a simple and unequivocal solution to the determination of equilibrium constants for solubilization processes [91–93]. Thus the two-site model is expected to be an excellent approximation for most solubilizates of interest. The fraction of solubilizate in the micelles can be written

$$p = \frac{C_M}{C_{tot}} = \frac{D_{free} - D_{obs}}{D_{free} - D_M} \qquad (8.26)$$

without further assumptions. Based on equations (8.26), different equilibrium constants can be given, for example

$$K = \frac{C_M}{C_{free}} = \frac{p}{1-p} \frac{V_{aq}}{V_M} \qquad (8.27)$$

where V_{aq} and V_M are the volumes of the aqueous pseudophase and of the micelles, respectively.

Stilbs [91–93] has, in studies of the partitioning of several compounds in anionic, cationic, and non-ionic surfactant systems, provided a detailed insight into the thermodynamics of solubilization. Very important observations were made in studies of alcohol solubilization in SDS solutions containing small amounts of hydrocarbon [92]. For example, on solubilizing butanol, hydrocarbon diffusion may become much more rapid than SDS diffusion. Thus hydrocarbon is no longer confined to the micelles. This micellar breakdown caused by the cosurfactant is of interest in connection with microemulsions. Stilbs *et al.* [94] have also extended the method to the study of partition coefficients in vesicle systems.

Information on the amplitudes and rates of molecular motion can be obtained in detail using static interaction and relaxation parameters, quadrupole splittings [95] and quadrupole relaxation rates [96] being most straightforward to apply. As a single example, we will mention the anisotropic reorientation of solubilized perdeuterated *trans*-decalin as studied by ^{2}H spin–spin and spin–lattice relaxation [97]. From several observations it could be ascertained that slow motions make an insignificant contribution to the observed relaxation rates. Therefore, it was relatively simple to deduce quantitative information on the correlation times for reorientation of solubilized trans-decalin around the three principal rotational diffusion axes. The analysis was facilitated further by the molecular rigidity and very low aqueous solubility of *trans*-decalin. The anisotropic reorientational behaviour of trans-decalin in micelles was found to be similar to that in simple solutions.

5.9   VESICLES AND LIPOSOMES

A great interest in NMR studies of lipid vesicles arose 20–30 years ago mainly from biological implications in that these vesicles show some important features of biological cells. Vesicles have, therefore, played a very important role as, quite appropriate, model systems for biological cells. A very important amount of work has concerned a detailed characterization of different molecular features of vesicle systems as well as a comparison between vesicles and biological membranes. This type of work has also been stimulated by the interest in using vesicles as carriers in pharmaceutical systems. Quite recently, the observation of apparently thermodynamically stable vesicle systems has created a renewed interest and broadened it towards the physical sciences. Another recent much studied topic is that of reversed vesicles, i.e. vesicles formed (for example by a mixture of one hydrophilic and one lipophilic amphiphile) with a hydrocarbon core in a hydrocarbon continuum (often with small amounts of water in the reversed bilayer).

Some key questions regarding vesicles and liposomes relate to the following parameters:

- Size and shape. Vesicles have generally been assumed to be spherical, often without any closer analysis but other shapes are certainly relevant as well.
- Vesicle structure, for example number of bilayers.

- Chain packing and dynamics.
- Inside-outside differences in packing etc.
- Trans-bilayer transport of water, ions, amphiphiles, etc.
- Counter-ion binding and hydration.
- Flocculation and fusion.
- Structural transitions such as the micelle–vesicle and vesicle–lamellar phase.

Dynamic NMR provides direct information on all these topics but we will here only dwell on one area of considerable current activity.

Studies on chain packing and dynamics are best performed using ^{2}H quadrupole relaxation on deuteriated surfactants. The general approach follows in relevant parts that outlined above for micellar and other solutions. During recent years there has been a renewed interest in the area and in particular relating to the influence of different long-range motions on the relaxation. Halle has presented an analysis that explains the frequency dependence of the NMR relaxation data from phospholipid vesicles [98]. In short, Halle considers the effect of vesicle tumbling and restricted tumbling (wobbling) of individual phospholipid molecules with respect to the bilayer normal. The same model can also account for the anisotropy of the relaxation rates obtained from aligned bilayers.

5.10   POLYMER–MICELLE SOLUTIONS

Mixed solutions of a polymer and a surfactant have become a much studied topic during the last few years. Dynamic NMR has so far been used only to a very limited extent but there are many possibilities such as:

- Quantifying the association between a surfactant and a polymer in solution on the basis of surfactant self-diffusion [99]. This is based on the typical situation that diffusion of the polymer–surfactant complex is much slower than that of either surfactant unimers or simple surfactant micelles.
- Characterizing the effect of association to a polymer on dynamics and order in surfactant self-assembly from relaxation [100]

The general aspects of these types of studies follow in main parts directly from those of simple surfactant solutions described above. One important complication in relaxation work is the need to introduce an additional slow motion term owing to the slow reorientation of the polymer–surfactant complex. The understanding of the dynamics of polymer–surfactant complexes would progress significantly with extended and detailed relaxation investigations directed to the polymer chain dynamics.

Here we will briefly introduce two novel types of investigations, i.e. relaxation studies aimed at investigating the effect of polymer on surfactant micelle size and studies of micellar diffusion in a polymer network.

Wong *et al.* [101] used ^{13}C relaxation of the surfactant molecules to study the effect and the resulting microstructural changes caused by the addition of a non-ionic polymer (poly(vinyl methyl ether), PVME) to micellar solution of CTAB and sodium salicylate. As illustrated in Figure 8.11 there are dramatic effects of polymer addition on relaxation, which could be interpreted in terms of a break-down of long thread-like micelles into small spherical micelles. In a detailed analysis of the T_1, NOE and differential line-broadening it was found that a three-step motional model has to be used, as previously observed for several systems of large micelles. The slow motion dominated by the reorientation of long micelles and the diffusion of surfactant molecules over the micelle surface are characterized by two correlation times. In the presence of polymer, on the other hand, the influence of the slowest motion is lost and a two-step model can be used to analyse the data. The slow motion correlation time is then close to that observed for simple surfactant solutions of spherical micelles. There is no appreciable influence on relaxation from the anisotropic environment arising from the polymer chains and therefore the association

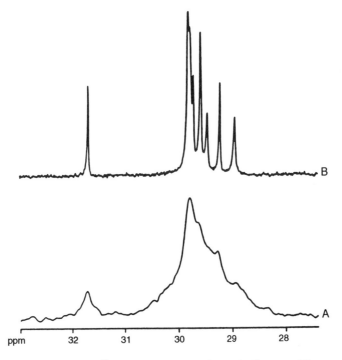

Figure 8.11 The decoupled ^{13}C spectra change dramatically on adding a polymer (poly(vinyl methyl ether)) to a solution of thread-like micelles (hexadecyltrimethyl-ammonium bromide–sodium salicylate mixtures) reflecting the transition to small spherical micelles) [10]. Reprinted with permission from Wong *et al.*, *Langmuir*, **8**, 460. Copyright (1992) American Chemical Society

must be rather loose. This is supported by the observation that T_1 and NOE of the ^{13}C signals of the polymer, and thus also polymer dynamics, are not affected by addition of micelles to a polymer solution. This study demonstrates the usefulness of detailed surfactant and polymer relaxation studies to elucidate central features of polymer–surfactant systems, which are now known to range from strongly associating to segregating.

The influence on micelle diffusion of the presence of polymer networks has recently been investigated in detail by Johansson and Löfroth [102–106] both experimentally and theoretically. This work, which provides very useful ways of accounting for obstruction effects, opens a new possibility of investigating polymer–micelle interactions which has been followed by others. The possibility of monitoring both attractive and repulsive interactions from micelle diffusion may be illustrated by a study on the interaction between octaethyleneglycol mono n-dodecyl ether ($C_{12}E_8$) micelles and ethylhydroxyethyl cellulose (EHEC). Micelle diffusion is influenced by the polymer in two ways, i.e. obstruction due to the polymer network and association of micelles to the polymer chains. The obstruction from the polymer chains has a large influence already at moderate polymer concentrations. As illustrated in Figure 8.12, the size and shape of the micelles are important factors and others are the polymer radius, the polymer volume fraction, and the polymer persistence length. The obstruction contributions from the polymer chains can for nonflexible polymer chains and spherical micelles be described by a model developed by Johansson et al. [102–106].

$$\frac{D_{mic}}{D_{aq}} = e^{-\alpha} + \alpha^2 e^{\alpha} E_1(2\alpha) \tag{8.28}$$

where $\alpha = \Phi_p(R_s + a)^2/a^2$, D_{mic} is the micellar self-diffusion coefficient in the polymer solutions in the absence of strong attractive interactions between the surfactant and the polymer, Φ_p the volume fraction of the polymer, R_s is the micellar radius, a is the polymer radius and E_1 is the exponential integral, i.e.

$$E_1(x) = \int_x^{\infty} du \, \frac{e^{-u}}{u} \tag{8.29}$$

Combining the obstruction effects from both surfactant micelles (volume fraction ϕ_s) and polymer chains leads to

$$D_{mic} = D_o(1 - 2\phi_s)(e^{-\alpha} + \alpha^2 e^{\alpha} E_1(2\alpha)) \tag{8.30}$$

Combining with a simple two-site model we obtain

$$D_{obs} = [D_o(1 - 2\phi_s)(e^{-\alpha} + \alpha^2 e^{\alpha} E_1(2\alpha))](1 - P_b) + P_b D_b \tag{8.31}$$

where P_b is the fraction of the surfactant micelles bound to the polymer and D_b the self-diffusion coefficient of the polymer–surfactant complex. It is demonstrated that from studies using polymers of different polarity, a

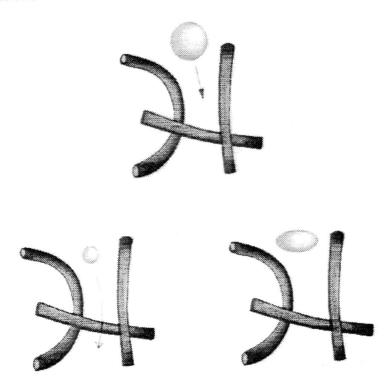

Figure 8.12 Micelle diffusion in a polymer network depends on association and obstruction effects. The latter are given by simple geometrical features like micelle size and shape and volume fraction and radius of polymer [107]. Reprinted with permission from Zhang *et al.*, *J. Phys. Chem.*, **98**, 2459. Copyright (1994) American Chemical Society

separation between obstruction and association effects can be achieved [107]. This is illustrated in Figure 8.13 where it is shown that micelle diffusion in the presence of a relatively polar polymer gives a retardation in diffusion which can be accounted for by obstruction alone. The additional reduction in diffusion in the presence of a more nonpolar polymer is attributed to association effects leading to a formation of polymer–surfactant complexes. This observation is consistent with phase diagram findings. Taking into account the obstruction from both micelles and polymer chains, the binding isotherm of $C_{12}E_8$ to the hydrophobic EHEC is calculated by using a two-site model. The association approximately follows a simple Langmuir isotherm.

5.11 ORDER–DISORDER TRANSITIONS OF SURFACTANT CONTINUOUS MICROSTRUCTURES

The formation of liquid crystalline phases at higher surfactant concentrations

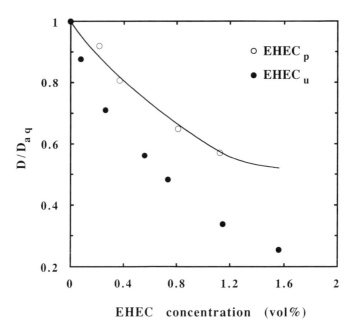

Figure 8.13 The self-diffusion of non-ionic micelles is much more retarded by a more hydrophobic polymer due to polymer–micelle association. The figure shows the micelle self-diffusion coefficient relative to that in the absence of polymer as a function of polymer concentration. The system shown is that of octaethylene glycol monodode-cylether and ethyl(hydroxiethyl) cellulose. Subscripts p and u refer to polar and nonpolar polymer samples [107]. Reprinted with permission from Zhang *et al.*, **98**, 2459. Copyright (1994) American Chemical Society

can often be considered as disorder–order phase transitions. For example, small micelles crystallize in a cubic lattice and long cylindrical micelles are ordered in a two-dimensional hexagonal array. Analogously, the multiply-connected bilayer structure, which is disordered in the liquid L_3 phase, crystallize in some systems into a bicontinuous cubic phase at higher surfactant concentrations. This phase transition was studied by Balinov *et al.* [51] in the AOT-brine system. They measured the water and surfactant self-diffusion in the L_3 and V_2 cubic phases and found that the self-diffusion varies smoothly across the transition. The data are shown in Figure 8.14. The results imply that the bilayer structures in the melted L_3 and the frozen cubic phases are very similar, and that the phase transition can be regarded as a simple order–disorder transition. Analogous results have been obtained in a ternary system with non-ionic surfactant [108] and the bilayer volume fraction at which the order–disorder transition occurs is thought to depend on the bilayer rigidity [109].

A similar transition was investigated in the cetylpyridinium chloride–sodium salicylate–water system, at a constant 1 : 1 molar ratio of surfactant to salt, by

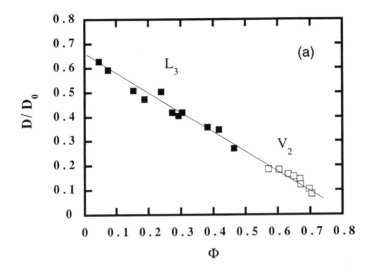

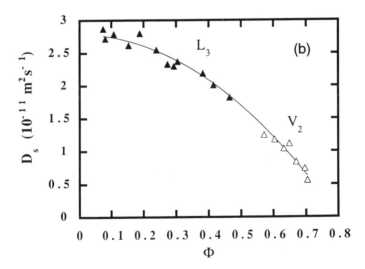

Figure 8.14 The relative self-diffusion coefficient (D/D_0) of water (a) and the absolute self-diffusion coefficient of the surfactant (b) plotted as a function of the surfactant volume fraction in the L_3 and cubic V_2 phases of the AOT–brine system. Note the continuous variation of the self-diffusion coefficients across the L_3–V_2 phase transition. Data taken from Ref. 51

Monduzzi *et al.* [110]. In this system, which behaves very similarly to the binary cetylpyridinium salycilate–water system, the classical micellar L_1 phase is at higher concentrations and higher temperatures in equilibrium with the bicontinuous cubic V_1 phase. This unusual phase behaviour results from an unusually low melting point of the hexagonal phase. The L_1–V_1 transition was studied by NMR self-diffusion and ^{14}N spin-relaxation of the surfactant ion. Similar to the L_3–V_2 transition described above, the self-diffusion coefficients and the relaxation rates varied smoothly across the L_1–V_1 transition. From these results, including the relatively high value of the surfactant self-diffusion coefficient, it could be concluded that the aggregate structure in the two phases is similar and that the surfactant aggregate has to be multiply connected also in the L_1 phase.

5.12   EMULSIONS

Emulsions are defined as dispersions of one liquid in another. The dispersed phase may be present as spherical droplets. This is an unstable situation, and therefore one stabilizes the system by adding one (or several) emulgators. These are often surfactants or polymers.

One important issue regarding emulsions is the size and distribution of sizes of these droplets. The NMR self-diffusion approach offers a unique tool for determining these parameters. The NMR approach to the problem rests on the fact that the dispersed phase (the liquid entrapped in the droplets) shows restricted diffusion. This was discussed in some detail above. It is clear that the equations pertaining to the echo decay from the dispersed phase (see equation (8.13) above) do indeed contain the droplet radius as a parameter, and it is thus possible to back out an estimate for the radius from the experimentally observed echo decays.

When applied to a real emulsion one has to consider the fact that the emulsion droplets in most cases are polydisperse with respect to their size. This can be taken into account if the molecules confined to the droplets are in a slow exchange situation, meaning that their lifetime in the droplet must be longer than Δ. For such a case, the echo attenuation is given by:

$$E_{\text{poly}} = \frac{\int_0^\infty R^3 P(R) E(R) \, dR}{\int_0^\infty R^3 P(R) \, dR} \tag{8.32}$$

where $P(R)$ represents the droplet size distribution function and $E(R)$ the echo attenuation according to equation (8.13) (or, within the SGP approximation, equation (8.12)) for a given value of R. From NMR data it is difficult (although in principle not impossible) to determine the actual form of $P(R)$.

However, given an analytical expression for $P(R)$ we may determine the parameters of that distribution function. The most commonly used form is the log-normal function as defined in equation (8.33), as it appears to be a reasonable description of the droplet size distribution of many emulsions. In addition, it has only two parameters which makes it convenient for modelling purposes.

$$P(R) = \frac{1}{2R\sigma\sqrt{2\pi}} \exp\left[-\frac{(\ln 2R - \ln d_0)^2}{2\sigma^2}\right] \tag{8.33}$$

In equation (8.33), d_0 represents the diameter median and σ is a measure of the width of the size distribution.

To illustrate the method and also discuss its accuracy we will use some recent results for margarines (or low calorie spreads) [111]. This system highlights some of the definite advantages of using the NMR method to determine emulsion droplet sizes, since other non-perturbing methods hardly exist for these systems.

Given in Figure 8.15 is the echo decay for the water of a low calorie spread containing 60% fat. These systems are W/O emulsions and as can be seen the water molecules do experience restricted diffusion (in the representation of Figure 8.15, the echo decay for free diffusion would be given by a Gaussian

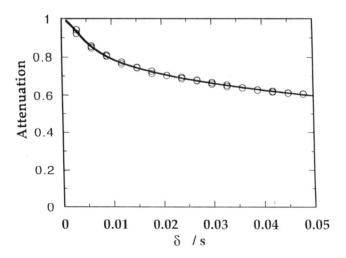

Figure 8.15 Echo intensity for the entrapped water in droplets formed in a low calorie spread containing 60% fat versus δ. The solid line corresponds to the predictions of equations (8.13), (8.32) and (8.33). The results from the fit are: $d_0 = 0.82$ μm and $\sigma_0 = 0.72$. Reproduced by permission of the American Oil Chemists' Society from Ref. 111

function). Also given in Figure 8.15 is the result of fitting equations (8.13), (8.32) and (8.33) to the data. As is evident, the fit is quite satisfactory and the parameters of the distribution function obtained are given in the figure caption. However, one might wonder how well determined these parameters are, given the fact that the equations describing the echo attenuation are quite complicated. To test this matter further, Monte Carlo error investigations were performed [111]. Thus random errors were added to the echo attenuation and the least squares minimization was repeated 100 times as described previously [112]. A typical result of such a procedure is given in Figure 8.16. As can be seen in Figure 8.16 the parameters are reasonably well determined, with an uncertainty in R and σ of about ±15%.

In conclusion, we summarize the main advantages of the NMR diffusion method as applied to emulsion droplet sizing. It is non-perturbing, requiring no sample manipulation (such as dilution with the continuous phase) and non-destructive which means that the same sample may be investigated many times, which is important if one wants to study long time stability or the effect of certain additives on the droplet size. It requires small amounts of sample (typically on the order of a few 100 mg). Moreover, it is a rapid technique.

So far, the discussion applies to the case where the molecules are confined to the droplets on the timescale of the experiment. This is a reasonable assumption for many emulsions, and it can in fact be tested by the NMR diffusion method by varying Δ. However, there are some interesting emulsion systems where this is not always the case. These are the so-called highly concentrated emulsions [113, 114], which may contain up to in excess of 99% dispersed

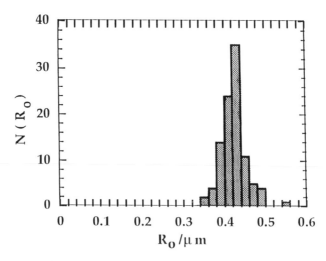

Figure 8.16 A Monte Carlo error analysis of the data in Figure 8.15 The value of the parameter d_0 in equation (8.33) is $d_0 = 0.82 \pm 0.044$ μm. Reproduced by permission of The American Oil Chemists' Society from Ref. 111

phase. Here the droplets are separated by a liquid film which may be very thin (on the order of 100 Å). If the dispersed phase actually crosses this film by some mechanism, the detailed nature of which does not need to concern us here, we are instead dealing with a system with permeable barriers, and the system can now be regarded as belonging to the general class of porous systems. The PGSE experiment as applied to this type of system has recently been treated by Callaghan *et al.* [29, 30]. These workers have shown that under certain conditions, one may actually obtain a peak in the plot of the echo-amplitude versus the gradient pulse duration, δ [115]. This is a surprising result at first sight, as we are accustomed to observe a monotonous decrease of the echo amplitude with δ, but it is actually a manifestation of the fact that the diffusion is no longer Gaussian. Such peaks can be rationalized by a formalism related to the one used to treat diffraction effects, and the analysis of the data may yield important information regarding not only the size of the droplets but also the permeability of the dispersed phase through the thin films as well as the long-term diffusion behaviour of the dispersed phase. We show in Figure 8.17 an example of such a diffraction like effect in a concentrated emulsion

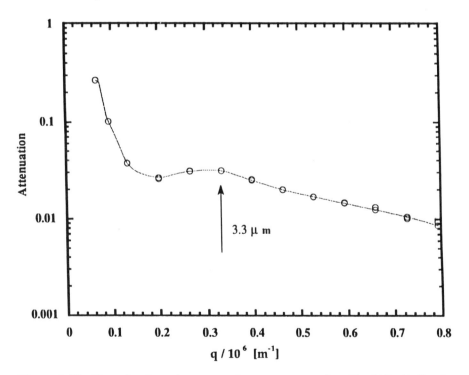

Figure 8.17 The echo intensity versus the parameter q $(q \equiv (\delta\gamma g)/(2\pi))$ for the water in a concentrated W/O emulsion consisting of a partially fluorinated surfactant, perfluorodecaline and water [116]. Reprinted with permission from Balinov *et al.*, *J. Phys. Chem.*, **98**, 393. Copyright (1994) American Chemical Society

system [116]. The particular example pertains to a concentrated emulsion based on a fluorinated non-ionic surfactant, where the continuous medium is a perfluorinated oil.

References

1. J. N. Israelachvili, *Intermolecular and Surface Forces*, 2nd edn, Academic Press, London (1991).
2. B. Lindman and H. Wennerström, *Topics in Current Chemistry*, vol. 1, **87**, 1 (1980).
3. G. J. T. Tiddy, *Phys. Rep.*, **57**, 1 (1980).
4. P. Ekwall, in H. Brown (ed.), *Advances in Liquid Crystals*, pp. 1, Academic Press, New York (1975).
5. D. F. Evans and H. Wennerström, *The Colloidal Domain: Where Physics, Chemistry, Biology, and Technology Meet*, VCH Publishers, Inc., New York (1994).
6. B. Lindman, O. Söderman and H. Wennerström, in R. Zana (ed.), *Surfactant Solutions, New Methods of Investigation*, pp. 295, Marcel Dekker, New York (1987).
7. C. Chachaty, *Prog. Nucl. Magn. Reson. Spectrosc.*, **19**, 183 (1987).
8. B. Lindman, O. Södermnan and H. Wennerström, *Ann. Chim. (Rome)*, **77**, 1 (1987).
9. B. Lindman, O. Söderman and P. Stilbs, in K. L. Mittal (ed.), *Surfactants in Solution*, pp. 1–24, Plenum, New York (1989).
10. J. Ulmius and H. Wennerström, *J. Magn. Res.*, **28**, 309 (1977).
11. U. Olsson, O. Söderman and P. Guéring, *J. Phys. Chem.*, **90**, 5223 (1986).
12. H. Wennerström and J. Ulmius, *J. Magn. Res.*, **23**, 431 (1976).
13. F. Noack, *Prog. Nucl. Magn. Reson. Spectrosc.*, **18**, 171 (1986).
14. W. Kühner, E. Rommel, F. Noack and P. Meier, *Z. Naturforsch.*, **42a**, 127 (1987).
15. H. Wennerström, G. Lindblom and B. Lindman, *Chemica Scripta·* **6**, 97 (1974).
16. H. Wennerström, B. Lindman, O. Söderman, T. Drakenberg and J. B. Rosenholm, *J. Am. Chem. Soc.*, **101**, 6860 (1979).
17. B. Halle and H. Wennerström, *J. Chem. Phys.*, **75**, 1928 (1981).
18. A. Abragam, *The Principles of Nuclear Magnetism*, Clarendon Press, Oxford (1961).
19. P. T. Callaghan, C. M. Trotter and K. W. Jolley, *J. Magn. Reson.*, **37**, 247 (1980).
20. P. Stilbs, *Prog. Nucl. Magn. Reson. Spectrosc.*, **19(1)**, 1 (1987).
21. O. Söderman and P. Stilbs, *Prog. Nucl. Magn. Reson. Spectrosc.*, **26**, 445 (1994).
22. E. L. Hahn, *Phys. Rev.*, **80**, 580 (1950).
23. E. O. Stejskal and J. E. Tanner, *J. Chem. Phys.*, **42**, 288 (1965).
24. D. W. McCall, D. C. Douglass and E. W. Anderson, *Ber. Bunsenges. Phys. Chem.*, **67**, 336 (1963).
25. R. L. Vold, J. S. Waugh, M. P. Klein and D. E. Phelps, *J. Chem. Phys.*, **48** (1968).
26. J. E. Tanner Thesis, Univ. of Wisconsin. Madison, Wis., USA, 1966.
27. P. Callaghan and A. Coy, *Phys. Rev. Lett.*, **68**, 3176 (1992).
28. P. T. Callaghan, *Principles of Nuclear Magnetic Resonance Microscopy*, Clarendon Press, Oxford (1991).
29. P. T. Callaghan, A. Coy, T. P. J. Halpin, D. MacGowan, J. K. Packer and F. O. Zelaya, *J. Chem. Phys.*, **97**, 651 (1992).
30. P. T. Callaghan and A. Coy, in P. Tycko (ed.), *NMR Probes of Molecular Dynamics*, Kluwer, Dordrecht, in press.

31. B. Balinov, B. Jönsson, P. Linse and O. Söderman, *J. Magn. Reson., A.*, **104**, 17 (1993).
32. D. C. Douglass and D. W. McCall, *J. Phys. Chem.*, **62**, 1102 (1958).
33. C. H. Neuman, *J. Chem. Phys.*, **60**, 4508 (1974).
34. J. S. Murday and R. M. Cotts, *J. Chem. Phys.*, **48**, 4938 (1968).
35. I. Lönnqvist, A. Khan and O. Söderman, *J. Colloid Interface Sci.*, **144**, 401 (1991).
36. B. Balinov, O. Urdahl, O. Söderman and J. Sjöblom, *Colloids Surf.*, **82**, 173 (1994).
37. X. Li, J. C. Cox and R. W. Flumerfelt, *AIChE J.*, **38**, 1671 (1992).
38. P. T. Callaghan, K. W. Jolley and R. Humphrey, *J. Colloid Interface Sci.*, **93**, 521 (1983).
39. J. C. Van den Enden, D. Waddington, H. Van Aalst, C. G. Van Kralingen and K. J. Packer, *J. Colloid Interface Sci.*, **140**(1), 105 (1990).
40. K. J. Packer and C. Rees, *J. Colloid Interface Sci.*, **40**(2), 206 (1972).
41. B. Jönsson, H. Wennerström, P. Nilsson and P. Linse, *Coll. Polym. Science*, **264**, 77 (1986).
42. P. N. Pusey, In D. Levesque, J. Hansen and J. Zinn-Justin (eds.), *Liquids, Freezing and the Glass Transition* (*Les Houches Session LI*); pp. 763, Elsevier: Amsterdam (1990).
43. M. Medina-Noyola, *Phys. Rev. Lett.*, **60**, 2705 (1988).
44. B. Lindman and H. Wennerström, *J. Phys. Chem.*, **95**, 6053 (1991).
45. P. G. Nilsson, H. Wennerström and B. Lindman, *J. Phys. Chem.*, **87**, 1377 (1983).
46. P. G. Nilsson, H. Wennerström and B. Lindman, *Chem. Scr.*, **25**, 67 (1985).
47. O. Söderman, E. Hansson and M. Monduzzi, *J. Colloid Interface Sci.*, **141**, 512 (1991).
48. U. Olsson and P. Schurtenberger, *Langmuir*, **9**, 3389 (1993).
49. D. M. Anderson and H. Wennerström, *J. Phys. Chem.*, **94**, 8683 (1990).
50. P. Ström and D. M. Anderson, *Langmuir*, **8**, 691 (1992).
51. B. Balinov, U. Olsson and O. Söderman, *J. Phys. Chem.*, **95**, 5931 (1991).
52. U. Olsson, B. Balinov and O. Söderman, in R. Lipowsky, D. Richter and K. Kremer (eds) *The Structure and Conformation of Amphiphilic Membranes. Springer Proceedings in Physics*; pp. 287, Springer, Berlin (1992); Vol. 66.
53. H. Wennerström and G. Lindblom, *Q. Rev. Biophys.*, **10**, 67 (1977).
54. G. Lindblom and H. Wennerström, *Biophys. Chem.*, **6**, 167 (1977).
55. P. O. Eriksson, L. B. Å. Johansson and G. Lindblom in K. L. Mittal and B. Lindman (eds), *Surfactants in Solution*, pp. 219, Plenum Press, New York (1984), Vol. 1.
56. P. T. Callaghan, M. A. LeGros and D. N. Pinder, *J. Chem. Phys.*, **79**, 6372 (1983).
57. K. Shinoda, *Prog. Colloid Polym. Sci.*, **68**, 1 (1983).
58. R. G. Laughlin, *Adv. Liq. Cryst.*, **3**, 41 (1978).
59. R. G. Laughlin, in T. Tadros (ed.), *Surfactants*; pp. 53, Academic Press, New York (1984).
60. N. O. Persson, K. Fontell and B. Lindman, *J. Colloid Interface Sci,*, **53**, 461 (1975).
61. A. Khan, K. Fontell, G. Lindblom and B. Lindman, *J. Phys. Chem.*, **86**, 4266 (1982).
62. E. Marques, A. Khan, M. G. Miguel and B. Lindman, *J. Phys. Chem.* **97**, 4729 (1993).
63. G. Chidichimo, N. A. P. Vaz, Z. Yaniv and J. W. Doane, *Phys. Rev. Lett.*, **49**, 1950 (1982).

64. K. Fontell, *Mol. Cryst. Liq. Cryst.*, **63**, 59 (1981).
65. T. Bull and B. Lindman, *Mol. Cryst. Liq. Cryst.*, **28**, 155 (1974).
66. G. Lindblom, K. Larsson, L. Johansson, K. Fontell and S. Forsén, *J. Am. Chem. Soc.*, **101**, 5465 (1979).
67. P. O. Eriksson, A. Khan and G. Lindblom, *J. Phys. Chem.*, **86**, 387 (1982).
68. L. B. Johansson and O. Söderman, *J. Phys. Chem.*, **91**, 5275 (1987).
69. O. Söderman, H. Walderhaug, U. Henriksson and P. Stilbs, *J. Phys. Chem,,* **89**, 3693 (1985).
70. P. O. Eriksson, G. Lindblom and G. Arvidson, *J. Phys. Chem.*, **89**, 1050 (1985).
71. B. Lindman and P. Stilbs, in S. Friberg and P. Bothorel (eds), *Microemulsions*, pp. 119, CRC Press, Boca Raton, FL (1986).
72. B. Lindman, M. C. Puyal, N. Kamenka, R. Rymdén and P. Stilbs, *J. Phys. Chem.*, **88**, 5048 (1984).
73. G. Gunnarsson, B. Jönsson and H. Wennerström, *J. Phys. Chem.*, **84**, 3114 (1980).
74. B. Lindman, M. C. Puyal, N. Kamenka, B. Brun and G. Gunnarsson, *J. Phys. Chem.*, **86**, 1702 (1982).
75. N. Kamenka, G. Haouche, B. Faucompré, B. Brun and B. Lindman, *J. Colloid Interface Sci.*, **108** (1985).
76. A. Ceglie, G. Colafemmina, M. Della Monica, U. Olsson and B. Jönsson, *Langmuir*, **9**, 1449 (1993).
77. O. Söderman, U. Henriksson and U. Olsson, *J. Phys. Chem.*, **91**, 116 (1987).
78. M. S. Leaver, U. Olsson, H. Wennerström and R. Strey, *J. Phys. II France*, **4**, 515 (1994).
79. R. Skurtveit and U. Olsson, *J. Phys. Chem.*, **96**, 8640 (1992).
80. B. Halle, *J. Chem. Phys.*, **94**, 3150 (1991).
81. H. Wennerström, *Chem. Phys. Lett.*, 18, 41 (1973).
82. L. Hwang, P. Wang and T. C. Wong, *J. Phys. Chem.*, **92**, 4753 (1988).
83. B. Lindman and P. Stilbs, in V. Degiorgio and M. Corti (eds), *Proc. S.I.F. Course XC*, pp. 94, North-Holland: Amsterdam (1985).
84. P. Stilbs and B. Lindman, *J. Phys. Chem.*, **85**, 2587 (1981).
85. D. Bratko and B. Lindman, *J. Phys. Chem.*, **89**, 1437 (1985).
86. M. Jansson and P. Stilbs, *J. Phys. Chem.*, **89**, 4868 (1985).
87. B. Lindman, in P. Laszlo (ed.), *NMR of Newly Accessible Nuclei*, pp. 193, Academic Press, New York (1983).
88. B. Lindman, H. Wennerström, H. Gustavsson, N. Kamenka and B. Brun, *Pure Appl. Chem.*, **52**, 1307 (1980).
89. P. G. Nilsson and B. Lindman, *J. Phys. Chem.*, **87**, 4756 (1983).
90. B. Halle Thesis, Lund, 1981.
91. P. Stilbs, *J. Colloid Interface Sci.*, **87**, 385 (1982).
92. P. Stilbs, *J. Colloid Interface Sci.*, **89**, 547 (1982).
93. P. Stilbs, *J. Colloid Interface Sci.*, **94**, 463 (1983).
94. P. Stilbs, G. Arvidson and G. Lindblom, *Chem. Phys. Lipids*, **35**, 309 (1984).
95. U. Henriksson, T. Klason, L. Ödberg and J. C. Eriksson, *Chem. Phys. Lett.*, **52**, 554 (1977).
96. U. Henriksson, T. Klason, E. Florin and J. C. Eriksson, *NATO Adv. Study Inst. Ser.*, **C61**, 681 (1980).
97. P. Stilbs, H. Walderhaug and B. Lindman, *J. Phys. Chem.*, **87**, 4762 (1983).
98. B. Halle, *J. Phys. Chem.*, **95**, 6724 (1991).
99. A. Carlsson, G. Karlström and B. Lindman, *J. Phys. Chem.*, **93**, 3673 (1989).
100. T. C. Wong, K. Thalberg and B. Lindman, *J. Phys. Chem.*, **95**, 8850 (1991).
101. T. C. Wong, C. Liu and C.-D. Poon, *Langmuir*, **8**, 460 (1992).

102. L. Johansson, U. Skantze and J. Löfroth, *Macromolecules*, **24**, 6019 (1991).
103. L. Johansson, C. Elvingson and J. Löfroth, *Macromolecules*, **24**, 6024 (1991).
104. L. Johansson and J. Löfroth, *J. Chem. Phys.*, **98**, 7471 (1993).
105. L. Johansson, P. Hedberg and J. Löfroth, *J. Phys. Chem.*, **97**, (1993).
106. L. Johansson Doctoral Thesis Thesis, Göteborg, 1993.
107. K. Zhang, M. Jonströmer and B. Lindman, *J. Phys. Chem.*, **98**, 2459 (1994).
108. U. Olsson, U. Würz and R. Strey, *J. Phys. Chem.*, **97**, 4535 (1993).
109. J. Bruinsma, *J. Phys. II France*, **2**, 425 (1992).
110. M. Monduzzi, U. Olsson and O. Söderman, *Langmuir*, **9**, 2914 (1993).
111. B. Balinov, O. Söderman and T. Wärnheim, *J. Am. Oil Chem. Soc.*, **71**, 513 (1994).
112. P. Stilbs and M. Moseley, *J. Magn. Reson.*, **31**, 55 (1978).
113. H. M. Princen, *J. Colloid Interface Sci.*, **91**, 160 (1983).
114. K. J. Lissant, *J. Colloid Interface Sci.*, **22**, 462 (1966).
115. P. T. Callaghan, A. Coy, D. MacGowan, K. J. Packer and F. O. Zelaya, *Nature (London)* **351**(**6326**), 467 (1991).
116. B. Balinov, O. Söderman and J. C. Ravey, *J. Phys. Chem.*, **98**, 393 (1994).

9 POLYMERS AND BIOPOLYMERS IN THE LIQUID STATE

Introduction

This chapter deals with the dynamics of large molecules in solution. It presents the basic NMR techniques and the theoretical treatments used for this purpose. The difficulties encountered in these studies are quite similar to those described in Chapter 8 for organized assemblies of medium size molecules, namely the increased importance of local dynamics as compared to overall motions. Distributions of correlation times would presumably be necessary to give an accurate description of internal motions; in fact, this description is most often reduced to simplified models involving two types of displacement only, considered to be either fast or slow relatively to each other. Anyway, molecular models and dynamical simulations are required to interpret NMR experiments which, in turn, permit the adjustment of the relevant structural and dynamical parameters used in these models. In this domain, there is no methodology free from any assumption. The confidence in the conclusions obtained in this way proceeds from comparisons with similar results obtained from independent experiments using other techniques than NMR. The text of this chapter is divided into two parts relating to either synthetic polymers or to biopolymers.

Part A   Local Dynamics of Large Molecules: Synthetic Polymers in Solution and in Melts

Part B   NMR and Dynamics of Biopolymers

Dynamics of Solutions and Fluid Mixtures by NMR
Edited by J.-J. Delpuech   © 1995 John Wiley & Sons Ltd

9A LOCAL DYNAMICS OF LARGE MOLECULES: SYNTHETIC POLYMERS IN SOLUTION AND IN MELTS

M.-A. Krajewski-Bertrand, F. Lauprêtre and L. Monnerie
Ecole Supérieure de Physique et Chimie Industrielles, Paris, France

Dynamics of Solutions and Fluid Mixtures by NMR
Edited by J.-J. Delpuech © 1995 John Wiley & Sons Ltd

A.1   Introduction

As shown by the numerous studies published in the literature, high-resolution ^{13}C and ^{1}H NMR have proven to be a most powerful tool for investigating the dynamics of organic compounds. Unlike fluorescence anisotropy or electron spin resonance techniques, high-resolution NMR does not require any labeling of the molecule under study. As a selective technique, it allows the observation of one signal per magnetically inequivalent nucleus, and, therefore, the dynamic behavior of each part of a molecule can be followed independently. Moreover, many NMR parameters are sensitive to molecular motions. They include the different relaxation times as well as the spectrum lineshape, the strength of the dipolar coupling and the chemical shift anisotropy. The available spectral windows depend on the type of measurement which is performed. They range from about 10^{-1} Hz for slow processes to several hundreds of MHz for very fast modes.

This chapter will be devoted to the local dynamics of synthetic polymers in a mobile state, that is, either in the bulk, at temperatures well above the glass transition temperature, or in solution. Owing to their size, macromolecules have a large spectrum of motional frequencies, ranging from the rapid librations and reorientations at the monomer unit scale to the much slower overall diffusion of the entire molecule. In the following, we will focus on the fastest modes of the polymer chains, involving only a few bonds. The characteristics of these very local motions are mainly determined by the chemical structure of the repeat unit. For long enough chains, they are independent of the molecular weight. This fast local dynamics can be investigated by determining the spin–lattice relaxation times, T_1, of the proton and carbon nuclei and the nuclear Overhauser enhancement of carbon spins under proton decoupling conditions, that probe modes in the Larmor frequency region.

In this chapter, we will first qualitatively describe the different local motions that may occur in a polymer chain either in solution or in the bulk at temperatures well above the glass transition temperature. We will review the orientation autocorrelation functions that have been proposed to account for these phenomena. The experimental tests of these functions will be discussed, using results from both fluorescence anisotropy decay and NMR relaxation experiments. Starting from the representation of the local dynamics thus obtained, the interpretation of NMR data derived from a large number of polymers will be performed and the main factors that govern the nature and frequency of local motions of polymers in solution or in the bulk above the glass transition temperature will be identified.

A.2 Description of the Local Motions of a Polymer Chain in Solution or in the Melt

A.2.1 QUALITATIVE DESCRIPTION OF SEGMENTAL CHAIN MOTIONS

The bond motions in a polymer chain are much more complicated than the Brownian motion of a small molecule. Clearly, the connectivity of the chain, which implies that no bond can be displaced independently of its neighbors, has to be taken into account.

Local chain motions can be schematically described in terms of conformational transitions involving short sequences of the chain. Let us consider a small segment located between the two chain tails P and Q and undergoing a conformational transition:

$$Ps_1 s_2 \ldots s_i Q \to P's_{1'} s_{2'} \ldots s_{i'} Q'$$

$s_1 s_2 \ldots s_i$ and $s_{1'} s_{2'} \ldots s_{i'}$ are the conformational states of the moving sequence at the start and end, respectively, of the transition. No conformational change is assumed to affect the P and Q moieties during the time interval where the above sequence undergoes the transition. However, depending on the motion of the sequence and owing to the connectivity of the polymer chain, P and Q will or will not have the same position at the start and end of the transition. According to the global motion of the chain tails attached to the moving unit, Helfand [1] has proposed a classification of conformational transitions in polymers in terms of three types of elementary processes. In the following, two factors will be considered: (i) the energy barrier associated with a conformational jump, (ii) the friction forces opposed by the environment to the motion of the macromolecule. Examples will be chosen among alkane chains whose conformational states are t, g^+ and g^-.

Type 1 transitions are such that they do not modify the positions of P and Q. Examples of type 1 transitions are the three-bond motion:

$$Ptg^+tQ \rightarrow Ptg^-tQ$$

and the crankshaft motion depicted in Figure 9A.1:

$$Pg^+tg^+tg^+tg^-Q \rightarrow Pttg^+tg^+ttQ$$

Such motions encounter a small viscous resistance since only a very limited part of the polymer chain is displaced in the medium. However, due to the simultaneous orientational change of the moving bonds, they require several barrier heights to be crossed simultaneously, for example an average of two barrier crossings for the above transitions.

Type 2 transitions are characterized by a small-amplitude translation of Q. Two examples of type 2 transitions are represented in Figure 9A.2:

$$PgttQ \rightarrow PttgQ'$$

$$PtttQ \rightarrow Pg^+tg^-Q'$$

The viscous friction accompanying the translation of Q is larger than the friction opposed to a type 1 transition. On the other hand, only one barrier height has to be crossed.

In type 3 transitions, the motion of Q due to the conformational changes of the small sequence is a rotation. Single conformational jumps belong to this

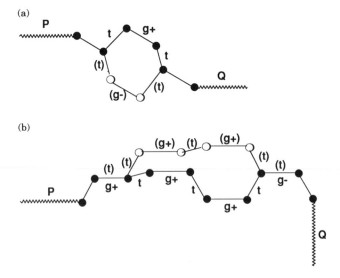

Figure 9A.1 Examples of type 1 transitions: (a) the three-bond motion: $P\ tg^+tQ \rightarrow P\ tg^-t\ Q$; (b) the crankshaft transition: $P\ g^+tg^+tg^+tg^-Q \rightarrow P\ ttg^+tg^+tt\ Q$

(a)

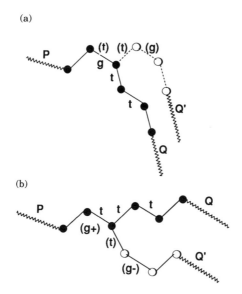

(b)

Figure 9A.2   Examples of type 2 transitions: (a) *P gtt Q → P ttg Q'*; (b) *P ttt Q → P g⁺tg⁻Q'*

type, as shown in Figure 9A.3:

$$PtQ \rightarrow PgQ'$$

These transitions are characterized by strong friction forces opposing the rotation of *Q*. On the other hand, they require only one barrier crossing.

These three types of elementary motions have interesting characteristics. For example, the first transition depicted in Figure 9A.2 does not modify the relative number of *trans* and *gauche* states. It can be described as a migration or diffusion of *gauche* conformations. The opposite phenomenon occurs in the second transition shown in Figure 9A.2. It can be considered as a loss of orientation and described in more general terms as a damping of the previous diffusion of bond orientation. Such a description holds for the three types of conformational transitions.

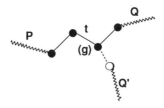

Figure 9A.3   Example of a type 3 transition: *P t Q → P g Q'*

Motions that induce a reorientation of a large part of the chain, such as transition types 2 or 3, are expected to be slow and molecular weight-dependent. These two points are in striking disagreement with the conclusions derived from the experimental study of local motions. In this respect, the very localized type 1 transitions that involve a minimum viscous energy may appear of particular interest. However, the activation energies determined experimentally are quite low. They are of the order of the energy barrier associated with a single conformational jump. Therefore, all the elementary motions that may occur in a polymer chain involve either several energy barriers or a relatively high chain friction in addition to the barrier height for one bond. This paradox has been elucidated by Skolnick and Helfand [2] by applying the approach of Kramers [3] to the *trans–gauche* conformational change for an internal bond in an alkane chain. The basic idea consists of taking into account the distortions of the bond lengths, valence angles and torsional angles of the units next to the bond undergoing a conformational change. By looking for the path of steepest descent required for the *trans–gauche* conformational change, it appears that the distortion resulting from the bond motion is spread over the neighboring units, without modifying the rest of the chain. Such an analysis shows that the deformation is mainly accommodated by distortions of the torsional angles of the neighboring bonds, and that some distortions of the valence angles may also occur. These distortions are of opposite sign with respect to the distortion of the bond undergoing the transition, in order to minimize the motion of the more remote parts of the chain.

A.2.2 DESCRIPTION OF SEGMENTAL MOTIONS AS OBTAINED FROM BROWNIAN DYNAMICS SIMULATIONS

Computer simulations provide a powerful tool for the exploration of the local dynamics of macromolecules. Molecular dynamics simulations use Newton's equation of motion:

$$m_i \vec{a}_i = \vec{F}_i \qquad (9A.1)$$

for a collection of i particles of mass m_i and of acceleration $\vec{a}_i$. $\vec{F}_i$ represents the sum of all external forces experienced by the particle i and is determined from the gradient of the potential energy U. U is a function of the coordinates of all the particles in the simulation. For a study of the local motions of a polymer in a dilute solution, this would include one polymer chain and hundreds of solvent molecules.

The Langevin equation is the starting point for stochastic simulations:

$$m_i \vec{a}_i = \vec{F}_i - \zeta_i \vec{v}_i + \overrightarrow{N_i(t)} \qquad (9A.2)$$

$\zeta_i \vec{v}_i$ and $\overrightarrow{N_i(t)}$ describe the interactions between the particles of interest, that is, the polymer atoms, and the solvent. The solvent is not represented explicitly. It damps the motion of the particles with friction terms $\zeta_i v_i$ proportional to the velocity v_i and supplies stochastic forces $\overrightarrow{N_i(t)}$ that mimic the collisions between the solvent molecules and the polymer. Often the high friction or diffusive limit of equation (9A.2) is used in simulations, which implies that the inertial term is negligible:

$$\zeta_i \vec{v}_i = \vec{F}_i + \overrightarrow{N_i(t)} \qquad (9A.3)$$

Such simulations are referred to as Brownian dynamics simulations. Usually, the contributions to the potential energy U that are taken into account, are those resulting from the distortion of bond lengths, bond angles and dihedral angles.

The first polymer investigated by Brownian dynamics simulations was polyethylene [4]. Results obtained by Helfand et al. on a cyclic chain of 200 bonds show that most of the transitions occurring belong to type 2 motions which lead to a small-amplitude translation of a part of the chain. The activation energy is slightly greater than the barrier height separating the trans and gauche states. However, the probability of such motions strongly depends on the conformation of the neighboring bonds. Moreover, these motions may occur either in an isolated way or a cooperative way. Indeed, such motions of the chain, which induce a translation of a part of the chain, and, therefore, a large dissipated energy resulting from viscous friction, are often associated with opposite motions of the second neighboring bond. These motions facilitate the displacement of the first bond, without involving any translation of a chain moiety. As regards the orientations of the bonds of the chains, this correlation of individual processes is equivalent to a diffusion of bond orientation along the chemical sequence. Therefore, from the results of the simulations carried out by Helfand et al., as well as from elementary considerations of motions of type 1, 2 and 3, it clearly appears that, concerning the orientations of the bond, the chain motion is equivalent to a random diffusion of the orientations along the chain, that is damped by isolated conformational jumps.

More recent Brownian dynamics simulations have been performed on both polyethylene and polyisoprene [5, 6]. For both polymers, the most striking result is that the distortions accompanying a conformational transition are highly localized, involving mainly the four carbons that are the closest to the bond undergoing a transition, as represented in Figure 9A.4. These distortions induce displacements of the positions of the nearest atoms, which consist mainly of variations in torsional angles, but can also affect bond angles and less probably bond lengths. This set of occurrences can be considered as cooperative motions through which the chain accommodates a conformational jump in order to localize the motion and to avoid the viscous resistance exerted on large amplitude motions. Moreover, more than two-thirds of the transitions occur as isolated transitions, the rest being correlated jumps. The second-neighbor motional coupling, which is important for localizing conformational transitions in polyethylene, does not occur in polyisoprene. The extent of

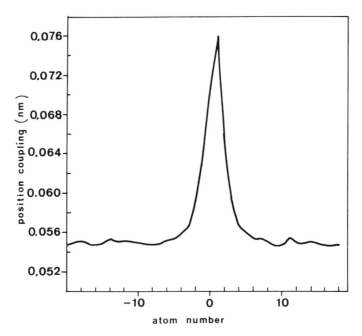

Figure 9A.4 Distortion of atomic positions accompanying a conformational transition between atoms 0 and 1 in simulated *cis*-1,4-polyisoprene [5]. Reprinted with permission from Adolf and Ediger, *Macromolecules*, **24**, 5834. Copyright (1991) American Chemical Society

correlated transitions seems to depend on the chemical structure of the monomer unit.

Another interesting feature can be extracted from the Brownian dynamics simulations. The polyisoprene histograms, indicating the probability that at time t the dihedral angle, $\phi(0)$, will have changed by $\phi(t)-\phi(0)$, are given in Figure 9A.5 for different values of t. They show that librations with a full-width half maximum of 50° take place in less than 1 ps from an arbitrary starting time, whereas conformational transitions are observed at much longer times [5].

A.2.3   POLYMER ORIENTATION AUTOCORRELATION FUNCTIONS

The interesting quantity in the NMR experiments described in this chapter is the orientation autocorrelation function:

$$G(t) = \langle P_2[\overrightarrow{u_n}(0).\overrightarrow{u_n}(t)]\rangle = \langle 3\cos^2\theta(t)-1\rangle/2 \qquad (9A.4)$$

where $\overrightarrow{u_n}$ is the unit vector associated with the nth bond of the chain. $\theta(t)$

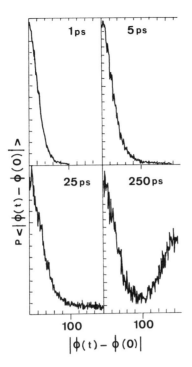

Figure 9A.5 Evolution of the distribution of dihedral angles for the C–CH$_2$ bond in *cis*-1,4-polyisoprene as a function of time [5]. Reprinted with permission from Adolf and Ediger, *Macromolecules*, **24**, 5834. Copyright (1991) American Chemical Society

describes the reorientation of the unit vector during the time t, and P_2 is the second Legendre polynomial. The brackets represent an ensemble average.

$J(\omega)$ is the spectral density defined by:

$$J(\omega) = \frac{1}{2} \int_{-\infty}^{+\infty} G(t)\exp(i\omega t)\, dt \tag{9A.5}$$

The mean correlation time, or effective correlation time, τ_c, is expressed as:

$$\tau_c = \int_0^{+\infty} \langle P_2[\overrightarrow{u_n}(0).\overrightarrow{u_n}(t)]\rangle\, dt = \int_0^{+\infty} \langle 3\cos^2\theta(t) - 1\rangle/2\, dt \tag{9A.6}$$

In spite of the complexity and variety of chain motions, the above results show that, at least in the simple case of alkane chains, the chain dynamics corresponds to a damped diffusion of bond orientation along the chemical sequence. These basic features are included in the following motional models.

Starting from the description of a chain constrained to a tetrahedral lattice and undergoing three- and four-bond motions (Figure 9A.6), Valeur *et al.* derived the following VJGM analytical expression [7]:

$$G_{\text{VJGM}}(t) = \exp(-t/\theta)\exp(t/\rho)\text{erfc}(t/\rho)^{1/2} \tag{9A.7}$$

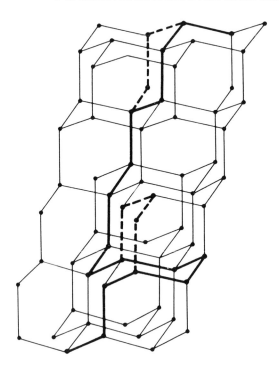

Figure 9A.6 Three-bond motion (top) and four-bond motion (bottom) on the tetrahedral lattice [7]. Reproduced by permission of John Wiley & Sons from Ref. 7

where erfc is the complementary error function, ρ is the characteristic time for three-bond motion and θ describes the damping of the orientation propagation along the chain or the orientation loss processes. However, such a function has an infinite slope at $t = 0$, which does not correspond to any physical reality.

By considering only three-bond motions and assuming that the effects of lattice-chain directional effects on the rearrangement kinetics are cut-off sharply rather than gradually as a function of distance from the central link of the crankshaft, Jones and Stockmayer have obtained the following JS expression [8]:

$$G_{JS}(t) = \sum_{k=1}^{s} G_k \exp(-t/\tau_k) \tag{9A.8}$$

with $s = (m+1)/2$, $\tau_k = 1/(4 \sin^2[k\pi/2(m+1)]\omega_1)$.

ω_1 is the probability per unit time that any particular three-bond *gauche* segment undergoes the rearrangement and $2m - 1$ is the number of bonds over which the kinetic coupling occurs. The G_k coefficients are given explicitly in [8]. $G_{JS}(t)$ is numerically similar to $G_{VJGM}(t)$.

More recently, Hall and Helfand [9] have derived a model for conformational dynamics based on the conformational correlations observed in their simulations and they have proposed the use of the same expression for the orientations. The HH orientation autocorrelation function is expressed as:

$$G_{HH}(t) = \exp(-t/\tau_1)\exp(-t/\tau_2)I_0(t/\tau_1) \qquad (9A.9)$$

where I_0 is the zero order modified Bessel function, τ_1 is the characteristic time responsible for the diffusion of bond orientation along the chain and τ_2 is the damping term corresponding to the orientation loss processes.

The spectral density is given by:

$$J_{HH}(\omega) = Re[(a + i\beta)^{-1/2}] \qquad (9A.10)$$

with $a = \tau_2^{-2} + 2\tau_1^{-1}\tau_2^{-1} - \omega^2$ and $\beta = -2\omega(\tau_1^{-1} + \tau_2^{-1})$.

The damping process consists either of nonpropagative specific motions or of distortions of the chain with respect to its most stable conformations. However, it must be pointed out that the molecular origin of the damping is still not well identified. Molecular processes such as specific isolated jumps, fluctuations of internal rotation angles, and chemical defects (head-to-head linking, stereochemical sequences, side-groups ...) are expected to affect the propagation of conformational changes along the chain sequence leading to a damping effect.

A more complete expression for the orientation autocorrelation function has been established by Dejean et al. [10]. The DLM orientation autocorrelation function describes the segmental motions as proposed by Hall and Helfand. It also takes into account independent librations of the internuclear vectors represented, as suggested by Howarth [11] by a random anisotropic fast reorientation of the CH vector within a cone of half-angle Θ. The resulting orientation autocorrelation function can be written as:

$$G_{DLM}(t) = (1 - a)\exp(-t/\tau_2)\exp(-t/\tau_1)I_0(t/\tau_1)$$
$$+ a\exp(-t/\tau_0)\exp(-t/\tau_2)\exp(-t/\tau_1)I_0(t/\tau_1) \qquad (9A.11)$$

where τ_1 and τ_2 are the correlation times defined in equation (9A.9) and τ_0 is the characteristic time of the libration. a is related to the half-angle Θ of the libration cone through the relation:

$$(1 - a) = [(\cos\Theta - \cos^3\Theta)/2(1 - \cos\Theta)]^2 \qquad (9A.12)$$

Assuming that τ_0 is much shorter than τ_1 and τ_2 (a situation which is often encountered), the second term in equation (9A.11) can be simplified and $G(t)$ written as:

$$G_{DLM}(t) = (1 - a)\exp(-t/\tau_2)I_0(t/\tau_1) + a\exp(-t/\tau_0) \qquad (9A.13)$$

The corresponding spectral density is written as:

$$J_{DLM}(\omega) = (1 - a)J_{HH}(\omega) + aJ_0(\omega) \qquad (9A.14)$$

with: $J_0(\omega) = \tau_0/(1 + \omega^2\tau_0^2)$.

A.2.4   SIDE-GROUP MOTIONS

All the models reported above focused on the segmental motions of the polymer backbone. However, some groups behave as internal rotators. Examples of such rotators are methyl and phenyl side-groups, phenyl rings belonging to the main chain or alkyl side-chains.

For a jump between two equilibrium positions, $G(t)$ is written as [12]:

$$G(t) = (A + C) + Be^{-t/\tau_i} \qquad (9A.15)$$

whereas, for a jump between three equilibrium positions:

$$G(t) = A + (B + C)e^{-t/\tau_i} \qquad (9A.16)$$

and for a stochastic process:

$$G(t) = A + Be^{-t/\tau_i} + Ce^{-4t/\tau_i} \qquad (9A.17)$$

where:

$$A = (3\cos^2\alpha - 1)^2/4, \quad B = 3\sin^2(2\alpha)/4, \quad C = 3\sin^4\alpha/4 \qquad (9A.18)$$

α is the angle between the rotation axis and the internuclear vector $C - H$ and τ_i is the internal correlation time.

For alkyl side-chains a model of multiple internal rotations about the successive carbon–carbon bonds has been developed [13]. The corresponding spectral density function is:

$$J_i(\omega) = \sum_{a,b,\dots,i} B_{ab}B_{bc} \dots B_{hi}B_{io} \frac{\tau_i}{1 + \omega^2\tau_i^2} \qquad (9A.19)$$

with:

$$\tau_i = (6D_0 + a^2D_a + b^2D_b + \dots + i^2D_i)^{-1} \qquad (9A.20)$$

The B matrix elements are geometric factors describing the alkyl chain [13], D_0 is the overall isotropic rotational diffusion constant for the molecule and $D_a, D_b, \dots, D_i$ are the rotational diffusion constants about the successive carbon-carbon bonds of the alkyl chain.

Fast internal motions can also be described in terms of a generalized order parameter, S, which is a measure of the spatial restriction of the motion, and an effective correlation time, τ_i, which is a measure of the rate of the motion [14]. The autocorrelation function is given by:

$$G_i(t) = S^2 + (1 - S^2)e^{-t/\tau_i} \qquad (9A.21)$$

For groups involved in independent motions of both the chain and side-groups, the resulting orientation autocorrelation function is the product of the corresponding individual orientation autocorrelation functions.

A.2.5 EXPERIMENTAL EVIDENCE OF THE SHAPE OF THE ORIENTATION AUTOCORRELATION FUNCTION FROM FLUORESCENCE ANISOTROPY DECAY STUDIES

Among the various spectroscopic techniques, only the fluorescence anisotropy decay provides a quasi-continuous sampling of the second moment orientation autocorrelation function of a vector which is, in this case, the transition moment of a fluorescent group covalently bound to the chain. From this point of view, the fluorescence anisotropy decay technique is a powerful tool for a critical discussion of the various motional models that have been proposed.

The absorption and emission of a fluorescent molecule can be described by considering the transition moment of the molecule, the direction of which is determined with respect to the geometry of the molecule. Under the action of a suitably polarized electromagnetic field, the absorption of light is proportional to the scalar product of the transition moment and the incident electric field. In the same way, the emission of light is proportional to the scalar product of the transition moment and the direction of the analyzer. Thus, the excitation of an isotropic population of fluorescent species by polarized light creates a temporary anisotropic population of excited molecules. Molecular motions progressively destroy this anisotropy and affect the polarization of the re-emitted light. At a time t after the excitation, the fluorescence anisotropy, $r(t)$, is defined as:

$$r(t) = [I_v(t) - I_h(t)]/[I_v(t) + 2I_h(t)] \qquad (9A.22)$$

where I_v and I_h are the fluorescence intensities for analyzer direction parallel and perpendicular, respectively, to the polarization of the incident beam. It can be shown that the evolution of $r(t)$ is directly proportional to the second moment of the orientation autocorrelation function:

$$r(t) \propto G(t) \qquad (9A.23)$$

The time interval during which $r(t)$ can be recorded depends on the fluorescence lifetime (10^{-10}–10^{-8} s). Experimentally, $G(t)$ can be determined up to approximately 100 ns.

Using such a technique to study polymer dynamics implies that the motion of the fluorescent label reflects the motion of the macromolecular chain. In the following example, labeling has been performed in the middle of the polystyrene chain using an anthracene derivative in such a way that the transition moment of this molecule lies along the local axis of the chain and cannot be involved in motional processes independent of those of the chain:

Experiments performed on this polymer using synchrotron radiation as the light source showed that it is indeed possible to differentiate the fits obtained from the various orientation autocorrelation functions [15]. For example, the VJGM function leads to a correct agreement at long times whereas, at short times, important differences with the experimental data are observed. The best fit is obtained with the HH orientation autocorrelation function. The presence of eventual librations cannot be detected by these experiments.

Similar studies were realized on bulk polymers at temperatures above the glass transition temperature [16]. Again, the HH orientation autocorrelation function was found to adequately represent the data. For polybutadiene in the bulk and in solution, the best fit was obtained with the same τ_2/τ_1 ratio. These experiments demonstrate that the motional models based on a damped diffusion of bond orientation lead to a satisfying description of the orientation autocorrelation function obtained from the fluorescence anisotropy decay technique.

A.3   Interpretation of NMR Experiments in Terms of Local Chain Dynamics. Test of Chain Motional Models

As indicated in the Introduction, this chapter focuses on the local dynamics of polymers, namely on the fastest motions of the polymer chain. In the following, we will concentrate on those NMR parameters that are sensitive to high frequency motions, that is, the spin–lattice relaxation times and the nuclear Overhauser effect for unlike spins.

In the case of two different groups of spins denoted A and X, the spin–lattice relaxation obeys the following differential equations [17]:

$$dM_A/dt = -(M_A - M_A^0)/T_1^{AA} - (M_X - M_X^0)/T_1^{AX} \qquad (9A.24)$$

$$dM_X/dt = -(M_A - M_A^0)/T_1^{XA} - (M_X - M_X^0)/T_1^{XX} \qquad (9A.25)$$

where M_A and M_X are the longitudinal magnetizations at time t of the spins A and X, respectively, and M_A^0 and M_X^0 are the longitudinal magnetizations at thermal equilibrium. Two characteristic time constants are necessary to describe the evolution of each group of spins. The complete description of relaxation in a multispin system requires a quantum mechanical density matrix formalism that is out of the scope of this presentation [18] (see also Chapter 2).

The coupled spin relaxation was studied in a few polymers using either proton [19–21] or ^{13}C [22–24] NMR. However, most experiments encountered in the polymer field deal with one of the two following situations:

- in ^{13}C experiments, the ^{13}C relaxation evolves under continuous proton decoupling conditions
- in ^1H experiments, the cross-relaxation between the different protons is negligible.

Under the above conditions, the equation governing the evolution of spin A simplifies to:

$$dM_A/dt = -(M_A - M_A^0)/T_1^{AA} \qquad (9A.26)$$

and the spin–lattice relaxation of A is described in terms of a unique relaxation time. It corresponds to an exponential relaxation of the spins A, which is observed experimentally.

The interpretation of the NMR experiments requires that the relaxation mechanism is well identified. The dipole–dipole interactions generally provide a major contribution to the proton and ^{13}C relaxations. This is of particular interest for the $^{13}CH_n$ groups where the dominant dipolar interaction is between the ^{13}C nucleus and the directly bonded protons. In this case, the dipolar relaxation of the ^{13}C spin is sensitive to the reorientation of the C–H bonds. The dipolar relaxation of the protons is ruled by $^1H–^1H$ interactions. Only the closest protons to the observed nucleus have to be taken into account. Depending on the chemical structure, one or several $^1H–^1H$ vectors will have to be considered.

It must be kept in mind that the chemical shift anisotropy, which is significant for sp^2 and sp carbons, may produce an additional relaxation mechanism. It is negligible at low and moderate magnetic field strengths. At high field strengths (11.74 T) it has been observed to contribute to up to 20% of the relaxation of protonated carbons [25].

A.3.1 EXPRESSIONS OF T_1 AND NOE

A.3.1.1 ^{13}C spin–lattice relaxation under continuous proton decoupling conditions

With the assumption of a purely $^{13}C–^1H$ dipolar relaxation mechanism, the spin-lattice relaxation time T_1 under ^1H decoupling conditions, and nuclear Overhauser enhancement (NOE) obtained from a ^{13}C experiment are given by the following expressions [26]:

$$(nT_1)^{-1} = (\hbar^2 \gamma_C^2 \gamma_H^2 / 10 r_{CH}^6)(J(\omega_H - \omega_C) + 3J(\omega_C) + 6J(\omega_H + \omega_C)) \qquad (9A.27)$$

$$NOE = 1 + \frac{\gamma_H}{\gamma_C} \frac{6J(\omega_H + \omega_C) - J(\omega_H - \omega_C)}{J(\omega_H - \omega_C) + 3J(\omega_C) + 6J(\omega_H + \omega_C)} \qquad (9A.28)$$

where n is the number of protons directly bound to the considered carbon, ω_H and ω_C are the ^1H and ^{13}C resonance frequencies and r_{CH} is the internuclear distance.

A.3.1.2 ^1H spin–lattice relaxation in the absence of cross-relaxation

If the cross-relaxation between the different protons is negligible, the spin–lattice relaxation time is expressed as:

$$(T_1)^{-1} = (3\hbar^2\gamma_H^4/10r_{HH}^6)[J(\omega_H) + 4J(2\omega_H)] \tag{9A.29}$$

where r_{HH} is the effective proton–proton internuclear distance.

A.3.2 TEST OF THE ORIENTATION AUTOCORRELATION FUNCTIONS BY SPIN–LATTICE RELAXATION EXPERIMENTS

Under extreme narrowing conditions, the mean correlation time or effective correlation time, τ_c (equation (9A.6)), can be calculated from the spin–lattice relaxation times without reference to any motional model. In contrast, when extreme narrowing conditions are not satisfied, the spin–lattice relaxation times are highly dependent on the nature of the motions in which the polymer units are involved and the NMR data can be used as a test for the orientation autocorrelation functions. In order to discriminate between the different motional models, extensive sets of data obtained at different experimental frequencies and temperatures are required.

Examples of the temperature dependences of nT_1 at three experimental frequencies and nuclear Overhauser enhancement are shown in Figures 9A.7 and 9A.8 for the methylene carbons (d) of polyisoprene in toluene solution [25]:

Apart from the high temperature region, the ^{13}C spin–lattice relaxation times are frequency dependent and the NOE values are significantly lower than their maximum value, indicating that extreme narrowing conditions are not satisfied. The experimental nT_1 minima have a very characteristic behaviour. They are much higher than the nT_1 minima calculated using either an isotropic diffusion model or the HH orientation autocorrelation function. For example, at 25 MHz, the nT_1 minimum of the methylene carbon (d) of polyisoprene is equal to 0.063 s, whereas the calculated values from the isotropic and HH models are 0.040 s and 0.050 s, respectively. The values of the nT_1 minima appear to be a very stringent test for the orientation autocorrelation function.

The variations of nT_1 and NOE as a function of reciprocal temperature, $1/T$, for the CH and CH_2 (c) carbons of bulk polyisoprene at temperatures well above the glass transition temperature, are depicted in Figures 9A.9 and 9A.10 [27]. As for the solution data, the most striking point is the fact that the nT_1 values at the minimum are significantly higher than those predicted by both isotropic and HH orientation autocorrelation functions. It is also important to

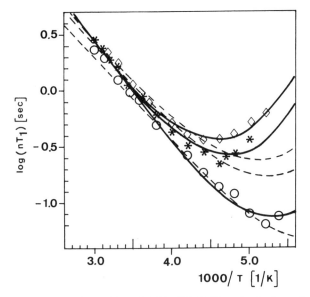

Figure 9A.7 125.8 ($\diamond$), 90.6 (*) and 25.2 ($\bigcirc$) MHz spin–lattice relaxation times, nT_1, of the methylene carbon (d) of polyisoprene in toluene-d_8 solution. Best fit of the Cole–Cole distribution (- - -) and of the biexponential autocorrelation function (———). Reprinted with permission from Gisser *et al.*, *Macromolecules*, **24**, 4270. Copyright (1991) American Chemical Society

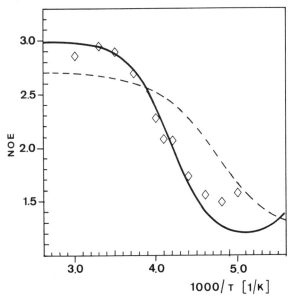

Figure 9A.8 Temperature dependence of the nuclear Overhauser enhancement, NOE ($\diamond$), at 125.8 MHz for the methylene carbon (d) of polyisoprene in toluene-d_8 solution. The lines are predicted by the best fit to T_1 data of the Cole–Cole distribution (- - -) and of the biexponential autocorrelation function (———). Reprinted with permission from Gisser *et al.*, *Macromolecules*, **24**, 4270. Copyright (1991) American Chemical Society

notice that the nT_1 rule does not apply to the data reported in Figures 9A.9 and 9A.10. The mean values of the T_{1CH}/T_{1CH2} ratios for the methylene carbon (c) are 1.49 at 62.5 MHz and 1.67 at 25.15 MHz. For the methylene carbon (d) of polyisoprene, these ratios are 1.35 at 62.5 MHz and 1.4 at 25.15 MHz.

The nT_1 values at the nT_1 minimum are listed in Table 9A.1 for a number of amorphous polymers in the bulk state at temperatures well above the glass transition temperature. For all the polymers considered, at the minimum, nT_1 is always higher than the values that are derived from the isotropic and HH models [10, 27–29]. The same conclusions apply for the VJGM and JS models. It must be noticed that these large discrepancies cannot be accounted for by internuclear distance imprecisions.

To reproduce the large deviations between calculated and experimental nT_1 values at the minimum, two different assumptions can be made. The first assumption is the existence, inside the polymer chain, of an additional motion

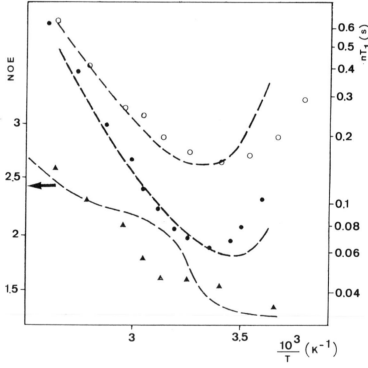

Figure 9A.9 Temperature dependence of the 62.5 ($\bigcirc$) and 25.2 ($\bullet$) MHz ^{13}C spin–lattice relaxation times, nT_1, and nuclear Overhauser enhancement, NOE ($\blacktriangle$) at 62.5 MHz, for the methine carbon (b) of *cis*-1,4-polyisoprene in the bulk state. Best fit of the DLM autocorrelation function (- - -) [27]. Reprinted with permission from Dejean de la Batie *et al.*, *Macromolecules*, **22**, 122. Copyright (1989) American Chemical Society

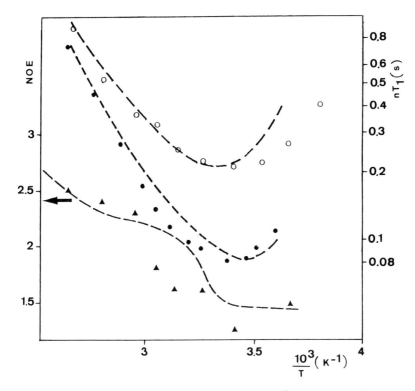

Figure 9A.10 Temperature dependence of the 62.5 (○) and 25.2 (●) MHz ^{13}C spin–lattice relaxation times, nT_1, and nuclear Overhauser enhancement, NOE (▲) at 62.5 MHz, for the methylene carbon (c) of *cis*-1,4-polyisoprene in the bulk state. Best fit of the DLM autocorrelation function (- - -) [27]. Reprinted with permission from Dejean de la Batie *et al.*, *Macromolecules*, **22**, 122. Copyright (1989) American Chemical Society

that is not taken into account in the previous models and that contributes to a partial reorientation of the CH vectors with a characteristic time that differs from the correlation time for orientational diffusion along the chain. The alternative explanation is that the chain motions have to be described by a distribution of correlation times.

If one considers the existence of an additional motion, the assignment of this additional motion to a process slower than the orientation diffusion along the chain associated with τ_1 must be eliminated. Indeed, results derived from fluorescence anisotropy decay in solution [15] and in the bulk [16] and NMR [30–32] on bulk polybutadiene have shown that such slow modes contribute only to a very small extent to the decay of the orientation autocorrelation function in this temperature range. Therefore, this additional motion has to be a faster mode and thus more local than the orientation diffusion process along the chain. A similar fast process has recently been observed by neutron scattering

Table 9A.1 Experimental nT_1 (second) values at the minimum in bulk polymers [10, 27–29]

| | Saturated CH | | CH_2 | |
	25.15 MHz	62.5 MHz	25.15 MHz	62.5 MHz
Poly(vinylmethyl ether)	0.070	0.177	0.070	0.177
Poly(propylene oxide)	0.068	0.168	0.096	0.250
Polyisobutylene			0.060	0.18

| | Unsaturated CH | | CH_2 | |
	25.15 MHz	62.5 MHz	25.15 MHz	62.5 MHz
cis-1,4-polyisoprene	0.064	0.154	0.080	0.21
			0.095	0.22
cis-1,4-polybutadiene	0.072	0.166	0.095	0.24

experiments in bulk polybutadiene [33]. It can be assigned to molecular librations, of limited extent, of the CH vectors about their equilibrium conformation and corresponds to oscillations inside a potential well.

For bulk polyisoprene, a fit of nT_1 values at the nT_1 minimum by the DLM orientation autocorrelation function, which is precisely based on segmental motions and independent librations, yields $a = 0.17$, 0.40 and 0.48, for the CH, CH_2 (c) and CH_2 (d), respectively, corresponding to $\Theta = 20°$, 33° and 37°. In terms of τ_1 and τ_2 correlation times, the fit has to be achieved simultaneously for the three carbons belonging to the CH and CH_2 groups. With the above a values and a τ_2/τ_1 ratio of the order of 40, the best fits obtained from equation (9A.12) are shown in Figures 9A.9 and 9A.10 for the same polymer [27]. As can be concluded from these figures, the overall agreement between the theoretical prediction and experimental data is very good. Deviations occur only in the low temperature range, where the measurements are less precise due to the broadening of the lines.

More generally, the DLM orientation autocorrelation function has led to satisfying fits of relaxation data for polymers in both the solution and bulk states at temperatures well above the glass transition temperature [10, 27–29]. The ability of the DLM function to describe the segmental motions in these two situations is a clear indication that the very nature of the motions involved in the magnetic relaxation at a given frequency is identical in the bulk at temperatures well above Tg and in solution.

As pointed out above, an alternative way to interpret the relaxation data in these systems is to use distributions of correlation times for the chain motions. As compared to the isotropic model, distributions predict broader and raised T_1 minima. In this regard, a systematic investigation was performed by Gisser *et al.* [25]. Representative distributions of correlation times calculated for the methylene carbon (d) of polyisoprene in toluene solution are drawn in Figure 9A.11. They are the unimodal symmetric Cole–Cole distribution and the bimodal distribution corresponding to a biexponential function. As a comparison, the distributions associated with the HH and DLM orientation autocorrelation functions are also plotted in Figure 9A.11. The best fit of the Cole–Cole distribution to spin–lattice relaxation data from the methylene carbon (d) of polyisoprene is shown in Figures 9A.7 and 9A.8. The height and shape of the T_1 minima cannot be represented quantitatively. More generally, all the unimodal distributions considered in ref. [25], either symmetric (Cole–Cole and Fuoss–Kirkwood distributions) or asymmetric (log χ^2 model), fail to reproduce these data. Besides, as shown in Table 9A.1, for all polymers except poly(vinylmethyl ether), the nT_1 values at a given experimental frequency are different for adjacent CH and CH_2 carbons of the polymer main chain. Therefore, a different unimodal distribution for the segmental correlation times would be required for each of these adjacent sites, which is not consistent with the concept of cooperative motions involving a few monomer bonds. As expected, the fit using the distribution associated with the HH orientation autocorrelation function is also unsatisfactory. Finally, only the distributions corresponding to the biexponential function with two very

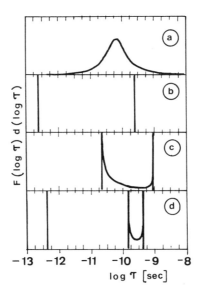

Figure 9A.11 Distribution of motional time constants calculated for methylene carbon (d) of polyisoprene at 250 K in toluene-d_8 solution; (a) Cole–Cole distribution; (b) biexponential; (c) HH model; (d) DLM model [25]. Reprinted with permission from Gisser *et al.*, *Macromolecules*, **24**, 4270. Copyright (1991) American Chemical Society

different correlation times (Figures 9A.7 and 9A.8), and to the DLM orientation autocorrelation function produce adequate fits.

The general conclusion of all the above tests of the orientation autocorrelation functions is that spin–lattice relaxation detects motions occurring on two well-separated timescales [25]. The slower class of motions corresponds to activated transitions, whose characteristics closely resemble those of conformational changes about backbone bonds. In agreement with results obtained from Brownian dynamics simulations, as well as from other experimental techniques, the faster class of motions is assigned to librations of the C–H bonds within potential energy wells. Both the biexponential and DLM orientation autocorrelation functions contain the features that are necessary to account for these two distinct components of the local chain motions.

A.4 Physical Interpretation of NMR Results. Influence of the Chemical Structure and Experimental Conditions on the Local Dynamics

A.4.1 INVESTIGATION OF THE LOCAL DYNAMICS IN DILUTE SOLUTIONS

A.4.1.1 Influence of the molecular weight

Figure 9A.12 shows the variation of the ^{13}C spin-lattice relaxation time of the backbone carbons of polystyrene as a function of molecular weight, M_w [34]. Above a molecular weight of about 10 000, corresponding to approximately 100 monomer units, T_1 is independent of the chain length. For M_w less than 10 000, T_1 increases as the molecular weight decreases. As, in this system, T_1 is an increasing function of the motional frequency, this result reflects a higher mobility of the low molecular weight compounds, consistent with the increasing contribution of the overall tumbling of the molecule to the relaxation.

A similar behaviour was observed for all the random coil polymers investigated ([35] and references cited therein). The striking result of these experiments is the relatively low molecular weight, corresponding to a critical size of 50 to 200 monomer units, above which the spin–lattice relaxation is size-independent. This observation demonstrates the local nature of the motions responsible for the spin–lattice relaxation in polymer chains. Above this critical size, the overall tumbling is much slower than the local motions and, therefore, its contribution to the relaxation is negligible.

On the contrary, polymers that have a well-defined rigid structure in solution show a size dependence of the spin–lattice relaxation up to higher molecular weights. For example, in poly(γ-benzyl-L-glutamate), T_1 varies with molecular weight up to a polymerization degree of at least 600 [36]. In this case, the local mobility is very restricted and the overall tumbling of the molecule plays a major role in the relaxation even for a relatively large chain size.

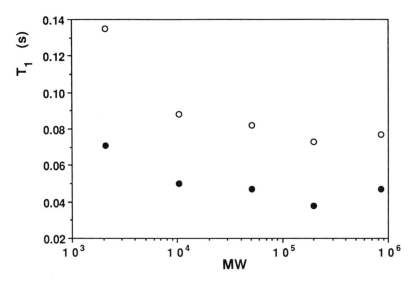

Figure 9A.12 Molecular weight dependence of the T_1 of the main-chain CH_2 carbons (●) and CH carbons (○) of polystyrene in C_2Cl_4 solution at 44 °C. Reproduced by permission of the American Institute of Physics from Ref. 34

A.4.1.2 Influence of the solvent

The friction forces encountered by the polymer atoms moving in a solvent are expected to strongly influence the dynamics. Indeed, early experiments have demonstrated that the relaxation data are solvent dependent and the correlation times follow an Arrhenius law [37]. To rationalize the role of the solvent on the local motions, Kramers' theory [3] was applied to conformational transitions in polymers. This theory is based on the passage of a particle over an energy barrier, E_a. The solvent is treated as a random frictional force opposing passage across the barrier. Any spatial or temporal correlations in the solvent are ignored. In the high friction limit, the rate constant, P, is expressed as:

$$P = k/\zeta \exp(-E_a/RT) \tag{9A.30}$$

where ζ is the friction coefficient of the solvent and k is a constant.

The friction coefficient is generally taken as proportional to the zero shear viscosity, η, of the solvent. Assuming that the segmental correlation time, τ_c, determined from NMR, is proportional to the reciprocal of the rate constant, it follows that:

$$\tau_c = A\eta \exp(E_a/RT) \tag{9A.31}$$

where A is a constant.

The temperature dependence of the solvent viscosity can be approximated by an Arrhenius law in a limited temperature range:

$$\eta = \eta_0 \exp(E_\eta/RT) \tag{9A.32}$$

which leads to:

$$\tau_c = B \exp[(E_a + E_\eta)/RT] \tag{9A.33}$$

where B is a constant independent of temperature and viscosity.

The apparent activation energy, E_{exp}, calculated from the temperature variation of τ_c, is therefore predicted to be the sum of the barrier height for conformational transitions and the viscous flow contribution:

$$E_{exp} = E_a + E_\eta \tag{9A.34}$$

There are only a few systematic studies performed over a wide range of solvent viscosities. Recently, the ^{13}C spin–lattice relaxation times were measured for polyisoprene in ten solvents [38]. Figure 9A.13 shows the viscosity dependence of the effective correlation time τ_c (equation (9A.6)) for the methylene carbon (d). As indicated in Figure 9A.13, τ_c is not proportional to η. It shows a η^α variation, with $\alpha = 0.41$, in disagreement with Kramers'

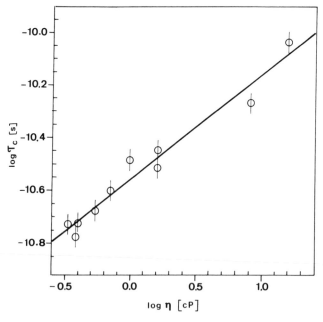

Figure 9A.13 Viscosity dependence of the apparent correlation time, τ_c, for methylene carbon (d) of polyisoprene in 10 wt% solution at 333K. The full line is least-squares fit to the data with the slope 0.41 [38]. Reprinted with permission from Glowinkowski et al., Macromolecules, **23**, 3520. Copyright (1990) American Chemical Society

theory. The activation energy is therefore expressed as:

$$E_{exp} = E_a + \alpha E_\eta \qquad (9A.35)$$

The activation energies, E_a, calculated using this expression, are independent of the solvent viscosity, as expected for an intramolecular potential barrier. These results were interpreted in terms of a generalized Kramers' theory [39] by allowing the solvent friction to be frequency dependent. The breakdown of the original Kramers' theory is likely due to the fact that the timescales for the polymer and solvent motions are not clearly separated [38]. Indeed, at high viscosities, the correlation times for both motions are very similar. Moreover, this investigation suggests that the comparison of activation energies requires an extensive study of the solvent dependence of the relaxation.

A.4.1.3 Influence of the chain tacticity and regularity

The influence of the microstructure on the chain mobility was detected for some stereoregular polymers. For example, the proton relaxation of the syndiotactic and isotactic poly(methyl methacrylate) chains [20, 21] indicated that the mobility is slightly higher for the isotactic chain than for the syndiotactic one. The same results were obtained for polypropylene and polybut-1-ene [40] whereas the reverse tendency was noticed for polystyrene [41]. In atactic polypropylene, differences between the mobility of the stereosequences were also observed [40]. This observation is a clear indication of the local character of the dynamics observed from T_1 measurements.

The role of the organization of the different groups in a chain on its dynamic behaviour was illustrated by the study of head-to-tail and head-to-head polystyrenes [42]. The ^{13}C spin–lattice relaxation times were analyzed in terms of a segmental motion using the VJGM orientation autocorrelation function and a twofold rotation of the phenyl group. Both compounds have the same correlation time, ρ, which describes the conformational jumps. On the contrary, the orientation propagation along the chain is damped much more quickly in the head-to-head polymer which also has a reduced internal mobility of the side-rings, owing to the location of these groups on vicinal carbons.

A.4.1.4 Influence of the chemical structure of the polymer on the segmental motions

Despite the large number of NMR relaxation studies on polymers in solution, the wide differences in experimental conditions and the analysis in terms of different motional models preclude an easy interpretation of the results. However, several ^{13}C relaxation data have been compared in ref. [35] using the isotropic rotational diffusion model to calculate an apparent correlation time for the motion of the chain backbone. Some significant results are reported in Table 9A.2.

Table 9A.2 ^{13}C spin-lattice relaxation times, nT_1, and apparent correlation time, τ_c, for main-chain carbons of several polymers in solution, at 25.14 MHz except as indicated [35]

Polymer	Solvent	Temperature (°C)	Concentration (wt %)	nT_1 (s)	τ_c (ns)
Polyoxymethylene	HFIP^a	30	3	0.60	0.082
Polyoxyethylene	C_6D_6	30	5	2.80	0.018
Polyethylene	ODCB^b	100	25	2.70	0.018
	ODCB	30^c	25	1.24	0.04
Poly-1-butene	CCl_3CHCl_2	100	10	0.34^d	0.14
Polystyrene	Toluene-d$_8$	30	15	0.10	0.49
cis-1,4-Polybutadiene	CDCl$_3$	54	20	3.00	0.016
trans-1,4-Polybutadiene	CDCl$_3$	54	20	2.38	0.021

^{a} Hexafluoro-2-propanol.
^{b} o-dichlorobenzene.
^{c} Extrapolated from high-temperature data with $E_a = 10.5$ kJ/mol.
^{d} At 22.6 MHz.

The polymers showing the highest segmental mobility are unsubstituted chains containing heteroatoms, like polyoxyethylene. This can be explained by the lower rotational barrier at heteroatoms. However, polyoxymethylene seems less mobile than polyoxyethylene. To interpret this result the conformations of both chains were investigated by Monte Carlo simulations which established that the preferential *gauche* conformations found in polyoxymethylene lead to a low probability of cooperative transitions and hence a restricted mobility [43]. This interpretation is in agreement with results obtained from recent coupled-spin relaxation studies of local dynamics in polyoxymethylene [23]. As observed by NMR experiments on polybutadiene, polydienes are also very flexible polymers [44]. Mobility associated with a *cis* double-bond is higher than with a *trans* one [44].

The presence of side-chains decreases the mobility of the backbone in the sequence polyethylene, polypropylene, polybut-1-ene, polystyrene, that is, the larger the substituent, the higher the motional restriction. In *ortho-*, *meta-* and *para*-substituted polystyrenes, ^{13}C spin–lattice relaxation studies permitted the investigation of the role of a substituent on both the chain and ring motions [45]. The *ortho* substituents decrease the mobility of the chain backbone and this effect is correlated with the substituent size: motions are slower in the bromo derivatives than in the chloro and fluoro ones. *Para* and *meta* substitutions only slightly modify the chain mobility.

Another example dealing with the role of substituents on the local motions is the series of polycarbonates represented in Figure 9A.14 [46–49]. Proton and ^{13}C spin-lattice relaxation times were measured at two field strengths and interpreted in terms of segmental motion using the HH orientation autocorrelation function, phenyl ring rotation and methyl rotation. The spin–lattice

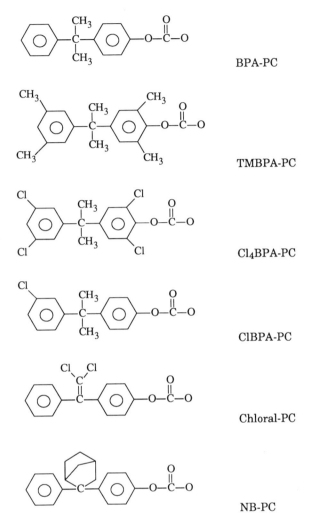

Figure 9A.14 Repeat units of polycarbonate derivatives and abbreviations. Reproduced by permission of John Wiley & Sons, Inc., from Ref. 49

relaxation times of the aromatic protons were used to characterize the segmental motion whereas the correlation times of the phenyl and methyl group rotation were determined from the spin-lattice relaxation times of the protonated aromatic and methyl carbons, respectively.

Results thus obtained show that the rate of the segmental motion depends on the number of phenyl ring substituents. The more substituted the polycarbonate, the slower the segmental motion. As indicated in Table 9A.3, the activation energies for the correlation times of the segmental motion increase

Table 9A.3 The glass transition temperatures, Tg, and activation energies, E_{exp}, for segmental motions (τ_1) measured for different polycarbonate derivatives in 10 wt % $C_2D_2Cl_4$ solutions [46–49]

Polymer	Tg (°C)	Activation energy, E_{exp}, for segmental motions (kJ/mol)
BPA-PC	150	16
Chloral-PC	164	18
ClBPA-PC	146	20
TMBPA-PC	203	34
Cl$_4$BPA-PC	225	34
NB-PC	232	30

with increasing substitution. The evolution of the activation energies parallels the variation of the glass transition temperature, Tg, with the chemical structure of the polycarbonate. The only exception is ClBPA-PC. The asymmetry of this molecule may hinder good packing of the polymer in the glassy state, thus leading to a decrease in the strength of the intermolecular interactions and a decrease in the Tg.

Backbone methyl group rotation varies very slightly in the different polycarbonates. On the contrary, the phenyl ring dynamics are strongly influenced by the molecular structure. Whereas internal rotation is observed for unsubstituted species, ring libration occurs in tetrasubstituted polymers. The behaviour of the monosubstituted polycarbonate depends on temperature. In this compound, both the unsubstituted and the substituted phenyl rings have restricted rotation below 40 °C, whereas they rotate freely above 40 °C. The unsubstituted ring is only slightly more mobile than the substituted one, and their motions seem to be coupled.

It is interesting to note that, for all the examples given in this section and as far as the intramolecular interactions are the dominant interactions in both the solution and the bulk states, the variation of the spin–lattice relaxation time as a function of the chemical structure roughly follows the variation of the glass transition temperature.

In all the above examples, all the chain carbons shared the same segmental dynamics. This may not be the case in more complex polymer systems. The detailed local dynamics of two aryl-aliphatic polyesters, differing only by the length of the alkyl spacer between the terphenyl ester groups:

with $m = 2$ or $m = 8$, have been studied by proton and ^{13}C relaxation measurements at two resonance frequencies [50].

Let us first consider the dynamic behaviour of the aromatic part. The aromatic proton relaxation is governed by the dipolar interaction between pairs of protons attached to vicinal carbons. Since the corresponding H–H vectors are parallel to the long symmetry axis of the terphenyl group, the aromatic proton relaxation is not sensitive to the internal rotation of the rings about this axis. It is well-described by an isotropic reorientation of the terphenyl group as a whole, consistent with the fact that the large terphenyl groups act as anchors for the rapid short-range processes affecting the aliphatic part of the molecule. The spin–lattice relaxation times and nuclear Overhauser enhancements of the aromatic protonated carbons indicate that the C-H aromatic vectors are involved in both the isotropic reorientation of the terphenyl group and a stochastic rotation of each phenyl group about its long symmetry axis. The internal rotation of the central phenyl ring is somewhat faster than the internal rotation of the outer ones.

The segmental motion of the aliphatic sequence was investigated in terms of the VJGM model. In the case of the central carbons, both correlation times and associated energy barriers, E_a as defined by equation (9A.34), depend on the length of the aliphatic segment. A longer aliphatic sequence not only induces an increase in flexibility in terms of the ρ correlation time (see equation (9A.7)), but also leads to a notable decrease in the measured activation energy. These activation energies are equal to 23.4 and 10.9 kJ mol^{-1} for the polymers with $m = 2$ and $m = 8$, respectively. They are of the order of two barrier crossings for $m = 2$ and of one barrier crossing for $m = 8$. In terms of the Helfand classification of local motions, they correspond, in the first case, to type 1 motions, which leave the chain tails at the same position at the start and end of the transition and require two barrier crossings, and, in the second case, rather to type 2 motions which accommodate the deformation on a larger scale and need roughly one energy barrier. Indeed, due to the large inertial effects of the terphenyl groups acting like anchors, type 1 motions are mainly expected to occur in the shorter flexible sequences ($m = 2$) whereas the longer ($m = 8$) flexible sub-chains have a behaviour close to that of the polyethylene oxide chain.

A.4.1.5 Influence of the chemical structure of the polymer on the librations

When the T_1 minimum is determined experimentally, the relative contributions of the segmental motions and librations can be separated, leading to a more detailed insight into the chain dynamics in terms of the DLM orientation autocorrelation function. In all the examples investigated, the correlation time, τ_1, that describes the segmental chain motion, and the apparent correlation time discussed in the previous section, show parallel dependences on the chemical structure. In simple homopolymers, whose monomer unit contains only a limited number of chain carbons, results were interpreted using the same correlation times τ_1 for the different chain carbons, consistent with the fact that all the chain units share the same segmental dynamics [10, 27–29]. However, at the present time, it seems more difficult to relate the variation of the damping term τ_2 to specific chemical structure effects. A good understanding of these parameters clearly need more experimental data.

As regards the libration, and, as explained in Section A.3.2, the occurrence of fast librations of the C–H bonds has been invoked to account for the high value of the T_1 minimum in ^{13}C relaxation experiments [10, 27–29].

Table 9A.4 Libration amplitudes of several polymers in solution

Polymer	Libration amplitude and chemical formula		Solvent	Reference
Polyvinylmethylether	33° 33° —CH$_2$—CH— OCH$_3$		CDCl$_3$	10
cis-1,4-polyisoprene	26° 19° —CH$_2$—CH=C—CH$_2$— CH$_3$		C$_2$D$_2$Cl$_4$	27
cis-1,4-polybutadiene	35° 26° —CH$_2$—CH=CH—CH$_2$—		CDCl$_3$	27
Polyisobutylene	CH$_3$ 23° —CH$_2$—C— CH$_3$		CDCl$_3$	29
Polyhydroxybutyrate	26° 31° —O—CH—CH$_2$—C— CH$_3$ O		CDCl$_3$	51
Polyhydroxyvalerate[a]	21° 29° —O—CH—CH$_2$—C— C$_2$H$_5$ O		CDCl$_3$	52
Polyvinylchloride	33° 30° —CH$_2$—CH— Cl		C$_2$D$_2$Cl$_4$	53

[a]Measured in poly(hydroxybutyrate-co-hydroxyvalerate).

Differences in libration amplitude can also explain the fact that the nT_1 rule does not hold for adjacent chain carbons in several polymers [27]. Table 9A.4 shows the values of the libration angles determined from ^{13}C spin–lattice relaxation data using the DLM motional model [10, 27, 29, 51–53].

When data are available, librations always appear as very fast modes, more than 100 times faster than the segmental motions. As regards the methylene carbons, the libration amplitude is similar in polyvinylmethylether (PVME) [10], polyhydroxybutyrate (PHB) [51] and polyvinylchloride (PVC) [53]. A smaller amplitude is observed for polyisobutylene [29], which is consistent with the larger steric hindrance induced by the two neighboring substituents. For the methine carbons, the libration amplitude decreases in the sequence PVME, PVC, PHB, polyhydroxyvalerate [52], that is, the more bulky the substituent, the more restricted the libration. The methine group of the highly sterically hindered poly(1-naphthyl acrylate) [54] does not show any libration. Methine groups in double bonds show a small libration amplitude which is consistent with the higher rigidity of the double bond. Surprisingly, in the case of vinylchloride-vinylidenechloride copolymers, the methylene group has a smaller libration amplitude in the vinyl chloride units than in the vinylidene chloride ones, which may originate from specific dipolar intermolecular interactions [55].

A.4.1.6 Motions of alkyl side-chains

Figure 9A.15 shows the spin–lattice relaxation times of the side-chain carbons of poly ((R,S)–3-methyl–1-octene) [56]. Interestingly, data are field dependent, even for the highly mobile terminal methyl group. The T_1 analysis was performed using equation (9A.21) combined with different models for the motion of the chain backbone. Results thus obtained show that both the order parameter and the correlation time decrease towards the side-chain end, indicating an increase in the mobility of the flexible side-chains far from the polymer main chain.

$$\begin{array}{ll} & -\!\text{CH}\!-\!\text{CH}_2\!- \\ & \qquad | \\ 0.241\text{s} & \text{CH}\!-\!\text{CH}_3 \\ & \qquad | \\ 0.536\text{s} & \text{CH}_2 \\ & \qquad | \\ 1.101\text{s} & \text{CH}_2 \\ & \qquad | \\ 1.819\text{s} & \text{CH}_2 \\ & \qquad | \\ 3.630\text{s} & \text{CH}_2 \\ & \qquad | \\ 8.159\text{s} & \text{CH}_3 \end{array}$$

Figure 9A.15 Experimental nT_1 values of poly((R,S)-3-methyl-1-octene) at 75.4 MHz in CDCl$_3$ solution at 40 °C [56]. Reprinted with permission from Perico *et al.*, *Macromolecules*, **23**, 4912. Copyright (1990) American Chemical Society

A.4.2 INVESTIGATION OF THE LOCAL DYNAMICS IN CONCENTRATED SOLUTIONS

It was observed on several polymers that the rate of segmental chain motions is independent of the concentration in dilute solutions up to 10–15 monomer mol%. Above this concentration range, the T_1 variation with increasing concentration indicates a decrease in the chain mobility ([57] and references therein).

An interesting example is the study of 2,2-propane diyl-bis(4-hydroxyphenyl) polyformal:

$$-\langle\bigcirc\rangle-\underset{\underset{CH_3}{|}}{\overset{\overset{CH_3}{|}}{C}}-\langle\bigcirc\rangle-O-CH_2-O-$$

by multifield ^{13}C and proton spin–lattice relaxation times [58]. The polymer concentration in $C_2D_2Cl_4$ solution was varied from 5 to 100 wt %. The segmental motions were described in terms of a Hall–Helfand orientation autocorrelation function. The methyl group threefold rotation, the phenyl ring rotation, and the formal group rotation, characterized by correlation times τ_{irm},

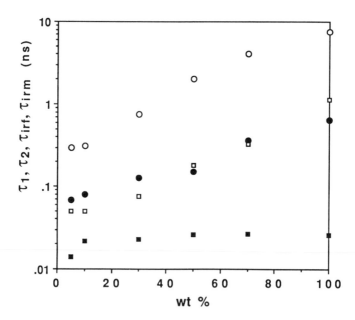

Figure 9A.16 Correlation times for segmental motions, τ_1 (●) and τ_2 (○), for formal group rotation, τ_{irf} (□), and for methyl group rotation, τ_{irm} (■) in 2,2-propane diyl-bis(4-hydroxyphenyl)polyformal in $C_2D_2Cl_4$ solutions at 130 °C, as a function of concentration. Reproduced from Ref. 58 by permission of the publishers Butterworth Heineman Ltd.

τ_{irp} and τ_{irf}, respectively, were also taken into account to represent the large set of T_1 data. Results obtained show that the segmental dynamics can be represented by the HH orientation autocorrelation function over the entire concentration range. The concentration dependence of the τ_1 and τ_2 correlation times is depicted in Figure 9A.16. It indicates a decrease in the segmental mobility with increasing concentration. The formal group rotation was described by double *trans-gauche* rotations about the C–O axes. As shown in Figure 9A.16, its rate is also a decreasing function of the concentration. In solutions up to 30 wt %, the internal motion of the phenyl ring is well represented by a stochastic diffusion. At higher concentrations, T_1 measurements are consistent with a restricted ring rotation, the rate and amplitude of which decrease with increasing concentration. On the contrary, no variation in the rate of the methyl group rotation was observed over the whole concentration range. Similar conclusions were drawn for the methyl group rotation in bisphenol-A-polycarbonate over the concentration range from 5 to 30 wt % [46]. It must be noticed that this detailed interpretation relies on the correlation time analysis and not on the T_1 data (note for instance that the T_1 of the methyl carbons decreases from 5 to 100 wt %).

The above study indicates that a motion involving a few atoms inside a small volume, such as the methyl group rotation, is not, or only slightly, modified by concentration effects. On the contrary, motions in which a group of atoms sweeps out a large volume, like the segmental motions or phenyl ring rotation, are restricted and/or slowed down.

A.4.3 INVESTIGATION OF THE LOCAL DYNAMICS IN THE BULK STATE AT TEMPERATURES WELL ABOVE THE GLASS TRANSITION TEMPERATURE

The polymers listed in Table 9A.5 were investigated in the bulk state at temperatures well above the glass transition temperature. The DLM orientation autocorrelation function gave good fits of the experimental data [10, 27–29]. As regards the libration motion, results obtained in the bulk are in complete agreement with the conclusions reached in solution. In the bulk state as well as in solution, the libration is a very fast process, the amplitude of which is

Table 9A.5 $T_{\text{ref/NMR}}$ reference temperatures (K) and abbreviated names of the polymers [61]

	$T_{\text{ref/NMR}}$	Abbreviated name
Poly(vinylmethyl ether)	344	PVME
Poly(propylene oxide)	270	PPO
Polyisobutylene	333	PIB
cis-1,4-Polyisoprene	297	PI
cis-1,4-Polybutadiene	234	PB

strongly related to the steric hindrance at the considered site; the larger the steric hindrance, the smaller the half-angle of the libration cone.

A.4.3.1 *Temperature dependence of the segmental dynamics at temperatures well above the glass transition temperature*

In the DLM description of local motions, τ_1 characterizes the segmental modes. In order to know whether these segmental motions observed by NMR in bulk at temperatures well above the glass transition temperature belong to the glass transition processes, it is of interest to compare the variations of τ_1 as a function of temperature with the predictions of the Williams–Landel–Ferry (WLF) equation [59]. The WLF equation describes the frequency dependence of the motional processes associated with the glass transition phenomena. It can be written as [60]:

$$\log[\tau(T)/\tau(T_g)] = -C_1^g(T - T_g)/(C_2^g + T - T_g) \qquad (9A.36)$$

where $\tau(T)$ is the viscoelastic relaxation time at temperature T, and $\tau(T_g)$ is the viscoelastic relaxation time at the T_g, which serves here as a reference. C_1^g and C_2^g can be written as:

$$C_1^g = \frac{B}{2.303 f_g} \qquad (9A.37)$$

$$C_2^g = \frac{f_g}{\alpha} \qquad (9A.38)$$

$$C_1^g C_2^g = \frac{B}{2.303 \alpha} \qquad (9A.39)$$

where f_g is the fractional free volume at the T_g, α is the thermal expansion coefficient of the free volume and $B \sim 1$.

Using the temperature $T_\infty = T_g - C_2^g$, that is, the temperature at which $\tau(T)$ tends to infinity, the WLF equation can be written:

$$\log[\tau(T)/\tau(T_g)] = -C_1^g + C_1^g C_2^g/(T - T_\infty) \qquad (9A.40)$$

T_∞ and the product $C_1^g C_2^g$ are constants characteristic of a given polymer. They do not depend on the reference temperature. Variation of $\log[\tau(T)/\tau(T_g)]$ as a function of $1/(T - T_\infty)$ is linear, and a change in the reference temperature only induces a translation of the line without any modification of its slope.

WLF coefficients for bulk *cis*-1,4-polyisoprene are $T_\infty = 146$ K, $C_1^g = 16.79$, and $C_1^g C_2^g = 900.2$ K for $T_g = 200$ K [60]. In Figure 9A.17 are plotted the variations of $\log[\tau(T)/\tau(T_g)] + A$, where A is an arbitrary constant, and of $\log(2\pi\tau_1)$, as a function of $(T - T_\infty)$ for bulk *cis*-1,4-polyisoprene. The slopes of the two lines are quite similar, which shows that the segmental motions associated with the τ_1 process have the same temperature dependence as the viscoelastic relaxation times. Therefore, the segmental motions described by τ_1

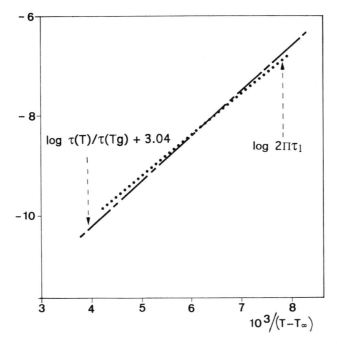

Figure 9A.17 $\log (2\pi\tau_1)(\ldots)$ and $(\log [\tau(T)/\tau(T_g)] + 3.04)$ (- — -) as a function of $10^3/(T - T_\infty)$ for *cis*-1,4-polyisoprene [27]. Reprinted with permission from Dejean de la Batie *et al.*, *Macromolecules*, **22**, 122. Copyright (1989) American Chemical Society

belong to the processes that are involved in the glass transition phenomena. Identical results have been obtained for poly(propylene oxide) [28], *cis*-1,4-polybutadiene [27] and poly(vinylmethylether) [10]. In the case of polyiso-butylene [29], the agreement between the variations of $\log[\tau(T)/\tau(T_g)] + A$ and of $\log (2\pi\tau_1)$ as a function of $(T - T_\infty)$ is not so good. In this case, the temperature dependence of the segmental motions, as observed by NMR, can be understood by considering both glass transition and secondary relaxation processes [29].

A.4.3.2 Reference temperatures and influence of the friction coefficient

Data plotted in Figure 9A.18 show that each polymer has its own dependence of $\log \tau_1$ as a function of $(T - T_\infty)$. Moreover, at a constant $(T - T_\infty)$ difference, the segmental mobility depends on the polymer considered. These results indicate that the differences in segmental mobility at a given temperature observed for the above polymers cannot be interpreted in terms of the differ-ences in the T_∞ temperatures. Similar results are observed when comparing the

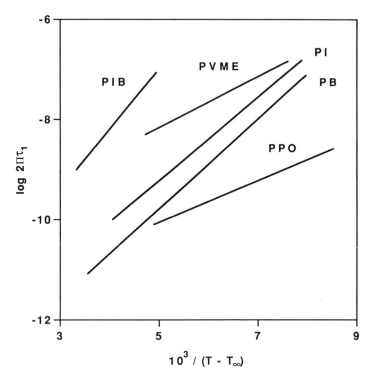

Figure 9A.18 $\log(2\pi\tau_1)$, derived from ^{13}C T_1 NMR experiments, as a function of $10^3/(T - T_\infty)$ for different polymers. Reproduced from Ref. 61 by permission of the publishers Butterworth Heineman Ltd

$\log \tau_1$ variations as a function of $(T - T_g)$ for the different polymers: each polymer has its own dependence of $\log \tau_1$ as a function of $(T - T_g)$. Although the τ_1 processes are controlled by the segmental motions of the polymer chains involved in the glass transition phenomena, the polymer matrices do not share the same local dynamics at a given $(T - T_\infty)$ or $(T - T_g)$ difference. Therefore, neither T_∞ nor the glass transition temperature T_g can be considered as good descriptors for rescaling the segmental motions in bulk polymers at temperatures well above T_g.

It is of interest to define a reference state in which all the polymers are in equivalent states from the point of view of their local mobility. In order to define such equivalent states, one has to look for a property in relation with the frequency window of each experiment. For the ^{13}C NMR relaxation experiments, the temperature, $T_{ref/NMR}$, at which the spin–lattice relaxation time reaches its minimum may constitute such an appropriate reference state. The T_1 minimum is directly related to the local mobility in the frequency range defined by $(\omega_H - \omega_C)$, (ω_C) and $(\omega_H + \omega_C)$ and can be easily determined experimentally. It is independent of the model used to describe the local dynamics. $T_{ref/NMR}$ reference temperatures have been obtained at 25.15 MHz for several polymers [61]. They are listed in Table 9A.5 and show that $(T_{\text{ref/NMR}} - T_g)$ strongly varies from one polymer to another. For example, $T_{\text{ref/NMR}} - T_g$ is 36° higher in PIB than in PI, which implies that the same mobility in terms of correlation time τ_1 is obtained at 36° higher in polyisobutylene (PIB) than in cis-1,4-polyisoprene (PI).

The next step of this approach is to relate the reference temperatures to data obtained from viscoelastic experiments. A quantity of particular interest for the local dynamics is the monomeric friction coefficient, ζ_0, which characterizes the resistance encountered by a monomer unit moving through its surroundings [62]. It has been shown to follow the WLF law. The variation of the monomeric friction coefficient with temperature is plotted in Figure 9A.19 for several polymers. PI and PIB data have been taken from ref. [60]. For cis-1,4-polybutadiene (PB), the approximation $\zeta_0 = \zeta_1$, where ζ_1 is the friction coefficient of a foreign molecule of like size, has been used. For a given $(T - T_\infty)$ difference, the value of ζ_0 strongly depends on the polymer considered. However, log ζ_0 values at $T_{\text{ref/NMR}}$ have the same order of magnitude for the polymers considered. Such a result implies that at these

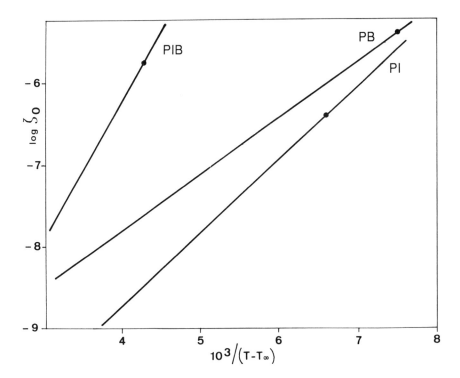

Figure 9A.19 Plots of log ζ_0 as a function of $10^3/(T - T_\infty)$ and log ζ_0 values at the $T_{\text{ref/NMR}}$ (•) reference temperatures of the polymer matrices. Reproduced from Ref. 61 by permission of the publishers Butterworth Heineman Ltd.

reference temperatures the polymers share a nearly common value of ζ_0. Therefore, the monomeric friction coefficient of the bulk polymer appears to be one of the main factors controlling the local dynamics in bulk polymers at temperatures well above the glass transition temperature. Extensive discussion concerning the role of ζ_0 can be found in ref. [61] where comparison between NMR and excimer fluorescence studies of local dynamics in bulk polymers at temperatures well above T_g shows that the amplitude of the local jumps observed by NMR depends on the precise chemical nature of the polymer chain, that is, on both intramolecular constraints and intermolecular interactions.

A.5 Conclusion

NMR experiments have led to significant progress in the detailed description of local polymer dynamics. For the main chain carbons, spin–lattice relaxation time measurements in the vicinity of the T_1 minimum region have demonstrated the existence of two types of motion. The faster motions are librations of the $^{13}C-^1H$ internuclear vectors within the potential energy wells. Their amplitude is mainly determined by the steric hindrance at the carbon site considered. The slower motions are segmental motions. They are well-represented by a damped diffusion of bond orientation along the chemical sequence. In addition, when they occur, the motions of groups that behave like internal rotators, such as phenyl ring or methyl rotations or alkyl side-chain motions, can also be investigated by NMR experiments.

In dilute solution, determinations of the spin–lattice relaxation times and nuclear Overhauser enhancements permit the characterization of the dynamic flexibility of the isolated chain. The correlation times of the polymer segmental motions are strongly dependent on the chemical structure of the monomer unit. Their variation as a function of the solvent viscosity is well-described in terms of a generalized Kramers' theory. In bulk polymers at temperatures well above the glass transition temperature, the segmental motions, as observed at the high frequency of the T_1 experiment, belong to the processes that are involved in the glass transition phenomenon. Their rate is mainly determined by the monomeric friction coefficient of the polymer.

The description of polymer dynamics, derived from NMR experiments, is consistent with results obtained from computer dynamics simulations. Both methods yield complementary information to the dynamics of polymer chains. Systematic studies using the two techniques should lead to a precise characterization of the geometry of local motions.

References

1. E. Helfand, *J. Chem. Phys.*, **54**, 4651 (1971).
2. J. Skolnick and E. Helfand, *J. Chem. Phys.*, **72**, 5489 (1980).

3. H. A. Kramers, *Physica*, **7**, 284 (1940).
4. E. Helfand, Z. R. Wasserman and T. A.Weber, *Macromolecules*, **13**, 526 (1980).
5. D. B. Adolf and M. D. Ediger, *Macromolecules*, **24**, 5834 (1991).
6. D. B. Adolf and M. D. Ediger, *Macromolecules*, **25**, 1074 (1992).
7. B. Valeur, J. P. Jarry, F. Gény and L. Monnerie, *J. Polym. Sci., Polym. Phys. Ed.*, **13**, 667, 675 and 2251 (1975).
8. A. A. Jones and W. H. Stockmayer, *J. Polym. Sci., Polym. Phys. Ed.*, **15**, 847 (1977).
9. C. K. Hall and E. Helfand, *J. Chem. Phys.*, **77**, 3275 (1982).
10. R. Dejean de la Batie, F. Lauprêtre and L. Monnerie, *Macromolecules*, **21**, 2045 (1988).
11. O. W. Howarth, *J. Chem. Soc., Faraday Trans. II*, **76**, 1219 (1980).
12. D. E. Woessner, *J. Chem. Phys.*, **36**, 1 (1962).
13. R. E. London and J. Avitabile, *J. Chem. Phys.*, **65**, 2443 (1976).
14. G. Lipari and A. Szabo, *J. Am. Chem. Soc.*, **104**, 4546 (1982).
15. J. L. Viovy, L. Monnerie and J. C. Brochon, *Macromolecules*, **16**, 1845 (1983).
16. J. L. Viovy, L. Monnerie and F. Merola, *Macromolecules*, **18**, 1130 (1985).
17. A. Abragam, *The Principles of Nuclear Magnetism*, Clarendon Press, Oxford, 1961.
18. A. G. Redfield, *Adv. Magn. Reson.*, **1**, 1 (1965).
19. F. Heatley and B. Wood, *Polymer*, **19**, 1405 (1978).
20. F. Heatley and M. K. Cox, *Polymer*, **21**, 381 (1980).
21. F. Heatley and M. K. Cox, *Polymer*, **22**, 190 (1981).
22. M. M. Fuson and D. M. Grant, *Macromolecules*, **21**, 944 (1988).
23. M. M. Fuson, D. J. Anderson, F. Liu and D. M. Grant, *Macromolecules*, **24**, 2594 (1991).
24. M. M. Fuson and J. B. Miller, *Macromolecules*, **26**, 3218 (1993).
25. D. J. Gisser, S. Glowinkowski and M. D. Ediger, *Macromolecules*, **24**, 4270 (1991).
26. A. Allerhand, D. Doddrell and R. J. Komoroski, *J. Chem. Phys.*, **55**, 189 (1971).
27. R. Dejean de la Batie, F. Lauprêtre and L. Monnerie, *Macromolecules*, **22**, 122 (1989).
28. R. Dejean de la Batie, F. Lauprêtre and L. Monnerie, *Macromolecules*, **21**, 2052 (1988).
29. R. Dejean de la Batie, F. Lauprêtre and L. Monnerie, *Macromolecules*, **22**, 2617 (1989).
30. J. P. Cohen-Addad and J. P. Faure, *J. Chem. Phys.*, **61**, 2440 (1974).
31. A. D. English and C. R. Dybowski, *Macromolecules*, **17**, 446 (1984).
32. A. D. English, *Macromolecules*, **18**, 178 (1985).
33. T. Kanaya, K. Kaji and K. Inoue, *Macromolecules*, **24**, 1826 (1991).
34. A. Allerhand and R. K. Hailstone, *J. Chem. Phys.*, **56**, 3718 (1972).
35. F. Heatley, in C. Booth and C. Price (eds), *Dynamics of Chains in Solution by NMR Spectroscopy*, Comprehensive Polymer Science, Pergamon Press, New York, vol. 1, p. 377 (1991).
36. P. M. Budd, F. Heatley, T. J. Holton and C. Price, *J. Chem. Soc., Faraday Trans. I*, **77**, 759 (1981).
37. F. Heatley and B. Wood, *Polymer*, **19**, 1405 (1978).
38. S. Glowinkowski, D. J. Gisser and M. D. Ediger, *Macromolecules*, **23**, 3520 (1990).
39. R. F. Grote and J. T. Hynes, *J. Chem. Phys.*, **73**, 2715 (1980).
40. T. Asakura and Y. Doi, *Macromolecules*, **16**, 786 (1983).
41. W. Gronski and N. Murayama, *Makromol. Chem.*, **179**, 1509 (1978).

42. F. Lauprêtre, L. Monnerie and O. Vogl, *Europ. Polym. J.*, **14**, 981 (1978).
43. F. Gény and L. Monnerie, *J. Polym. Sci.*, **17**, 147 (1979).
44. W. Gronski and N. Murayama, *Makromol. Chem.*, **177**, 3017 (1976).
45. F. Lauprêtre, C. Noël and L. Monnerie, *J. Polym. Sci.*, **15**, 2143 (1977).
46. J. J. Connolly, E. Gordon and A. A. Jones, *Macromolecules*, **17**, 722 (1984).
47. J. J. Connolly and A. A. Jones, *Macromolecules*, **18**, 906 (1985).
48. A. K. Roy and A. A. Jones, *J. Polym. Sci., Polym. Phys. Ed.*, **23**, 1793 (1985).
49. J. A. Ratto, P. T. Inglefield, R. A. Rutkowki, K.-L. Li, A. A. Jones and A. K. Roy, *J. Polym. Sci., Polym. Phys. Ed.*, **25**, 1419 (1987).
50. P. Tékély, F. Lauprêtre and L. Monnerie, *Macromolecules*, **16**, 415 (1983).
51. M. E. Nedea, R. H. Marchessault and P. Dais, *Polymer*, **33**, 1831 (1992).
52. P. Dais, M. E. Nedea and R. H. Marchessault, *Polymer*, **33**, 4288 (1992).
53. T. Radiotis, G. R. Brown and P. Dais, *Macromolecules*, **26**, 1445 (1993).
54. A. Spyros and P. Dais, *Macromolecules*, **25**, 1062 (1992).
55. H. Menge and H. Schneider, *Polymer*, **34**, 4208 (1993).
56. A. Perico, A. Altomare, D. Catalano, M. Colombani and C. A. Veracini, *Macromolecules*, **23**, 4912 (1990).
57. F. Heatley, *Prog. Nucl. Magn. Reson. Spectrosc.*, **13**, 47 (1979).
58. C. C. Hung, J. H. Shibata, A. A. Jones and P. T. Inglefield, *Polymer*, **28**, 1062 (1987).
59. M. L. Williams, R. F. Landel and J. D. Ferry, *J. Am. Chem. Soc.*, **77**, 3701 (1955).
60. J. D. Ferry, in *Viscoelastic Properties of Polymers*, 3rd edn., Wiley, New York, 1980.
61. F. Lauprêtre, L. Bokobza and L. Monnerie, *Polymer*, **34**, 468 (1993).
62. F. Bueche, *J. Chem. Phys.*, **20**, 1959 (1952).

9B NMR AND DYNAMICS OF BIOPOLYMERS

L. Y. Lian and I. L. Barsukov
University of Leicester, Leicester, UK

B.1 Introduction

For proteins and nucleic acids to function, flexibility is essential. It is this flexibility that, in part, accounts for the diversity in their functional roles. For proteins in particular, this diversity can be illustrated by their different functions—as enzymes, structural proteins, transport proteins, receptors, etc. Structurally, a protein folds in such a way that key amino acids are strategically located in well-defined active sites, where groups can interact in a coordinated manner with ligands or other proteins, etc. Nucleic acids are polymers of

Dynamics of Solutions and Fluid Mixtures by NMR
Edited by J.-J. Delpuech   © 1995 John Wiley & Sons Ltd

mononucleotides and can either adopt compact structures like tRNA's or have extended coil conformations like many large DNA's. Although both the proteins and nucleic acids adopt some specific fold, their structures are by no means static. In fact, the functional diversity of proteins and nucleic acids reflects on the large amount of information that can be stored in these molecules. The exact nature of some of the motions present in these macro-molecules are now known, primarily through improved experimental methods of detection and more sophisticated theoretical studies. An understanding of the nature and timescales of the molecular motions allows more detailed studies of the macromolecules; for example, it helps with the description of mechanistic pathways of enzymes, and clarifies the nature of ligand binding and molecular associations between a protein and its receptor, or a DNA.

Within a protein these motions range from torsional oscillations of rigid groups, such as aromatic rings and peptide bond (CONH), to rapid local motions of individual groups, to slower distortion of large regions within the molecule. Many of these motions occur at different timescales, ranging from 10^{-14} to 10^1 s. Table 9B.1 shows examples of the typical timescales involved in the motions of proteins, together with examples of the biophysical techniques that can be used for studying these dynamic processes. As evident from the table, no one biophysical method alone will enable detection and detailed

Table 9B.1 Typical motions in proteins and methods of detection

	Log_{10} of characteristic times (s)	Methods of detection
Local denaturation and opening of secondary structure	3 to −1	^3H isotope exchange; ^2H–^1H exchange NMR; ^1H–^1H exchange; NMR chemical shift averaging
Aromatic side-chain rotation	1 to −5	NMR chemical shift averaging
Conformational changes and allosteric transitions	0 to −5	^1H–^1H exchange; NMR chemical shift averaging
Diffusion	−6 to −9	NMR relaxation; ESR
Segmental motion	−7 to −9	NMR relaxation, fluorescence depolarization
Aliphatic side-chain rotation	−8 to −9	NMR relaxation
Vibrational and torsional modes	−11 to −13	X-ray, IR and Raman

studies of all these motions. In addition, theoretical methods such as dynamics simulations are increasingly being used for more in-depth analysis of experimental data and for prediction.

As far as NMR is concerned, motions and dynamics are reflected by five different NMR parameters: chemical shifts, spin-spin coupling constants, the area enclosed by a resonance, relaxation times and Nuclear Overhauser Effect. NMR data can provide both qualitative and quantitative measurements of flexibility and exchange rates.

This part of Chapter 9 describes several basic experimental analytical NMR techniques that are frequently used for the qualitative and quantitative analysis of dynamic and exchange processes, focusing on protein systems; the same approach can be applied to most biological macromolecules. The analysis of data for dynamic processes, such as the determination of rate and binding constants, and of relaxation times, can be complicated; hence, a large proportion of this section is devoted to the theoretical aspects of chemical exchange and relaxation time measurements, since an understanding of the main features of these experiments is crucial for the interpretation of the NMR data. Also discussed in some detail are the hydrogen exchange experiments, these experiments being one of the most widely used for detecting both local and large-scale motions in proteins and for assessing protein stability.

B.2   Theoretical Aspects of Chemical Exchange and Relaxation Measurements

B.2.1   CHEMICAL EXCHANGE

In the context of NMR, 'chemical exchange' refers to any process in which a nucleus exchanges between two or more environments in which its NMR parameters (e.g., chemical shift, scalar coupling or relaxation) differ. These may be intramolecular or intermolecular processes.

Intramolecular exchange processes include:

- motions of side-chains in proteins [1]
- helix-coil transitions of nucleic acids [2]
- unfolding of proteins [3]
- conformational equilibria [4]

Intermolecular exchange processes include:

- binding of small molecules to macromolecules [5]
- protonation/deprotonation equilibria of ionizable groups [6]
- isotope exchange processes (notably exchange of labile protons of a macromolecule with solvent) [7]
- enzyme-catalysed reactions [8]

In all these cases, the effect of the exchange process on the NMR spectrum depends upon its rate, *relative to the magnitude of the accompanying change in NMR parameters* [9, 10].

Thus if an ionization process, for example, was accompanied by a change in chemical shift of 250 Hz, an exchange rate of $>1000 \text{ s}^{-1}$ would be regarded as *fast*, one of $\sim\!250 \text{ s}^{-1}$ as *intermediate* and one of $<100 \text{ s}^{-1}$ as *slow* on the 'NMR timescale'. Each of these regions of exchange leads to characteristic effects on the NMR spectrum (Figure 9B.1). In slow exchange, separate resonances are seen for the nucleus of interest in each of the two states. By contrast, in fast exchange, a single averaged resonance is observed. In the intermediate exchange region, there is a complex series of changes in appearance of the spectrum as the two separate signals coalesce into one. Bearing in mind the typical values of NMR parameters, the range of rates which can be studied is approximately $0.05\text{--}5000 \text{ s}^{-1}$. This range can be extended to faster rates by the use of nuclei with larger chemical shift ranges than ^1H, notably ^{19}F and ^{13}C; within this faster range, analysis of the spectrum can provide estimates of the rate constants of exchange.

We shall describe the NMR methods for qualitative and quantitative evaluation of exchange effects in terms of the situation in which the chemical shift of the nucleus is the main parameter differing between the states. Because of the typical magnitudes of scalar couplings (for $^1\text{H}\text{--}^1\text{H}$ couplings, $0\text{--}20 \text{ s}^{-1}$) and relaxation rates ($1\text{--}300 \text{ s}^{-1}$), these are commonly seen to be in the fast

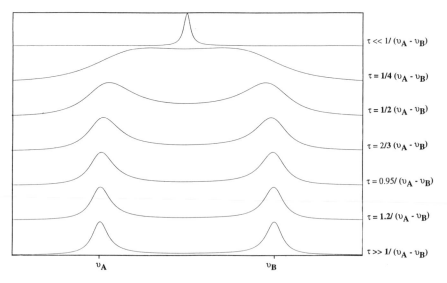

Figure 9B.1 Change in chemical shifts and linewidths in the presence of chemical exchange between two equally populated environments. Bottom to top: increasing exchange rates as denoted in the figure. Reproduced by permission of Oxford University Press from Ref. 9

exchange regime—that is, as averages. We shall also restrict discussion largely to exchange between two states, while indicating in general terms how the methods can be extended to more complex processes.

The first step in any study of chemical exchange by NMR is to establish to which region of change the spectrum (or, more correctly, the resonances of interest, since different resonances can show different exchange behaviour) corresponds. We therefore begin by summarizing some relevant definitions of the different exchange regions, concentrating in detail on the second-order ligand binding process since this is one of the more common situations in which chemical exchange studies can be informative.

For a two-site first-order exchange where A and B are interconverting forms of a molecule,

$$A \underset{k_{-1}}{\overset{k_{+1}}{\rightleftharpoons}} B \tag{9B.1}$$

Lifetime of state A, $\tau_A = 1/k_{+1}$

Lifetime of state A, $\tau_B = 1/k_{-1}$

For a two-site second-order exchange, such as the binding of a ligand, L, to an enzyme, E,

$$E + L \underset{k_{-1}}{\overset{k_{+1}}{\rightleftharpoons}} EL \tag{9B.2}$$

For a nucleus in the ligand molecule:

$$\text{Lifetime in state } EL, \ \tau_{EL} = 1/k_{-1} \tag{9B.3}$$

$$\text{Lifetime in state } L, \ \tau_L = 1/(k_{+1}[E]), \tag{9B.4}$$

and for a nucleus in the enzyme:

$$\text{Lifetime in state } EL, \ \tau_{EL} = 1/k_{-1}$$

$$\text{Lifetime in state } E, \ \tau_E = 1/(k_{+1}[L]) \tag{9B.5}$$

where $[E]$ is the concentration of *free* enzyme, $[L]$ the concentration of the *free* ligand and $[EL]$ the concentration of the protein–ligand complex. The total concentrations of protein and ligand are denoted by $[E_T]$ and $[L_T]$ respectively:

$$[E_T] = [EL] + [E] \tag{9B.6}$$

$$[L_T] = [EL] + [L] \tag{9B.7}$$

The equilibrium constant for the reaction is given by

$$K_d = k_{-1}/k_{+1} = [E][L]/[EL] \tag{9B.8}$$

From equations (9B.6) to (9B.8), the concentration of the protein–ligand complex $[EL]$ can be expressed in terms of the measurable protein and ligand concentrations and the equilibrium constant:

$$[EL] = \tfrac{1}{2}\{[E_\mathrm{T}] + [L_\mathrm{T}] + K_\mathrm{d} - (([E_\mathrm{T}] + [L_\mathrm{T}] + K_\mathrm{d})^2 - 4[E_\mathrm{T}][L_\mathrm{T}])^{1/2}\} \quad (9\mathrm{B}.9)$$

From equations (9B.4), (9B.5) and (9B.8), the lifetimes of states L and E can in turn be expressed in terms of the mole fraction of the free and bound ligand to give

$$\tau_\mathrm{L} = p_\mathrm{L}/(k_{-1}p_\mathrm{LE}) \qquad\qquad (9\mathrm{B}.10)$$

$$\tau_\mathrm{E} = p_\mathrm{E}/(k_{-1}p_\mathrm{EL}) \qquad\qquad (9\mathrm{B}.11)$$

where p_L and p_E are the fraction of the ligand and the enzyme, respectively, in the free state, given by

$$p_\mathrm{L} = [L]/[L_\mathrm{T}] \quad \text{and} \quad p_\mathrm{E} = [E]/[E_\mathrm{T}]$$

and p_LE and p_EL are the fraction of the ligand and the enzyme, respectively, in the bound state, given by

$$p_\mathrm{LE} = [EL]/[L_\mathrm{T}] \quad \text{and} \quad p_\mathrm{EL} = [EL]/[E_\mathrm{T}] \qquad (9\mathrm{B}.12)$$

In many discussions of exchange effects, a single lifetime τ is used to characterize the process defined by equation (9B.2); for the ligand:

$$1/\tau = 1/\tau_\mathrm{EL} + 1/\tau_\mathrm{L} = k_{-1}(1 + p_\mathrm{LE}/p_\mathrm{L}) \qquad (9\mathrm{B}.13)$$

The relationship between the lifetime τ and the NMR parameters of chemical shift, relaxation time and scalar coupling constant determines the appearance of the resulting NMR spectrum (in practice, the coupling constant timescale is rarely used in proteins, since the linewidths are of the same order as proton-proton coupling constants). In the following, we summarize this relationship, focusing on the ligand resonance, although analogous equations will of course apply to a nucleus on the enzyme. It should be noted that τ, and hence the appearance of the spectrum, depends upon the ligand concentration, as is evident from equation (9B.13).

B.2.1.1   Slow exchange

In the system defined by equation (9B.2), separate resonances for the free and bound ligand will be observed if the slow exchange condition is met, that is, if the rate of equilibration, $1/\tau$, is *slower* than the chemical shift separation, $\Delta\delta$ (in rad s^{-1}), of the resonances of the bound and free ligand:

$$1/\tau \ll \Delta\delta, \quad \text{where } \Delta\delta = |(\delta_\mathrm{EL} - \delta_\mathrm{L})| \qquad (9\mathrm{B}.14)$$

Slow-exchange on the relaxation timescale is defined as

$$1/\tau << |(R_{2,\text{EL}}) - (R_{2,\text{L}})|$$

or

$$1/\tau << |(R_{1,\text{EL}}) - (R_{1,\text{L}})| \tag{9B.15}$$

where R_1 and R_2 are, respectively, the longitudinal and transverse relaxation rates, the subscript indicating the species concerned. When equation (9B.15) is satisfied,* the relaxation rates of resonances in the bound and free ligand are given by

$$R_{1,2\text{EL}}|_{\text{obs}} = R_{1,2\text{EL}} + k_{-1}$$

$$R_{1,2\text{L}}|_{\text{obs}} = R_{1,2\text{L}} + k_{-1}\frac{p_{\text{LE}}}{p_{\text{L}}} \tag{9B.16}$$

B.2.1.2   Fast exchange

When the rate of equilibration $1/\tau$ is *faster* than the chemical shift separation and *faster* than the difference in relaxation rates,

$$1/\tau >> \Delta\delta, \tag{9B.17}$$

and

$$1/\tau >> |R_{2,\text{EL}} - R_{2,\text{L}}| \tag{9B.18}$$

a single resonance is observed for the ligand† at a 'weighted average' chemical shift, δ_{obs}, given by

$$\delta_{\text{obs}} = \delta_{\text{L}}p_{\text{L}} + \delta_{\text{EL}}p_{\text{LE}} \tag{9B.19}$$

The transverse relaxation rate of this single resonance is given by

$$R_{2,\text{obs}} = p_{\text{LE}}R_{2,\text{EL}} + p_{\text{L}}R_{2,\text{L}} + p_{\text{LE}}p_{\text{L}}\tau(2\pi\Delta\delta)^2 \tag{9B.20}$$

The magnitude of the third, 'exchange', term in equation (9B.20) depends upon the magnitude of $\tau(\Delta\delta)^2$ and also on the ligand concentration, being

---

*Regardless of whether the conditions of equations (9B.14) or (9B.17) are satisfied, the rate of equilibration may be faster or slower than the difference between the respective relaxation rates in the two environments (equations (9B.15), (9B.18)). Thus, the slow and fast exchange conditions must be applied separately to chemical shifts and relaxation rates. Under the slow exchange conditions described by equation (9B.15), separate resonances will always be present for the free and bound ligand regardless of whether these resonances are in fact resolved. If the resonances are not resolved, the observed spectrum will be the superpositions of two resonances of different linewidths. In situations when an excess of ligand is present, that is $p_{\text{EL}} << p_{\text{L}} \sim 1$, the free ligand resonance will be the only observable signal, with a transverse relaxation rate $R_{2,\text{L,obs}} = R_{2,\text{L}} + k_{-1}p_{\text{LE}}$. If on the other hand $\Delta\delta >> 1/\tau >> |R_{2,\text{EL}} - R_{2,\text{L}}|$, two resonances, separated by $\Delta\delta$, will be observed, with transverse relaxation rates given by equation (9B.20).

†For a nucleus on the protein, under these same conditions, the resonance position is given by $\delta_{\text{obs}} = \delta_{\text{E}}p_{\text{E}} + \delta_{\text{EL}}p_{\text{EL}}$.

greatest at $p_{EL} \sim 1/3$ (see ref. [7]). When $\tau(\Delta\delta)^2$ is small, the third term is negligible and the relaxation rate is a weighted average

$$R_{2,obs} = p_{LE}R_{2,EL} + p_L R_{2,L} \qquad (9B.21)$$

The longitudinal relaxation rates are not affected by the chemical shift difference $\Delta\delta$, and, under fast exchange conditions, R_1 for the averaged resonance is always a weighted average, provided that $1/\tau >> |R_{1,EL} - R_{1,L}|$.

B.2.1.3   Intermediate exchange

Intermediate exchange occurs when the lifetime is of the order of $\Delta\delta$ or the difference in relaxation rates:

$$1/\tau \simeq \Delta\delta \qquad (9B.22)$$

and

$$1/\tau \simeq |R_{2,EL} - R_{2,L}| \qquad (9B.23)$$

Resonances in the spectrum broaden markedly, often becoming unobservable; hence these spectra are generally difficult to analyse. When spectra of sufficient quality is available, lineshape analysis can be used to obtain quantitative estimates of the exchange rates [11].

Some of the spectral features described above are is illustrated in Figure 9B.2 which shows the 2D spectrum of RCAM-bovine pancreatic trypsin inhibitor (BTPI), where 14–38 disulphide bridge is reduced and the corresponding cysteine residues are protected by carboxamido methylation. In the spectrum at 10 °C (Figure 9B.2a), for example, five separate signals are observed for the five aromatic protons of Phe 45: 8.23 ppm (Hε1), 7.67 ppm (Hζ), 7.60 ppm (Hε2), 7.53 ppm (Hδ1), and 7.31 ppm (Hδ2). All five aromatic protons appear to experience different magnetic environments. This spectrum at 10 °C indicates that Phe 45 side-chain rotates at a rate slower than the chemical shift separation between the pairs of δ protons or ε protons ($k << 10^2$ s^{-1}). As the temperature is raised, some of these signals begin to shift. As evident in the spectrum at 36 °C (Figure 9B.2b) the δ resonances now appear as a *sharp* signal at 7.405 ppm and the ε resonances as a *very broad* signal centred around 7.90 ppm. The ζ resonance remains at about 7.70 ppm. The spectrum at 36 °C illustrates another feature, that is, in the same spectrum, signals can appear to be in both fast exchange (Phe 45 δ protons) or intermediate exchange (Phe 45 ε protons). Similarly, for the same temperature range, the chemical shifts of the resonances from many of the other aromatic residues indicate that these latter side-chains appear to be in fast exchange on the chemical shift time-scale, with the shifts of two pairs, δ and ε protons appearing as two 'averaged' single peaks. From the known crystal structure of BPTI, Phe 45 is situated in the interior of the protein where rotation of the aromatic side-chain is restricted.

It is essential as a first step in studying chemical exchange by NMR to distinguish between fast, intermediate and slow exchange. Table 9B.2 summarizes the definition of exchange rates on the NMR timescale. Once the chemical exchange rate regime for the exchange process has been established using the above procedure (together with other biochemical data), the second stage of analysing these exchange processes can be progressed. This latter stage includes obtaining kinetic information using quantitative analysis of either the lineshape or magnetization transfer data [9, 10].

In the NMR studies of large macromolecules, the most widely used method for obtaining chemical exchange rates by magnetization transfer is the two-dimensional experiment [12]. A major advantage of the 2D experiment is its ability to provide a 'map' of the exchange process(es). All the sites are 'labelled' in a single experiment without previous knowledge of the spectrum, thus allowing all exchange pathways to be observed simultaneously. Problems associated with 1D selective excitation in crowded spectra are completely avoided. Formally and in practice, the two-dimensional experiment used for studying magnetization transfer due to the Nuclear Overhauser Effect and due to chemical exchange are identical. The 2D method is most suitable for the slow exchange regime (see Chapter 3, Section 4.4).

For a quantitative analysis, it is possible to view the intensities of cross-peaks in the 2D NOESY spectrum of an exchanging system in the absence of scalar coupling as the solution to the Bloch-McConnell equations:

$$\mathbf{M} = \mathbf{M}_0 \exp(\mathbf{R}t_m) \qquad (9B.24)$$

where $\mathbf{M}$ is the matrix comprising the peak volumes of the 2D spectrum, $\mathbf{R}$ is a matrix, of dimensions equal to the number of sites between which exchange is occurring, containing all the exchange rate constants and spin–lattice relaxation rates and t_m is the mixing time. For the exchange between two sites A and B, $\mathbf{R}$ is given by:

$$\mathbf{R} = \begin{pmatrix} -k_{-1} - R_A & k_{+1} \\ k_{-1} & -k_{+1} - R_B \end{pmatrix} \qquad (9B.25)$$

(see also Chapter 3, Section 4.4.1 and equation (3.122)).

Table 9B.2 The NMR time scale

	Exchange rate		
Timescale	Slow	Intermediate	Fast
Chemical shift, δ^a	$k \ll \|\delta_A - \delta_B\|$	$k \approx \|\delta_A - \delta_B\|$	$k \gg \|\delta_A - \delta_B\|$
Coupling constant, J^b	$k \ll \|J_A - J_B\|$	$k \approx \|J_A - J_B\|$	$k \gg \|J_A - J_B\|$
T_2 relaxation^c	$k \ll \|1/T_{2,A} - 1/T_{2,B}\|$	$k \approx \|1/T_{2,A} - 1/T_{2,B}\|$	$k \gg \|1/T_{2,A} - 1/T_{2,B}\|$

^{a} $(\delta_A - \delta_B)$ is typically in the order of hundreds of hertz.
^{b} J is typically in the order of 1–10 Hz.
^{c} $1/T_{2,A} - 1/T_{2,B}$ is typically in the order of 1–20 Hz for protons; the linewidth at half-height, $\Delta = 1/\pi T_2$.

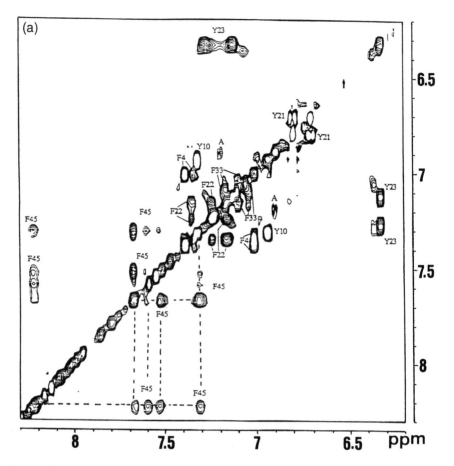

Figure 9B.2 The aromatic region of the TOCSY spectrum of RCAM BPTI at (a) 10 °C and (b) 36 °C in H_2O with the aromatic resonances of the Phe 45 spin system linked by dotted lines

When the exchange rates dominate over the relaxation rates, the 2D spectrum is in effect a a graphic display of the exchange pathways. The very appearance of an exchange cross-peak at ω_A, ω_B is sufficient proof that exchange is taking place between the sites i and j (Figure 9B.3a).

The integrated intensity I_{ij} of the 2D absorption peak at ω_A, ω_B is given by:

$$I_{AB}(t_m) = M_{0A} \exp(-\mathbf{R}t_m)_{ij} \tag{9B.26}$$

For the simplest case of first-order two-site exchange with equal populations ($M_{0A} = M_{0B} = M$, $k_{+1} = k_{-1} = k$) and equal relaxation rates ($R_A = R_B = R$), this

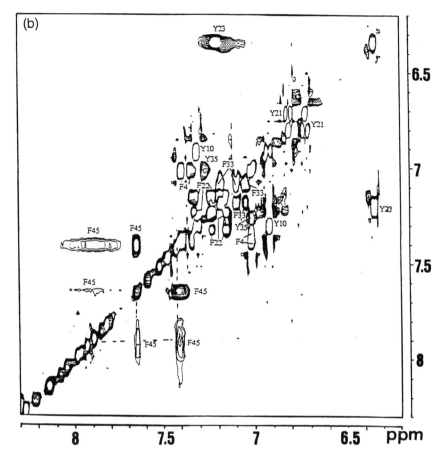

Figure 9B.2 (*continued*)

has the solution:

Diagonal peaks

$$I_{AA} = I_{BB} = \{[1 + \exp(-2kt_m)]\exp(-Rt_m)\}/2 \qquad (9B.27)$$

Cross-peaks

$$I_{AB} = I_{BA} = \{[1 - \exp(-2kt_m)]\exp(-Rt_m)\}/2 \qquad (9B.28)$$

In this simple two-site exchange system, the exchange rate can be determined from the initial rate of build-up of cross-peak intensity at short mixing times (Figure 9B.3b)

$$(\delta I_{AB}/\delta t_m)i = M_0 k \qquad (9B.29)$$

(a)

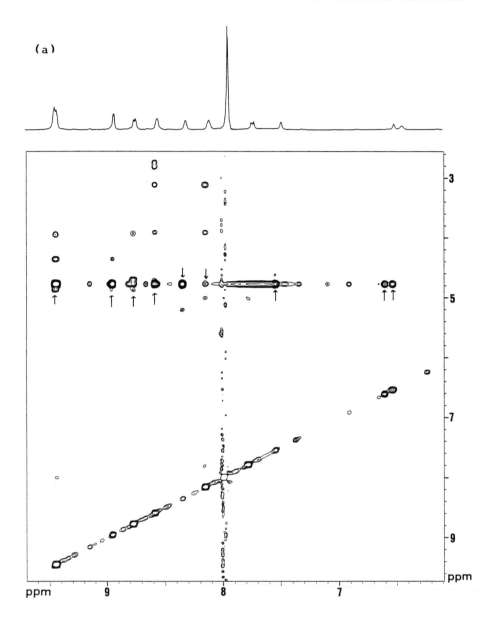

Figure 9B.3 (a) A phase-sensitive contour plot of a 2D exchange experiment showing the exchange of the amide protons of the peptide viomycin with water at 310 K, mixing time 210 ms. The NH–H₂O exchange cross-peaks are indicated by arrows. (b) Representative plot of the variation of peak intensity with mixing time t_m: (●) and (■) for diagonal- and cross-peaks, respectively. Reproduced by permission of Academic Press from Ref. 13

(b)

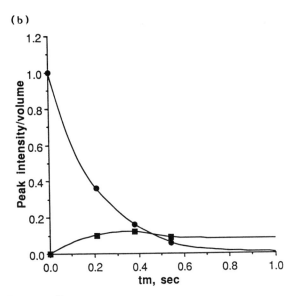

Figure 9B.3 (*continued*)

or directly from the ratio of the cross and diagonal peak intensities at short mixing times

$$I_{AA}/I_{AB} = (1 - kt_m)/kt_m \qquad (9B.30)$$

However, the cross-peak intensity may be weak at mixing times short enough to satisfy the initial rate assumption. More generally, it is necessary to record a series of 2D exchange spectra with different t_m values and to fit the dependence of diagonal peak and cross-peak intensity on mixing time to equations (9B.27) and (9B.28). When the exchange process is more complex than the very simple two-site case, the equivalent of equations (9B.27) and (9B.28) can be derived from equation (9B.24) (with the matrices of appropriate dimensions); the general solution is given in reference [12]. This can then be fitted to the peak intensities as a function of t_m in a similar way; the number of parameters is likely to be such that the intensity data for several cross-peaks will have to be analysed together.

The data shown in Figure 9B.3 allow the amide proton exchange rates with the solvent water of the nonapeptide viomycin to be determined [13]. The rates of exchange for each amide proton (k_n) observed by the 2D method at pH 4 and 28 °C can be grouped as:

$$k_{11}, k_{12} \gg k_3, k_5, k_7 > k_2, k_8, k_9 \gg k_4, k_{13}$$

These results are in agreement with previously obtained crystal structure in so far as those amides which are either involved in intramolecular hydrogen bonding, in particular amide of residue 4, or are protected from the solvent (residues 13, 8) have the slowest amide exchange rates.

B.2.2   HETERONUCLEAR RELAXATION MEASUREMENTS

Since the heteronuclear relaxation rates (^{15}N and ^{13}C) depend chiefly on the dynamics of the heteronuclei-proton bond vectors with respect to the external magnetic field, the motions of these bond vectors can be characterized, allowing the dynamics of the protein to be monitored using a set of ^{15}N and/or ^{13}C relaxation parameters. This approach is feasible today since the availability of isotopically ^{13}C and ^{15}N-labelled proteins combined with the current advances in NMR methodology mean that these experiments do not suffer from sensitivity problems. Strong dependence of chemical shifts in proteins upon spatial structure allows one to resolve signal from nuclei at different positions of the polypeptide chain and thus relate relaxation information to the particular part of the molecule. The interpretation of relaxation measurements for heteronuclei in terms of motional characteristics is much more straightforward than for protons. This is due to the fact that relaxation behaviour of protonated ^{13}C and ^{15}N is mainly influenced by the chemically attached proton(s) and chemical anisotropy of the electron density, that is, it can be related to the internal motion of H–N or H–C vector.

Relaxation data are usually analysed in the framework of the so-called 'model-free' approach [14, 15]. Although this approach does not use any model for intramolecular motion, it is based upon an oversimplified model for a correlation function which contains several adjustable parameters. These parameters are evaluated from the results of relaxation measurements and are interpreted within specific motional models to give geometrical properties of the internal dynamics. The most widely used parameter in the 'model-free' approach is an order parameter S^2, which is associated with the degree of restriction of internal motion. As more complex systems were studied, the originally proposed 'model-free' correlation function was found to be inadequate to describe the experimental results in several cases and had to be expanded [16, 17]. Further improvement in the interpretation of relaxation data may come from expanding the basic set of measured relaxation parameters in order to directly obtain values of the spectral density function at discrete frequency points from the relaxation data [18, 19]. This latter approach leaves out any assumption on the shape of spectral density function.

In the following section the analysis of protein relaxation is described for ^{15}N data, since these are most widely used. The models of correlation function with increasing complexity are gradually introduced to account for deviation between experimental and theoretical values. The results obtained with direct evaluation of correlation function are described as well. Interpretation of protein dynamics using models of internal motion concludes the section.

B.2.2.1   Relaxation parameters

Relaxation of heteronuclei which have one or more directly attached protons is dominated by dipole–dipole interaction with the proton(s) and chemical shift anisotropy. In such cases the usually measured ^{15}N relaxation rates T_1, T_2, $T_{1\rho}$

and NOE are given by [20]:

$$\frac{1}{T_1} = d^2\{J(\omega_A - \omega_X) + 3J(\omega_X) + 6J(\omega_A + \omega_X)\} + C^2 J(\omega_X)$$

$$\frac{1}{T_2} = \frac{1}{2}d^2\{4J(0) + J(\omega_A - \omega_X) + 3J(\omega_X) + 6J(\omega_A + \omega_X)\}$$

$$+ \frac{C^2}{6}\{3J(\omega_X) + 4J(0)\} \tag{9B.31}$$

$$\frac{1}{T_{1\rho}} = \frac{1}{2}d^2\{4J(\omega_e) + J(\omega_A - \omega_X) + 3J(\omega_X) + 6J(\omega_A + \omega_X)\}$$

$$+ \frac{C^2}{6}\{3J(\omega_X) + 4J(\omega_e)\}$$

$$\text{NOE} = 1 + \left[\frac{\gamma_A}{\gamma_X}d^2\{6J(\omega_A + \omega_X) - J(\omega_A - \omega_X)\}T_1\right]$$

where $d^2 = \gamma_A^2\gamma_X^2\hbar^2/(10)\langle 1/r_{AX}^3\rangle^2$ and $C^2 = 2/(15)\omega_x^2(\sigma_\| - \sigma_\perp)^2$; $\hbar$ is Planck's constant; γ_X and γ_A are gyromagnetic ratios of ^{1}H and ^{15}N; ω_A and ω_X are ^{1}H and ^{15}N Larmor frequencies in the constant magnetic field, B_0; $\omega_e = B_{eff}\gamma_x$ is Larmor frequency in the effective spin-lock field B_{eff}; r_{AX} is NH bond length; $\sigma_\|$ and $\sigma_\perp$ are the parallel and perpendicular components of the axially symmetric ^{15}N chemical shift tensor; $J(\omega)$ is the spectral density function for the motion of the NH vector. The equation for $T_{1\rho}$ corresponds to the on-resonance limit [21]. The spectral density function contains the information on the motion of the NH vector and can be calculated as real Fourier transform-ation of the correlation function $g(t)$:

$$J(\omega) = \int_0^\infty g(t)\cos(\omega t)\, dt \tag{9B.32}$$

As can be seen from equation (9B.31), NMR relaxation measurements allow one, at best, to sample spectral density function only at discrete points, which are combinations of Larmor frequencies of two interacting nuclei. This greatly restricts information about the correlation function $g(t)$. The positions of relaxation sampling points in relation to correlation functions of different types of motion in proteins are shown in Figure 9B.4. The slowest is the chemical exchange process, which normally occurs on a millisecond timescale. This only has an influence on $T_{1\rho}$ and T_2 values. The overall motion of a protein and fast internal motion contribute to all relaxation rates and these contributions depend both on timescale and degree(s) of freedom of motion. Very fast internal motion gives equal contribution at all frequencies sampled and only the degree of freedom for this motion can be derived.

Factors that need to be considered when analysing the relaxation data from NMR are the effects of *chemical exchange, overall motions (isotropic or* anisotropic), and internal motions (fast and slow). The influence of each of these factors is described next.

B.2.2.1.1 Chemical exchange

Increase of ^{15}N transverse relaxation rate due to chemical exchange was found to be common for proteins. Chemical exchange analysis using lineshapes and magnetization transfer effects are discussed in Section B2.1. Here we deal with its implication to heteronuclear relaxation measurements.

When chemical exchange between states with different chemical shift of ^{15}N is present, the equations for T_2 and $T_{1\rho}$ have to be modified. In the case of fast exchange between two sites with populations p_A and p_B the expressions are [22]:

$$\frac{1}{T_2} = \frac{1}{T_2^0} + p_A p_B \delta\omega^2 J_e(0)$$

$$\frac{1}{T_{1\rho}} = \frac{1}{T_{1\rho}^0} + p_A p_B \delta\omega^2 J_e(\omega_r)$$

$$(9B.33)$$

where $\delta\omega$ is chemical shift difference between two sites (in rad s^{-1}) and $J_e(\omega)$ is spectral density function for the exchange process, and $\omega_r = |\gamma B_1|$. Variation of the intensity of the spin-lock field offers the possibility to map the low-frequency part of the spectral density function. In the range normally available for the radio-frequency field, spectral densities corresponding to molecular motion are frequency independent (Figure 9B.4); all changes in $T_{1\rho}$ with the change of radio-frequency can be attributed to the chemical exchange process. For the case of random exchange between two sites, the correlation function has a Lorentzian form [23]:

$$J_e(\omega_r) = \frac{\tau_{ex}}{1 + (\omega_r \tau_{ex})^2}$$

$$(9B.34)$$

where the correlation time of exchange process is $\tau_{ex}^{-1} = \tau_A^{-1} + \tau_B^{-1}$ (see Chapter 3, equation 3.142). Equation (9B.33) is strictly valid under conditions of fast exchange $\delta\omega\tau_{ex} \ll 1$ [23] but, as was shown in reference [22], can in addition be a good approximation in the case of intermediate exchange as long as single resonance is observed. To obtain parameters of exchange process, radio-frequency field dependence of $T_{1\rho}$ has to be measured and fitted into equation (9B.34). The method was used by Szyperski et al. [22] to detect two intramolecular exchange processes in BPTI. The first one with $\tau_{ex} = 1.3$ ms was associated with the isomerization of the chirality of the disulphide bond. The second process, with the exchange rate too fast to be evaluated from spin-lock experiments, was attributed to local segmental motion. When the spin-lock field satisfy the condition $\omega_r\tau_{ex} \gg 1$, contribution of chemical exchange into

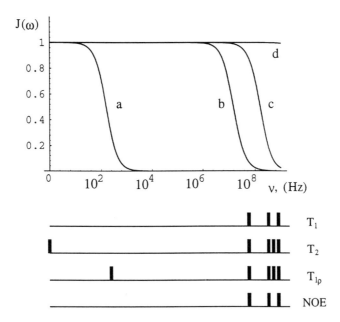

Figure 9B.4 Spectral density functions, calculated for different types of protein motions assuming exponential decay of correlation function and normalized to give unit value at zero frequency, $J(\omega) = 1/(1 + (\tau\omega)^2)$. (a) Chemical exchange with correlation time $\tau = 10^{-3}$ s; (b) overall isotropic tumbling with $\tau = 10^{-8}$ s, corresponding to protein of molecular weight ~15–20 kDa; (c) fast internal motion with $\tau = 10^{-9}$ s; (d) very fast internal motion with $\tau = 10^{-10}$ s. Positions of frequencies 0, ω_r, ω_N, $\omega_H + \omega_N$ and $\omega_H - \omega_N$ (from left to right), contributing to ^{15}N relaxation rates at 600 MHz are marked with vertical bars

$T_{1\rho}$ becomes negligible, and $T_{1\rho} = T_{1\rho}^0 \approx T_2^0$. The value $T_{1\rho}^\infty$ measured at high radio-frequency field is equivalent to the T_2 value in the absence of chemical exchange. In the following discussion of relaxation parameters, T_2 represents the transverse relaxation time in the absence of chemical exchange, normally measured as $T_{1\rho}^\infty$.

B.2.2.1.2 Motion

The standard set of relaxation parameters T_1, T_2 and NOE (equation (9B.31)) is not enough to derive the values of spectral density function independently of the model. The set of parameters can be extended to include the measurement of relaxation of two-spin orders [18, 19], thus offering the possibility of obtaining discrete values of the spectral density function directly from NMR experiments. However, even with known discrete values of the spectral density function, it is difficult to obtain the form of the appropriate correlation function, let alone derive a detailed picture of molecular motion. Thus, NMR relaxation parameters can only be used to reject correlation functions which fail to reproduce either the combination of discrete spectral density function values in the case of using T_1, T_2 and NOE, or the exact values of the spectral density function when the expanded set of relaxation parameters was measured; these correlation functions can be obtained either from theoretical analysis of motional models or by using molecular dynamics simulations.

The established approach for studying molecular mobility within a protein using NMR includes subdividing the motion of the particular internuclei vector into the slow motion of the molecule as a whole and faster independent local internal motion(s). The correlation function is then expressed as:

$$g(t) = g_0(t)^* g_I(t) \qquad (9B.35)$$

where $g_0(t)$ is the correlation function of *overall* motion and $g_I(t)$ is the correlation function of internal motion. Different models for the correlation function are described below followed by experimental results. It starts with the simplest correlation function corresponding to protein diffusion as a rigid body. This correlation function is then gradually modified to account for relaxation data that does not fit a simple model. The number of adjustable parameters is increased until it is equal to the number of relaxation values measured. A flowchart of a commonly adopted approach for the analysis of ^{15}N relaxation data for proteins is summarized in Figure 9B.5.

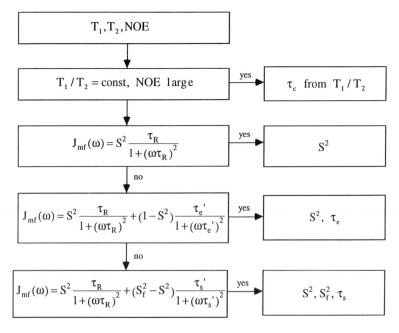

Figure 9B.5 Flowchart of ^{15}N relaxation data analysis

Overall isotropic motion. Isotropic tumbling of a macromolecule in solution corresponds to an exponential correlation function

$$g_o(t) = \exp(-t/\tau_R) \tag{9B.36}$$

and the spectral density function has a Lorentzian form

$$J(\omega) = \frac{\tau_R}{1 + (\omega \tau_R)^2} \tag{9B.37}$$

The shape of the spectral density function is shown in Figure 9B.4 for the standard set of NMR parameters of a small protein. It is characterized by fast decrease of intensity with values at high frequencies ω_H, $\omega_H + \omega_N$ and $\omega_H - \omega_N$ being very small in comparison to $J(0)$ and $J(\omega_N)$. The values of relaxation parameters T_1, T_2 and NOE calculated for isotropic overall motion of a protein are shown in Figure 9B.6. As can be seen from this figure, the rigid body isotropic motion model requires not only the relaxation parameters to be the same for all nuclei, but also puts restriction on possible combinations of relaxation parameters. Clear disagreement of such predictions with experimental results requires the extension of the model. One of the reasons for this disagreement can be residual chemical exchange broadening. Transverse correlation time T_2 monotonically increases with the decrease of correlation time (Figure 9B.6). Thus, the use of single overall correlation time for the whole molecule sets a lower limit for the expected T_2 due to mobility. The decrease of T_2 below this limit is usually interpreted as chemical exchange broadening $1/T_2^{ex}$.

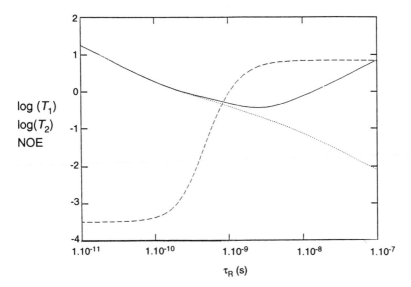

Figure 9B.6 Dependence of ^{15}N T_1 (————), T_2 ($\cdots\cdots$) and ^{1}H–^{15}N NOE (- — — — -) on the correlation time of overall motion for rigid body isotropic model

Overall anisotropic motion. If the shape of a protein is substantially non-spherical, motion of the individual internuclear vector depends strongly on its orientation relative to the main ellipsoid axes. For axially symmetric motion the overall spectral density function of equation (9B.37) has to be replaced with:

$$J(\omega) = \frac{A_1 \tau_1}{1 + (\omega \tau_1)^2} + \frac{A_2 \tau_2}{1 + (\omega \tau_2)^2} + \frac{A_3 \tau_3}{1 + (\omega \tau_3)^2} \qquad (9B.38)$$

where geometrical parameters A_1, A_2, A_3 depend on the angle φ between the HN vector and symmetry axis; correlation times τ_1, τ_2, τ_3 depend on the diffusion coefficients parallel and perpendicular to the symmetry axis. When protein conformation is known, equation (9B.38) can be used to estimate the contribution of anisotropy of motion into relaxation. For most of the proteins studied so far, estimated anisotropy was small enough to be neglected. Anisotropy of motion in solution was systematically studied for *calmodulin* [24], which according to X-ray structure has dumb-bell type conformation with a high degree of anisotropy. Experimental data showed that angular dependence of effective correlation time is much smaller than expected for the rigid structure. This was interpreted as an indication of the flexibility of central α-helix which connects the two parts of the dumb-bell.

Fast internal motion. The failure of spectral density function of equation (9B.37) with a single correlation time to reproduce experimental results is most striking for NOE values. The prediction of the rigid body model for proteins with correlation time >8 ns is an NOE value close to the theoretical maximum of 0.82 (Figure 9B.6). The experimental values for some residues were found to be small or negative. The main source of errors which gives rise to this discrepancy is the negligibly small high-frequency model values of the spectral density function in comparison to low frequency ones. The correlation function has to be modified to increase the relative values of the spectral density function at high frequencies. A convenient and easily interpretable model of the correlation function was proposed by Lipari and Szabo in their 'model-free' approach [14, 15]. In this model the additional internal fast motion was assumed to be present with a correlation function:

$$g_I^A(t) = S^2 + (1 - S^2)\exp(-t/\tau_e) \qquad (9B.39)$$

with order parameter S^2 representing the degree of spatial restriction and τ_e, the effective correlation time of internal motion. Being an approximation of the unknown internal correlation function $g_I(t)$, equation (9B.39) reproduces some of its characteristics, namely:

$$S^2 = g_I(\infty)$$

and

$$\tau_e(1 - S^2) = \int_0^\infty (g_I(t) - S^2)\, dt \qquad (9B.40)$$

which can be used to calculate S^2 and τ_e for different models of motion. The correlation function of equation (9B.39) accounts for two processes: fast partial loss of coherence owing to spatially restricted internal motion followed by slower total loss of coherence due to isotropic tumbling.

From equations (9B.32), (9B.35), (9B.36) and (9B.39) the 'model-free' spectral density function $J_{mf}(\omega)$ is now:

$$J_{mf}(\omega) = S^2 \frac{\tau_R}{1 + (\omega\tau_R)^2} + (1 - S^2) \frac{\tau'_e}{1 + (\omega\tau'_e)^2} \qquad (9B.41)$$

where $\tau'_e = (1/\tau_R + 1/\tau_e)^{-1}$. As is expected from general principles, introduction of faster motion increases the high-frequency values of the spectral density function. This has a different influence on relaxation parameters, as shown in Figure 9B.7. Transverse relaxation rate $1/T_2$ shows small dependence on τ_e especially at large S^2, because $J(0)$ is the main contributor to the relaxation rate. Longitudinal relaxation rate $1/T_1$ depends on high-frequency values of the spectral density function, which can be comparable for overall and internal motions. This creates a complex behaviour with small dependence of $1/T_1$ on S^2 at certain values of τ_e (~1 ns for $\tau_R = 10$ ns, Figure 9B.7). NOE values also demonstrate complex dependence on τ_e and S^2 since they are influenced by the ratio of high-frequency terms.

When internal motion is extremely fast ($\omega\tau'_e \ll 1$) and the order parameter is not too small [$S^2\tau_R \gg (1 - S^2)\tau'_e$], the spectral density function can be simplified to:

$$J_{mf}(\omega) \approx S^2 \frac{\tau_R}{1 + (\omega\tau_R)^2} \qquad (9B.42)$$

In this approximation, according to equation (9B.42), the ratio T_2/T_1 depends only on overall correlation time and is constant for nuclei with internal motion which satisfy extremely fast conditions (Figure 9B.7). In this approximation the NOE value is also independent of internal motions and close to the maximum value. It is proposed that the T_2/T_1 ratio can be used for evaluation of the overall correlation time when a substantial proportion of protein signals have similar values for this ratio with an additional criteria of large NOE values [25]. As is illustrated in Figure 9B.7, T_1/T_2 ratio remains constant for $S^2 \geqslant 0.6$ and $\tau_e \leqslant 20$ ps. A decrease of S^2 and an increase of τ_e lead to a decrease of the T_1/T_2 ratio.

Slow internal motions. The 'model-free' approach, as proposed originally, was found to be adequate to describe relaxation data for most of ^{15}N in proteins; however, in some cases, combinations of relaxation parameter were found which could not be accounted for within the error of the experimental data [16, 17]. Additional internal motion had to be assumed to achieve agreement

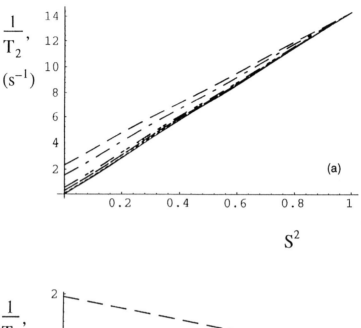

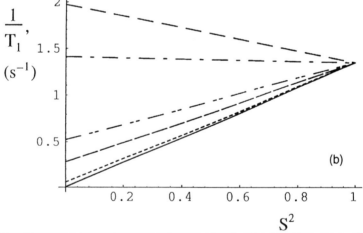

Figure 9B.7 Dependence of ^{15}N relaxation parameters (a) $1/T_2$; (b) $1/T_1$; (c) T_1/T_2; and (d) $^{1}H-^{15}N$ NOE on the order parameter S^2 in the 'model-free' approximation calculated for $\tau_R = 10$ ns, $\omega = 2\pi \times 600$ rad s^{-1} and different values of correlation time of the fast internal motion τ_e: 10^{-9} ns (— — —), 5×10^{-10} ns (—·—·—), 10^{-10} ns (—··—··—), 5×10^{-11} ns (··········) and 10^{-12} ns (———)

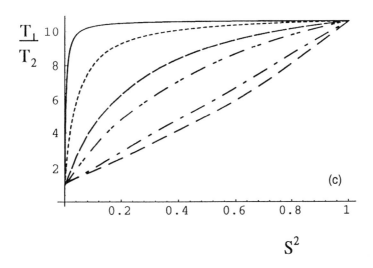

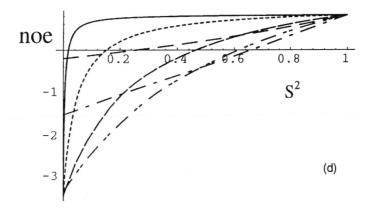

Figure 9B.7   (*continued*)

with experimental data. This led to an extended 'model-free' approach with a spectral density function [16, 17]:

$$J_{mf}^e(\omega) = S^2 \frac{\tau_R}{1+(\omega\tau_R)^2} + (S_f^2 - S^2) \frac{\tau_s'}{1+(\omega\tau_s')^2} + (1 - S_f^2) \frac{\tau_f'}{1+(\omega\tau_f')^2} \qquad (9B.43)$$

where subscripts s and f correspond to slow and fast internal motions, respectively. In the case of independent axially symmetric fast internal motion, $S^2 = S_f^2 S_s^2$. The number of adjustable parameters in equation (9B.43) is larger than the number of normally measured relaxation values (T_1, T_2 and NOE) and therefore the function has to be simplified to allow meaningful interpretation. The simplest assumption is for extremely fast additional motion, thus leading to a spectral density of

$$J_{mf}^e(\omega) = S^2 \frac{\tau_R}{1+(\omega\tau_R)^2} + (S_f^2 - S^2) \frac{\tau_s'}{1+(\omega\tau_s')^2} \qquad (9B.44)$$

The changes introduced by extended correlation functions to the values of relaxation parameters are discussed in reference [17]. When the generalized order parameter S^2 is kept unchanged, introduction of fast internal motion ($S_f^2 < 1$) can lead to a substantial increase in NOE value, while T_1 and T_2 remain unchanged. This is accompanied by the increase of τ_s in comparison with the case of single internal motion ($S_f^2 = 1$). Having three adjustable parameters guarantees that all combinations of T_1, T_2 and NOE can be accounted for when using the spectral density function equation (9B.44).

B.2.2.2   Strategy of relaxation data analysis

A block diagram of the currently adopted procedure for the analysis of relaxation data is shown in Figure 9B.5. It is based upon the correlation function approximations discussed above. For resonances with large NOE values the ratio T_1/T_2 is calculated. If a substantial number of the ^{15}N nuclei have similar values for the T_1/T_2 ratio, this ratio is used for evaluation of the overall correlation time. If large variations in the T_1/T_2 ratio are observed, nuclei for which this ratio is largest, or those which are expected to have most restricted internal motion (i.e. in regular elements of secondary structure) are used for τ_R evaluation. The overall correlation time is then fixed and further motional parameters are obtained. At later stages τ_R can be further optimized using a 'model-free' approximation of spectral density function, but in all published cases the final τ_R was very close to the initial estimation.

After finding the value of overall correlation time, a suitable model of the spectral density function has to be chosen. It has to predict relaxation parameters with the accuracy within experimental errors and contain a minimum number of adjustable parameters. One possible approach is to gradually increase the complexity of the model, until agreement between calculated and experimental data is achieved, as shown in Figure 9B.5. The T_1/T_2 ratio and NOE values can be used as initial criteria for choosing the model. When the T_1/T_2 ratio is *close* to the value used for τ_R evaluation, and NOE is close to the maximum value expected for the deduced τ_R, the simplest correlation function of equation (9B.42) is appropriate. When the T_1/T_2 ratio is *larger* than the value used for τ_R evaluation, but the NOE is still close to maximum, an additional exchange term R_{ex} has to be added for the transverse relaxation:

$$1/T_2^{obs} = 1/T_2 + R_{ex} \qquad (9B.45)$$

and R_{ex} is optimized together with S^2. Relaxation data which cannot be accounted for by the above model are analysed using the 'model-free' spectral density function of equation (9B.41) and, if necessary, the exchange term is included. If the agreement between calculated and experimental parameters is still not achieved, an extended 'model-free' spectral density function of equation (9B.44) is applied and R_{ex} can be used as well. As was pointed out by Powers *et al.* [26], there is an overlap region, when the data are described equally well by both 'model-free' function of equation (9B.41) and extended 'model-free' function of equation (9B.44). For such data τ_e normally has value of 0.5–1.0 ms. Two types of internal motions may still be present but impossible to resolve because the correlation time difference for these motions is less than 5–10-fold. The overall order parameter S^2 is generally the same for both models [16].

The second approach is to use Monte Carlo simulations to find an appropriate model of the spectral density function [27]. Two equations are normally checked separately with simulated relaxation parameters of each residue: 'model-free' function with a single correlation time of equation (9B.41) and the fast approximation of 'model-free' function of equation (9B.42) together with exchange term of (9B.45). If the value of τ_e and/or R_{ex} obtained from the analysis of simulated data is found to be non-zero (within 95% confidence limit), then the corresponding parameter is included into the spectral density function for final analysis. Data for which the agreement between calculated and experimental parameters cannot be achieved are analysed using an extended 'model-free' function. This statistical approach has the advantage of being more systematic in checking different models.

Non-linearity of the equations for relaxation parameters requires the minimization of target function:

$$R^2 = \left[\frac{T_1^{exp} - T_1^{cal}}{T_1^{cal}}\right]^2 + \left[\frac{T_2^{exp} - T_2^{cal}}{T_2^{cal}}\right]^2 + \left[\frac{NOE^{exp} - NOE^{cal}}{NOE^{cal}}\right]^2 \qquad (9B.46)$$

in order to obtain 'model-free' parameters. A suitable function is substituted for the spectral density, as described above. Errors in calculated parameters are estimated statistically by generating artificial data sets with the values of relaxation parameters randomly chosen within experimental errors from measured ones. Minimization of the target function (equation (9B.46)) is repeated with generated data and 'model-free' parameters thus obtained are used to estimate errors [28]. It was found that accuracy in determination of S^2 is much higher than for τ_e.

B.2.2.3   Experimental results and examples

Pulse sequences for measuring relaxation parameters T_1 and T_2 are described in references 19, 29, 30 and NOEs in references 19 and 25 and in Chapter 2 of this book. They are shown in Figure 9B.8. The sequences are based upon INEPT transfer of magnetization from ^{1}H to ^{15}N and back to increase sensitivity of the experiments. For T_2 measurements CPMG sequence [20] is applied in the delay T to reduce the effect of chemical exchange and zero quantum magnetization exchange on T_2 values; alternatively, heteronuclei spin-lock pulse can be used for these purposes [19]. The time interval between 180° pulses in CPMG or spin-lock field strength B_{s1} sets the upper limit on the lifetime of the exchange process which can still contribute into the linewidth. This was found to be approximately 10 ms for δ approximately 0.5 ns [16]. The repetition rate of CPMG sequence has also to be fast enough to prevent exchange between inphase and antiphase heteronuclear coherences. ^{1}H 180° pulses are applied every 5–10 ms during the delay T in T_1 and T_2 sequences to suppress the effect of ^{1}H–^{15}N dipolar/CSA cross-correlation [29, 30]. For the T_2 sequence a 180° ^{1}H proton pulse is applied at the spin-echo maximum. When present, dipolar/CSA cross-correlation can cause as much as 30% error in T_2 values [30]. Errors in T_1 due to cross-correlation are much smaller, but can still be significant. Relaxation delay in the NOE experiment has to be long enough for heteronuclear magnetization to reach equilibrium. It is normally 4–6 s, depending on the size of the protein. Proton saturation can be achieved by the application of a train of 120° pulses or broad-band decoupling. The saturation time has to be at least 3s for the NOE to reach maximum value. Relaxation delays (RD) in T_1 and T_2 experiments are shorter than in the NOE experiment since they are determined by the recovery of proton magnetization.

Results of relaxation analysis for different proteins are summarized in Table 9B.3. In almost all the proteins analysed, a majority of the backbone ^{15}N had very similar values of T_1/T_2 ratios for residues in the protein core. These ratios were used to determine correlation times of the molecules and in some cases were further refined with appropriate models for the spectral density function. One exception was the (30–51) intermediate of BPTI, for which T_1/T_2 ratio varied significantly for different residues due to the lack of a rigid core [31]. In this case the largest T_1/T_2 ratio was used to evaluate correlation time of the molecules, assuming that the chemical exchange effect was totally suppressed.

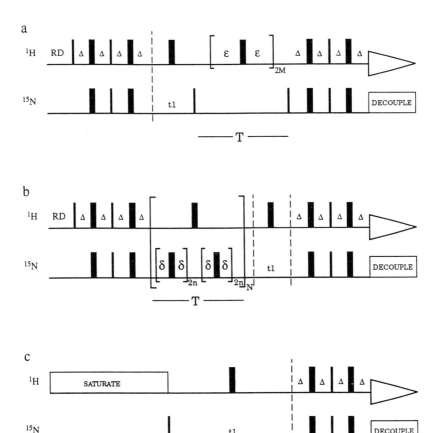

Figure 9B.8   Pulse sequences for measuring heteroatom (a) T_1 relaxation time; (b) T_2 relaxation time and (c) $^1\text{H}-^{15}\text{N}$ NOE. Thin and thick vertical bars represent 90° and 180°, pulses, respectively. For polarization transfer, $\Delta = 1/(4J_{NH})$. ^1H pulses are applied every 5–10 ms in order to suppress the effect of $^1\text{H}-^{15}\text{N}$ dipolar/CSA cross-correlation. The delay between ^{15}N 180° pulses in CPMG sequence is ~0.5 ms

   Relaxation data for most of the residues in the protein core were found to be in agreement with either the simplest model of extremely fast internal motion (equation (9B.42)) or the model with single correlation time (equation (9B.41)) under conditions when internal motion is still very fast, $\tau_e \leqslant 50$ ps, see Table 9B.3. Practically all residues in well defined elements of secondary structure belong to this category and reported values of the order parameter S^2 for α-helices and β-sheets are close to 0.85 ($\pm$0.04) (see references to Table 9B.3). At the same time a high proportion of residues which are not involved in the formation of secondary structure also have large order parameters and, thus, highly restricted fast motions. Smaller order parameters and/or the

Table 9B.3 Summary of ^{15}N relaxation data analysis for different proteins

Protein	Number of residues	τ_R (ns)	Model[a]	S_f^{2} [b]	S_s^2	τ_e [c] (ns)	$\Theta°$	$\varphi°$	Reference
Interleukin-1β	153	8.3	[1] 96	0.83±0.5			20.7±3.3		16
			[3] 32	0.79±0.05	0.25–0.95	0.5–4	20.7±3.3	28±14	
Ribonuclease H	134	10.4	[1] 29	0.88±0.06			17±5		26
			[2] 48	0.88±0.06		0.029–0.84	17±5		
			[3] 30	0.7–0.95	0.45–0.95	≤3.9	28–11	63–15	
Calbindin D$_{9k}$	75	4.25	[2] 65	0.84±0.03		0.015±0.01	19±2		33
			[3] 6	0.66–0.79	0.38–0.7[d]	1.1–2.7	30–23	65–40	
Glucose permease IIA domain	162	6.24	[1] 63	0.75–0.95		0.01–0.14	25–11		27
			[2] 65	0.75–0.95	0.34–0.53[d]	0.5–1.3	25–11		
			[3] 8	0.71–0.8			27–22	69–52	
Interleukin-4	129	7.6	[1] 82	0.9		0.04±0.02	15		33
			[2] 8	0.83			20		
			[3] 23	0.79±0.08	0.75±0.14	1.85±0.73	22±5	34±14	
FK506 binding protein	107	9.2	[1] 23	0.88±0.06			16±4		34
			[2] 71	≤0.66		0.0–0.45	>30		
(30–51) BPTI intermediate	58	12.2	[2] 38	0.06–0.9		~1	69–15		31

[a] Model [1]–[3] correspond to correlation functions, given by equations (9B.42), (9B.41) and (9B.44), respectively. Models are followed by number of residues analysed by the model. ^{15}N nuclei with additional chemical exchange broadening are also included.
[b] In the case of model [1] and [2] S_f^2 corresponds to order parameter S^2.
[c] In the case of model [3] τ_e corresponds to τ_s.
[d] S_s^2 of one residue is less than 0.25; this disagrees with two-site jump model.

presence of slower internal motion was found in the loop regions and N- and C-terminal parts of proteins. These regions are considered to be flexible and normally correspond to poorly defined regions in X-ray and NMR structures. Large values of exchange broadening, indicating internal motions on milli- and microsecond timescales, are found in many regions of protein structure. The correlation between slow internal motions and secondary structures appears to be better than for the fast motions. Residues in well defined secondary structure have small values of exchange broadening, while loops often exhibit larger scale slow motions.

A 500 ps molecular dynamic simulation (MD) was performed for interleukin-1β, surrounded by bulk water molecules [35] and the results of the calculations were compared with experimental data of reference [16]. Correlation functions for all backbone N–H vectors show ultrafast subpico-second decay to a value of ~0.85. The behaviour of correlation functions can be subdivided into three classes. Firstly, for the majority of vectors, correlation function decays a little further and stabilizes at the value ~0.8, corresponding to the calculated value of S^2. These backbone NHs participate in stable hydrogen bonds, which were not broken during simulation. Fast small amplitude mobility, found in MD simulation, for the residues, is in agreement with ^{15}N relaxation measurements. The second type of H–N vectors demon-strates larger amplitude orientational changes on the 10–50 ps timescale with S^2 values as low as 0.2. These residues are localized in loop 3 with no specific secondary structure. Predicted flexibility of this region is consistent with X-ray temperature factors but NMR relaxation parameters do not register increased mobility in the region. The third kind of behaviour is observed for some residues in loop 1. Here infrequent transitions occur between different configurations of hydrogen bonds. This results in N–H vectors jumps between states with well-defined orientation with small angle fluctuation between transitions. The rates of the transitions cannot be quantified from MD trajectory owing to insufficient time span. The same residues required the use of an extended two correlation time 'model free' function to facilitate analysis of the relaxation data. MD simulations failed to predict enhanced mobility in loop 1 near residue 50, which was ascribed to the shortness of calculated trajectory.

Direct evaluation of spectral densities.

Analysis of experimental data described above was based upon the assumption about the shape of correlation function. A different approach was proposed in references [18] and [19] and applied to the small protein *Eglin c*, a 70-residue proteinase inhibitor. The results obtained are briefly described here. Measure-ment of additional two-spin relaxation rates allows one to derive directly spectral density function values at discrete frequencies. It was found that low-frequency components of spectral density function $J(0)$ and $J(\omega_N)$ are the most sensitive to the presence of internal motion and decrease significantly in flexible regions. The values of high-frequency components $J(\omega_N + \omega_H)$, $J(\omega_H)$ and $J(\omega_H - \omega_N)$ showed no correlation with the flexibility, estimated from the precision of structure determination. Comparison of the 'model-free' prediction of spectral density function and its direct measurement showed that when $\tau_e < 25$ ps, equation (9B.41) was adequate to reproduce spectral density values within the experimental error. Discrepancy increased when the internal motion was found to be slower. The extended model is probably required in such cases. The main difference between 'model-free' prediction and direct estimation was in the high-frequency values. The values $J(\omega_H)$ and $J(\omega_H - \omega_N)$, evaluated directly, were systematically larger than $J(\omega_N + \omega_H)$, while 'model-free' function predicts monotonic decrease with the increase of frequency. A monotonic decrease of spectral density is the requirement of a general diffusion model of molecular motion [15]. Thus the discrepancy observed is most likely to be due to systematic experimental errors, as was pointed out in reference [18]. The accuracy of direct spectral density evaluation remains to be justified.

B.2.2.4 Motional model interpretation of relaxation data

Correlation functions derived from different motional models or calculated from molecular dynamics simulations are in general agreement with the form adopted by 'model-free' approach. The most popular model now is the 'diffusion in a cone'. This model assumes that the motion of interproton vector is restricted by a cone of angle Θ with uniform distribution of orientation within the cone [36, 37]. The order parameter is related to Θ by:

$$S^2 = [0.5 \cos \Theta (1 + \cos \Theta)]^2 \tag{9B.47}$$

The spectral density function of extended 'model-free' approach can be interpreted by superposition of fast diffusion within a cone and slow jump between two states as illustrated in Figure 9B.9. Order parameter S_f^2 (see equation (9B.43)) is related to semi-angle Θ of the fast motion cone by equation (9B.47). Under the simplest assumption of equal population of two states, the angle φ_s between average positions of an HN vector in each of them is related to the order parameter of slow motion by [16]:

$$S_s^2 = (1 + 2 \cos^2 \varphi_s)/4 \tag{9B.48}$$

Reported values of φ and Θ for some proteins are presented in the Table 9B.3. As expected, the most restricted motions with $\Theta \sim 20°$ were found for the residues in the regular secondary structure elements. High mobility with large values of Θ are normally characteristic for N- and C-terminal and unstructured loops. In some cases slow motion was also found in such regions.

An alternative model relates fast internal motion to the variation in the effective bond length. This model gives an adequate description of

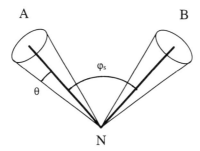

Figure 9B.9 Model for the internal motion on two time scales: very fast (<20 ps) and fast (0.5–4 ns). The slower motion corresponds to a jump of HN vector between two states A and B with the change of an orientation by angle φ_s. Faster motion is represented by a free diffusion in a cone with semi-angle Θ, centred about the average orientation of the vector in each state. NH vector in each state is represented by a thick bar

experimental results. The maximum change in the effective bond length required by this model was reported to be ~10% [16].

B.2.2.5 Homonuclear relaxation

Dipole–dipole interaction between protons is the main mechanism for homonuclear relaxation. Equations describing proton relaxation parameters are analogous to equations (9B.31) and can be found in reference [10]. The main difference to the heteronuclear case is in the contribution of several protons to the relaxation of a proton of interest. Distances to surrounding protons are often known with a low degree of accuracy and correlation functions associated with each interproton vector can be rather different. This makes proton relaxation data difficult to interpret in terms of motions of interproton vectors. The only relaxation parameter which can be directly associated with the individual interproton vector is the cross-relaxation rate and this is measured in 2D NOESY experiments. However, the problem of unknown interproton distance remains, so cross-relaxation rates can be most successfully used only if distance between protons is chemically fixed. Nevertheless, some dynamics information can be extracted as illustrated by the following examples.

The usual approximation for the correlation function is a 'model-free' form under conditions of extremely fast internal motion, equation (9B.42). Under the condition $\omega\tau_R \gg 1$, which is normally valid for proteins with molecular weight $\geqslant 5$ kDa, $J(0)$ gives the main contribution to proton relaxation rates. The other simplified assumption used when considering relaxation times T_1 and T_2 is that the spectral density function is associated with a proton, rather than interproton vector. This allows one to separate geometrical and motional terms.

Relaxation time T_2 of aromatic residues and resolved high-field methyl groups was used in [38] to estimate mobility of different domains in the multidomained protein, *urokinase*. Using the simplifications mentioned above, spin–spin relaxation of proton i is now given by the equation:

$$\frac{1}{T_{2i}} = \frac{1}{4} (\gamma_h^2 \hbar)^2 S^2 \tau_R \sum_j \left(\frac{1}{r_{ij}^3}\right)^2 \qquad (9\text{B}.49)$$

where r_{ij} is interproton distance and summation is over all protons contributing to the relaxation. The unknown geometrical parameter $\Sigma_j (1/r_{ij}^3)^2$ was assumed to be close for different aromatic ring protons or protons in methyl groups for residues inside protein molecules and estimated from X-ray structure of lysozyme. T_2 values were evaluated from proton linewidth in 1D and 2D spectra and known coupling constants. Chemical exchange contribution was considered negligible. Parameter $\tau_{\text{eff}} = S^2 \tau_R$ was calculated using equation (9B.49) and systematic difference was found between protons in different domains with much smaller deviations for protons in the same domain. By using average order parameter values S^2 of 0.6 and 0.8 for methyl groups and aromatic ring protons, respectively, as estimated from molecular dynamics calculations, overall correlation times were obtained from τ_{eff}. These were found to be 15.3 ns for the kringle domain and 26.0 ns for the protease domain of urokinase with good agreement between the data obtained for different proton types. The presence of an extensive independent domain motion was concluded. This example demonstrates how a large number of assumptions have to be used to analyse homonuclear relaxation data and the limited dynamics information available.

When the interproton distance is fixed chemically or conformationally, cross-relaxation rate σ can be used to estimate proton intramolecular mobility. With only one relaxation parameter directly associated with the motion of interproton vector being measurable, spectral density function has to be simplified as described above. In reference [39] order parameter S_{ij}^2 was associated with each interproton vector of DNA dodecamer duplex and adjusted to achieve the best agreement between calculated and experimental time dependence of NOESY cross-peak intensities. For most proton pairs, the order parameter was found to be close to 1.0, while some interproton vectors had order parameters as low as 0.6. Similar results were obtained for DNA octamer [40] using a combination of NMR and molecular dynamics simulation.

B.3   Slow Large-scale Exchange. Hydrogen–Deuterium Exchange

Hydrogen exchange rates, as detected by NMR, monitor the internal motions and flexibility of folded proteins since it is widely agreed that fluctuations are required to allow exchange of labile hydrogens which are completely or partially buried within a protein structure. From the hydrogen exchange studies of a number of small proteins, it is concluded that the hydrogen exchange rate can be influenced by a wide variety of dynamical features, ranging from local side-chain motions to cooperative global unfolding. Hence, one of the most important features of amide hydrogen exchange rate determination is the vast range of exchange rates that can be detected in a single protein; this timescale can vary from being very fast ($>10^{-3}\,\mathrm{s}^{-1}$) to several months, although the timescale for the slow exchanging amide protons varies from protein to protein. It is important to distinguish between the use of amide exchange rate to deduce protein stability and protein motions from its use for studying protein unfolding processes. Since this chapter is concerned with the dynamics of intact proteins, the discussion below will focus on the use of amide exchange rates for the former purpose rather than for studying the unfolding/folding process although the dynamics of a protein and its unfolding/folding characteristics are related.

B.3.1   MECHANISMS OF HYDROGEN EXCHANGE IN A NATIVE PROTEIN

The hydrogen exchange mechanisms described here will focus on those that are applicable to the intact native protein. Despite the many studies available, there is still little agreement on how hydrogen exchange occurs in proteins and what dynamic processes are responsible for the exchange of interior protons. The wide variation in exchange rates of the different amide protons within a protein ($>10^{-2}$ to $10^{-9}\,\mathrm{min}^{-1}$) implies that it does not occur entirely by infrequent transient unfolding of the entire protein. Two models are currently used: the local unfolding model and the non-cooperative fluctuations or (penetration) model [41].

The first model involves the concept of local unfolding or 'breathing' of the protein. This idea was initially due to Linderstrom-Lang and includes the presence of a variety of transient 'open' conformations of the protein, with the exchanging groups exposed to the solvent:

$$\text{folded} \underset{k_{-1}}{\overset{k_1}{\rightleftharpoons}} \text{'open'} \overset{k_{ex}}{\rightleftharpoons} \text{hydrogen exchange}$$

A variety of 'open' states need to be postulated in order to account for the different rates of exchange of the different sites.

The rates of the process are generally considered in terms of whether the rate-limiting step is the opening of the structure, the EX_1 mechanism, or the exchange reaction, the EX_2 mechanism. In only the latter case is the amide exchange rate sensitive to factors that are known to affect the intrinsic exchange reaction in model compounds, in particular pH. An EX_1 mechanism requires that $k_{ex} > k_{-1}$, hence the experimentally derived rate of amide exchange gives directly the value of k_1, the rate of opening of the protein structure. An EX_2 mechanism will apply if $k_{ex} < k_{-1}$, so that the observed amide exchange rate is $K_1 k_{ex}$ (with $K_1 = k_1/k_{-1}$), that is, the equilibrium constant for the breathing process (opening) of the protein is obtained. Proteins demonstrate the EX_2 type of exchange under most conditions; in only a few cases, particularly over a narrow range of high pH and temperature, where the exchange is very rapid, is the EX_1 mechanism observed, as was shown for the slowest exchanging protons of BPTI [42]. The hydrogen exchange rate constant, k_{ex}, for the chemical exchange step of a peptide NH group fully exposed to solvent (with no contribution from protein conformation) is given by:

$$k_{ex} = k_H[H^+] + k_{OH}[OH^-] + k_0$$

where k_H and k_{OH} are pseudo first-order rate constants for acid and base-catalysed exchange, k_0 is the rate constant for uncatalysed exchange with water. Base-catalysed exchange proceeds through an N-protonation mechanism [43]. Acid catalysed exchange involves an O-protonation mechanism [44, 45]. Both acid- and base-catalysed exchange mechanism are significant, with the slowest rate occurring at pH 3 where acid- and base-catalysed processes are equal in rate; the rate increases ten-fold for each unit change in pH away from the minimum value. The activation energy for the intrinsic chemical exchange step is between 71 and 83 kJ mol^{-1}; that is a threefold increase in the exchange rate for every 10° increase in temperature.

The alternative model for amide proton exchange in the folded native protein is the penetration model in which clusters of the NH groups are not exposed to the solvent by a single, cooperative open/close transition. Rather, the transient exposure to solvent of buried groups in folded proteins is mediated by small fluctuations of less that 1 Å of many atoms about their average positions. The hydrogen bonds present are envisaged to break and reform rapidly as the donor and acceptor atoms fluctuate. Occasionally, the donor and acceptor groups form transient hydrogen bonds with migrating water molecules, resulting in hydrogen exchange. In the penetration model, exchange takes place in a conformation similar to the crystal structure. The specific mechanisms suggested for such non-cooperative fluctuations are defects in the protein packing and transient formation of channels from bulk solvent to interior sites. This model predicts that neighbouring protons will have very different exchange rates and activation energies; that is, the exchange is highly localized and this has been illustrated in BPTI, and lysozyme. For example, in *lysozyme*, non-helical but adjacent amide protons exchange at very different rates, and in some helices, the exchange rates are more rapid on the solvent side of the helix and slower on the packed side, these observations being inconsistent with a local unfolding mechanism [46].

B.3.2 METHODS FOR MEASURING AMIDE EXCHANGE RATES BY NMR

For the purpose of studying protein flexibility and mobility, hydrogen–deuterium exchange rates are measured by following a time-course of the amide exchange for a period ranging from several days to months. The fastest exchanging amide protons ($>10^{-3}$ s^{-1}) are identified simply by direct comparison of the 2D spectra obtained with and without water suppression. In the absence of cross-relaxation, any attenuation of amide resonances with presaturation implies that the corresponding amide proton is exchanging relatively rapidly with water. For the more slowly exchanging amide protons, the exchange rates are determined by measuring the intensities of the amide proton resonances at several time intervals after dissolution of the protein sample in D_2O. The time intervals for making these measurements and the total time-course are governed by the range of exchange rates to be measured; generally, the rate of exchange is slowest at around pH 3. There is also a temperature dependence of the amide exchange rates; in some cases, the temperature can be altered to make some of the rates more amenable for exchange rate measurements.

B.3.3 INTERPRETATION OF AMIDE EXCHANGE RATES

The amide exchange rates of the small protein bovine pancreatic trypsin inhibitor (BPTI, M_w 6.5 kDa) have been the most thoroughly determined and characterized by NMR [47, 48, 49]. Other proteins include lysozyme [46, 50], cytochrome c [51, 52], ribonuclease A [53], barnase [54], staphylococcal nuclease [55] and α-lactalbumin [56]. In the case of **BPTI**, 10 min after dissolving the protein in D_2O at pH 3.5 and 30 °C, about 34 out of the 58 backbone amide protons are observed. The most rapidly exchanging amides, with rates $>10^{-2}$ min^{-1}, have been identified to be from residues found mainly on the protein surface, exposed to the solvent and not hydrogen bonded in the crystal structure: 1–4, 8–9, 11–13, 15, 17, 25–26, 37, 40, 42–43, 46–50 and 58 [49]. These residues are also either at the N or C-termini, or in unstructured loop regions in the three-dimensional structure of the protein. For residues 51–56 within the α-helical region, the exchange rates span a range of 2.4×10^{-4} to 5.1×10^{-5} min^{-1}. The residues at the ends of this helix, 47–49 and 57, exchange faster than those in the centre. The slowest exchanging amide protons have amide hydrogen exchange rate constants in the range of 3×10^{-6} to 2×10^{-9} min^{-1} and have been identified as residues 18, 20–24, 29, 31, 33, 35, 44 and 45. The amide protons of these residues are observed 60 days after dissolution in D_2O. Most of these 'slow exchanging' residues are located within the β-sheet region (which spans from residues 16 to 36) and are involved in the typical interstrand hydrogen bonding network in a β-sheet. Exchange is slower towards the centre of the β-strands than at either ends.

Amides protons that exchange with the slowest rates are from residues Y21, F22 and Y23 with an exchange rate of approximately $2 \times 10^{-9} \, min^{-1}$. These residues, together with the side-chains of residues 30, 33, 43, 45 and the backbone of residues 20–24, 30–33 and 45 make up the core of BPTI. They are predominantly hydrophobic aromatic rings, and in the crystal structure these residues have the lowest b factors. This slow exchanging core may well be the protein folding core, that is the slowest exchanging amides will be the first to be protected from exchange during the refolding process.

From the above example of BPTI, the differential rates of amide exchange must be explained in terms of the hydrogen bonding patterns and the accessibility of the exchanging amide proton to both the surrounding water and the catalyst of exchange. The presence of hydrogen-bonding alone cannot account entirely for the slow rate of amide exchange because in proteins, all NH's are hydrogen bonded either intramolecularly or intermolecularly to water molecules. It is often that the slowest exchanging protons are the ones that are both most buried in rigid regions of the protein and are at the same time involved in hydrogen bonding. From the many amide exchange rate studies of proteins by NMR, one of the most useful results has been the identification of the slowest exchanging protons as the protein folding core; that is, the last hydrogens to exchange can be used to mark the region where hydrogens are first protected during the refolding process. Between the fastest exchanging surface amide protons and the slowest exchanging core, the intermediate amide proton exchange rates are used to identify the regular secondary structure and the loop regions, with amide protons in the loop region exchanging faster than those in the regular secondary structures. The manner in which the amide exchange rates is distributed throughout the polypeptide backbone will also give an indication of the cooperativity of the local fluctuations; that is whether amide protons from an entire region (either sequentially along the polypeptide backbone or in spatial proximity) exchange at similar rates.

Is there a correlation between the global stability and the rate of amide exchange from the folded state? One of the main problems in trying to address this question is finding a suitable set of exchangeable protons that serve as indicators of global stability. The conventional method has been to identify slow exchanging protons in the molecule which are not involved in residual structure after the major unfolding transition. Because these protons exchange as a result of global fluctuations, their exchange with the solvent can be correlated with thermal stability. However, using the single-site mutant approach described by Woodward [48] in which the amide exchange characteristics of single site mutants of BPTI are examined, a different result can be obtained. For example, in the less destabilized mutants G37A and N44G, either there are no effects or rates are increased for the NH in the vicinity of the mutation; in the more destabilized mutants, Y35G and N43G, there is widespread but variable acceleration of rates. In other words, there is no general correlation between global stability and amide proton exchange rates from a folded protein. Hence, although there is correlation between local protein stability and amide hydrogen exchange rates, this may not be generally true for the whole molecule.

B.4 Conclusions

For many years, NMR has been used in various ways to study and elucidate dynamic properties of macromolecules. However, many of the earlier studies have been limited, predominantly due to the inability both to perform experiments and to analyse the results in sufficient detail to yield the desired dynamic information. In recent years, some of these problems have in part been solved, as a result of three major advances. First, many approaches are now available to afford sequence specific assignments of the NMR resonances; these techniques have stemmed from the ability to make large amounts of recombinant proteins expressed in *E. coli* and include stable isotope-labeling and site-directed mutagenesis. Secondly, in combination with the previous set of developments, spectrometer hardware is currently capable of performing many types of experiments more efficiently and effectively than have previously been possible. The ability to resolve and assign many of the NMR resonances means that the amount of NMR information that can be obtained can be 'adjusted' accordingly, depending on the biological questions to be addressed. For example, specific sites can be targeted if the interest is in the active site of the protein, rather than in the whole protein system. The ability to investigate individual sites also means that more systematic approaches can be undertaken to unravel the complexity of the dynamical behaviour present in a macromolecule. As a result of the latter, the many theoretical models for analysing NMR relaxation data can be developed, tested, and improved, thereby allowing more meaningful information to be obtained. The third significant development in recent years is the availability of low-cost powerful computers which make it more practical to perform many of the calculations required in order to extract dynamic information.

We have attempted to show in this section the different approaches that can be adopted in order to study dynamics at different timescales in macromolecules. The methods discussed are relevant to situations ranging from when only a small part of the macromolecule system is of interest (for example the chemical exchange experiments) to when the whole macromolecule is studied from the point of view of defining flexible functional regions (relaxation and amide hydrogen exchange measurements).

Acknowledgment

We thank Dr G. H. E. Scott for making Figure 9B.2 available to us.

References

1. I. D. Campbell, C. M. Dobson, G. R. Moore, S. J. Perkins and R. J. P. Williams, *FEBS Letters*, **70**, 96 (1976).
2. C. W. Hilbers and D. J. Patel, *Biochemistry*, **14**, 2656 (1975).
3. P. E. Evans, R. A. Kautz, R. O. Fox and C. M. Dobson, *Biochemistry*, **28**, 362 (1989).
4. A. Gronenborn, B. Birdsall, E. I. Hyde, G. C. K. Roberts, J. Feeney and A. S. V. Burgen, *Molecular Pharmacology*, **20**, 145 (1981).
5. E. I. Hyde, B. Birdsall, G. C. K. Roberts, J. Feeney and A. S. V. Burgen, *Biochemistry*, **19**, 3746 (1980).
6. J. A. Lenstra, B. G. J. Bolscher, S. Stob, J. J. Beintema and R. Kaptein, *European Journal of Biochemistry*, **98**, 385 (1979).
7. J. L. Leroy, E. Charretier, M. Kochoyan and M. Gueron, *Biochemistry*, **27**, 8894 (1988).
8. K. V. Vasadava, J. I. Kaplan and B. D. Nazeswara Rao, *Biochemistry*, **23**, 961 (1984).
9. L. Lian and G. C. K. Roberts, in G. C. K. Roberts (ed.), *Effects of Chemical Exchange on NMR Spectra. A Practical Approach*, IRL Press, Oxford (1993).
10. O. Jardetzky and G. C. K. Roberts, *NMR in Molecular Biology*, Academic Press, New York (1981).
11. B. D. Nageswara Rao, N. J. Oppenheimer and T.L. James (eds) *Nuclear Magnetic Resonance, Line-Shape Analysis and Determination of Exchange Rates*, vol. 176, Academic Press, London (1989).
12. R. R. Ernst, G. Bodenhausen and A. Wokaun, *Principles of Nuclear Magnetic Resonance in One and Two Dimensions*, Clarendon Press, Oxford (1987).
13. C. M. Dobson, L. Y. Lian, C. Redfield and K. D. Topping, *J. Magn. Reson.*, **69**, 201 (1986).
14. G. Lipari and A. Szabo, *J. Am. Chem. Soc.*, **104**, 4559 (1982).
15. G. Lipari and A. Szabo, *J. Am. Chem. Soc.*, **104**, 4546 (1982).
16. G. M. Clore, P. C. Driscoll, P. T. Wingfield and A. M. Gronenborn, *Biochemistry*, **29**, 7387 (1990).
17. G. M. Clore, A. Szabo, A. Bax, L. E. Kay, P. C. Driscoll and A. M. Gronenborn, *J. Am. Chem. Soc.*, **112**, 4989 (1990).
18. J. W. Peng and G. Wagner, *Biochemistry*, **31**, 8571 (1992).
19. J. W. Peng and G. Wagner, *J. Magn. Reson.*, **98**, 308 (1992).
20. A. Abragam, *The Principles of Nuclear Magnetism*, Clarendon Press, Oxford (1961).
21. J. W. Peng, V. Thanabal and G. Wagner, *J. Magn. Reson.*, **94**, 82 (1991).
22. T. Szyperski, P. Luginbuhl, G. Otting, P. Guntert and K. Wuthrich, *J. Biomol. NMR*, **3**, 151 (1993).
23. C. Deverell, R. E. Morgan and J. H. Strange, *Molec. Phys.*, **18**, 553 (1970).
24. G. Barbato, M. Ikura, L. E. Kay, R. W. Pastor and A. Bax, *Biochemistry*, **31**, 5269 (1992).
25. L. E. Kay, D. A. Torchia and A. Bax, *Biochemistry*, **28**, 8972 (1989).
26. R. Powers, G. M. Clore, S. J. Stahl, P. T. Wingfield and A. Gronenborn, *Biochemistry*, **31**, 9150 (1992).
27. M. J. Stone, W. J. Fairbrother, A. G. Palmer, J. Reizer, M. H. Saier and P. E. Wright, *Biochemistry*, **31**, 4394 (1992).
28. A. G. Palmer, M. Rance and P. E. Wright, *J. Am. Chem. Soc.*, **113**, 4371 (1991).
29. A. G. Palmer, N. J. Skelton, W. J. Chazin, P. E. Wright and M. Rance, *Molec. Phys.*, **75**, 699 (1992).

30. L. E. Kay, L. K. Nicholson, F. Delaglio, A. Bax and D. A. Torchia, *J. Magn. Reson.*, **97**, 359 (1992).
31. C. P. M. van Mierlo, N. J. Darby, J. Keeler, D. Neuhaus and T. E. Creighton, *J. Molec. Biol.*, **229**, 1125 (1993).
32. J. Kordel, N. J. Skelton, M. Akke, A. G. Palmer and W. J. Chazin, *Biochemistry*, **31**, 4856 (1992).
33. C. Redfield, J. Boyd, L. J. Smith, R. A. G. Smith and C. M. Dobson, *Biochemistry*, **31**, 10431 (1992).
34. J. W. Cheng, C. A. Lepre, S. P. Charnbers, J. R. Fulghum, J. A. Thomson and J. M. Moore, *Biochemistry*, **32**, 9000 (1993).
35. I. Chandrasekhar, G. M. Clore, A. Szabo, A. M. Gronenborn and B. R. Brooks, *J. Molec. Biol.*, **226**, 239 (1992).
36. G. Lipari and A. Szabo, *Biophys. J.*, **30**, 489 (1980).
37. G. Lipari and A. Szabo, *J. Chem. Phys.*, **75**, 2971 (1981).
38. U. K. Nowak, L. Xiang, A. J. Teuten, R. A. G. Smith and C. M. Dobson, *Biochemistry*, **32**, 298 (1993).
39. J. D. Baleja and B. D. Sykes, *J. Magn. Reson.*, **91**, 624 (1991).
40. T. M. G. Koning, R. Boelens, G. A. van der Marel, J. H. van Boom and R. Kaptein, *Biochemistry*, **30**, 3787 (1991).
41. T. E. Creighton, *Proteins: Structures and Molecular Properties*, W. H. Freeman, New York (1984).
42. H. Roder, G. Wagner and K. Wuthrich, *Biochemistry*, **24**, 7396–7407 (1985).
43. C. Perrin, E. Johnston, C. Lollo and P. A. Kobrin, *J. Am. Chem. Soc.*, 4691 (1981).
44. C. Perrin and C. Lollo, *J. Am. Chem. Soc.*, **106**, 2754 (1984).
45. E. Tuchsen and C. Woodward, *J. Molec. Biol.*, **185**, 421 (1985).
46. T. G. Pedersen, N. K. Thomsen, K. V. Andersen, J. C. Madsen and F. M. Poulsen, *J. Molec. Biol.*, **230**, 651 (1993).
47. G. Wagner and K. Wuthrich, *J. Molec. Biol.*, **160**, 343 (1982).
48. K. S. Kim, J. A. Fuchs and C. K. Woodward, *Biochemistry*, **32**, 9600 (1993).
49. K. S. Kim and C. K. Woodward, *Biochemistry*, **32**, 9609 (1993).
50. T. Pedersen, B. Sigurskjold, K. Andersen, M. Kjaer, F. Poulsen, C. M. Dobson and C. Redfield, *J. Molec. Biol.*, **218**, 413 (1991).
51. M. F. Jeng, S. W. Englander, G. Elove, A. Wand and H. Roder, *Biochemistry*, **29**, 10433 (1990).
52. H. Roder, G. A. Elove and S. W. Englander, *Nature*, **335**, 700 (1988).
53. A. D. Robertson and R. L. Baldwin, *Biochemistry*, **30**, 9907 (1991).
54. J. Clarke, A. M. Hounslow, M. Bycroft and A. R. Fersht, *Proc. Natl. Acad. Sc.*, **90**, 9871 (1993).
55. S. N. Loh, K. E. Prehoda, J. F. Wang and J. L. Markley, *Biochemistry*, **32**, 11022 (1993).
56. C. L. Chyan, C. Wormald, C. M. Dobson, P. A. Evans and J. Baum, *Biochemistry*, **32**, 5681 (1993).

10 LIQUID-LIKE MOLECULES IN RIGID MATRICES AND IN SOFT MATTER

Introduction

In this last chapter, a further degree of complexity is reached in the systems probed by NMR spectroscopy. These are essentially heterogeneous materials where, however, the mobility of the examined nuclei permits high-resolution NMR or magnetic resonance imaging. The general theme of the chapter concerns the observation of liquid-like molecules which are contained as inclusions or adsorbates in solid materials, or else which are constituting what is presently named soft matter. In this respect, Chapter 8 and, in part, Chapters 4 and 5 of this book were also dealing with (micro) heterogeneous solutions, in which, however, the viscosity was still of the same order of magnitude as that encountered in liquids. The objective of this chapter is to examine either fluid microsystems in a rigid matrix—itself not detected by NMR—or viscoelastic materials such are swollen polymers, gels, or biomaterials. These systems lie at the frontier between liquid and solid states so that methodologies of solid-state NMR are necessarily called to mind in some sections of this chapter.

Another objective is to show the great versatility of NMR methodologies, which are equally prone to yield unvaluable information on amorphous, structurally ill-defined, materials. There is, in fact, a wide panel of applications to illustrate this potential of NMR: medical imaging, agricultural science, soil chemistry, food science.... Three topics have been chosen which typically show the wealth of information brought by NMR spectroscopy in these fields. The first part of this chapter is devoted to swollen polymers and gels obtained by cross-linking polymeric chains, with emphasis on the restriction brought about over local microdynamics by the gelation process. The second topic concerns porous solid materials, unexpectedly using the chemical shifts in enclosed fluid molecules to probe the structure of the rigid matrix, as the result of molecular collisions onto the surface of solid catalysts or the walls of cavities in zeolites. The importance of dynamical considerations is stressed in the last part of this chapter, reporting recent developments of magnetic resonance imaging or relaxometry of water molecules to study biological tissues in food science.

Part A   Swollen Polymers and Gels

Part B   Fluids in and on Inorganic Materials

Part C   Magnetic Resonance in Food Science

10A SWOLLEN POLYMERS AND GELS

J. P. Cohen-Addad
Laboratoire de Spectrométrie Physique, University of Grenoble I,
France

A.1 Introduction

This chapter deals with the NMR observation of dynamic fluctuations which occur in polymeric systems observed above or near the glass transition temperature. The linkage of monomeric units induces necessarily a collective dynamical behavior of skeletal bonds which propagates along every chain segment. The strong interaction determined by the linkage originates two properties specific to polymer chains. The first one concerns the broad spectrum

Dynamics of Solutions and Fluid Mixtures by NMR
Edited by J.-J. Delpuech   © 1995 John Wiley & Sons Ltd

of relaxation times associated with internal fluctuations which affect every chain. The linear structure of macromolecules induces a hierarchy of random motions which occur within every chain. This hierarchy is quantitatively described by the Zimm model in dilute solution or by the Rouse model in a low molecular weight polymer melt [1,2]. The second property is the topological hindrance which has no equivalent in ordinary liquids; contrary to ideal chains, real macromolecular segments cannot intersect. Consequently, a length of uncrossable chain contour must be assigned to each macromolecule to account for both the finite size of skeletal bonds and their linkage [3]. This hindrance is specific to polymer chains which overlap; it occurs in semi-dilute or concentrated solutions or in a melt. It induces the effect of *dynamic screening*, namely the chain molecular weight independence of dynamic fluctuations, observed within short segments which pertain to long chains. The description of the dynamical behavior of one chain in a polymeric medium is focused on two crucial aspects.

(i) The first one deals with the characterization of the elementary random mechanism which underlies all chain fluctuations and which arises from collective rotational isomerizations of a few monomeric units. The characterization is usually based on the assumption that there exists a mean friction coefficient (see also Chapter 9A, Sections A.2.2 and A.4.3.2); the friction property is described by taking the *free volume* concept into consideration [4]. The friction coefficient is independent of the chain molecular weight. In this chapter, the presence of a free volume effect is illustrated by varying both the temperature and the concentration of polymeric solutions. The free volume is probed from solvent molecules or chain segments; it is also probed in permanent polymeric gels by detecting random motions of free chains throughout network structures.

(ii) The second main feature, about dynamic chain fluctuations, concerns the progressive onset of collective random motions of monomeric units which involve longer and longer segments and correlatively, longer and longer timescales of observation. Collective internal motions give rise to the translational and rotational diffusions of one chain considered as a whole; these random processes are characterized by one terminal relaxation spectrum which is strongly molecular weight dependent [5]. The longest relaxation time determines the diffusion coefficient of one macromolecule. The diffusion is a semi-macroscopic property which accounts for the ensemble of internal chain fluctuations; it is conveniently observed from NMR or light scattering.

Macroscopic dynamical properties of polymeric systems, observed above the glass transition temperature, are usually referred to the dynamic relaxation modulus, $G(t)$; for long chains, this viscoelastic quantity exhibits three characteristic domains of time evolution which reveal the hierarchy of internal chain fluctuations. Segmental motions are reflected from the evolution of $G(t)$, detected within short time intervals ($<10^{-6}$ s); in this domain, $G(t)$ is independent of the chain molecular weight, in agreement with the property of dynamic screening. Then, an effect of temporary elasticity appears as a plateau; its length is molecular weight dependent and it may last over about 10^{-4} s. The existence of a plateau gives macroscopic evidence for the dynamic screening effect. Finally, long-range fluctuations are observed over 10^{-2} s or more; they exhibit a strong chain length dependence. One of the most baffling problems, about chain fluctuations, concerns the understanding of the profound origin of the dynamic screening effect.

Details about macromolecular motions cannot be given within this short chapter, consequently short illustrations will be presented in order to show how NMR can be sensitive to several types of random motions. Segmental motions are probed from spin–lattice relaxation rates while the dynamic screening effect and semi-local fluctuations are disclosed from the transverse relaxation of the nuclear magnetization.

A.2   Principle of NMR Approaches

Nuclear magnetic properties observed in polymeric systems show a marked axial symmetry. By analogy with viscoelastic properties of molten polymers, it may be considered that the nuclear magnetization exhibits a dual relaxation behavior. The longitudinal component of nuclear spins presents a relaxation process analogous to that of the longitudinal magnetization observed in ordinary liquids whereas the transverse component exhibits a relaxation behavior analogous to that of nuclear spins embedded in a solid.

A.2.1   LONGITUDINAL RELAXATION

The spin–lattice relaxation is induced by a quasi-resonant exchange which occurs between the Zeeman energy of the spin system and the thermal energy of molecules which carry nuclear magnetic dipoles. The condition of quasi-resonant exchange of energy is used to calibrate the timescale of relaxation of molecular processes; the Larmor frequency plays the role of a reference. The exact calculation of relaxation rates necessitates an accurate description of random molecular motions which are nowadays well described in the case of small molecules in an ordinary liquid. In the case of polymeric systems, the calculation of relaxation rates comes against the description of segmental motions; the difficulty arises from the necessity to take collective rotational isomerizations of monomeric units into consideration. Several correlation times must be introduced to account for the collective random monomeric rotations. As a consequence of the existence of a spectrum of relaxation instead of a single correlation time, the spin–lattice relaxation rate is usually found to depend on the Larmor frequency on both sides around the maximum observed by varying the temperature of the polymeric system. A few models of segmental motions which involve several skeletal bonds have been proposed until now [6,7]. They are discussed in Chapter 9A of this volume. In the present chapter, experimental results are reported only to give a short illustration of the analysis which has been already proposed elsewhere [8]. Measuring the T_1^{-1} spin-lattice relaxation rate of protons or of ^{13}C nuclei attached to polymer chains, the function defined by the product $\omega_0 T_1^{-1}$ is considered. The analysis is based on the assumption that the ratio ω_0/T_1 obeys a property of homogeneity with respect to the variable determined by the product of the Larmor frequency ω_0 and one characteristic correlation time. Such a property implies that the spectrum of correlation times associated with random segmental motions represents multiples of one characteristic time whatever its exact definition; let τ_1^c denote this correlation time. The ratio ω_0/T_1 is written as a function F of the product $\omega_0 \tau_1^c$:

$$\omega_0/T_1 = F(\omega_0 \tau_1^c) \qquad\qquad (10A.1)$$

where τ_1^c may depend on both the polymer concentration ϕ and the temperature of the polymeric system. Equation (10A.1) shows that a frequency–temperature or a frequency–concentration scaling can be applied to the analysis of experimental results. Such an analysis has been already extended to several solvent–polymer systems [9]. In this chapter, the temperature–concentration scaling is illustrated by keeping the Larmor frequency as constant.

A.2.2   TRANSVERSE MAGNETIC RELAXATION: RESIDUAL INTERACTION

The transverse magnetization of nuclei attached to long chains exhibits a dynamical behavior very specific to molten polymers or to concentrated solutions. It is now well established that topological constraints exerted on long chains, in the liquid state, induce a pseudo-solid behavior of the transverse nuclear magnetization [10]. This pseudo-solid relaxation results from a residual spin–spin interaction due to segmental motions which appear necessarily as non-isotropic diffusional rotations when the timescale of NMR observation is too short compared with the time interval required to achieve the full rotation of one chain. The correlation time of random rotation of a high polymer, considered as a whole, is usually much longer than 1 ms when it is observed about 100 K above the glass transition temperature; consequently, the timescale of NMR measurements is too narrow a window compared with the broad spectrum of chain relaxation. A time break of observation of chain fluctuations must be necessarily associated with NMR measurements performed on high polymer melts or on concentrated solutions. The effect of a time break of observation is enhanced by the screening property inherent in the dynamical behavior of long chains. A space scale is correlatively associated with the effect of a time break; it corresponds to the chain segment formed from a mean number of skeletal bonds. Let N_v denote this mean number; it defines one so-called NMR submolecule. It has not been proved until now that the length of the chain segment associated with the dynamic screening effect, as it is observed from viscoelastic measurements, is in coincidence with the NMR submolecule. Nevertheless, segmental fluctuations which occur within different submolecules are supposed to be stochastically independent from one another. Consequently, a residual dipole–dipole interaction of nuclear spins attached to polymer chains can be ascribed to any submolecule. It is a function of the N_v number and the end-to-end vector $\mathbf{r}_v$ which characterizes the elongation of a given submolecule. Let $\mathbf{H}_D^R$ denote the residual dipole–dipole interaction; it is expressed by the key equation

$$\mathbf{H}_D^R = 0.3\{2(\rho_v^z)^2 - (\rho_v^y)^2 - (\rho_v^x)^2\}\Lambda(a/\sigma_v)^2\mathbf{H}_D^0 \qquad (10A.2)$$

with $(\sigma_v)^2$, the mean square end-to-end distance of one submolecule; a is the mean skeletal length. The normalized ρ_v vector is defined by $\rho_v = \mathbf{r}_v/\sigma_v$ and Λ is called the second order stiffness because it involves angular correlations of three skeletal bonds along one segment. The $\mathbf{H}_D^0$ spin–spin interaction reads

$$\mathbf{H}_D^0 = \sum_{p,p'} A_{pp'}B_{pp'} \qquad (10A.3)$$

where the sum is extended to all p, p' nuclear spins located on one submolecule; spin-operators $A_{pp'}$ are supposed to commute with the Zeeman energy and numerical factors $B_{pp'}$ result from transformations of spherical harmonics from monomeric unit frames to the reference frame ascribed to the elongation vector $\mathbf{r}_v$ of one submolecule. Details about these numerical factors are without importance, here. The main feature about equation (10A.2) is the presence of a reduction factor $(a/\sigma_v)^2$. For a Gaussian segment, $(\sigma_v)^2 = N_v a^2$ with $N_v \simeq 10^2$; the reduction factor is about equal to 10^{-2}. This means that the strength of the residual interaction is about equal to 10^3 rad s^{-1}, in agreement with experimental values of spin–spin relaxation rates.

A.2.3 TRANSVERSE RELAXATION: TIME SCALING

Then, the transverse magnetization is written as a product

$$M_x(t) = M_x^R(t)\Phi_R(t) \qquad (10A.4)$$

where

$$M_x^R(t) = \mathrm{Tr}(\mathbf{M}_x^R(t)\mathbf{M}_x)/\mathrm{Tr}(\mathbf{M}_x^2) \qquad (10A.5)$$

and

$$\mathbf{M}_x^R(t) = \exp(i\mathbf{H}_D^R t)\mathbf{M}_x \exp(-i\mathbf{H}_D^R t) \qquad (10A.6)$$

$\mathbf{M}_x$ is the spin-operator which represents the x-component of the nuclear magnetization. Properties of the $M_x^R(t)$ function are closely related to the time break of observation of fluctuations. The $\Phi_R(t)$ function results from the existence of fast but non-isotropic rotational motions of skeletal bonds; the nature of $\Phi_R(t)$ is analogous to that of any spin–spin relaxation function defined in ordinary liquids except for the non-isotropic character of random motions. For short correlation times, $\Phi_R(t)$ is nearly equal to unity. It is clearly seen from Equations (10A.2) and (10A.6) that the main property of the $M_x^R(t)$ function is induced by the presence of the term $(a/\sigma_v)^2$; σ_v is a physical quantity which reflects directly statistical properties of chain segments in the polymeric medium. A time scaling is induced whenever the term is varied by changing physical properties of the polymeric system. Such an NMR approach applies conveniently to investigations into the dynamic screening effect, also detected from viscoelastic measurements. Without any lack of generality, the time break relaxation function is written as M_x^R $(\Delta_G \Lambda (a/\sigma_v)^2 t)$, Δ_G is the strength of the dipole–dipole interaction in the absence of any random motion. The mathematical structure of this function is expected to be invariant when the only quantity which is varied is σ_y. Transverse relaxation functions must obey a property of superposition by applying an appropriate factor to the time-scale

$$M_x^R[\Delta_G \Lambda (a/\sigma_v')^2 t] = M_x^R(\Delta_G \Lambda [\sigma_v''/\sigma_v']^2 (a/\sigma_v'')^2 t) \qquad (10A.7)$$

the timescale factor is defined by

$$s = [\sigma_v''/\sigma_v']^2 \qquad (10A.8)$$

A.2.4 PSEUDO-SOLID SPIN-ECHOES

The presence of a residual spin–spin interaction is easily disclosed experimentally by forming pseudo-solid spin-echoes; these result from a $90°/x - \tau/2 - 180°/x - \tau/2 - 90°/y$ pulse sequence. These echoes look like solid echoes observed from nuclear spins embedded in any solid system; however, they spread over a long time interval because the strength of the residual interaction observed above the glass transition temperature is much weaker than the interaction corresponding to an ordinary density of spins in a solid. Also, polymeric systems usually present strong diamagnetic heterogeneities; consequently, $180°/x$ pulses must be used to focalize the magnetization during the observation of pseudo-solid echoes. These echoes obey four specific properties but two of them are currently used to characterize the spin–spin relaxation process.

(i) Considering one pseudo-solid echo formed at a time τ_1 and another one formed at a time τ_2, they must intersect at a time $\tau_1 + \tau_2$.

(ii) Let $M_x^R(t)$ still denote the time break relaxation function and $M_x^{R^*}(t, \tau)$ denote the analogous function for pseudo-solid echo formed at a time τ, the following relationship between derivatives must be satisfied

$$\frac{dM_x^R(\tau)}{dt} = -\frac{\partial M_x^{R^*}(\tau, \tau)}{\partial t} \qquad (10A.9)$$

The observation of these two properties is used to demonstrate that the transverse relaxation is governed by a pure spin–spin residual interaction. The above equation no longer applies to the observed relaxation function when the function $\Phi_R(t)$ plays a non-negligible role. The derivative of the transverse relaxation function is expressed as

$$\frac{dM_x(\tau)}{dt} = \frac{dM_x^R(\tau)}{dt} \, \Phi_R(\tau) + M_x^R(\tau) \, \frac{d\Phi_R(\tau)}{dt} \qquad (10A.9')$$

while the derivative of the pseudo-solid echo function is written as

$$\frac{dM_x^*(\tau)}{dt} = -\frac{dM_x^R(\tau)}{dt} \, \Phi_R(\tau) + M_x^R(\tau) \, \frac{d\Phi_R(\tau)}{dt} \qquad (10A.9'')$$

The addition of Equations (10A.9′) and (10A.9″) to each other, divided by the $M_x(t)$ relaxation function, yields the logarithmic derivative of the $\Phi_R(t)$ function; the logarithmic derivative of $M_x^R(t)$ is obtained by subtracting equations (10A.9′) and (10A.9″) from each other and by dividing by the $M_x(t)$ relaxation function.

A.3 Concentrated Polymer Solutions

In this section, the effect of addition of solvent on the polymeric dynamics is analysed, using several timescales of observation. Rates of random dynamical processes are increased while the dynamic screening effect is expected to extend over a longer space scale.

A.3.1 SEGMENTAL MOTIONS

It is considered that segmental motions which are observed from the spinlattice relaxation rate involve several monomeric units. More precisely, it is assumed that collective rotational isomerizations of a small number of successive monomeric units are converted into a translational random motion of a short segment. An effective friction coefficient of one monomeric unit can be associated with this translational motion: let $\zeta_f(T)$ denote this friction coefficient. It corresponds to an effective correlation time $\theta_f(T) = b^2 \zeta_f / k_B T$; b is the mean length of one skeletal bond, and k_B is the Boltzmann constant.

The property of homogeneity of the ω_0/T_1 function has been already illustrated for polybutadiene chains, observing ^{13}C or ^1H nuclei [8]. It has been supposed that the product of the relaxation rate and the Larmor frequency can be expressed as a function F the variable x (see equation (10A.1)):

$$\omega_0/T_1 = F(x)$$

with $x = \omega_0 \theta_f(T_i) a(T_i, T)$. T_i is a reference temperature. The shift fact or $a(T_i, T)$ is in agreement with results of viscoelastic measurements according to the Williams, Landel and Ferry theory [4] which governs all the processes related to glass transition phenomena. The WLF law reads (see also equation (9A.36) in Chapter 9A):

$$\text{Log}[a(T_i, T)] = -5.78 \, \frac{T - T_i}{T - T_i + 94.8} \tag{10A.10}$$

where $T_i = T_g + 55$ K. This equation applies to any chain microstructure provided the T_i reference temperature is defined with respect to the T_g glass temperature corresponding to the observed polybutadiene chain [11]. Polybutadiene is a flexible polymer whose viscoelastic properties have seen extensively studied. The experimental ω_0/T_1 curve can be drawn as a function of $T - T_g$ or $T - T_i$, for a given Larmor frequency; it must be shifted towards high temperatures when the Larmor frequency is increased. The first criterion of application of this analysis is the observation of a constant value of the maximum of the ω_0/T_1 function, whatever the Larmor frequency. Then, the second criterion of application is the possibility to superpose all ω_0/T_1 functions recorded by using several Larmor frequencies. Considering two Larmor frequencies ω_0' and ω_0'', the property of homogeneity reads

$$\omega_0' a(T' - T_i) = \omega_0'' a(T'' - T_i)$$

where T' and T'' are determined from any two chosen equal amplitudes of experimental curves.

The ω_0'/ω_0'' ratio gives the relative evolution of the $a(T - T_i)$ shift factor. A similar approach has been applied to polyisoprene in dilute solution in toluene [9]. The spin–lattice relaxation time of the methylene carbon C4, has been measured as a temperature function, for three Larmor frequencies. The three T_1/ω_0 curves superpose very well over the entire temperature range of observation; the $\theta_0 \langle \sigma(T) \rangle$ reduced variable is used, with the $\langle \sigma(T) \rangle$ time constant expressed as

$$\langle \sigma(T) \rangle = A \eta^{0.41} \exp(13.10^3/RT)$$

η is the viscosity. Such a result shows that all motional correlation times which govern the spin–lattice relaxation process are proportional to a single basic time constant. This way of analysis of experimental results has been extended to the spin–lattice relaxation measured from ^{13}C nuclei attached to Aroclor® molecules, in polybutadiene-Aroclor solutions [12]. Although the ^{13}C spectrum is very complicated, T_1/ω_0 curves have been found to obey both a property of frequency-temperature superposition (for $\omega_0 = 90.6$ and 25.2 MHz) and a property of temperature–concentration superposition for three polybutadiene concentrations. In this section, the property of homogeneity is illustrated from a concentration–temperature scaling observed on polyisoprene in solution in toluene; the Larmor frequency is kept constant ($\omega_0/2\pi = 40$ MHz) (Figure 10A.1a). The minimum value of the relaxation time is found to be independent of polymer concentrations ranging from 0.58 to 1 (w/w); it is equal to 50 ms.

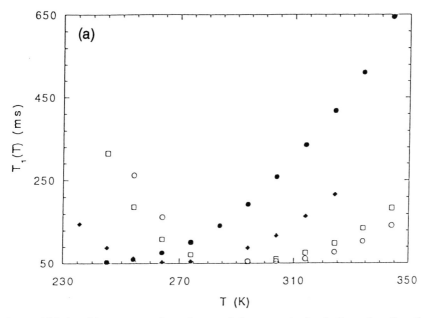

Figure 10A.1a Temperature dependence of the proton spin–lattice relaxation time of polyisoprene in deuterated toluene solutions. Polymer concentrations (w/w) are: 1.00 (○), 0.94 (□), 0.78 (♦), 0.58 (●). From Ref. 25

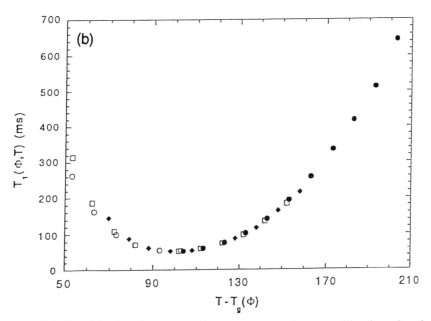

Figure 10A.1b All relaxation curves obey a property of superposition by using the $T_g(\phi)$ temperature variable. From Ref. 25

This result shows that the number of interacting protons does not depend on the polymer concentration; interactions between protons located on different chain segments can be neglected. The property of superposition of all relaxation curves is well observed by using the variable $T - T_g(\phi)$ (Figure 10A.1b); ϕ is the polymer concentration and $T_g(\phi)$ is the glass transition temperature of the solution characterized by (ϕ). The broad shape of the relaxation curve near the minimum indicates that a spectrum of correlation times is involved in segmental motions; nevertheless, the maximum of efficiency corresponds to correlation time values near 4×10^{-9} s.

The experimental observation of the property of homogeneity of spin–lattice relaxation rates with respect to the Larmor frequency and one basic dynamical segmental parameter is interpreted as reflecting average fluctuations which occur within short chain segments rather than a specific rotational isomerization process of skeletal bonds. Helical turns are known to form and to dissociate rapidly in any homopolymer; the average fluctuation probably spreads over about 10 monomeric units since the spin–lattice relaxation rate is not affected by crosslinks between segments unless the contour length defined by two consecutive junctions is reduced down to 20 skeletal bonds [13]. Finally, it may be worth emphasizing that spin–lattice relaxation rates usually observed on high polymers are independent of the chain molecular weight. This property reflects the effect of dynamic screening; it does not hold when polymer chains are too short. Then, variations of the polymer molecular weight correspond to variations of the concentration of chain ends which plays the role of solvent molecules.

To conclude, ω_0/T_1 may be considered, in most cases, as a homogeneous function with respect to the product of the Larmor frequency and one characteristic molecular time constant; this property is very useful to disclose relative variations of this time constant as a function of the temperature or the concentration. The coincidence of maximum values of the ω_0/T_1 ratio with one another, indicates that the mean square dipole–dipole interaction is hardly dependent on the temperature or the polymer concentration.

A.3.2   SOLVENT DIFFUSION

Local motions which occur in macromolecular systems can be probed from the diffusion process of small molecules in concentrated polymeric solutions. The translational diffusion is detected from NMR over a timescale which may vary from about 1 to 100 ms. Such a time interval corresponds to a very large number of elementary collisions and a long random path; consequently, details about mechanisms of molecular jump are not disclosed from this NMR approach. However, the dynamical behavior of small solvent molecules, immersed in a polymer melt and observed over a long time interval, permits the determination of characteristic parameters of the diffusion process.

According to the well-known Einstein definition, the diffusion constant is written as

$$D = k_B T 6 \gamma m \qquad (10A.11)$$

where γ^{-1} is the correlation time of the solvent molecular velocity and m is its mass [14]. Using a method originally devised by Markoff and described by Chandrasekhar, the D coefficient is also written as

$$D = \langle a^2 \rangle / 6\tau \qquad (10A.12)$$

where $\langle a^2 \rangle$ is the mean square jump length and τ^{-1} is the average frequency of jumps [15,16]. Finally, applying Langevin's equation, D is defined as

$$D = k_B T / \zeta_0 \qquad (10A.13)$$

where the ζ_0 friction coefficient is governed by Stokes' law

$$\zeta_0 = 6\pi a \eta_0$$

a is the molecular diameter and η_0 is the liquid viscosity [17]. The purpose of investigations into diffusion properties is twofold.

(i) Starting from a free volume law which characterizes a pure molten polymer, it is of interest to analyse the evolution of the free volume description upon addition of solvent.

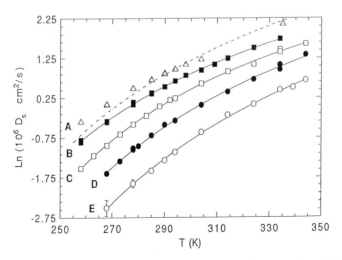

Figure 10A.2 Temperature dependence of the self-diffusion of cyclohexane in polybutadiene. Solvent concentrations are (w/w) 0.05 (○), 0.19 (●), 0.29 (□) and 0.375 (■). Experimental data were analyzed using the WLF approach and a reference temperature equal to 298 K [21]. Reprinted with permission from *Macromolecules*, **26**, 3946. Copyright (1993) American Chemical Society

(ii) The limiting value of the ζ_0 coefficient obtained when the solvent concentration goes to zero, gives a good estimate of the monomeric friction coefficient.

Since the pioneer work of Stejskal, the pulsed field gradient method is currently used to characterize the diffusion process of small molecules or of macromolecules in dilute or semi-dilute solutions [18–20]. In this chapter, the NMR approach is illustrated from the self-diffusion of cyclohexane molecules through polybutadiene. Typical variations of the D_s self-diffusion coefficient of cyclohexane in polybutadiene are reported in Figure 10A.2. as a temperature function; several concentrations are considered [21].

The WLF method of analysis (equation 10A.10) has been extended to polymeric solutions assuming that polymer and solvent fractional free volumes add linearly to each other [22]. More recently, a different method of analysis has been proposed by Vrentas and Duda; it includes several parameters such as a jumping unit volume [23].

It is worth emphasizing that these two approaches become equivalent to each other when parameters introduced in the Vrentas–Duda analysis have appropriate values. This is the case of the cyclohexane–polybutadiene system. Measurements performed by varying both the polymer concentration and the temperature lead to the determination of free volume parameters with a good accuracy. For the pure polymer the fractional free volume f_p and the thermal expansion coefficient α_p are:

$$f_p = 0.124 \pm 0.005 \text{ at } 298 \text{ K} \quad \text{and} \quad \alpha_p = (7.0 \pm 0.5) \times 10^{-4} \text{K}^{-1}$$

while for the pure solvent the corresponding quantities are:

$$f_s = 0.223 \text{ at } 298 \text{ K and } \alpha_s = 11.8 \times 10^{-4} \text{K}^{-1}$$

There is an interesting feature about the trace diffusivity of solvent determined from experimental curves when the solvent concentration is set equal to zero. The $D_s(0, T)$ diffusion coefficient is found to obey the free volume law of the pure polymer when the above numerical values are assigned to parameters. Furthermore, the $\zeta_s(0, T)$ translational friction coefficient, derived from the $k_B T/D_s(0, T)$ ratio, is equal to $(1.64 \pm 0.06) \times 10^{-10}$ kg s^{-1}, at 298 K; this value is in good agreement with the result of viscoelastic measurements performed on polybutadiene (1.8×10^{-10} kg s^{-1}).

The trace diffusivity clearly shows that there is no discontinuity when the solvent concentration goes to zero; in other words, solvent molecules can be used to probe directly local properties of polymer melts.

A.3.3   EVIDENCE FOR A TEMPORARY GEL-LIKE BEHAVIOUR

The most striking feature about NMR investigations into dynamical properties of high polymer systems is the observation of a gel-like behaviour of the transverse magnetic relaxation. This is illustrated in Figures 10A.3a and 10A.3b where the proton transverse relaxation and pseudo-solid echoes observed on high molecular weight polybutadiene chains $(M_w = 16.10^4)$, in a melt, at room temperature, are compared with the relaxation curve and echoes observed at 150 °C, on a vulcanized polybutadiene $(M_w = 7.7.10^4$ and sulfur concentration: 0.01 g/g). Similar NMR properties are observed on these two systems. This result shows clearly that there exists a time break of chain fluctuations which occurs in a melt as in a permanent gel. During this time break, it may be considered that any given chain in a melt is divided into submolecules; each submolecule has fixed ends during this time break. It is like assuming that there exists a temporary network structure which is detected from NMR. It is worth emphasizing that the search for a characterization of dynamical properties leads actually to the observation of non-isotropic rotations of monomeric units, resulting from the time break of chain fluctuations. The time break, as disclosed from viscoelastic properties, is usually associated with a temporary network structure built from chain segments determined by a mean number of skeletal bonds. Let N_e denote this mean number; N_e is defined by analogy with the description of the property of elasticity of permanent gels

$$N_e = \rho_p RT / G_N^0 M_m \qquad (10A.14)$$

ρ_p is the pure polymer density while G_N^0 is the modulus of temporary elasticity; M_m is the mean molar weight of one skeletal bond. The addition of solvent is known to induce a dilation of the mesh size of the temporary network structure. This effect can be investigated from NMR by applying the time-scaling analysis of the transverse relaxation, in the following way.

(i) Without asserting that the time break observed from NMR is also associated exactly with the N_e mean number of skeletal bonds, it is assumed that the N_v number related to NMR properties is proportional to N_e: $N_v = \beta N_e$.

(ii) The temporary network structure is supposed to be dilated by the addition of a good solvent, in agreement with the variation of the modulus of temporary elasticity

$$G_N(\phi) = G_N^0 \phi^{-\alpha} \qquad (10A.15)$$

with α varying from 2 to 2.2, for different polymers [3].

(iii) The dilation is expected to be reflected by a variation of the timescale of the transverse relaxation function. Correspondingly, relaxation curves must obey a property of superposition, by applying an appropriate factor to the timescale of observation.

Relaxation curves recorded from polybutadiene solutions have been shown to satisfy such a property, provided measurements are performed at a temperature higher than a threshold temperature which will be called T_s,

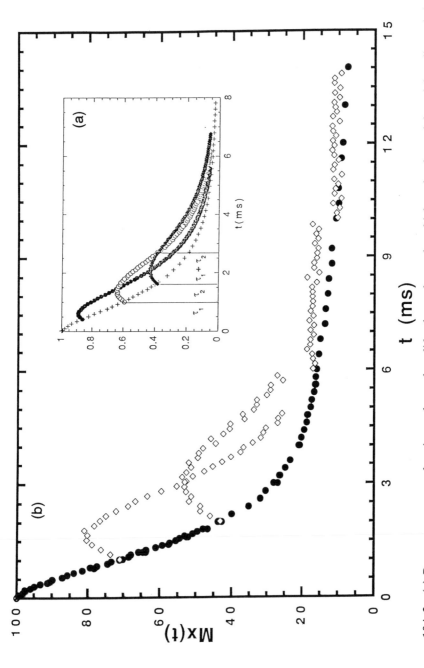

Figure 10A.3 (a) Proton transverse relaxation and pseudo-solid echoes observed on high molecular weight polybutadiene chains, at 300 K. From Ref. 25. (b) Proton transverse relaxation and pseudo-solid echoes observed on vulcanized polybutadiene chains at 150 °C. ($M_w = 0.77 \times 10^5$ g/mole, the sulfur concentration is 0.01 g/g.) [25]. Reprinted with permission from *Macromolecules*, **25**, 6855. Copyright (1992) American Chemical Society

hereafter; T_s is higher than the T_g glass transition temperature of the polymeric system ($T_s - T_g \simeq 80$ K). Above this threshold temperature, the mathematical structure of the relaxation function is kept invariant except for the timescale. The property of superposition is illustrated in Figure 10A.4a, corresponding to pure polybutadiene (number average molar weight $M_n = 1.9 \times 10^5$); temperatures of observation are 294 and 334 K, respectively. Because of the property

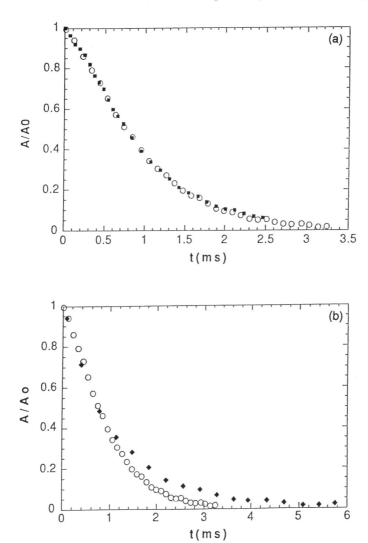

Figure 10A.4 Proton transverse relaxation of pure polybutadiene. (a) Relaxation curves recorded at 294 ($\bigcirc$) and 334 K ($\blacksquare$) exhibit a property of superposition. (b) Relaxation curves recorded at 254 ($\blacklozenge$) and 294 K ($\bigcirc$) do not obey a property of superposition. From Ref. 25

of superposition, normalized relaxation curves can be characterized by choosing a given amplitude A and by measuring the time interval corresponding to this amplitude

$$\tau_A = (\Delta_G \Lambda)^{-1}(\sigma_v/a)^2 M_x^{-1}(A) \tag{10A.16}$$

where $M_x^{-1}(A)$ is the inverse function of $M_x(t)$ For Gaussian segments, the dependence of τ_A on σ_v^2 is really a dependence on the N_v number of skeletal bonds, in dynamic screening segments. For polybutadiene in concentrated solutions, it has been shown that

$$\tau_A = \left(1 + \frac{4.5}{G_N^0(x_{1,2})\phi^{2.2}}\right)(T - T_g(\phi, x_{1,2}) - 50).8 \times 10^{-3} \tag{10A.17}$$

the time interval is given in milliseconds and G_N^0 is expressed in dynes cm^{-2} (10^{-1} kg m^{-1} s^{-2}) divided by 10^6. The above equation applies to different chain microstructures obtained by varying the $x_{1,2}$ concentration of monomeric units in the vinyl-1,2 conformation. The $T_g(\phi m\, x_{1,2})$ glass transition depends on both the polymer concentration and the $x_{1,2}$ concentration of vinyl monomeric units. The presence of a temporary network structure manifests itself through the first factor. It comes from the $M_x^R(t)$ component of the relaxation function [24]. The second factor results from fast non-isotropic motions of monomeric units which give rise to the $\Phi_R(t)$ function defined in equation (10A.4). Clearly, the reference temperature $T_i' = T_g(\phi, x_{1,2}) + 50$ which appears in NMR properties, is not very different from the T_i reference temperature introduced from viscoelastic measurements (equation (10A.10)). The second factor reflects a free volume effect; it is necessarily a conjugate variable of an expansion coefficient which is about equal to 10^{-3} K^{-1}. Then, the order of magnitude of the τ_A time interval must be given by the inverse of the reduced dipole–dipole interaction: $10^5/N_v$ rad s^{-1} = 500 rad s^{-1} ($N_v = 200$), this must be compared with the estimate equal to 125 rad s^{-1}, according to equation (10A.17). It is seen that NMR properties, specific to polymeric systems, are not due to the absence of motions but to the non-isotropic character of these motions. Equation (10A.17) applies to polyisobutylene solutions or to polyisoprene by using different values of parameters. A sharper analysis of relaxation functions is obtained by using equations (10A.9') and (10A.9''). They yield an exponential time function for $\Phi_R(t)$ whereas $M_x^R(t)$ is found to hardly depend on temperature variations [25].

A.3.4   NEAR THE GLASS TRANSITION

The property of superposition no longer holds when relaxation curves are recorded at temperatures lower than the T_s threshold temperature. This is illustrated in Figure 10A.4b, for pure polybutadiene; temperatures of observation are 254 and 294 K, respectively. It is considered that the progressive enhancement of the chain stiffness, induced by lowering the temperature, is mainly due to a progressive change of asymmetry of random rotations of monomeric units. The great sensitivity of NMR to rotational isomerizations leads to a change of the shape of the relaxation curve. Near the glass transition temperature, two-dimensional Fourier transform methods are used to characterize molecular motions; geometric information that is obtained is independent of model assumptions [26]. Non-exponential correlation functions and non-Arrhenius behaviour of correlation times are currently disclosed by applying these methods to polymeric systems. Although no satisfactory interpretation of the glass transition has been given until now, the stretched exponential Kohlrausch–Williams–Watts function [31] yields good fits to relaxation measurements.

A.4 Free Chains in Polymeric Gels

Polymeric network structures are usually formed by long chains crosslinked by covalent bonds located at random along any chain; they can also result from reactions which occur between appropriate chemical functions, attached to chain ends, prior to the gelation process. The dynamic behaviour of polymeric gels, observed in a macroscopic gel above the glass transition temperature, is again characterized by a stress relaxation modulus. Segmental motions are not sensitive to the presence of crosslinks, provided the concentration of covalent coupling junctions is not too high. The average number of monomeric units considered along one chain between two consecutive crosslinks must be higher than about 20 to observe no effect on the spin–lattice relaxation rate of protons attached to chain segments. The presence of crosslinks gives rise to two main properties specific to polymeric gels: these can be highly strained but they cannot flow. At long times, the stress relaxation modulus exhibits a constant finite value. Therefore, the viscoelastic behaviour of a network structure diverges from that of an uncrosslinked polymer. The storage modulus has a constant finite limiting value when it is observed at zero frequency whereas the loss modulus corresponds to infinite viscosity.

In this section, attention is focused on internal dynamical properties of a polymeric gel probed from polymer chains moving through the network structure. The main question which is addressed about free chains moving in polymeric network structures concerns the restrictions of fluctuations induced by the presence of crosslinks. Free chains must turn around nodes to move randomly; also, loops of a given free chain, formed around network chain segments, probably induce more hindrance than those formed around other free chains. It is shown that the presence of crosslinks can be accounted for by variations of the monomeric friction coefficient which governs the dynamics of free chains.

The NMR approach is illustrated from polydimethylsiloxane (PDMS) which is a flexible polymer [27]. Most chains play the role of strands used to build the polymeric structure while others move randomly throughout this structure.

A.4.1   DYNAMICAL CONTRAST

The main feature about the NMR approach concerns the dynamical contrast detected from the relaxation of the transverse magnetization of protons linked to PDMS chains. Relaxation functions associated with chain segments which participate in the network structure usually spread over a few milliseconds; the relaxation is governed by residual dipole–dipole interactions which result from topological constraints created by the presence of crosslinks. Relaxation functions corresponding to free chains, vary over a time interval about equal to 400 ms. These two timescales clearly discriminate properties of the network structure from those of free chains when these are short enough.

A.4.2   NATURE OF SYSTEMS

Three types of modified PDMS chains are mixed up with one another to synthetize network structures; let A, B and C denote these three types. Silane functions are randomly located along A chains (number average molar weight: $M_n = 7650$ g mol^{-1}; average concentration of SiH groups: 1.22×10^{-3} g^{-1} mol) while ends of all B chains carry vinyl functions (number average molar weight: $M_n = 12\,500$ or 15 700 g mol^{-1}); C chains bear no reactive functions (number average molar weight: $M_n = 13\,300$ g mol^{-1}). Platinum is used as a catalyst to synthetize network structures between A and B chains; these structures are swollen by chemically inert C chains when the gelation process is completed. The concentration of free chains is easily varied by changing initial concentrations of C chains in mixtures prior to the gelation process. The concentration of chemically inert chains is called ϕ_1(w/w).

A.4.3   NATURE OF THE SPIN-SYSTEM RESPONSES

Spin-system responses are characterized according to the principle of analysis described in Section A.2. Pseudo-solid spin-echoes are well observed from relaxation curves associated with the polymeric network structure whereas no such specific echoes can be formed from relaxation curves associated with chains moving randomly through polymeric structures. Consequently, a pure dynamical property is detected from the relaxation process of protons linked to free chains. The relative amplitude of the spin-system response associated with free chains is experimentally found to be equal to ϕ_1.

A.4.4   PROPERTY OF SUPERPOSITION

Normalized relaxation curves associated with free chains present a property of superposition when the ϕ_1 polymer concentration is varied. A σ_t superposition

factor is applied to the timescale of each curve except for the curve corresponding to the pure melt which is chosen as a reference. Variations of the σ_t
superposition factor are illustrated in Figure 10A.5a. The property of
superposition shows clearly that the mathematical structure of the relaxation

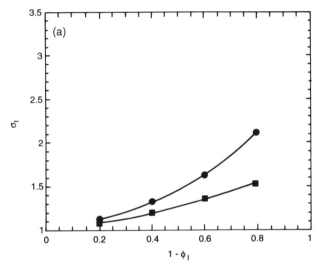

Figure 10A.5a Variations of the superposition factor of relaxation curves as a
function of the gel fraction (for chains with vinyl ends $M_n = 12\,500$ (●) and
$M_n = 15\,700$ (■)) [27]. Reprinted with permission from *Macromolecules*, **25**, 593.
Copyright (1992) American Chemical Society

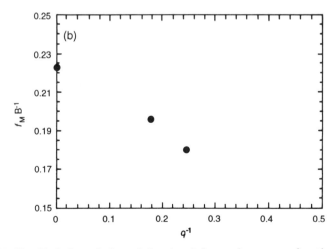

Figure 10A.5b Variation of the gel fractional free volume as a function of the q
factor of swelling (equation (10A.19)). The fraction free volume of the pure molten
polymer is also reported [27]. Reprinted with permission from *Macromolecules*, **25**,
593. Copyright (1992) American Chemical Society

function is kept invariant under variations of the concentration of free chains in the polymeric network structure. The invariance of the relaxation function implies that a single basic dynamical parameter governs the chain dynamics even though a broad spectrum of correlation times is implied in the description of this random motion. All correlation times are probably multiples of this single dynamical parameter which is supposed to be a monomeric friction coefficient.

A.4.5 POLYMERIC GEL-FREE VOLUME

Variations of the σ_t superposition factor are conveniently analyzed by assuming that a fractional free volume f_g specific to a polymeric gel can be assigned to any monomeric unit which participates in the network structure. Then, the fractional free volume f_M probed by free chains is supposed to be a linear combination of f_g and f_1, the fractional free volume of pure molten PDMS. The fractional free volume of the mixture is expressed as:

$$f_M = (1 - \phi_1)f_g + \phi_1 f_1 + \kappa \phi_1 (1 - \phi_1) \qquad (10A.18)$$

the quadratic term in equation (10A.18) accounts for a small deviation from linearity with respect to the ϕ_1 variable. The f_g value is expected to be smaller than f_1 because conformational fluctuations of network chain segments are restricted by the presence of crosslinks. The experimental value of f_g/B is found to be equal to 0.18 at room temperature when the molar weight of PDMS chains which carry vinyl groups is 12 500 g mol^{-1}; the f_g/B value is 0.2, at room temperature, when the molar weight is equal to 15 700 g mol^{-1}. Correspondingly, experimental values of κ/B are 0.02 and 0.01, respectively. The B term for polymers is usually set equal to unity. The fractional free volume of pure PDMS is set equal to 0.223, at room temperature [4].

The fractional free volume of a given network structure can be related to the property of maximum swelling induced by the osmotic pressure of a good solvent. The swelling property is observed after extracting all free chains. The swelling ratio Q_m is obtained by dividing the volume V_m of the swollen gel by the volume V_D of the dry gel. The Q_m ratio is found to vary as a linear function of the weight ratio of free chains:

$$Q_m = q\left(1 + \frac{m_1}{m_g}\right) \qquad (10A.19)$$

with $q = 4.1$, for $M_n = 13\ 500$ g mol^{-1} and $q = 5.7$ for $M_n = 15\ 700$ g mol^{-1}; m_1 and m_g are the weights of free chains and crosslinked chains, respectively. Then, the q factor is a measure of the available volume determined by the presence of free chains prior to their extraction. The fractional free volume f_M is found to be a linear function of q^{-1} which is illustrated in Figure 10A.5b. This result is in accordance with observations made on epoxy-gels or on polyester-gels, using dilatometric measurements [28, 29].

A.4.6 CONCLUSION

The NMR approach to dynamical properties of polymer chains, moving randomly through a polymeric network structure, shows that internal fluctuations are described within the framework of the Rouse model [2]. There is a continuous evolution from the behaviour of chains moving in a pure polymer to that of chains moving in a gel. The assignment of a free volume specific to the gel structure accounts for the complex topological interaction of one free chain with covalent nodes. It is considered that the nature of fluctuations is kept invariant except for their timescale.

A similar law of superposition is observed on short polyethylene-oxide (PEO) chains moving in long polymethylmethacrylate (PMMA) chains, in a blend [30]. The basic process which governs dynamical properties of the mixture reflects a linear combination of fractional free volumes of PEO and PMMA.

A.5   Conclusion

The description of dynamical fluctuations of polymeric systems, observed above the glass transition temperature, impinges on two main effects: free volume and dynamic screening. The thermal energy gives rise to rotational isomerizations of monomeric units which induce a collective segmental behaviour; this generates a free volume temperature dependence of dynamical parameters. Macromolecules swollen by other chains retain the ability to exhibit a hierarchy of fluctuations because these linear molecular systems comprise thousands of skeletal bonds. Huge numbers of conformations are associated with long chains; they confer a fractal character which is detected from NMR. The longer the chain segment which is observed, the higher the symmetry of random rotations of monomeric units. The dynamic screening effect is detected from NMR as well as from the temporary elasticity displayed from viscoelastic measurements. Both NMR and viscoelastic responses of polymers manifest a dual behaviour; this appears as liquid-like and solid-like magnetic relaxations while it corresponds to viscosity and elasticity for flowing systems. The NMR approach to the terminal relaxation of polymers was not addressed in this short chapter; it is based on a molecular weight scaling of timescales of magnetic relaxation [10]. Properties of homogeneity of relaxation functions which underlie most interpretations, come from the very nature of macromolecules.

References

1. B. H. Zimm, *J. Chem. Phys.*, **24**, 269 (1956).
2. P. E. Rouse, *J. Chem. Phys.*, **21**, 1272 (1953).
3. W. W. Graessley and S. F. Edwards, *Polymer*, **22**, 1329 (1981).
4. J. D. Ferry, *Viscoelastic Properties of Polymers* (3rd edn), J. Wiley, New York (1980).

5. P. G. De Gennes, *Scaling Concepts in Physics*, Cornell University Press, Ithaca (1979).
6. R. Dejean de la Batie, F. Lauprêtre and L. Monnerie, *Macromolecules*, **22**, 122 (1989).
7. C. K. Hall and E. J. Helfand, *J. Chem. Phys.*, **77**, 3275 (1982).
8. A. Guillermo, R. Dupeyre and J. P. Cohen-Addad, *Macromolecules*, **23**, 1291 (1990).
9. D. J. Gisser, S. Glowinkowski and M. D. Ediger, *Macromolecules*, **24**, 4270 (1991).
10. J. P. Cohen-Addad, in J. W. Emsley, J. Feeney and L. H. Sutcliff (eds), *Progress in Nuclear Magnetic Resonance Spectroscopy*, Pergamon Press, **25**, 1 (1993).
11. J. M. Carella, W. W. Graessley and L. J. Fetters, *Macromolecules*, **17**, 2775 (1984).
12. D. J. Gisser and M. D. Ediger, *Macromolecules*, **25**, 1284 (1992).
13. G. C. Munie, J. Jonas and I. J. Rowland, *J. Polym. Sci.*, **18**, 1061 (1981).
14. A. Einstein, *Ann. d. Physik*, **17**, 549 (1905).
15. A. A. Markoff, *Wahrscheinlichkeitsrechnung* (Leipzig) (1912).
16. S. Chandrasekhar, in Nelson Was (ed.), *Noise and Stochastic Process*, Dover Publications, New York (1954).
17. P. Langevin, *C. R. Acad. Sci. (Paris)*, **146**, 530 (1908).
18. E. O. Stejskal and J. E. Tanner, *J. Chem. Phys.*, **42**, 288 (1965).
19. E. D. von Meerwall, J. Grigsby, D. Tomich and R. Van Antwerp, *J. Polym. Sci. Polym. Phys. Ed.*, **20**, 1037 (1982).
20. P. T. Callaghan and D. N. Pinder, *Macromolecules*, **17**, 431 (1984).
21. A. Guillermo, M. Todica and J. P. Cohen-Addad, *Macromolecules*, **26**, 3946 (1993).
22. H. Fujita and A. Kishimoto, *J. Chem. Phys.*, **34**, 393 (1961).
23. J. S. Vrentas and J. L. Duda, *J. Polym. Sci. Polym. Phys. Ed.*, **15**, 403 (1977).
24. A. Labouriau and J. P. Cohen-Addad, *J. Chem. Phys.*, **94**, 3242 (1991).
25. M. Todica, Thesis, University of Grenoble (1994).
26. H. W. Spiess, *Chem. Rev.*, **91**, 1321 (1991).
27. J. P. Cohen-Addad and O. Girard, *Macromolecules*, **25**, 593 (1992).
28. K. Shibayama and Y. Suzuki, *J. Polym. Sci.*, **26**, 3 (1965).
29. Y. G. Won, J. Galy, J. P. Pascault and J. Verdu, *Polymer*, **79**, 32 (1991).
30. C. Brosseau, A. Guillermo and J. P. Cohen-Addad, *Polymer*, **33**, 2076 (1992).
31. G. Williams and D. C. Watts, *Trans. Faraday Soc.*, **60**, 80 (1970).

10B FLUIDS IN AND ON INORGANIC MATERIALS

J. P. Fraissard
University of Paris VI, France

Dynamics of Solutions and Fluid Mixtures by NMR
Edited by J.-J. Delpuech © 1995 John Wiley & Sons Ltd

B.1   Introduction: Physical and Chemical Adsorption

In a solid the particles which make up its lattice (ions, atoms or molecules) are subjected in the deep layers to forces which equilibrate, but in the superficial layers the particles of the solid can only be subjected to dissymmetric forces which are expressed at its periphery by an attractive force field. This field has a range limited to distances of the order of magnitude of the atomic dimensions, i.e. a few angstroms, but it is enough to attract fluids, liquids or gases, located in the immediate vicinity of the interface. These are the forces which cause the bonding of molecules to the surfaces, a phenomenon which is called 'adsorption'.

The characteristics of adsorption depend very much on the nature of the solid (adsorbent), of the fluid (adsorbate) and the experimental conditions, particularly the temperature. Physical adsorption or 'physisorption' occurs preferentially at low temperature and involves primarily the physical properties of the adsorbate. At higher temperatures (higher must be taken in a relative sense) chemical adsorption or 'chemisorption' involves a chemical association between the surface and the adsorbed species rather than a simple physical interaction.

Physisorption involves physical bonding forces, therefore weak, of the van der Waals type. It can be considered as a sort of condensation of the gas at the surface of the adsorbent. For a given temperature this pseudo-condensation occurs at a pressure much less than the saturated vapour pressure, since the superficial force field of the adsorbent attracts the gas molecules which are in the immediate vicinity. This explains why physisorption only occurs at fairly low temperatures, far from the critical temperature of the fluid concerned. Moreover, since the surface-physisorbed molecule bonds are weak, the electron charge distributions in the adsorbate and the adsorbent are practically the same as those of the two isolated systems. Consequently, the physisorbed molecules have the same properties as the free molecules and they can move freely over the surface of the solid.

Chemisorption corresponds to a true chemical reaction between the adsorbate and certain sites of the adsorbent. It therefore concerns large bond energies and leads to a profound modification of the electron distribution of the chemisorbed molecule, which can even be dissociated.

Let us consider the series of events likely to arise when a gas molecule collides with the crystalline surface (Figure 10B.1):

- Elastic reflection;
- Physical adsorption and diffusion of the molecule at the surface of the solid;
- If the chemical and experimental conditions are suitable, chemisorption (dissociate or not) of the physisorbed molecule on an active site of the surface.

These phenomena can be visualized on a Lennard-Jones diagram [1] representing the evolution of the potential energy of the adsorbate–adsorbent system in terms of the molecule-surface distance. Figure 10B.2 shows a physisorbed state and a dissociated chemisorbed state. The distance between the two asymptotes for $r \to \infty$ corresponds to the dissociation enthalpy of the molecule (for example, 430 kJ/mole for $H_2 \Leftrightarrow 2H$). The activation energy E_a necessary to pass from physisorption to chemisorption depends both on the values of the enthalpies ΔH_p and ΔH_c of physisorption and chemisorption and on the positions of the corresponding curves (r_C and r_P). E_a can therefore be very small. Consequently, if the chemisorbed and physisorbed phases coexist at equilibrium, there can be rapid exchange between the corresponding species.

Porous solids such as silica, titania, alumina, zeolites play an important role in many processes in the chemical industry where they are used as catalysts or as adsorbents. The state of adsorbed molecules, which is in several respects intermediate between the liquid and the solid states, has been the subject of numerous investigations, mainly for two reasons:

(i) It is clear that the determination of reaction mechanisms in heterogeneous catalysis requires knowledge about the state of the chemisorbed complex;
(ii) But chemisorbed or physisorbed molecules can also be used to investigate the chemical and physical properties of solid surfaces.

Excellent review articles have been published on NMR studies of adsorption [2–12]. For a deeper understanding these publications should be consulted.

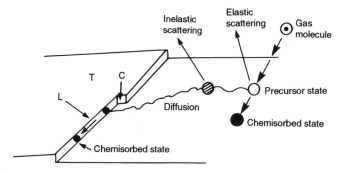

Figure 10B.1 Adsorption of a gas molecule. ⊙: gas molecule. ○: physisorbed molecule. ●: chemisorbed molecule

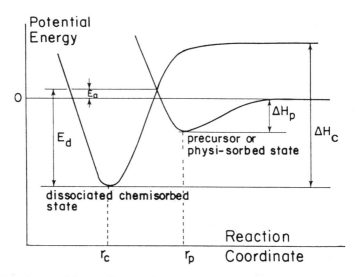

Figure 10B.2 Lennard-Jones diagram for dissociative chemisorption: E_a, activation barrier for transition between physisorption and chemisorption; E_d, activation energy for desorption; ΔH_p and ΔH_c, enthalpy of physisorption and chemisorption, respectively

This chapter presents, with the help of a few examples, various applications of physical adsorption in NMR studies of the properties of solid surfaces or of chemisorption.

B.2   Experimental Conditions and Difficulties

B.2.1   SENSITIVITY

Since the appearance of Fourier transform spectrometers the same spectrum can be accumulated a very large number of times. The resulting sensitivity increase makes it possible to detect many different nuclei in homogeneous media. Despite this, in the case of adsorbed phases, NMR signals can be detected only when the adsorbent has a high specific area (>10 m^2 g^{-1}). Moreover, only nuclei with spin of $1/2$ are easily suitable. The characteristics of some nuclei which are particularly interesting for the study of adsorption are listed in Table 10B.1.

The minimum number N_{min} of nuclei per cm^3 necessary to produce a signal-to-noise ratio of 10 after z accumulations is given by [7]:

$$N_{min} \sim 10^{18} I_r^{-1} \left(\frac{\nu_{100}}{100}\right)^{1/2} \left(\frac{T}{300}\right)^{3/2} . B_0^{-3/2} \left(\frac{T_1}{z} \delta\omega.\Delta\nu\right)^{1/2} V_c \qquad (10B.1)$$

where I_r is the relative NMR intensity (Table 10B.1), ν_{100} is the resonance

Table 10B.1 NMR characteristics of some nuclei with spin 1/2 often used for the study of adsorbed phases

	^{1}H	^{13}C	^{15}N	^{129}Xe
Resonance frequency (MHz) for $B = 2.34869$ Tesla	100	25.14	10.13	27.66
Relative intensity of the NMR signal for a given value of B_0	1	1.59×10^{-2}	1.04×10^{-3}	2.12×10^{-2}
Natural abundance of the isotope (%)	99.985	1.108	0.365	26.4
Product of relative intensity and natural abundance (I_r)	100	1.8×10^{-4}	3.8×10^{-6}	5.6×10^{-3}
Chemical shift range (isotropic values) (ppm)	10	300	500	7500
Typical NMR linewidths of adsorbed molecules (Hz) at room temperature	500	100	100	100

frequency in MHz if $B_0 = 2.34869$ Tesla, T is the sample temperature in K, T_1 is the longitudinal relaxation time in s, $\delta\omega$ is the observed half-width of the NMR signal in rad s^{-1}, and $\Delta\nu$ is the bandwidth of the whole receiver system in Hz and V_c is the volume of the receiver coil in cm^3. As an example we consider measurements of ^{13}C NMR signals for adsorbed molecules having a natural abundance of ^{13}C nuclei (1.1%).

Taking $\nu_{100} = 25.1$ MHz, $T = 300$ K, $V_c = 1$ cm^3 and $I_r = 1.8 \times 10^{-4}$ (Table 10B.1) it follows:

$$N_{min}(^{13}C) = 8 \times 10^{20} \left(\frac{T_1}{z} \, \delta\omega \, \Delta\nu \right)^{1/2} \tag{10B.2}$$

Typical values for the remaining parameters are $T_1 = 0.1$ s, $\delta\omega = 100$ rad s^{-1} (~20 Hz), $\Delta\nu = 500$ Hz and $z = 10^4$ accumulations, leading to

$$N_{min}(^{13}C) = 5.6 \times 10^{20}$$

without nuclear Overhauser effect enhancement.

B.2.2   NUCLEAR OVERHAUSER EFFECT (NOE)

The signal enhancement due to magnetization transfer from the coupling protons (index S) to nuclei investigated (C, N ... index I) is generally less than the optimum factor:

$$NOE = 1 + \eta = 1 + \frac{\gamma_S}{\gamma_I} \frac{T_{1I}}{T_{1S}} \sim \begin{cases} 3 \text{ for } ^{13}C \\ -3.9 \text{ for } ^{15}N \end{cases} \tag{10B.3}$$

with γ_s and γ_I the gyromagnetic ratios, T_{1I} the longitudinal relaxation time and

T_{IIS} the cross-relaxation time due to the dipole–dipole coupling between I and S spins. This reduction is due to the restricted molecular mobility [13] and the predominance of other relaxation mechanisms.

B.3   The Intrinsic Chemical Shift within Adsorbed Phases. Magnetic Susceptibility Correction Factors

One of the first high resolution NMR studies carried out around 1963 on adsorbed molecules indicated that proton resonance frequencies deviate from free molecule values [14]. These variations may represent a disturbance of the electron distribution in the molecule during adsorption and consequently allow the determination of the electron distribution in the chemisorbed complex. The intrinsic shift varies with the nature of the substrate and, for a given sample, with the nature of the functional group. Indeed, the adsorption process can affect the electron environment of a single type of nucleus when there is preferential orientation of the molecule with respect to the surface of the solid.

In liquids and gases the chemical shift is proportional to B_0 and changes with temperature and pressure as a result of intermolecular effects (hydrogen bonds [15]; van der Waals bonds [16, 17], proton exchanges [18. 20]). This shift also depends on other factors such as the local magnetic and electric fields [21–25].

However, the experimental shift is never simply due to the influence of the electron environment of the adsorbed molecule. Several corrections have to be made. One of the most important in the case of light nuclei such as ^{1}H is the volume magnetic susceptibility correction.

There are several methods for measuring NMR chemical shifts [19]. However, for a variety of reasons (wide lines, competitive adsorption, etc.), only the method of substitution can be used for adsorbed phase studies. The reference and the sample (adsorbate–adsorbent system) are put into identical glass tubes and are studied successively. It is necessary therefore to correct the observed resonance shift, δ_{obs}, of the sample in order to obtain the real (corrected) resonance shift, δ_{real} [26, 27]. If B_0 is the magnetic field applied to a solid of volume susceptibility χ_{V}, the resulting field can be written as [19,26,27]:

$$B = B_0 + b_1 + b_2 + b_3 + b_4 \qquad (10\mathrm{B}.4)$$

b_4 corresponds to the influence of the electron environment surrounding each nucleus. The first correction, $b_1 = (4\pi/3)\chi_{\mathrm{V}}B_0$, is the field due to the spherical Lorentz surface [28]. The following term, $b_2 = -\alpha\chi_{\mathrm{V}}B_0$, is the demagnetizing field. The parameter α depends on the sample shape and on the B_0 direction, i.e. $4\pi/3$ for a sphere, 2π for a cylinder (provided the height is at least five times the diameter), the axis of which is perpendicular to B_0, or zero for the same cylinder parallel to B_0. The correction b_3 is the field which depends on the distribution of magnetic dipoles inside the Lorentz sphere. It can be shown that this field is zero for an isotropic or cubic distribution. In general it is represented as a function of χ_V:

$$b_3 = k\chi_V B_0$$

Geshke [15] considers that b_3 is zero when all the particles are distributed randomly, which is indeed the case for an amorphous polycrystalline adsorbent. On the other hand, Gradsztajn et al. have shown that in the special case of adsorption on carbons, b_3 has a more noticeable effect than either b_1 or b_2 [29]. The microscopic susceptibility effect depends on the anisotropic and granular structure and on the porous nature of the samples. However, b_3 is generally negligible.

Using the same geometry for the reference and the sample, the correction is:

$$\Delta\delta = \delta_{real} - \delta_{obs} = \left[\frac{4\pi}{3} - \alpha - k\right][\chi_{V(ref)} - \chi_{V(sample)}] \qquad (10B.5)$$

The chemical shift of the adsorbed molecules is generally determined relative to that of the isolated molecules and therefore the best reference sample is the corresponding gas at very low pressure [27]. In this case $\chi_{V(ref)} \sim 0$. Further, according to Wiedeman's rule, the total volume susceptibility $\chi_V(sample)$ can be written as the weighted average:

$$\chi_{V(sample)} = \chi_{V(sp)} + F_a[\chi_{V(admolecule)} - \chi_{V(sp)}] \qquad (10B.6)$$

where F_a is the volume fraction of the adsorbed molecules and $\chi_{V(sp)}$ the bulk susceptibility of the solid powder.

Finally, if the gas is only physically adsorbed, there is no variation of b_4, and $\delta_{real} \rightarrow 0$. Then the extrapolation of δ_{obs} values to zero coverage yields the susceptibility correction for the solid powder:

$$\delta_{obs} = \left[\frac{4\pi}{3} - \alpha - k\right]\chi_{V(sp)} \qquad (10B.7)$$

This method, first proposed by Bonardet et al. [26, 27], leads to good agreement with direct measurements of volume magnetic susceptibilities on the basis of an analysis of proton resonance shifts. We may note that the sum $(b_1 + b_2)$ is zero for a spherical sample. This property proves sometimes useful when the origin of the observed shift is investigated.

Measuring the chemical shift by a substitution method is, in our opinion, the most realistic approach. Another method, however, using an internal reference, consists of simultaneously adsorbing the substrate and the reference. As noted in several cases [30, 33], this method turns out to be rather unsound as a result of (i) the preferential adsorption of either gas; (ii) the need to adapt the reference to the experimental conditions; (iii) the presence of intermolecular interactions or possible reactions between the two types of adsorbed molecules.

The usual corrections for ^1H NMR with diamagnetic solids vary approximately between 0.1 and 2 ppm. The magnetic susceptibility corrections must be made preferably by ^{1}H NMR. With heavier atoms the van der Waals interactions may not be negligible [32]. For example, with ^{13}C, the above method is valid for sufficiently large molecules such as n-butane, cyclohexane or TMS. For small molecules such as methane the chemical shift owing to van der Waals interactions is greater than that due to the volume susceptibility.

The volume susceptibility correction is important only in the case of hydrogen resonance, since it is of the same order of magnitude as the chemical shifts b_4 due to the modification of the electron environment by chemisorption. With heavy atoms this correction is generally negligible. Moreover, in this latter case, the influence of purely physical adsorption on b_4 is usually not negligible. We shall see in the case of ^{129}Xe-NMR of adsorbed xenon how the van der Waals interactions can be used for testing the solid surface properties.

B.4 NMR Features of Chemisorbed Phases in Fast Exchange with an Excess of the Physisorbed Phase

Physically adsorbed molecules are generally very mobile at the surface of solids (if the temperature is not too low). The spectral components are then bell-shaped and fairly narrow. Consequently, as in the case of liquids, the isotropic value of the shielding tensor can be measured. Conversely, when the temperature is not very high, the mobility of strongly chemisorbed molecules is low. The characteristic spectra are then generally very broad, which makes it impossible to measure the chemical shifts. To determine them it is necessary to narrow the spectral components. Two methods can be used: (i) rapid exchange between chemisorbed and physisorbed molecules; (ii) MAS-NMR.

In addition to the line broadening, chemical shift anisotropy can be measured for molecules which are not free to move: adsorbed molecules at low temperatures [33, 34] or located in small pore [35, 36] etc. In this case also the above techniques can be used to reduce the linewidth and to measure the isotropic chemical shift [36].

The new NMR line-narrowing techniques have not resolved all the problems of reducing linewidth. For example, with MAS-NMR a line cannot be narrowed unless the sample rotation speed is at least equal to the linewidth (in Hz) in the absence of rotation. This technique therefore will not work when the width of the components characteristic of the chemisorbed complexes is greater than the maximum sample rotation speed (about 5000 Hz with sealed tubes). This is, for example, important in the case of catalysis of molecules chemisorbed on paramagnetic sites.

Theoretically, the spectrum of an adsorbate–adsorbent system must have as many components as there are regions where the molecules can be found: chemisorption sites of different force and/or type, physical adsorption, intercrystallite space, etc. However, this is only true if the residence time of the molecules at each site is sufficiently long (on the NMR timescale). Generally (except at very low temperature), these residence times are short; consequently, because of exchange the spectrum consists of a few fairly narrow components due to the coalescence of the above, their chemical shifts being linear combinations of the chemical shifts associated with the various states weighted by their concentrations. The most important problem is then to deduce, from the dependence of these shifts on the number of adsorbed molecules, the shifts characteristic of the chemisorbed molecules. Generally, these values are not deduced directly from experiment, since the experimental shifts extrapolated to zero coverage are not necessarily identical to those of the chemisorbed complexes.

As an example, Figure 10B.3 shows the surface coverage dependence of the NMR spectrum of ethanol adsorbed on alumina [37]. The preferential broadening of the OH component when the degree of coverage θ decreases indicates that the CH_3-CH_2-OH molecules are attached to the surface by this

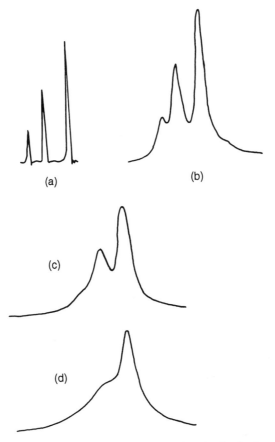

Figure 10B.3 High resolution NMR spectra of ethanol adsorbed at various degrees of coverage; (a) pure liquid; (b) $\theta = 20$; (c) $\theta = 10$; (d) $\theta = 3$

group. Even at low θ the CH_3 component remains relatively narrow and detectable because of the CH_3 rotation.

Long before the appearance of the MAS-NMR on commercial spectrometers, the exchange technique was proposed for adsorption on diamagnetic solids by Bonardet *et al* [38, 39], developed for paramagnetic solids by Borovkov *et al.* [40] and Enriquez and Fraissard [41], then expressed in another mathematical form by Michel *et al.* [42, 43]. Let us consider only one sort of chemisorption site, A (total number N_A) and the equilibrium:

$$P + A \rightleftharpoons Ch - A \qquad (10B.8)$$

where P represents the physisorbed molecules (number N_P), and Ch the chemisorbed complexes (number N_{Ch}). The equilibrium constant is given by:

$$K = \frac{N_{Ch}}{N_P(N_A - N_{Ch})} = \frac{N_{Ch}}{(N - N_{Ch})(N_A - N_{Ch})} \qquad (10B.9)$$

N is the total number of adsorbed molecules $(N = N_P + N_{Ch})$ and $N_A - N_{Ch}$ is the number of unoccupied sites.

In the case of fast exchange (generally $N \gg N_{Ch}$), the experimental chemical shift δ of the single line observed is given by:

$$\delta = \frac{N_{Ch}}{N} \delta_{Ch} + \frac{N - N_{Ch}}{N} \delta_P \qquad (10B.10)$$

where δ_{Ch} and δ_P are the chemical shifts of chemisorbed and physisorbed molecules, respectively.

B.4.1   FIRST METHOD

Generally, for light nuclei, the chemical shift of the physisorbed phase, δ_P, is negligible. $\delta_P \sim \delta_{reference\ gas} \sim 0$. Equation (10B.10) becomes:

$$\delta = \frac{N_{Ch}}{N} \delta_{Ch} \qquad (10B.11)$$

When the chemical and physical adsorptions are sufficiently different, the variation, $\delta = f(1/N)$, of equation (10B.11) shows a linear section as soon as N_{Ch} is constant, therefore when the number of chemisorbed molecules is maximal and equal to the total number of chemisorption sites, $(N_{Ch})_{max}$.

From the slope of the straight line, we can deduce the value of δ_{Ch} if $(N_{Ch})_{max}$ can be determined by other means [40, 41].

B.4.2   SECOND METHOD

By a proper combination of equations (10B.9) and (10B.11) we obtain the equations:

$$Y = \frac{X}{\Delta_{m_0}} - N_A \frac{1}{\Delta_{m_0}^2} \tag{10B.12}$$

where

$$Y = N \left(\frac{\delta}{\delta_{m_0}}\right)^2 \left(1 - \frac{\delta}{\delta_{m_0}}\right)^{-1} \tag{10B.13}$$

and

$$X = N \frac{\delta}{\delta_{m_0}} \times \left(1 - \frac{\delta}{\delta_{m_0}}\right)^{-1} \tag{10B.14}$$

δ_{m_0} is the maximum shift value obtained from a simple extrapolation of δ for $N \to 0$. Δ_{m_0} is defined as

$$\Delta_{m_0} = \delta_{m_0} / \delta_{Ch}$$

The quantities X and Y are given by the experiment. From the slope and the intercept of the straight line representing equation (10B.12), the shift, δ_{Ch}, and the number of chemisorption sites, N_A, can be determined [42, 43].

We give in Sections B.5 and B.6 some simple examples of the use of the rapid exchange techniques.

Note: Gedeon *et al.* [44] have shown that one can in the same way determine the population-weighted chemical shifts for molecules physically adsorbed on two fairly different sites, given a judicious choice of references.

B.5   Diamagnetic Systems: Study of the Brönsted Acidity of Catalysts

Of the various possible techniques, proton chemical shift determination would appear to be a direct means of determining the Brönsted acidity, as it is related to the electron density of the H atoms. Before high resolution techniques for solids were available (MAS, CRAMPS), the only means of determining the isotropic chemical shift δ_{iso} of the OH group located at the solid surfaces was the method of line-narrowing by rapid exchange, in order to mobilize the OH group protons so as to obtain a sharp NMR signal, whereas when they are static they give a broad signal [38, 39].

B.5.1 ^{1}H-NMR

Let us consider a surface S consisting of a certain number of acidic groups, denoted S–OH. Because of the strong dipolar interaction of spins, the NMR lines of these OH groups are very wide (of the order of 10^{-1} Tesla). On this surface let us adsorb a molecule AH (base) containing at least one other nucleus A which can be detected by NMR, e.g. ^{15}N or ^{17}O. AH can capture a surface proton by equilibrium:

$$S-OH + AH \rightleftharpoons S-O^- + AH_2^+ \qquad (10B.15)$$

If rapid exchange occurs between the surface proton, S–OH, and those of the adsorbed molecule, AH, the acid proton must affect the chemical shift of the adsorbed phase. The ^{1}H spectrum should contain only one line at δ_{obs} owing to the coalescence of the lines at δ_{AH}, $\delta_{AH_2^+}$ and δ_{OH}. Thus:

$$\delta_{obs}^H = P_{OH}\delta_{OH} + P_{AH}\delta_{AH}^H + P_{AH_2^+}\delta_{AH_2^+}^H \qquad (10B.16)$$

By the same reasoning

$$\delta_{obs}^A = P'_{AH}\delta_{AH}^A + P'_{AH_2^+}\delta_{AH_2^+}^A \qquad (10B.17)$$

where P_i and P'_i are the concentrations of H and A nuclei in the group i. Knowing δ_{AH}^A and $\delta_{AH_2^+}^A$, the relative concentrations, P'_{AH} and $P'_{AH_2^+}$, the dissociation coefficient of S–OH in the presence of AH can be calculated from equation (10B.17). By equation (10B.16), it is then possible to calculate δ_{OH}^H which could not be measured directly.

In fact, the study is more complex than is indicated by the above analysis which assumes that all the AH molecules are adsorbed on the S–OH sites. At high coverage, physical adsorption is important and gives rise to many equilibria such as:

$$S-OH \ldots AH \rightleftharpoons S-O^- + AH_2^+ \qquad (10B.18)$$

$$(AH-H)^+ + AH' \rightleftharpoons (AH-H')^+ + AH \qquad (10B.19)$$

$$S-OH \ldots AH + AH' \rightleftharpoons S-O \ldots AH' + AH \qquad (10B.20)$$

Furthermore, several types of site for chemisorbing AH can coexist on the surface (for example, Brönsted and Lewis acid sites). The problem then becomes above all a chemical one since, amongst this more or less complex set of equilibria, the characteristics of reaction (10B.15) alone have to be determined. Using the adsorption of AH = ^{15}NH$_3$ Bonardet et al. have found [38, 39]:

δ_{OH} (referred to gaseous TMS) = 2 and 4-5 ppm in silica gel and silica-alumina, respectively.

B.5.2 ^{13}C-NMR

Gay [45] proposed a method based on the same principle to determine the surface concentration of the acidic OH groups of a solid. This author studied the ^{13}C chemical shift of pyridine adsorbed on a silica gel (containing a monolayer of Si–OH groups) in the presence of an increasing amount of HCl.

The results show a linear dependence of the shift of each ^{13}C in the adsorbed phase on the HCl/pyridine ratio up to a value of unity. Above this value the shift no longer varies. This behaviour expresses the variation of the relative concentrations of pyridine molecules and pyridinium ions formed in the reaction:

$$py + HCl \underset{b}{\overset{a}{\rightleftharpoons}} pyH^+ + Cl^- \qquad (10B.21)$$

according to the equation:

$$\delta_{obs}^k = (1 - P_I)\delta_M^k + P_I\delta_I^k \qquad (10B.22)$$

where, M = molecules; I = ion; $k = C_{(2)}$, $C_{(3)}$ or $C_{(4)}$; $P_I = n_I/n$, n = total number of adsorbed molecules (py + pyH$^+$).

Unfortunately, only the resonance lines of $C_{(2)}$ and $C_{(3)}$ are in general well separated in the ^{13}C spectrum. This is why Gay proposed the equations:

$$\delta_{obs}^{C(2)} - \delta_{obs}^{C(3)} = (1 - P_I)(\delta_M^{C(2)} - \delta_M^{C(3)}) + P_I(\delta_I^{C(2)} - \delta_M^{C(3)}) \qquad (10B.23)$$

The use of such an experimental quantity is more convenient because the constant contributions to the resonance frequencies, such as bulk susceptibility or intermolecular interactions, are eliminated.

This method for determining the concentration of acidic OH groups is easily applicable in the case of very high acidity, i.e. when reaction (10B.21) is complete. If this is not the case, the dissociation coefficient of the OH groups in the presence of pyridine must be taken into account. This coefficient can be determined by studying another nuclear spin such as ^{15}N.

B.6   Chemisorption on Paramagnetic Solids. Electron Transfer

The spectral lines characterizing molecules chemisorbed on paramagnetic sites are too broad for methods such as MAS-NMR to be used. This is why, even today, fast exchange between chemisorbed and physisorbed species is particularly important in this case.

B.6.1   ADSORPTION OF OLEFINS ON PARAMAGNETIC CENTRES

Kazansky and coworkers [40] have applied the fast exchange method (Section B.4) to study the adsorption of various molecules such as olefins, cycloalkanes, benzene, etc. on paramagnetic centres supported on aerosil. The physical model is very close to that which is used in the chemistry of complex compounds in solution. When adsorbed molecules enter the coordination sphere of a paramagnetic ion, a spin density can arise in their nuclei as a result of contact interaction, this leads to a shift in the spectral components. Dipole–dipole interaction may be another source of paramagnetic shifts. For example, the

NMR spectrum (at 223 K) of ethylene on Ni^{2+} (trigonally coordinated) supported on silica consists of one line shifted upfield with respect to the spectrum of the same molecules condensed in pores of the adsorbent. According to equation (10B.11) this shift increases linearly with the reciprocal adsorption when the total number of adsorbed molecules is much higher than the number of superficial Ni^{2+}. The negative spin density detected in the proton is easily explained by spin polarization in the C–H σ bond, induced by a positive spin density transferred into the π-orbital of the molecule by formation of a complex with the metal ion.

This example does not offer many alternatives for interpretation. However, in more complicated cases, in order to deduce the mechanism of ligand–metal bond formation, one has to be able to interpret correctly the mechanism of spin correlation, which causes the measured spin densities. Such an interpretation very often requires

- measurement of spin densities of all the nuclei,
- calculation of the MOs of the complexes formed,

the most sophisticated quantum mechanical methods being used to take account of the effect of electron correlation.

B.6.2   DECOMPOSITION OF FORMIC ACID ON ELECTRON-DONOR CENTRES

By vacuum treatment at different temperatures T_t, electron-donor centres are released or created at the surface of TiO_2 [46, 47]. For example, above 250 °C, oxygen vacancies are created and, at the same time Ti^{3+} ions, which can be assayed by the now-classical method of adsorbing an electron-acceptor such as TCNE, and detecting the $TCNE^-$ signal by ESR [48, 49].

The decomposition of formic acid used as a test reaction in catalysis on TiO_2 is almost exclusively a dehydration and the corresponding rate constant is observed to vary with T_t in the same way as the number of donor centres: the latter can therefore be assumed to play an important role in the catalysis reaction. It can be shown, moreover, that HCOOH and Ti^{3+} interact, since the Ti^{3+} ESR signal is shifted when the acid is adsorbed. It is not possible, however, to detect any hyperfine coupling since the natural ^{13}C and ^{17}O concentrations are too small. In fact, the final results show that even with a sufficiently high concentration of these nuclear spins the components due to electron-nucleus coupling could not be resolved, the Ti^{3+} ESR signal of the Ti^{3+}–HCOOH complex being too broad.

Application of NMR has confirmed the ESR results but has made it possible to define better the form of the chemisorbed complex [49]. We summarize here the results obtained by ^1H and ^{13}C NMR of the CH group of formic acid adsorbed on amorphous TiO_2 treated under vacuum at 400 °C. Experiments were carried out at high surface coverage. The number n of adsorbed molecules is

$$3 \times 10^{18} < n = [\text{HCOOH}]\text{ads}/\text{m}^2 < 11 \times 10^{18}$$

For comparison, the number E of electron-donor centres is

$$[E] = [\text{Ti}^{3+}] = 10.75 \times 10^{16} \text{ spin}/\text{m}^2 \ (<<n)$$

The existence of a single signal for both ^{1}H and ^{13}C shows that chemically and physically adsorbed molecules exchange rapidly. The observed shifts $\delta(^1\text{H})$ and $\delta(^{13}\text{C})$ (relative to the gas phase) are linear functions of $(1/n)_{T < 363K}$ (Figure 10B.4) and of $(1/T)_n$. These variations show that the chemisorbed molecules interact strongly with paramagnetic centres. Consequently according to equation (10B.11):

$$\delta = \frac{S}{n} \delta_{\text{Ch}}$$

S is the number of chemisorption sites per m^2 and δ_{Ch} is the chemical shift of a nucleus of the chemisorbed molecule.

Assuming that the only paramagnetic electron centres are the electron-donor centres assayed by ESR ($E = S$) it is possible to calculate δ_{ch}:

$$\delta_{\text{ch}}(^1\text{H}) = -75 \text{ ppm (upfield)}$$
$$\delta_{\text{ch}}(^{13}\text{C}) = 3934 \text{ ppm (downfield)}$$

$\delta_{\text{ch}}(^1\text{H})$ and $\delta_{\text{ch}}(^{13}\text{C})$ are also linear functions of $(1/T)_n$ [49].

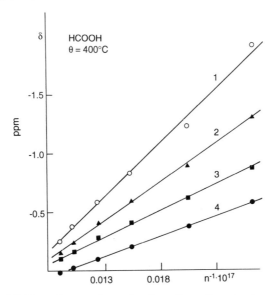

Figure 10B.4 $\delta(^1\text{H})$ against reciprocal adsorption. Sample temperatures T_t: (1) 300 K; (2) 320 K; (3) 340 K; (4) 358 K. Adapted from Ref. 49

The authors have checked that these chemical shifts are due simply to Fermi–contact interaction. Whence the hyperfine coupling constants:

$$a = 2.85 \text{ MHz or } 1.019 \times 10^{-4} \text{ Tesla}$$

$$a = -37.30 \text{ MHz or } -13.33 \times 10^{-4} \text{ Tesla}$$

The above results can only be explained as follows. *Assume firstly that the HCOOH molecules remain planar after adsorption.* Formation of a (Ti–HCOOH) π-complex transfers positive electron spin density ρ_π directly into the π^*-MOs of the molecule. By polarization of the spins along the σ (C–H) bond, this density, ρ_π^C, near the carbon, induces a spin density which is positive on ^{13}C and consequently negative on ^{1}H (Figure 10B.5). From the value of the coupling constant a it is possible to calculate ρ_π^C in the π^*-MO at the carbon atom by using McConnell and Chesnut's relationship [50]:

$$a(10^{-4} \text{Tesla}) = -25 \, \rho_\pi^C \tag{10B.24}$$

whence:

$$\rho_\pi^C = 4 \times 10^{-2}$$

From this result and the quantum mechanical study of $(HCOOH)^-$ one deduces that an electronic charge density of about 0.1 e is transferred from Ti^{3+} into the π^*-MO of HCOOH.

However, when the carbon atom is sp hybridized (planar configuration) and the position and negative spin densities on C and H, respectively, are due solely to spin polarization along the σ (C–H) bond induced by the electron density ρ_π^C, the absolute values of the coupling constants $|a_H|$ and $|a_c|$ are of the same order of magnitude [51, 52]. The fact that $|a_c|$ is much greater than $|a_H|$ proves that a part of the electron spin density on the π^*-MO is directly transferred into the 2s (C) orbital, and therefore, that the adsorbed species

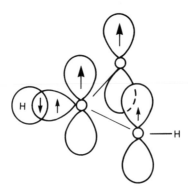

Figure 10B.5 Spin polarization in the (C–H) bond induced by the transferred spin density in the HCO_2H π^*-MO. Adapted from Ref. 49

$(HCOOH)^{\varepsilon-}$ is no longer planar. The deviation from planarity, α, can be calculated from the following equation [53]:

$$a_c(\alpha)(10^{-4}\,\text{Tesla}) = [a_c(0) + 1190 \times 2\tan^2\alpha]\rho_\pi^C \qquad (10\text{B}.25)$$

where: $a_c(\alpha)$ is the true coupling constant; $a_c(0)$ is the coupling constant corresponding to the theoretical planar structure; α is the angle between the bonds and the plan normal to the C_3 symmetry axis, treating $(HCOOH)^{\varepsilon-}$ as a CH_3 group.

The value of $a_c(0)$ for the planar structure can be calculated from Karplus and Fraenkel's equation [53] and $\sigma-\pi$ spin polarization constant [51]:

$$a_c(0) = 0.57 \times 10^{-4}\,\text{Tesla}$$

and, consequently:

$$\alpha = 6.5°$$

These calculations are only approximate. However, they give with sufficient precision the order of magnitude of the distortion of $(HCOOH)^{-0.1}$ relative to the theoretical plane.

Finally, by studying the ^1H NMR signal of C–H and OH, it has been possible to compare the distances, r, between Ti^{3+} and each of the H atoms, and to detail the form of the chemisorbed complex (Figure 10B.6).

The wave mechanical study explains why dehydration of HCOOH is facilitated by chemisorption of the molecule on electron-donor centres. When $(HCOOH)^-$ is formed by the capture of an electron, all the bond energies decrease but most importantly those of the two C–O bonds. Moreover, the C–OH bond is the weakest and, consequently, the easiest to break [52].

The calculated electron spin density distribution is in complete agreement with the experimental results deduced from NMR. The two methods prove that when HCOOH is chemisorbed on an electron-donor centre it loses its planar structure.

B.7 ^{1}H-NMR Study of Hydrogen Chemisorbed on Metals. Application to the Dispersion

The physical and electronic properties of metal particles, in particular their magnetism, must depend on their size, at least when they are sufficiently

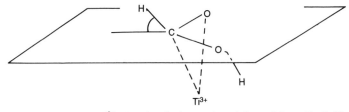

Figure 10B.6 $HCO_2H - Ti^{3+}$ chemisorbed complex. Adapted from Ref. 49

small. For this reason it could be interesting to use NMR to study these properties.

In 1973, Knight [54] predicted that the nuclear spins of even and odd particles would have different chemical shifts. The NMR spectra of small copper particles show a line which is less shifted and broader than that of the metal, and which could be the envelope of the two lines predicted by Kobayashi *et al.* [55]. The form of the NMR spectrum of metal nuclei should therefore depend on particle size. Moreover, at very low temperature it should be possible to observe two lines characteristic of even or uneven particles.

NMR study of the supported metal can, indeed, give direct information about the dispersion. But it involves considerable theoretical and technical problems. In addition, it cannot be generalized to all metals used in catalysis. For this reason, Fraissard *et al.* [56, 57] had the idea of studying a metal surface by NMR by using a gas, such as hydrogen, which could be sorbed by all metals and therefore used as a universal probe. Hydrogen has also the advantage of being very sensitive to NMR detection and one of the gases most used in catalysis.

Interest of these studies has since been confirmed by other authors both with platinum and other metals [58–63].

We summarize below the case of Pt and Pd.

B.7.1　HYDROGEN CHEMISORBED ON PLATINUM

After pretreatment at 400 °C under 10^{-5} torr and prior to any hydrogen adsorption, the ^{1}H-NMR spectrum consists of a single symmetrical line, *a*, whose chemical, δ, is close to zero, more or less intense depending on the nature of the support. This signal has the advantage of being able to serve as an internal reference for the chemical shifts. But this signal also has the disadvantage of washing a part of the adsorbate spectrum. This problem must be overcome:

- either by subtraction of the OH spectrum recorded before absorption; but this solution is not always reliable as the OH relaxation changes with the chemisorption on the metal:
- or by prior complete OH–OD exchange with D_2O and D_2. In this case the first H atoms to be chemisorbed exchange with the OD nearest the particles; but in all cases the signal of these OH groups is much narrower and smaller than the initial signal.

After adsorption at 26 °C of a very small amount of hydrogen, the NMR spectrum consists of two lines: the previous one, *a*, which has not moved, and a second, *b*, very much shifted upfield (δ_b negative), which is characteristic of H_2 chemisorbed on Pt [56]. When the number, N, of H_2 molecules/g increases, line *b* grows but its chemical shift, δ_b, remains about the same up to a coverage, $\theta_{H,26}$ of the order of 0.5–0.6 (depending on the techniques used for

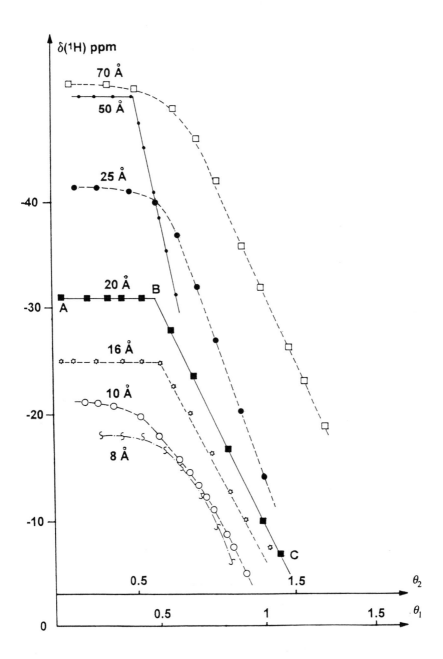

Figure 10B.7 The ^{1}H chemical shift, δ(ppm), of hydrogen adsorbed on Pt, at 26 °C, versus the coverage θ_H. Influence of the particle size d. Adapted from Refs 57–59, 63

the determination of the monolayer) if the sample is monodispersed, that is if the particle size is homogeneous (segment AB, Figure 10B.7). Beyond this latter value, δ_b varies linearly with $\theta_{H,26}$ (segment BC). As shown in Figure 10B.7, $| \delta_b |$ is increasing with the particle diameter d, at any coverage θ_H.

The proton shift is generally small (0–10 ppm) and positive. The negative and often very large shift, δ_b, can only be of the Knight shift. The Pauli paramagnetism of Pt is mainly due to unpaired spins in the d band. Thus covalent bonding of each H atom localizes a metal d electron and leads to a decrease of the net spin density. This concept is confirmed by UPS studies which indicate that substrate d electrons are involved in the surface bond [64]. The linear decrease of $| \delta_b |$ with increasing θ_H (>0.5) was quite easy to explain. The Knight shift is proportional to the paramagnetic susceptibility of Pt which, in turn, is proportional to the overall spin density; this must decrease linearly when θ_H increases, as is shown experimentally (segment BC).

Furthermore, the relative variation ($\Delta\delta_b/\delta_b$) when θ_H goes from 0 to 1 appears to increase with increasing particle size. This seems to indicate that the spin density change is restricted to the zone of Pt–H bonds (one or two atom layers). Consequently, δ_b must be close to zero at $\theta_{H,26} = 1$, for the mono and diatomic layers. It is observed to be the case for the most highly dispersed samples. Slichter et al. have confirmed these points by ^{195}Pt-NMR [65].

It is somewhat surprising to find a plateau for about $0 < \theta_H < 0.5$. For a well defined metal dispersion δ_b varies only with θ_H. The constant value of δ_b proves therefore, that along the horizontal AB, hydrogen is chemisorbed at room temperature with a constant coverage, $\theta_{H,26}$, corresponding to point B. This result can only be explained as follows: when the hydrogen enters the NMR tube containing the sample, it is chemisorbed on the first particles it encounters. This chemisorption defines two zones in the samples, one (A) towards the bottom of the tube which contains only bare particles, the upper one (B) containing particles with a constant coverage $\theta_{H,26} = \theta_B$ (Figure 10B.8). When N increases along AB, zone B extends at the expense of zone A. The coverage is the same for all particles for $N = N_B$. This two-zone distribution is confirmed by the variation of the intensity of line b when the sample is moved relative to the receiver coil. Consequently, at 26 °C even when H_2 concentration is very low, the coverage of the few particles which have chemisorbed hydrogen is quite high. Beyond θ_B, the H–Pt interaction becomes weaker (mainly due to the repulsive interactions between chemisorbed H atoms) and the H_2 distribution on the particles is more homogeneous. In addition the diffusion of H_2 from one particle to another is rapid on the NMR timescale. In this case $| \delta_b |$ decreases linearly with increasing θ_H, as was stated above.

In the case of Pt supported on zeolite, one can also have such a distribution of the chemisorbed phase inside the zeolite crystallite. If the sample has two well-differentiated particle distributions, two ^{1}H-NMR signals are detected

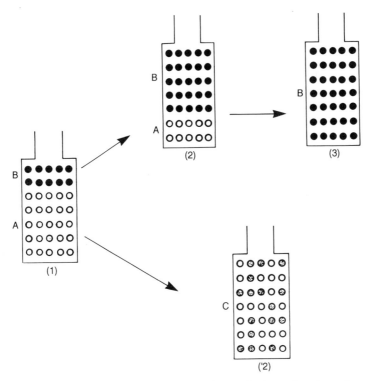

Figure 10B.8 Distribution of bare Pt particles and those bearing chemisorbed hydrogen. (1) (2) (3): distribution of H_2 chemisorbed at 26 °C. Variation with $[H_2]$. (2'): distribution of H_2 after homogenization of (1) at 400 °C. Adapted from Refs 63 and 66

upon the H_2 adsorption, each depending on θ_H as above [66]. Finally, this horizontal is difficult to detect in the following cases: wide particle size distribution or sample monodispersed but with large particles.

The distribution of the chemisorbed phase also depends on the chemisorption temperature. For example, when this is performed at 400 °C, the hydrogen is of course distributed throughout the tube and the variation of δ_b with the concentration N measured at 26 °C corresponds to a more homogeneous distribution (Figure 10B.9, EFGHI). At this temperature and at very low concentration the probability of there being more than 2H per particle is negligible. There is then in the NMR tube a random distribution of bare particles and particles with 2H. δ_E (point E) corresponds to shift δ_b when for a given dispersion, there are 2H on the particles. The number of H_2 molecules chemisorbed at point F corresponds roughly to the number of particles in the sample [63]; whence the determination of the number of Pt atoms per particle, given the metal concentration of the catalyst.

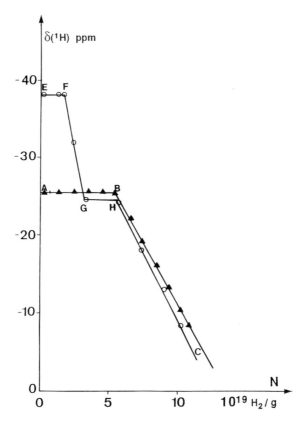

Figure 10B.9 The ^{1}H chemical shift, δ_H (ppm), versus the coverage θ_H. line ABC: chemisorption at 26 °C; line EFGH: chemisorption at 26 °C followed by homogenization in sealed tube at 400 °C. Adapted from Refs 63 and 662

B.7.2   HYDROGEN SORPTION ON PALLADIUM

We have just seen that for certain supported metals M, such as platinum, ^{1}H-NMR of chemisorbed hydrogen gives us some information about

- the nature of the Pt–M bond,
- the effect of particle size and hydrogen coverage on the NMR shift of hydrogen,
- the temperature and coverage dependence of the distribution of the chemisorbed phase on the metal particles.

The ^{1}H-NMR of hydrogen sorbed on supported palladium gives very different results which make it possible to identify several species in the presence of the metal: chemisorbed hydrogen, H_S, and hydrogen in α- and β-hydrides [62] (H_α and H_β, respectively).

From a thermodynamic point of view, *for a given particle* the five phases (H_2, Pd, Pd–H_α, Pd–H_β and surface Pd–H_S) cannot coexist since the variance, v, of the system would be negative ($v = c + 2 - \phi$, where c is the number of independent components, Pd and H_2, and ϕ the number of phases). Since H_2 and Pd–H_S are always present we deduce that Pd–H_α and Pd–H_β cannot coexist at the same time as Pd. At low H_2 concentration we have therefore Pd and Pd–H_α, at high concentration Pd–H_α and Pd–H_β. For these two cases $\phi = 4$ and $v = 0$. However, experiment shows that $v = 2$ since the system evolves with the hydrogen pressure and temperature. There must therefore also be a relationship between [H_2], [H_S] and [H_α] and between [H_2] [H_S] and [H_β]

$$H_2 \rightleftharpoons H_S \rightleftharpoons H_\alpha \qquad\qquad (10B.26)$$

$$H_2 \rightleftharpoons H_S \rightleftharpoons H_\beta \qquad\qquad (10B.27)$$

[H_2] and [H_S] are related by the classical dissociative chemisorption isotherms. The H_α and H_β concentrations must depend on the extent of superficial particle coverage.

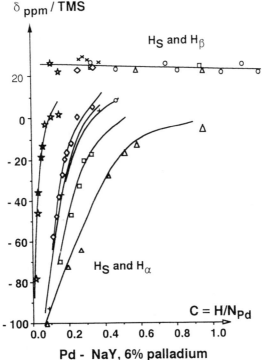

Figure 10B.10 Palladium supported on NaY zeolite at 26 °C. ^{1}H- NMR chemical shift dependence on the hydrogen concentration. Adapted from Ref. 62

^{1}H-NMR verifies totally these thermodynamic conclusions. At low hydrogen concentration, N (in moles per mole of metal), the spectra show a signal S_1, due to hydrogen in contact with the metal, whose NMR shift δ_1 is negative (upfield), with a high absolute value $|\delta_1|$ (Figure 10B.10). This first line appears at about -100 ppm, but its exact position depends on the dispersion of the metal. Shift δ_1 increases algebraically with the hydrogen concentration. However, this variation is not regular. At the beginning of hydrogen addition, δ_1 increases linearly with N. The $\delta_1 = f(N)$ plot then curves with a decrease in the gradient. The onset of this change occurs in a concentration range and for a shift which depends on the dispersion D. For example, it occurs at about $N = 0.005$, $\delta_1 \approx 0$ ppm for a sample whose dispersion is 0.05 and at about $N = 0.6$, $\delta_1 \approx -15$ ppm for a sample whose dispersion is 0.50. For the highest values of N, δ_1 becomes constant and close to 0 ppm.

In parallel with the first change in the slope of the $\delta_1 = f(N)$ plot, a second signal S_2 appears at about $+26$ ppm; its intensity increases with N at the expense of that of S_1 which finally disappears.

At low hydrogen concentration the curves for the NMR shift variation, $\delta_1 = f(N)$, of samples with different dispersions are not superposable. The slope of the linear section of these plots decreases when the dispersion increases. For a given hydrogen concentration the shift therefore increases algebraically with the particle size.

However, when the shift δ_1 of the hydrogen is expressed relative to the coverage, the $\delta_1 = f(\theta)$ curves for the various samples are observed to coincide (Figure 10B.11). The single curve presents two parts: the first linear section (AB) has a slope of about 100 ppm/unit of coverage; the second part (CD) is a horizontal, linked to the former by a short curve. The extensions of those two linear sections intersect at E which corresponds to a θ slightly greater than unity. This is exactly the value of the coverage for which S_2 appears, at about $+26$ ppm. The variation of $\delta_2 = f(N)$, decreasing slightly with increasing θ, is also independent of the sample dispersion.

Signals S_1 and S_2 characterize the two systems (H_S, H_α) and (H_S, H_β), respectively. In each case, the presence of a unique signal proves that at 26 °C there is a rapid exchange between either H_S and H_α or H_S and H_β. The chemical shifts vary according to the relationships

$$\delta_1 = \frac{N_S\delta_S + N_\alpha\delta_\alpha}{N_\alpha + N_S} \quad \text{and} \quad \delta_2 = \frac{N_S\delta_S + N_\beta\delta_\beta}{N_S + N_\beta} \tag{10B.28}$$

where N_S, N_α and N_β are the numbers of hydrogen atoms at the particle surface and in the α- and β-hydride phases, respectively.

The effect of the H_S concentration on that of H_α is also shown by the experiment on the coadsorption of carbon monoxide. When increasing amounts of CO are adsorbed on the sample which already contains a small amount of hydrogen, δ_1 decreases algebraically, the intensity of signal S_1 decreases, and

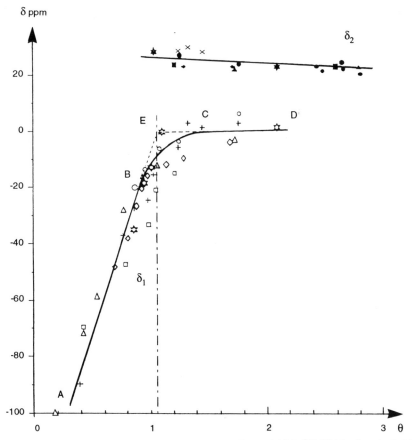

Figure 10B.11 Palladium supported on NaY zeolite at 26 °C. ^{1}H-NMR chemical shift dependence on the hydrogen coverage θ. (θ is determined by chemisorption). Adapted from Ref. 62

δ_1		δ_2	
(1) D = 0.0492	☆		★
(4) D = 0.0145	◇		◆
(5) D = 0.0195	+		×
(6) D = 0.24	⊙		●
(7) D = 0.247	□		■
(8) D = 0.5	△		▲

it is no longer detected when the adsorbed CO exceeds that of the initial hydrogen. According to

$$H_\alpha \rightleftharpoons H_S$$

the decrease in N_S due to the presence of CO leads to a reduction of N_α by a displacement of the above equilibrium. The α-hydride can therefore disappear in the presence of CO, despite the hydrogen pressure.

Finally, for a given value of the hydrogen concentration, N, δ_1 decreases when T increases. Increasing the temperature therefore shifts the above equilibrium to the right. This result confirms the interest of measuring hydrogen chemisorption isotherms at high temperature to determine the metal dispersion.

In conclusion, the ^{1}H-NMR spectra of the Pd-H system show

- a signal S_1, with negative shift for low hydrogen concentration, due to rapid exchange between chemisorbed hydrogen H_S and the H_α hydrogen of the Pd–H_α hydride. At 26 °C the $\delta_1 = f(N)$ plot becomes horizontal for $\theta = 1$ and the maximum concentration of Pd–H_α (Pd–$H_{0.035}$).
- a signal S_2, downfield, for hydrogen concentrations corresponding to a coverage greater than unity, due to Pd–H_β hydride in equilibrium with H_S.

In addition:

- The $\delta_1 = f(\theta)$ plots are independent of the dispersion. The continuous variation of δ_1 with θ proves that the H_α and H_S concentrations increase homogeneously throughout all the particles.
- H_S concentration controls the formation of the Pd–H_α and Pd–H_β phases.
- The equilibrium $H_\alpha \rightleftharpoons H_S$ is displaced in the forward direction when the temperature is raised.

B.8   NMR of Physisorbed Xenon used as a Probe

B.8.1   INTRODUCTION

The central idea of the pioneers [67, 68] of this general research was to find a molecule, non-reactive, particularly sensitive to its environment, to physical interactions with other chemical species and to the nature of adsorption sites, which could be used as a probe for determining in a new way solid properties which are difficult to detect by classical physico-chemical techniques. In addition, this probe should be detectable by NMR since this technique is particularly suitable for investigating electron perturbations in rapidly moving molecules.

Xenon is the ideal probe because it is an inert gas, monoatomic, with a large spherical electron cloud. From the NMR point of view, the ^{129}Xe isotope has a spin of one-half. Its natural abundance in xenon is 26 per cent and its sensitivity of detection relative to the proton is about 10^{-2}. The recent development by Pines et al. [69, 70] of xenon optical polarization techniques have increased the sensitivity for detection of adsorbed xenon by a factor of 10^3. The high polarizability of the xenon atom makes it very sensitive to its environment. Small variations in the physical interactions with the xenon atom cause marked perturbations of the electron cloud which are transmitted directly to the xenon nucleus and greatly affect the NMR chemical shift.

B.8.2   CHEMICAL SHIFT OF XENON ADSORBED IN A ZEOLITE

In pure xenon gas, the ^{129}Xe chemical shift can be expressed by a virial expansion of the xenon density ρ [71, 72]:

$$\delta(T, \rho) = \delta_0 + \delta_{1,Xe}(T).\rho_{Xe} + \delta_{2,Xe}(T).\rho_{Xe}^2 + \delta_{3,Xe}(T).\rho_{Xe}^3 + ... \quad (10B.29)$$

where $\delta(T, \rho)$ is the resonance shift at a temperature T and a density $\rho. \delta_0$ corresponds to its value extrapolated at an infinitely low density; $\delta_{i,Xe}(T)$ are the virial coefficients of the shift in density.

For mixtures of xenon and another gas, A. Jameson et al. showed that in the linear range [71, 72]:

$$\delta = \delta_0 + \delta_{1,Xe}(T).\rho_{Xe} + \delta_{1,A}(T).\rho_A \quad (10B.30)$$

where ρ_{Xe} and ρ_A are the densities of Xe and 'solvent' molecules A; $\delta_{1,Xe}$ and $\delta_{1,A}$ characterize Xe–Xe and Xe–A interactions, respectively.

As in the gas phase, the main information has been obtained from the analysis of the variation of the chemical shift against the xenon concentration, generally at 26 °C. The amount of xenon adsorbed is expressed as the number, N, of atoms per gram of anhydrous zeolite or the number of atoms, n_s, per cage (zeolites Y, ZK-4, erionite, etc.).

Fraissard et al. have shown that the chemical shift of adsorbed xenon is the sum of several terms corresponding to the various perturbations it suffers [68, 73, 74]:

$$\delta = \delta_{ref} + \delta_S + \delta_{Xe} + \delta_{SAS} + \delta_E + \delta_M \quad (10B.31)$$

δ_{ref} is the reference (gaseous xenon at zero pressure); δ_S arises from interactions between xenon and the surface of the zeolitic pores, assuming that the solid does not contain any electrical charges. In this case it depends only on the dimensions of the cages or channels and on the ease of xenon diffusion. $\delta_{Xe} = \delta_{1,Xe}.\rho_{Xe}$ corresponds to Xe–Xe interactions; it increases with the local density of adsorbed xenon and becomes predominant at high xenon pressure. When the Xe–Xe collisions are isotropically distributed (large spherical cage) the relationship $\delta = f(N)$ is a straight line (Figure 10B.12(1)). The slope, $d\delta/dn$, is proportional to the local xenon density and, therefore, inversely proportional to the 'void volume'. If the Xe–Xe collisions are anisotropically distributed (narrow channels) $d\delta/dn$ increases with N (Figure 10B.12(2)).

When there are strong adsorption sites (SAS) in the void space interacting with xenon much more than the cage or channel walls, each xenon spends a relatively long time on these SAS, particularly at low xenon concentration. The corresponding chemical shift δ will be greater than in the case of a non-charged structure (influence of δ_{SAS}, Figure 10B.12(3)). When N increases, δ must decrease if there is fast exchange of the atoms adsorbed on SAS with those adsorbed on the other sites. When N is high enough the effect of Xe–Xe interactions becomes again the most important and the dependence of δ on N is

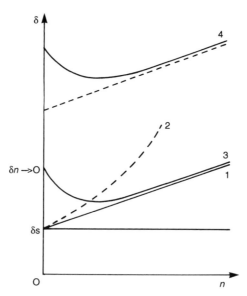

Figure 10B.12 Variation of the chemical shift of xenon adsorbed in a porous solid (see the text)

then similar to that of Figure 10B.12(1). In this case $\delta_{N \to 0}$, which is the chemical shift extrapolated to zero concentration, depends on the nature and number of these strong adsorption sites. Often, these SAS are more or less charged and sometimes with paramagnetic cations. The theoretical curve, 12 (3), is displaced downfield (Figure 10B.12(4)). The difference between 12(1) and 12(3) expresses the effect, δ_E, of the electrical field and, if it exists, δ_M due to the magnetic field created by these cations [68].

B.8.3   INFLUENCE OF THE PORE STRUCTURE ON δ_S. APPLICATION TO POROSITY

The variation $\delta = f(N)$ is characteristic of the zeolite structure when the Si/Al ratio is very high and when there are no strong adsorption sites accessible to xenon. Figure 10B.13 illustrates the sensitivity of this technique to the pore structure. The slopes of the curves depend on the void volume of the pores; this is logical since, for a given number of Xe atoms per gram, the local density and thus the Xe–Xe interactions depend on this volume. The chemical shift δ_S at zero concentration is related to the structure: the smaller the channels or cavities, or the more restricted the diffusion, the greater δ_S becomes. It is assumed that at 26 °C the experimental chemical shift is the average of the shift of xenon in rapid exchange between a position A on the pore surface

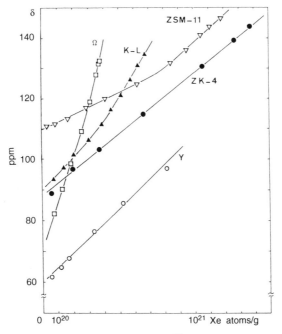

Figure 10B.13 Chemical shift δ of adsorbed ^{129}Xe versus number of xenon atoms per gram of zeolite (Ω, L, ZSM-11, ZK-4, Y)

(denoted by δ_a) and a position in the volume V of the cavity or channel (denoted by $\langle \delta_V \rangle$). (Figure 10B.14):

$$\delta = \frac{N_a \delta_a + N_V \langle \delta_V \rangle}{N_a + N_V} \tag{10B.32}$$

where N_a and N_V are the number of xenon atoms in each state. This equation is valid whatever the xenon concentration, then for $N \to 0$ $\delta \to \delta_S$. $\langle \delta_V \rangle$ is a function of δ_a and the distance travelled by the xenon atom between two successive collisions with the pore wall. In fact, $\delta_V = \delta_a$ when the xenon leaves the surface. δ_V then decreases during the journey between two collisions, whence the need to determine a mean value, $\langle \delta_V \rangle$.

In order to obtain by this technique precise data on the void space of a zeolite of unknown structure and on the dimensions of structural defects, Fraissard *et al.* have calculated the 'mean free path', $\bar{l}$, of xenon imposed by the structure and determined the dependence of δ_S on $\bar{l}$ either by calculation (sphere, infinite cylinder) or experimentally [75, 76].

$$\delta = \delta_a \frac{a}{a + \bar{l}} \tag{10B.33}$$

where a is a constant. Equation (10B.33) can be deduced from equation

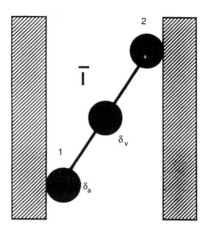

Figure 10B.14 Fast exchange model. δ_a and δ_v are the chemical shifts of xenon adsorbed on the wall and in the free space

(10B.32). The large variation of δ_S with $\bar{l}$ (Figure 10B.15) proves that ^{129}Xe-NMR is very sensitive to pore dimensions [75, 76].

Derouane and Nagy [77] have been able to relate δ_a to the curvature of the adsorbing surface, therefore also to the pore diameter. Finally, Ripmeester *et al.* [78] have continued this study by calculating the Lennard–Jones potential curves for a Xe atom and a spherical layer representing the atoms of a cage. They deduced that in small cages Xe is 'solid like' and in larger cages 'gas like', and that consequently it is difficult to imagine a single correlation between the xenon chemical shift and the dimensions of the adsorption zones.

The symmetry of the adsorption site can sometimes be reflected by the form of the signal when the interactions to which the xenon atom is subjected are very anisotropic. For example, Springuel-Huet and Fraissard [36] have shown that the ellipsoid-shaped channels of SAPO-11 and AlPO$_4$-11 cause chemical shift anisotropy effects to be observed in the ^{129}Xe-NMR spectra of adsorbed xenon, owing to the similarity between the small axis of the pore section (6.7 × 3.9 Å) and the diameter of the xenon atom (4.4 Å). Similar results have been obtained by Ripmeester *et al.* [79] on Xe/clathrates systems.

When the zeolite structures include several adsorption zones, the xenon spectrum must contain an equal number of components characteristic of these zones, provided that the diffusion of the xenon atoms is not too rapid on the NMR timescale. Thus, the spectrum of Na-mordenite [80, 81] shows two signals at 26 °C arising from xenon in the main channels and in the side-pockets. However, because of rapid exchange of xenon between these two regions, there is only one signal for H$^+$-exchanged mordenite. The spectrum of ferrierite [82] at 26 °C shows two signals characteristic of two types of channel. In this case it has even been possible to detect the presence of a mordenite intergrowth [82]. Two signals have also been detected for the Xe/H-

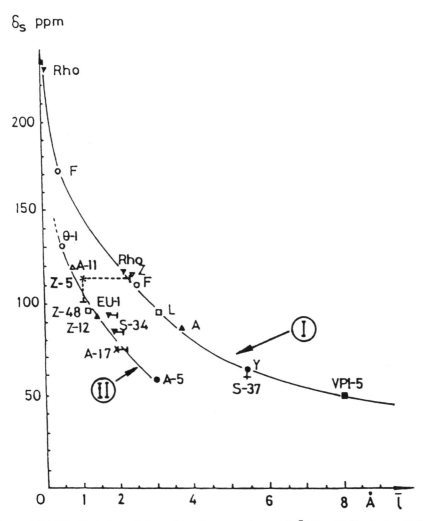

Figure 10B.15 Variation of δ_s against the mean free path $\bar{l}$. Rho; F: Ferrierite; Z-5, 48, 12: ZSM-5, 48 and 12; A-5, 11, 17: AlPO$_4^{-5}$, 11 and 17; S-34, 37: SAPO-34 and 37. Adapted from Refs 75 and 76

Rho system, but only at a temperature as low as $-50\,°C$ to avoid rapid exchange of atoms located in the octagonal prisms and the large cavities. In Cs-Rho, only the signal corresponding to the large cavities is detected, since all the prisms are occupied by Cs$^+$ [83, 84].

It is clear from these examples that this technique can be used to identify all the different microporous adsorption zones, namely: structure defects of zeolite [68], heteropolyanions [85], solid polymers [86], clathrates [73, 74, 79], etc.

B.8.4   INFLUENCE OF TEMPERATURE

The study of the temperature dependence is another way of checking the size of the pores in which the xenon is adsorbed [87]. The variation in temperature changes the relative residence times of xenon on the surface and in the free space, which in turn influence the xenon chemical shift. The variation of δ with the temperature then depends on $\langle\delta_V\rangle$. It is very small, by a few ppm only, if $\langle\delta_V\rangle \sim \delta_a$ (ZSM-5) or, conversely, very large, by about 40 ppm from $-100°$ to $+100\,°C$, if $\langle\delta_V\rangle \ll \delta_a$ (Na Y).

From a quantitative point of view, Chen *et al*, have shown that [87]:

$$\frac{d \, \text{Ln}(\delta_a - \delta_S)/(\delta_S - \langle\delta_V\rangle)}{dT} = -\frac{\Delta E}{RT^2} \qquad (10\text{B}.34)$$

ΔE is the energy difference between the xenon in the two states A and V. The fast exchange model is valid in the temperature range where the $\text{Ln}[(\delta_a - \delta_s)/(\delta_s - \langle\delta_V\rangle)] = f(1/T)$ plot is a straight line.

B.8.5   EFFECT OF THE Si/Al RATIO OF ZEOLITES

The relationship between δ_S and the porous structure is strictly valid only for an electrically neutral surface. The chemical shift of xenon must therefore depend on the chemical composition of the zeolite, in particular on the number of charges, i.e. on the Si/Al ratio. This will be all the more pronounced when the temperature is low and the pore diameter is small (increase of the effect of δ_a with the contact time of xenon at the surface).

In the case of faujasite either with Na^+ cation or decationized, the $\delta = f(N)$ relationships are almost parallel straight lines at $26\,°C$, and $\delta_{N\to0}$ decreases monotonically by 4 ppm when Si/Al increases from 1.28 to 54 [68]. For ZSM-5 and ZSM-11 (narrow channels) the dependence of $\delta_{N\to0}$ on [Al] is greater than for faujasite (large cavities). In addition, this variation shows a break at about [Al] = 2 Al/unit cell (u.c.) which demonstrates that the Xe-wall interaction changes for this concentration [88]. Then Al is not randomly distributed in the lattice but its distribution depends on its concentration. Moreover, we mention finally the difference between ZSM-5 and ZSM-11 for [Al] $\geqslant 2$ Al/u.c., despite the similarity of their structure, which points out again the sensitivity of ^{129}Xe-NMR technique.

B.8.6   INFLUENCE OF CATIONS

We explained in Section B.8.2 why the $\delta = f(N)$ curve goes through a minimum when there are strong adsorption sites in the zeolite. This has been shown for X and Y zeolites with divalent cations such as Mg^{2+}, Ca^{2+}, Zn^{2+}, Cd^{2+}, and Ni^{2+} [68, 89, 90], where xenon adsorption becomes saturated within the given pressure range.

With the help of a fast exchange model analogous to that of Fraissard *et al.* [76], but taking into account the possible presence of SAS, Cheung *et al.* were able to give an analytical expression of these curves in the form [91]:

$$\delta = \frac{a}{\rho} + \delta_g . \rho \qquad (10B.35)$$

ρ is the overall xenon density in the zeolite; δ_g is a coefficient analog to $\delta_{1,Xe}$ in equation (10B.29); a depends on the distribution of N_a and N_v as defined in equation (10B.30). The $1/\rho$ dependence results in an initial decrease in δ with coverage before the term linear in ρ begins to dominate. Let us consider, for example, Mgλ Y and Caλ Y zeolites, dehydrated under vacuum at 500 °C, where λ denotes the degree of cation exchange with Na$^+$ [68]. When C^{2+} cations are in the sodalite cage or in the hexagonal prisms ($\lambda < 55\%$), δ is a linear function of [Xe], identical with that for NaY, whatever the extent of dehydration of the sample. When some C^{2+} cations are situated in the supercages ($\lambda > 55\%$), one observes values of δ for CλY, compared to NaY, which are greatest when λ is high, especially at low xenon density. According to Fraissard and Ito the experimental value of $\delta_{N \to 0}$ for $N = 0$ is roughly proportional to the square of the electric field at the nuclei of Xe atoms adsorbed on C^{2+} cation [68].

In the opinion of Cheung *et al.* the high value of δ is not due solely to the high polarization of xenon and the distortion of the xenon electron cloud by the strong electric fields created by the 2+ cations. They suggest the formation of a partial bond between these two species formed by donation of a xenon 5p electron to the empty s-orbital of the 2+ cation [91]. A similar model concerning electron transfer from Xe to Pt was proposed by Ito *et al.* to explain the high shift of δ in platinum supported on NaY [92].

The effect of the dehydration and rehydration of zeolites on the influence of C^{2+} cations can also be studied by this technique [68]. For example, the chemical shift, δ, and the signal width, $\Delta \omega$, of xenon adsorbed on Mg$_{70}$ Y increase with the extent of dehydration of the solid. Their values are greatest when dehydration is complete. Inversely the spectra evolve with rehydration: for a given xenon pressure, line **a** due to xenon in the supercages containing only bare Mg^{2+} decreases in favour of line **b** corresponding to Mg^{2+} surrounded by water molecules. This technique therefore makes it possible to follow the diffusion of an adsorbate in a zeolite crystallite.

The problem is more difficult in the case of paramagnetic cations, especially when the extent of exchange is so high that the magnetic term, δ_M, in equation (10B.31), becomes large, as has been shown by Bansal and Dybowski [89] and by Gedeon *et al.* [90]. However, Scharf *et al.* have succeeded, using this technique, to follow the reduction and the reoxidation of Ni$_{7.5}$–NaY zeolite desorbed under vacuum at 350 °C [93]. In the sample reduced at the lower temperatures (≈ 100 °C) two types of environment for the xenon atoms are evident, one corresponding to xenon in the nickel-exchanged material, the other to xenon in contact with NaY or HY. Reduction at higher temperatures produces an upfield shift of the first resonance, indicating that this environment becomes more like the environment of xenon in NaY zeolite. For the highest temperature, 370 °C, only the line corresponding to xenon in NaY is detected. These results prove that nickel ions are removed from the supercages upon reduction. In the same way these authors have shown that the reoxidation in 600 torr of oxygen at various temperatures does not reverse the process of reduction.

A particularly interesting case is that of faujasite-type zeolites containing Ag^+ or Cu^+ cation [94] whose external electron structure is nd^{10}. Figure 10B.16 shows the ^{129}Xe isotropic chemical shifts of xenon in NaX and in the fully silver-exchanged zeolite AgX. The shifts in dehydrated and oxidized AgX are distinctly lower than that of NaX over the range of concentration studied. Most remarkably, the shifts decrease with concentration down to negative values in the range −40 to −50 ppm at low xenon concentration. In contrast to these results, the samples reduced at 100 °C and 300 °C show high-frequency shifts with respect to NaX. After reduction at 100 °C, δ increases steadily with the number n_S of Xe per supercage from +100 ppm to +170 ppm for about $0.3 < n_S < 4$, whereas after reduction at 300 °C the shift values are between +140 and 160 ppm for $0.1 < n_S < 1$ with a shallow minimum at about $n_S = 0.7$. In this latter case, the $\delta = f(N)$ variation has the classical shape of zeolite supported metals. The very unusual upfield shift in the case of the dehydrated and oxidized sample is due to a specific interaction of xenon with Ag^+ cations in the supercages. This shielding is due to $d_\pi - d_\pi$ back-donation from Ag to Xe involving the silver 4d and xenon 5d-orbitals.

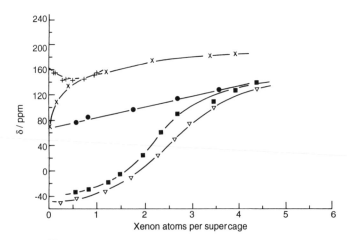

Figure 10B.16 ^{129}Xe NMR chemical shift versus number of xenon atoms per supercage for the zeolites: NaX (100 dehyd. ●); AgX (400 dehy. ■), AgX (450 ox. Δ); AgX (100 red. ×); AgX (300 red. +). Adapted from Ref. 94

Similar studies have confirmed that Cu^{2+} is autoreduced to Cu^+ in the dehydration of CuY at 400 °C and shown that there are only Cu^+ ions in the supercages [94]. The Cu^{2+} are located in the sodalite cavities or the prisms.

B.8.7   CATION EXCHANGE BETWEEN DIFFERENT ZEOLITES

^{129}Xe-NMR of adsorbed xenon makes it possible not only to follow the exchange of cation sites in the zones accessible to xenon in a given crystallite, but also to visualize the exchange of cations between different zeolites. For example, the two peaks in Figure 10B.17(a), (I) 144 and (II) 89 ppm, correspond to xenon adsorbed in dehydrated RbNaX and NaY zeolites, respectively. At the same xenon equilibrium pressure, RbNaX has a higher xenon chemical shift than NaY due to its higher cation concentration inside the zeolite and, more importantly, to the much higher polarizability of Rb^+ than Na^+ cations [95]. Upon mechanical mixing at 300 K (sample b) the intensities of the two signals decrease and a third broad signal (III) appears at about 141 ppm (Figure 10B.17(b)). After treatment of the mixture at 673 K and 10^{-5} torr (sample c) spectrum (b) changes to (c) with much greater decrease of the

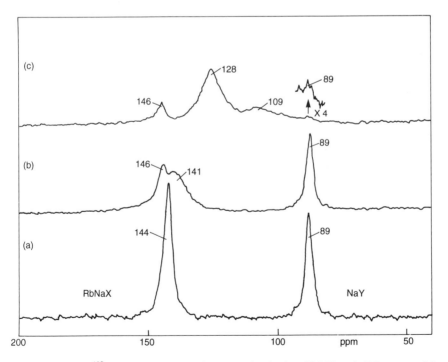

Figure 10B.17 ^{129}Xe-NMR spectra of xenon adsorbed at 300 K and 600 torr on (a): samples treated separately at 673 K under 10^{-5} torr; (b): samples (a) mixed mechanically at 300 K under vacuum; (c): sample (b) treated again at 673 K under 10^{-5} torr

intensities of the RbNaX and NaY signals, the disappearance of signal (III) and the detection of two additional broader signals at 128 (IV) and 109 (V) ppm.

The two xenon resonance signals corresponding to the two pure zeolites can be restored upon cooling sample (b) to 200 K, which proves that the third signal in Figure 10B.17(b) is due to coalescence of two signals. The mixture obtained directly in the tube by shaking is very inhomogeneous as regards to the distribution of the crystallites of the two zeolites. There are three types of zone, namely pure RbNaX, pure NaY and RbNaX–NaY mixture. Because of diffusion, the three corresponding lines are distinguished at 300 K. Consequently, the information obtained by Xe-NMR at 300 K is characteristic of macroscopic zones, i.e. containing several crystallites. The extent of these zones depends on the mean lifetime of the xenon atom in the crystallites compared to the NMR timescale. These zones can be reduced by lowering the temperature of the NMR experiment so as to obtain more and more localized information. The rate of exchange of xenon atoms between one crystallite and another must depend also on the crystallite size and above all on the barrier for diffusion to the external surface; this barrier is related to the size of the windows of the cavities and channels, other things being equal.

The low temperature ^{129}Xe-NMR spectra of xenon adsorbed on (c), show that, in contrast to the case of (b), down to 200 K, there are still four well resolved peaks, two of them corresponding to the original RbNaX and NaY zeolites. The other two signals located between those of RbNaX and NaY should correspond to $Rb_{58-x}Na_{24+x}X$ and $Rb_xNa_{56-x}Y$ zeolites, resulting from ion exchange between the two original solids.

B.8.8   DISTRIBUTION OF THE ADSORBED PHASES

Gedeon et al. [96] showed that the values of the slope, $d\delta/dN$, and of the intercept $\delta_{N\to0}$ of the $\delta = f(N)$ plots for each degree of NaY hydration can be used to distinguish the water located in zones accessible or inaccessible to xenon and to measure the volume of water in the pores. Conversely, during adsorption, the spectrum of the xenon probe is characteristic of the concentration gradient of the adsorbate in the sample tube and even within the zeolite crystallite.

Chmelka et al. [97] confirmed the interest of this technique by applying it to the study of the adsorption of various compounds: benzene, trimethylbenzene and hexane. The results were confirmed by multiple quantum NMR spectroscopy.

B.8.9   DEALUMINATION OF HY ZEOLITE. DEACTIVATION BY
COKING

The dealumination of decationized zeolites by steaming leads to the presence of extraframework aluminium, Al_{NF}, in the pores. These species carry mean charges ($\leqslant 3+$) depending on their state of complexation with the oxygen atoms and other Al_{NF}. They generally play the role of attractive centres stronger than the OH with respect to the adsorbed xenon. However, it is often necessary to reduce the experiment temperature to observe the resulting minimum in the $\delta = f(N)$ variation. Figure 10B.18 is related to a HY sample obtained by calcination of a NH_4Y zeolite at 773 K in a stream of air and slightly

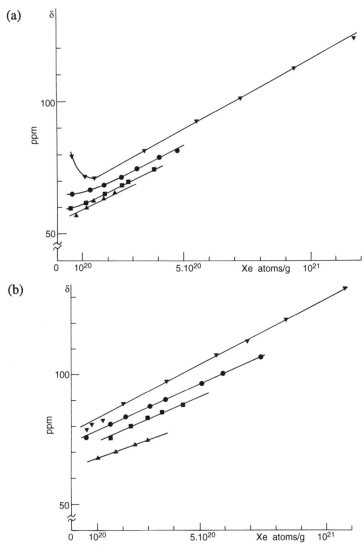

Figure 10B.18 (a) ^{129}Xe-NMR chemical shift as a function of sorbed xenon (atoms. g^{-1}) for HY sample at: (▼) 273 K, (●) 300 K, (■) 319 K, (▲) 338 K. (b) ^{129}Xe-NMR chemical shift as a function of sorbed xenon (atoms. g^{-1}) for sample HY-3% at: (▼) 273 K, (●) 300 K, (■) 319 K, (▲) 338 K. Adapted from Ref. 98

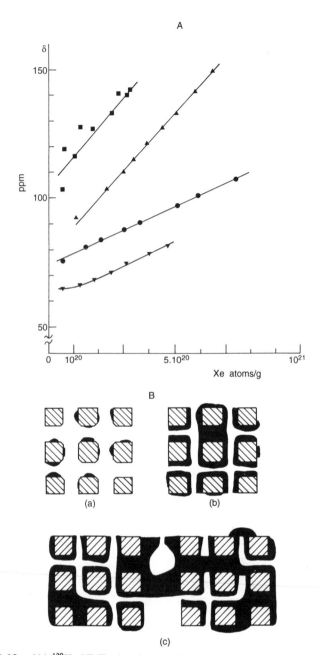

Figure 10B.19   (A) ^{129}Xe-NMR chemical shift as a function of sorbed xenon (atoms. g^{-1}) at 300 K for samples: (▼) HY; (●) HY-3%; (▲) HY-10.5%; (■) HY-15%. (B) Schematic representation of coke distribution. Scheme a: in HY-3%; Scheme b: in HY-10.5%; Scheme c : in HY-15%. Adapted from Ref. 99

dealuminated during its preparation [98]. The presence of Al_{NF} in this sample is unambiguously detected by Xe-NMR at 273 K. At 338 K on the other hand, the $\delta = f(N)$ plot is a straight line practically superposable on that of non-dealuminated HY.

The deactivation of zeolite by coking is a crucial problem in industry, in particular during cracking reactions. Xe-NMR can provide information about the distribution of coke during the reaction. For example, the $\delta = f(N)$ variation of the HY-3% sample slightly coked (3% w/w) by cracking n-heptane at 723 K is linear over the entire xenon concentration range regardless of the experiment temperature between 273 and 338 K (Figure 10B.18). This result proves that the coke is first deposited on the Al_{NF} species. At the same time the zeolite loses more than half its catalytic activity, which proves that these species are important to the cracking activity of the zeolite.

The linear section obtained at 273 K for HY and HY-3% are parallel, which is logical in view of the very small decrease in the volume due to the coke. However, the shift $\delta_{n \to 0}$ at [Xe] = 0 for HY-3% is 78 ppm, 15 ppm greater than the value obtained for HY under the same conditions. This increase in $\delta_{n \to 0}$ without change of slope corresponds to a decrease in the diffusion of xenon from one cage to another. This proves that right at the beginning of coking the coke is located at the windows between supercages and, therefore, that the Al_{NF} species were also located at this position. Figure 10B.19 shows the evolution of the $\delta = f(N)$ curves with the extent of coking and the schematic correspondence of the coke distribution in the zeolite [99].

B.8.10   EFFECT OF THE EXTRAFRAMEWORK ALUMINIUM

It is well known that extraframework aluminium Al_{NF} plays an important role in the catalytic properties of zeolites. The presence of these species is usually checked by ^{27}Al-NMR on the dehydrated samples normally used in catalytic reactions but the nature of these species is still unclear. Chen et al. have recently studied the nature of the Al_{NF} in MFI type zeolite by ^{129}Xe and ^{27}Al-NMR [100].

Table 10B.2   Aluminium concentration per unit cell; Al_T: total; Al_F: framework; AL_{NF}: non-framework

Sample	Al_T/u.c.	Al_F/u.c.	Al_{NF}/u.c.
R	4.0	4.0	0
A	8.0	4.0	4.0
B	3.4	2.9	0.5
C	2.2	1.6	0.6

^{27}Al-NMR shows that the reference sample R does not contain any extraframework aluminium (Table 10B.2). On the contrary, for those prepared in fluoride medium, there are always some Al_{NF} present in the sample. The quantity of Al_{NF} in the samples depends on the synthesis conditions.

For the four samples studied by [129]Xe-NMR (Table 10B.2), the xenon chemical shift δ dependences on the xenon loading N are different (Figure 10B.20). In the case of sample R, δ increases monotonically with N. For the samples synthesized in fluoride medium, δ first decreases and then increases with N. This indicates the presence of some strong adsorption sites inside the channels of the samples. These sites can only be more or less charged Al_{NF} species. Comparison of the number of Al_{NF} and the $\delta_{N \to 0}$ value shows that the charge of Al_{NF} increases in the order A < B < C. The average charge on each Al_{NF} depends upon the amount of Al_{NF} as well as the Al_{NF}/Al_F ratios inside the zeolite.

B.8.11 ADSORPTION ON OTHER MICROPOROUS SYSTEMS

[129]Xe NMR represents, of course, an efficient means for studying the microporosity and the symmetry of all materials analogous to zeolites, that is,

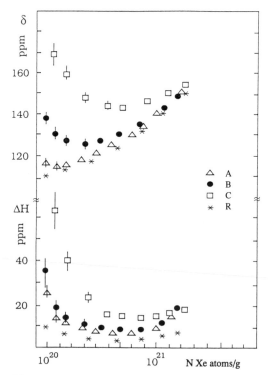

Figure 10B.20 [129]Xe-NMR chemical shift, δ and linewidth, ΔH, as a function of sorbed xenon for samples R, A, B and C (see Table 10B.2)

crystalline and containing micropores, tungstoheteropolyanions for example; but this technique is also important for many other compounds. Ripmeester and coworkers pioneered the study of clathrates and clathrate-hydrates. They have provided interesting details about the nature and the symmetry of the cages where the xenon is located [101, 102].

This technique is also important in the case of polymers:

- microscopic heterogeneity of PVC [103];
- detection of subregions in amorphous polyethylene owing to minor difference in density and geometry [104];
- effect of cross-linking (by sulphur curing at 160 °C for 20 min) on the amorphous structure of solid EPDM [105].

Xe-NMR can also characterize the pore structure of materials without internal cavities but having large surface area and porosity

- pore structure created by compression of aerosils [106];
- porosity of pillared clays [107].

However, in these last cases, exchange with the gas phase is not negligible and it is necessary to operate at low temperature. Cheung has shown that the non-linear dependence of the chemical shift at 144 K on the concentration of xenon adsorbed on silica, silica-alumina or alumina, was due to the distribution of the pore sizes of these solids and expressed the successive adsorption of xenon in pores of increasing size [108].

Finally, we should note studies on the porosity of Illinois coal [109] and on the dynamics of xenon adsorbed on graphite [110].

B.8.12 CONCLUSION

By means of ^{129}Xe-NMR of adsorbed xenon used as a probe it is possible to:

(1) determine the dimensions and the forms of the internal free volumes of zeolites without knowing their structures;
reveal structure defects produced, for example, by dealumination, and determine their characteristics;
calculate the short-range crystallinity, as opposed to that determined by X-rays;
(2) follow the mechanism of the synthesis and crystallization of zeolites during their preparation;
(3) locate cations in the zeolite structure and follow their migration or change in their environment depending on various factors;
(4) locate any 'encumbering' species, for example: adsorbed molecule, extraframework species, coke formed during catalytic cracking reactions, etc...

(5) determine the dispersion of metals (particularly when the particles are too small to be detected by electron microscopy), and the distribution of the molecules adsorbed on the particles (as opposed to the mean coverage);
(6) compare the diffusion of species as a function of the size of the zeolite 'windows'.

This list of applications is not exhaustive. Indeed, this technique an also be applied to any microporous system (amorphous solids, polymers, clays, etc.) provided one works at low temperature.

B.9   General Conclusion

Physical adsorption involves physical interactions of the van der Waals type between the adsorbate and the adsorbant. Physical adsorption isotherms recorded at low temperature are often used to determine the textural properties of a solid (surface area, porosity, etc.). Because the relevant forces are weak, physisorbed molecules are generally very mobile at the surface of the solid. This mobility, which depends on the experimental conditions, greatly affects the NMR spectrum of the adsorbate. ^{129}Xe-NMR has shown that when the adsorbed xenon diffuses rapidly, the chemical shift can be related to the porosity of the adsorbant. However, the zone 'examined' per unit time (on the NMR timescale) decreases with the temperature of the experiment; the measured shift can then be characteristic of certain adsorbent sites which interact specifically with the probe.

Physical adsorption is also a precursor of chemical adsorption, essential to heterogeneous catalysis. The chemisorbed molecules interact strongly with the adsorption sites and their mobility is generally restricted under normal conditions for detecting NMR spectra, which are therefore very broad. Line-narrowing by high resolution NMR techniques for solids is not always possible (strong dipolar interactions, paramagnetic or metal systems). Fortunately, when there is both chemical and physical adsorption (generally, after saturation of the chemisorption sites) the activation energy for exchange between physis-orbed and chemisorbed molecules is low. Whence a certain 'animation' of the latter, which increases with the concentration of the physisorbed species. Thanks to this site exchange process, the NMR characteristics of the chemis-orbed phase can then be obtained. This makes it possible to work back to the superficial chemical properties of the adsorbent and often to the mechanisms of heterogeneous catalysis.

References

1. J. E. Lennard-Jones, *Trans. Faraday Soc.*, **28**, 333 (1932).
2. D. E. O'Reilly, *Adv. Catal.*, **12**, 31 (1960).
3. H. Winkler, *Bull. Ampere*, 10th Year, Special Edition, p. 219 (1961).
4. K. J. Packer, *Progr. Nucl. Magn. Reson. Spectrosc.*, **3**, 87 (1967).

5. H. A. Resing, *Advan. Mol. Relaxation Processes*, **1**, 109 (1968); **3**, 199 (1972).
6. E. G. Derouane, J. Fraissard, J. J. Fripiat and W. E. E. Stone, *Catal. Rev., Sci. Eng.*, **7**, 121 (1972).
7. H. Pfeifer, *NMR: Basic Principles and Progress*, Vol. 7, Springer, New York, p. 53 (1972); *Phys. Rept.*, **26C**, 293 (1976).
8. J. H. Lunsford, *CRC Critical Rev. Solid State Sci.*, **6**, 337 (1976).
9. J. J. Fripiat, *J. Physique Colloq. C4*, **38**, 44 (1977).
10. W. N. Delgass, G. L. Haller, R. Kellerman and J. H. Lunsford, *Spectroscopy in Heterogeneous Catalysis*, Academic Press, New York, Ch. 7 (1979).
11. J. Tabony, *Progr. Nucl. Magn. Reson. Spectrosc.*, **14**, 1 (1980).
12. T. M. Duncan and C. Dybowsky, *Surf. Sci. Rep.*, **1**, No. 4, North-Holland, November 1981.
13. D. Michel, H. Pfeifer and J. Delmau, *J. Magn. Reson.*, **45**, 30 (1981).
14. J. Fraissard, R. Caillat, J. Elston and B. Imelik, *J. Chim. Phys.*, 1017 (1963).
15. D. Geschke, *Z. Naturforsch.*, A, **23**, 5 (1968).
16. W. T. Raynes, A. D. Buckingham and H. J. Bernstein, *J. Chem. Soc.*, **36**, 3481 (1962).
17. S. Gordon and B. P. Dailey, *J. Chem. Phys.*, **34**, 1084 (1961).
18. H. M. McConnell, *J. Chem. Phys.*, **38**, 430 (1958).
19. J. A. Pople, W. G. Schneider and H. J. Bernstein, *High Resolution Nuclear Magnetic Resonance*, McGraw-Hill, New York, 1959.
20. C. P. Slichter, *Principles of Magnetic Resonance*, Harper & Row, New York, 1963.
21. T. W. Marshall and J. A. Pople, *Molec. Phys*, **1**, 199 (1961).
22. A. D. Buckingham, *Can. J. Chem.*, **38**, 300 (1960).
23. A. D. Buckingham and E. G. Lovering, *Trans. Faraday Soc.*, **68**, 2077 (1962).
24. A. D. Buckingham and K. A. McLauchlan, *Proc. Chem. Soc.*, 144 (1963).
25. J. M. Deutch and J. S. Waugh, *J. Chem. Phys.*, **43**, 2568 (1965).
26. J. L. Bonardet, A. Snobbert and J. Fraissard, *C.R. Acad. Sci., Paris, Ser. C*, **272**, 1836 (1971).
27. J.L. Bonardet and J.P. Fraissard, *J. Magn. Reson.*, **22**, 543 (1976).
28. H. A. Lorentz, *Theory of Electrons*, Teubner, Leipzig, p. 138 (1909).
29. S. Gradsztajn, J. Conard and H. Benoit, *J. Phys. Chem. Solids*, **31**, 1121 (1970).
30. G. M. Muha and D. J. C. Yates, *J. Chem. Phys.*, **49**, 5073 (1968).
31. T. A. Egerton and R. D. Green, *Trans. Faraday Soc.*, **67**, 2699 (1971).
32. D. Michel, W. Meiler, A. Gutsze and A. Wronkowki, *Z. Phys. Chemie* (Leipzig), **261**, 953 (1980).
33. S. Kaplan, H. A. Resing and J. S. Waugh, *J. Chem. Phys.*, **59**, 5681 (1973).
34. H. Ernst, D. Feuzke and H. Pfeifer, *Ann. Physik*, **38**, 257 (1981).
35. E. O. Stejskal, J. Schaefer, J. M. S. Henis and M. K. Tripodi, *J. Chem. Phys.*, **61**, 2351 (1974).
36. M. A. Springuel-Huet and J. Fraissard, *Chem. Phys. Lett.*, **154**, 299 (1989).
37. E. G. Derouane, *Bull. Soc. Chim. Belg.*, **78**, 89 (1969).
38. J. L. Bonardet, Thesis, University Pierre et Marie Curie, Paris, 1974.
39. J. L. Bonardet, L. C. de Menorval and J. Fraissard, in H. A. Resing and C. G. Wade (eds), *Magnetic Resonance in Colloid and Interface Science*, ACS Symposium Series 34, San Francisco, p. 248 (1976).
40. V. Yu Borovkov, G. M. Zhidomirov and B. V. Kazansky, *J. Struct. Chem.*, **6**, 3084 (1975).
41. M. A. Enriquez and J. Fraissard, *J. Catal.*, **77**, 96 (1982).
42. A. Michael, W. Meiler, D. Michel, H. Pfeifer, D. Hoppach and J. Delmau, *J. Chem. Soc. Faraday Trans. 1*, **82**, 3053 (1986).

43. D. Michel, A. Germanns and H. Pfeifer, *J. Chem. Soc., Faraday Trans.*, **1**, 237 (1982).
44. A. Gedeon, J. L. Bonardet and J. Fraissard, *J. Chim. Phys.*, **85**, 872 (1988).
45. J. Gay, *J. Catal.*, **48**, 430 (1977).
46. M. A. Enriquez, J. Fraissard and B. Imelik, *C.R. Acad. Sc. (Paris), Série C*, **270**, 238 (1971).
47. M. A. Enriquez and J. Fraissard, *J. Catal.*, **71**, 89 (1982).
48. B. D. Flokhart, C. Naccache, J. A. N. Scott and R. C. Pink, *J. Chem. Soc. Chem. Comm.*, 238 (1965).
49. M. A. Enriquez and J. Fraissard, *J. Catal.*, **74**, 89 (1982).
50. H. M. McConnell and D. B. Chesnut, *J. Chim. Phys.*, **34**, 842 (1961).
51. C. N. La Mar, W. D. Horrocks and R. H. Holm, *NMR of Paramagnetic Molecules*, Academic Press, New York, 1973.
52. B. Bigot, M. A. Enriquez and J. P. Fraissard, *J. Catal.*, **73**, 121 (1982).
53. M. Karplus and G. Fraenkel, *J. Chem. Phys.*, **35**, 1312 (1961).
54. W. D. Knight, *J. Vol. Sci. Technol.*, **10** (1973).
55. S. Kobayashi, T. Takahashin and N. Sasaky, *J. Phys. Soc., Jap. Suppl.*, **31**, 1442 (1971).
56. J. L. Bonardet, J. Fraissard and L. C. De Menorval, in G. C. Bond, P. B. Wells and F. C. Tompkins (eds), *Proc. 6th International Congress on Catalysis*, London, July 12–16 1976; The Chemical Society. Vol 2, p. 633 (1977).
57. L. C. De Menorval and J. Fraissard, *Chem. Phys. Lett.*, **77**, 309 (1981).
58. T. Sheng and J. D. Gay, *J. Catal.*, **71**, 119 (1981); **77**, 53 (1982).
59. N. Reinecke and R. Haul, *Ber. Bunsenges. Physik. Chem.*, **88**, 1232–38 (1984).
60. J. Sanz and J. H. Rojo, *J. Phys. Chem.*, **89**, 4974 (1985) and references therein.
61. X. Wu, B. C. Gerstein and T. S. King, *J. Catal.*, **118**, 238 (1989); **121**, 271 (1990); **123**, 43 (1990).
62. M. Polisset and J. Fraissard, *Colloids and Surfaces. A: Physico Chemical and Engineering Aspects*, **72**, 197 (1993).
63. D. Rouabah and J. Fraissard, *Solid State NMR*, **3**, 153 (1994).
64. W. Eberhardt, F. Greuter and E. W. Plummer, *Phys. Rev. Lett.*, **46**, 1085 (1981).
65. H. E. Rhodes, P. K. Wang, H. T. Stokes, C. P. Slichter and J. H. Sinfelt, *Phys. Rev.*, **B26**, 3559 (1982).
66. D. Rouabah, Thèse Université Pierre et Marie Curie, Paris.
67. T. Ito and J. Fraissard, *Proc. Fifth Int. Cong. on Zeolites*, Naples, June 1980, p. 510; Heyden, 1980.
68. J. Fraissard and T. Ito, *Zeolites*, **8**, 350 (1988) and references therein.
69. D. Raftery, L. Reven, H. Long, A. Pines, P. Tang and J. A. Reiner, *J. Phys. Chem.*, **97**, 1649 (1993).
70. D. Raftery and B. Chmelka, in B. Blümich and R. Kosfeld (eds), *NMR Basic Principles and Progress*, Springer-Verlag, Berlin, in press.
71. A. K. Jameson, C. J. Jameson and H. S. Gutowsky, *J. Chem. Phys.*, **53**, 2310 (1970).
72. C. J. Jameson, A. K. Jameson and S. M. Cohen, *J. Chem. Phys.*, **59**, 4540 (1973).
73. P. Barrie and J. Klinowski, *Prog. Nucl. Magn. Reson. Spectrosc.*, **24**(2), 91 (1992).
74. C. Dybowski, N. Bansal and T. Duncan, *Ann. Rev. Phys. Chem.*, **42**, 433 (1991).
75. J. Demarquay and J. Fraissard, *Chem. Phys. Lett.*, **136**, 314 (1987).
76. M. A. Springuel-Huet, J. Demarquay, T. Ito and J. Fraissard, 'Innovation in Zeolite Material Science', in P. J. Grobet *et al.* (eds), *Studies in Surface Science and Catalysis*, Elsevier Science, vol. 37, 183 (1988).
77. E. G. Derouane and J. B. Nagy, *Chem. Phys. Lett.*, **137**, 341 (1987).

78. J. A. Ripmeester, C. J. Ratcliffe and J. S. Tse, *J. Chem. Soc., Faraday Trans. 1*, **84**, 3731 (1988).
79. J. A. Ripmeester, J. S. Tse, C. J. Ratcliffe and B. M. Powell, *Nature*, **325**, 135 (1987).
80. J. Ripmeester, *J. Magn. Reson.*, **56**, 247 (1984).
81. T. Ito, L. C. de Ménorval, E. Guerrier and J. Fraissard, *Chem. Phys. Lett.*, **111**, No 3, 271 (1984).
82. T. Ito, M. A. Springuel-Huet and J. Fraissard, *Zeolites*, **9**, 68 (1989).
83. T. Ito and J. Fraissard, *Zeolites*, **7**, 554 (1987).
84. M. L. Smith, D. R. Corbin, L. Abrams and C. Dybowski, *J. Phys. Chem.*, **97**, 7793 (1993).
85. J. L. Bonardet, G. B. McGarvey, J. Moffat and J. Fraissard, *Colloids and Surfaces A: Physicochemical and Engineering Aspects*, **72**, 191 (1993).
86. S. K. Brownstein, J. E. L. Roovers and D. J. Worsfold, *J. Magn. Reson.*, **26**, 392 (1988).
87. Q. J. Chen and J. Fraissard, *J. Phys. Chem.*, **96**, 1809 (1992).
88. Q. J. Chen, M. A. Springuel-Huet, J. Fraissard, D. Corbin and C. Dybowski, *J. Phys. Chem.*, **96**, 10914 (1992).
89. N. Bansal and C. Dybowski, *J. Phys. Chem.*, **92**, 2333 (1988).
90. A. Gedeon, J. L. Bonardet, T. Ito and J. Fraissard, *J. Phys. Chem.*, **93**, 2563 (1989).
91. T. T. P. Cheung, C. M. Fu and S. Wharry, *J. Phys. Chem.*, **92**, 5170 (1988).
92. T. Ito, L. C. De Menorval and J. Fraissard, *J. Chim. Phys.*, **80**, 7–8 (1983).
93. E. W. Scharf, R. W. Crecely, B. C. Gates and C. Dybowski, *J. Phys. Chem.*, **90**, 9 (1986).
94. A. Gedeon and J. Fraissard, *Chem. Phys. Lett.*, **219**, 440 (1994), and references therein.
95. Q. J. Chen and J. Fraissard, *Chem. Phys. Lett.*, **169**(6), 595 (1990).
96. A. Gedeon, T. Ito and J. Fraissard, *Zeolites*, **8**, 376 (1988).
97. B. F. Chmelka, J. G. Pearson, S. B. Liu, R. Ryoo, L. C. de Menorval and A. Pines, *J. Phys. Chem.*, **95**, 303 (1991) and references therein.
98. M. C. Barrage, J. L. Bonardet and J. Fraissard, *Catalysis Lett.*, **5**, 143 (1990).
99. T. Ito, J. L. Bonardet, J. Fraissard, J. B. Nagy, C. André, Z. Gabelica and E. Derouane, *Applied Catalysis*, **43**, 5 (1988).
100. Q. J. Chen, J. L. Guth, A. Seive, P. Caullet and J. Fraissard, *Zeolites*, **11**, 798 (1991).
101. J. A. Ripmeester and D. W. Davidson, *J. Molecular Structure*, **75**, 67 (1981).
102. J. A. Ripmeester and C. J. Ratcliffe, *J. Phys. Chem.*, **94**, 8773 (1990) and references therein.
103. M. D. Sefcik, J. Schaefer, J. A. E. Desa and W. B. Yelon, *Polymer Prepr.*, **24**(1), 85 (1983).
104. I. R. Stengle and L. K. Williamson, *Macromolecules*, **20**, 1428 (1987).
105. G. K. Kennedy, *Polymer Bulletin*, **20**, 605 (1990).
106. W. C. Conner, E. L. Weist, T. Ito and J. Fraissard, *J. Phys. Chem.*, **93**, 4138 (1989).
107. G. Fetter, D. Tichit, L. C. de Menorval and J. Figueras, *Applied Catalysis*, **65**, 1 (1990).
108. T. J. P. Cheung, *J. Phys. Chem.*, **93**, 7549 (1989).
109. P. C. Wernett, J. W. Larsen and H. J. Hue, *Energy and Fuels*, **4**, 413 (1990).
110. G. Neue, *Z. Phys. Chem.*, **7**, 271 (1987).

10C MAGNETIC RESONANCE IN FOOD SCIENCE

B. P. Hills
Institute of Food Research, Norwich

Dynamics of Solutions and Fluid Mixtures by NMR
Edited by J.-J. Delpuech   © 1995 John Wiley & Sons Ltd

C.1   Introduction

The last decade has seen a dramatic increase in the application of NMR and MRI to food science. This has been motivated, in part, by the increasing versatility and availability of NMR spectrometers and, in the case of imaging, by the realization that many of the powerful techniques developed for observing anatomical details in clinical applications are also applicable to foods. Indeed, the first images of fruit and vegetables demonstrated the same beautiful anatomical detail familiar in clinical imaging. Nevertheless, in most foods there is little advantage in non-invasive imaging of structure over invasive techniques such as light or electron microscopy since foods can be physically sliced with impunity. Moreover, spatial resolutions obtainable with MRI can never compete with resolutions routinely available in light and electron microscopy. Food quality control in a factory environment is, perhaps, one obvious application where non-invasive food structure determination could be important, but this awaits further technical development. The true potential of food imaging lies not in static structure determination but in following, non-invasively, in real time, dynamic changes such as mass and heat transport as foods are processed, stored, packaged and distributed. This dynamic information is contained in real time changes in maps of the NMR parameters causing image contrast. Depending on the imaging protocol these are spatial maps of spin number density, relaxation times, magnetization transfer rates, chemical shifts, self-diffusion coefficients and velocity flow rates, each of which will be considered in the following sections.

The time-dependent maps of mass and heat transport obtained by imaging can be used to develop theoretical computer models provided the appropriate transport mechanism such as diffusion or capillary flow has been identified. The theoretical model is perhaps the most valuable end product of the whole imaging process because it enables food processing conditions to be optimized numerically without repetitive trial-and-error experimentation.

We begin the discussion by considering imaging protocols suitable for obtaining quantitative maps of NMR parameters in heterogeneous food materials. This will be followed by a discussion of the interpretation of the NMR parameter maps in terms of mass and heat transport and associated physico-chemical changes in the food material. This leads naturally onto an analysis of the effect of food microstructure on water proton relaxation and diffusion and of the molecular mechanisms of water proton relaxation in biopolymer systems. Finally, high resolution NMR protocols suitable for following composition changes in food materials will be briefly discussed. Throughout the chapter an attempt will be made to present a broad overview of the current, and potential, applications of NMR and MRI to food science. This approach necessarily means some loss of detail so references to the literature will be given wherever appropriate.

C.2   Quantitative Imaging Protocols for Food Materials

It is instructive to consider some of the difficulties which arise if attempts are made to do quantitative food imaging with the spin-echo pulse sequence shown in Figure 10C.1. For simplicity the phase encoding gradients have been neglected since they are not essential to the analysis. The image intensity, $I(\mathbf{r})$ and the NMR signal, $S(\mathbf{r})$, are Fourier conjugates of each other such that

$$S(\mathbf{k}) = \int d\mathbf{r} I(\mathbf{r}) \exp(i2\pi \mathbf{k}.\mathbf{r}) \qquad (10C.1)$$

and

$$I(\mathbf{k}) = \int d\mathbf{k} S(\mathbf{k}) \exp(-i2\pi \mathbf{k}.\mathbf{r}) \qquad (10C.2)$$

where $\mathbf{k}$ is the wavevector $(2\pi)^{-1}\gamma G\delta$ and $\mathbf{G}$ is the applied imaging gradient [1]. In systems characterized by heterogeneity on a (macro)molecular distance scale, such as gelatinized starch, it is straightforward to obtain quantitative spin density maps, $M_0(\mathbf{r})$, and longitudinal or transverse relaxation time maps, $T_1(\mathbf{r})$ and $T_2(\mathbf{r})$, from the image intensity $I(\mathbf{r})$ because relaxation and diffusive dephasing are simply factorizable such that,

$$I(\mathbf{r}) = M_0(\mathbf{r})[1 - \exp(-TR/T_1(\mathbf{r}))].R(t_a, t_b, \mathbf{r}).\mathrm{DIFF}(t_a, t_b, \mathbf{r}) \quad (10C.3)$$

where $R(t_a, t_b, \mathbf{r})$ and $\mathrm{DIFF}(t_a, t_b, \mathbf{r})$ are the attenuating effects of transverse relaxation and diffusion respectively. The time delays t_a and t_b are identified in Figure 10C.1. TR is the repeat time for the whole sequence during spectral accumulation. For a homogeneous sample

$$R(t_a, t_b) = \exp{-2(t_a + t_b)/T_2} \qquad (10C.4)$$

and

$$\mathrm{DIFF}(t_a, t_b) = \exp[-\gamma^2 G^2 D t_a^2 (2(t_a + t_b) - 4t_a/3)] \qquad (10C.5)$$

Equation (10C.5) expresses the signal attenuation due to diffusion between the two read-out gradient pulses (Figure 10C.1). These equations show that the spin density map, $M_0(\mathbf{r})$, can be obtained as the intercept of a plot of $\ln I(\mathbf{r})$ against the echo time TE, defined as $2(t_a + t_b)$, provided $t_a \ll t_b$ so that the term

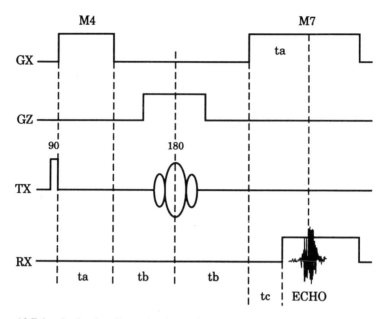

Figure 10C.1 A simple slice-selective spin-echo imaging sequence with no phase encoding gradients

in $4t_a/3$ can be neglected. The same result is obtained if TE $\ll T_2$, but this condition is not always possible to fulfil if the sample has short relaxation times. The transverse relaxation time can be obtained from the slope of this plot and the longitudinal relaxation time by varying TR.

Unfortunately in materials having spatial heterogeneity on the 1–1000 micron distance scale, which includes most foods, this simple protocol does not give quantitative relaxation times. Diffusion becomes restricted and coupled to bulk and surface relaxation so that it is no longer generally possible to factorize diffusion and relaxation as in equation (10C.3). Moreover the relaxation is usually multiple exponential whenever different compartments or phases in the food are associated with different intrinsic relaxation times. Heterogeneous samples also generate significant internal magnetic field gradients which arise from discontinuities in the magnetic susceptibility across local phase boundaries. Diffusion through these local field gradients causes rapid signal attenuation through dephasing. Clearly these complications prevent the determination of quantitative relaxation time maps in biological and heterogeneous food materials with a sequence such as that in Figure 10C.1. Nevertheless, it is still possible to derive values for the spin density map, $M_0(\mathbf{r})$ as the extrapolated origin in a plot of log $I(\mathbf{r})$ against TE when $t_a \ll t_b$.

To measure quantitative T_2- or T_1-weighted images a more general imaging protocol is needed whereby relaxation contrast is first generated in a non-spatially resolved preparation sequence which is immediately followed by a spatially resolving imaging sequence (see Figure 10C.2). For example, T_1 contrast can be generated using a standard inversion recovery preparation sequence; while T_2 contrast can be generated with a standard CPMG preparation sequence. Figure 10C.3 shows a simple imaging sequence that could be used with these preparation sequences. This minimizes dephasing by diffusion in susceptibility gradients and between the read-out gradient pulses by placing the initial read-out gradient pulse on the other side of the slice-selective 180 degree pulse (with a change of sign) and by fixing the delays t_a and t_b at their minimum values.

An additional complication arises because the transverse relaxation times in food samples usually depend on the 180–180° CPMG pulse spacing for reasons to be discussed later. This means that T_2 maps are best obtained from the signal attenuation caused by progressively increasing the number of echoes in the CPMG preparation sequence while keeping the pulse spacing fixed,

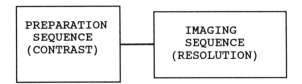

Figure 10C.2 The preparation and imaging sequences generate contrast and spatial resolution respectively

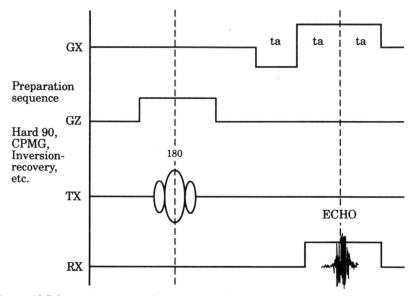

Figure 10C.3 A simple one-dimensional, slice selective imaging sequence which minimizes dephasing

instead of the conventional method of simply stepping out the echo time by increasing the pulse spacing. By choosing the appropriate preparation sequence in Figure 10C.2 it is possible to weight the image with many other NMR parameters such as the rotating frame relaxation time $(T_{1\rho})$, magnetization transfer rates, self-diffusion coefficients, stimulated echo contrast or even by coherence transfer using multiple quantum filters.

C.3   Fast Linear and Radial One-dimensional Imaging

Although two-dimensional imaging sequences with slice selection and phase encoding can be used to obtain two (or three)-dimensional spin density and relaxation time maps, most foods can be cut or moulded into a rectangular or cylindrical shape. Whenever this is the case, one-dimensional linear and/or radial imaging offers many advantages over multi-dimensional imaging. One-dimensional imaging is necessarily faster because no phase-encoding gradients or slice selection are needed and sufficient signal is usually obtained in just one or two scans enabling rapid changes to be followed non-destructively in real time, even with long recycle delays of a few seconds. This is important since many processes such as baking and cooking cause very rapid changes. In principle the rapid two-dimensional imaging techniques developed in clinical research such as echo planar imaging [2] or snapshot flash imaging [3] could be applied, but the price paid for increased speed in those protocols is loss of spatial resolution, which is not the case in one-dimensional imaging if the processing changes only occur in one dimension.

In rectangular geometry it is therefore necessary to arrange for the processing operation, such as drying, to cause transport along only one of the principal axes. The profile along the direction of interest can then be obtained at any instant of time by applying a linear gradient along this direction followed by one-dimensional Fourier transformation, FT{ }, without slice selection or phase encoding. A similar process can be used when the transport occurs in the radial direction towards the curved surface of a cylindrically symmetric sample [4, 5]. In this case the gradient is applied at right angles to the axis of the cylinder (i.e. in the x direction) and the radial profile, radial $(\mathbf{r})$, is obtained, first by Fourier transformation, which gives an image projection, proj( ), (an ellipsoid for a uniform cylinder), followed by an Inverse Abel transform, IA{ }, which converts the image projection into the desired radial profile (a top-hat function for a uniform cylinder). These relationships can be summarized as

$$\text{radial}\,(\mathbf{r}) = \text{IA}[\text{proj}(x)] = \text{IA}[\text{FT}(\text{FID}(\mathbf{k}))] \qquad (10\text{C.6})$$

Here fid$(\mathbf{k})$ is the free induction decay or echo acquired in the field gradient; proj(x), is the image projected along the x-axis, and $\mathbf{k}$ is the wavevector $(2\pi)^{-1}\gamma \mathbf{G}t$. In addition to the speed and simplicity of one-dimensional imaging it is considerably easier to develop and solve transport models in one spatial variable than in two or three, and once the model has been successfully tested in one spatial variable it is merely a computational exercise to generalize it to other sample geometries in two or three dimensions.

C.4 Time-dependent Volumetric and Gravimetric Fluid Content Maps

In most food applications it is not the time dependence of the spin density map, $M_0(\mathbf{r})$, itself that is of major interest but a quantitative map of volumetric or gravimetric water (or lipid) content as the sample undergoes some process such as drying or rehydration. The changing volumetric water (or oil) content is easy to obtain since $M_0(\mathbf{r})$ is necessarily linearly proportional to it. However, if the drying or rehydration process is associated with sample swelling, shrinkage or changes in packing density, the spin density map will no longer be simply related to the gravimetric water (or oil) content. In such cases it may be better to deduce the gravimetric water content indirectly from maps of the longitudinal or transverse relaxation times rather than $M_0(\mathbf{r})$. To do this it is first necessary to calibrate the system by plotting the relaxation rate against the reciprocal of the gravimetric water content measured using conventional (i.e. non-spatially resolved) NMR on uniform samples of known water content. This calibration curve can then be used to change the experimental relaxation time maps directly into gravimetric water (or lipid) content maps.

C.5 Interpreting Time-dependent Water (or Oil) Content Maps

C.5.1 DIFFUSIVE PROFILES

The time course of moisture profiles have been measured by imaging techniques for a variety of food materials during surface air drying. The examples include the drying of gels [6], apple tissue [7], potato [8], ears of corn [9] and extruded starch products. Because these materials are extremely fine grained (gels, corn, extruded starch) and/or have membrane barriers to water transport (apple and potato tissue) it is reasonable to analyse the moisture profiles using Fickian diffusion theory. There are, however, a number of subtleties in the application of diffusion theory to these drying problems. First, the diffusion coefficient, D, is dependent on the water content, W. This can be treated by specifying some functional dependence on water content, $D\{W\}$ and determining the parameters in this function by fitting the image profiles. A second difficulty is the treatment of sample volume changes during processing such as the non-uniform shrinkage that usually occurs during drying. Such shrinkage is best treated numerically by solving the diffusion equations with finite element methods. For example, the sample size can be reduced at each time step by an amount determined by the average shrinkage measured for the entire drying period. This correctly describes the overall shrinkage but not the correct shrinkage distribution. A more subtle approach is to reduce the number of voxels associated with each region of the sample by an amount determined by the local moisture content. A third problem is the non-isothermal nature of most real drying process, which means that simultaneous heat and mass transfer need to be considered, including cross-phenomena such as thermal diffusion and the Dufour effect [10]. This requires some estimate of sample temperature profiles during drying, usually by insertion of thermocouples in repeat experiments outside the imaging coils. An exciting alternative is to use the temperature dependence of the NMR parameters themselves as a probe of the temperature profile, a possibility that will be discussed in a later section.

C.5.2   CAPILLARY–GRAVITATIONAL PROFILES

Many water-rich food materials have a coarse-grained, granular structure over a variety of distance scales ranging from one to several hundred microns. Whenever this is the case moisture transport during drying and rehydration is no longer purely diffusive in character but controlled by capillary and gravitational forces. Consider, for example, drying a model system such as a randomly packed bed of glass beads by blowing dry air over the surface of an initially water-saturated sample [11]. In this system there is no sample shrinkage so the image intensity can be used to calculate a map of the degree of saturation, $S(\mathbf{r}, t)$, in the sample. The degree of saturation, $S(\mathbf{r}, t)$, in each volume element located at $\mathbf{r}$, is defined as the ratio of the volume of water in the unsaturated sample over the volume of water in the saturated sample, and ranges from 0 to 1. In isothermal drying or rehydration $S(\mathbf{r}, t)$ will obey an equation of the form

$$\partial S/\partial t = \nabla \cdot [D_\mu(S) + D_c(S)]\nabla S - (g\rho K_0/\phi\eta)\nabla K_r(S) \qquad (10C.7)$$

where

$$D_\mu(S) = l \cdot \phi^{-1} \cdot (\partial\mu/\partial t) \qquad (10C.8)$$

and

$$D_c(S) = K_0 \cdot K_r(S) \cdot \phi^{-1} \cdot \eta^{-1} \cdot (\partial P_c/\partial S) \qquad (10C.9)$$

Here L is an Onsager coefficient for diffusion, ϕ the porosity, μ the chemical potential, K_0 the Darcy flow permeability, K_r the relative Darcy flow permeability, η the shear viscosity and P_c the capillary pressure. The first term in equation (10C.7) describes Fickian diffusion under a chemical potential gradient. This term is negligible in coarse-grained materials such as the glass-bead bed. The second term in $D_c(S)$ describes capillary transport; while the third term corresponds to the gravitational force. If the drying process is not isothermal, additional heat transport terms must be included. Provided the drying is slow these equations can be solved by assuming a steady-state condition such that the moisture profile is determined by a balance between capillary and gravitational forces, and the resulting capillary-gravitational profiles have been observed in imaging experiments on beds of glass beads having a diameter of about 400 micron [11, 12] and an example is reproduced in Figure 10C.4. However, when particle sizes fall in the region of 1–20 μm as, for example, with beds of potato starch granules, gravitational forces can be neglected compared to the enormous capillary suction potentials. In this case the moisture profiles observed during drying can assume a highly irregular character because air replaces moisture in regions of low packing density and bubbles can be formed by the high suction pressures. Two-dimensional images of the moisture profiles during the displacement of one fluid by another (one of which could be air) in these coarse-grained materials also provide novel views of the phenomenon of percolation, which is an aspect that has received only passing attention in the imaging literature [11].

C.5.3 VAPOUR PHASE DIFFUSION

Capillary and gravitational forces can only drive water (or oil) transport as long as the liquid phase is continuously connected throughout the hetero-

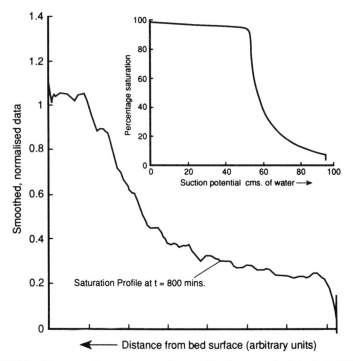

Figure 10C.4 The experimental saturation profile for a packed bed of 400 micron diameter glass beads after surface drying for 800 minutes. For comparison a plot of saturation versus capillary suction pressure is shown since, according to capillary theory the two profiles are equivalent [12]

geneous material. At sufficiently low liquid contents the connectivity throughout the liquid phase is lost and mass transport proceeds only through the gas
phase. In packed beds of non-porous spheres the point at which all liquid-phase
connectivity is broken corresponds to the 'pendular' state, where the liquid is
located only at the points of contact between the spheres. Vapour phase
diffusion is therefore expected to be an important mass transport mechanism in
low moisture content foods, especially in processes such as the baking of
biscuits and breads. Imaging studies of these processes are still in their
exploratory stages and no literature reports are yet available.

C.5.4   THE DYNAMICS OF SURFACE MOISTURE CONTENT

Real-time changes in surface moisture contents of foods during processing and
storage are of considerable importance since they affect surface texture and the
rate of growth of surface spoilage organisms and food-borne pathogenic
bacteria. MRI offers a new and rapid method for the real-time monitoring of
surface moisture contents during processing. The experimental decrease in
surface moisture content in the randomly packed bed of 400 micron glass
beads as it is surface dried in a current of dry air is plotted in Figure 10C.5.
The asymptotic value in the degree of surface saturation corresponds to the
pendular state, and, in this case, the decrease in time is roughly exponential. In

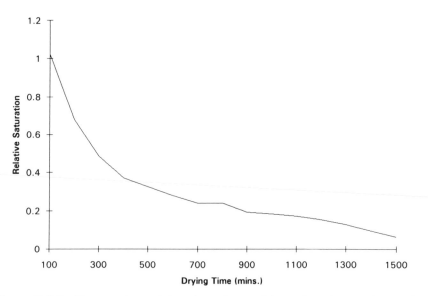

Figure 10C.5   The experimental time dependence of the surface degree of saturation
for the glass bead bed of Figure 10C.4 [12]

general the time dependence of the surface moisture content will depend on the difference in moisture fluxes to and from the surface, which, in turn, depend on the drying conditions and the nature of the material. Much experimental and theoretical work remains to be done in measuring and modelling surface moisture contents in heterogeneous food materials during processing.

C.6 Interpreting Time-dependent Relaxation Time Maps

Proton transverse, longitudinal and rotating frame relaxation times depend not only on water (or oil) content but also on molecular mobility and this makes them especially useful in following the kinetics of phase changes such as crystallization of water or lipid, phase separation, gelation and/or macromolecular aggregation and denaturation. Table 10C.1 lists a number of applications of relaxation time maps in food science. Low concentrations of paramagnetic ions dramatically shorten water proton relaxation times so the transport of these ions in, for example, plant tissue can be monitored [15–17]. Fat and water also have different proton relaxation times allowing fat and water distributions in meat [18, 19] and fish [14] to be measured separately. Table 10C.1 includes a number of miscellaneous reports where imaging has been used to monitor the quality of fruits.

Ice has a very short proton transverse relaxation time of a few microseconds so it is invisible in conventional imaging experiments on partly frozen foods. This permits imaging studies of the kinetics of ice formation in cryobiology [26] since the ice appears as a black region surrounded by the high intensity unfrozen regions. Somewhat surprisingly the frozen core of raw potatoes undergoing freeze-drying appears as a bright, high intensity region inside the

Table 10C.1   Relaxation weighted imaging of foods

	Reference
Relaxation contrast in courgettes	13
Crystallization of trimyristin-water emulsions	14
Freezing peach halves	14
Syneresis of cheese curds	14
Transport of paramagnetic ions in plant systems	15, 16, 17
Fat and water distribution in meat	18, 19
Fat and water distribution in fish	14
Oil content in French salad dressing	20
Oil/water volume fractions in emulsions	21
Bruising in apples, onions and peaches	22
Deterioration of apples in storage	23
Oxygen-dependent core breakdown in Bartlette pears	24
Watercore distribution in apples	25

potato. This apparent contradiction arises from the presence of unfrozen water associated with the cell walls and starch granules inside the frozen potato cells. It is therefore possible to follow freeze-drying kinetics with MRI whenever there are significant amounts of unfrozen water [27].

C.7 Magnetization Transfer Contrast (MTC) Imaging

In clinical spin-warp imaging it is the differing transverse and longitudinal water proton relaxation times that is the basis for most of the observed tissue contrast. The differences in water proton relaxation times in these tissues arise predominantly from the transfer of magnetization between the water and the pool of protons associated with the more rigid macromolecules such as proteins and polysaccharides, characterized by much shorter transverse relaxation times. The molecular mechanism of this important transfer process will be discussed in Section C.13. Here it is sufficient to note that it is, in principle, possible to generate maps weighted according to the effective magnetization transfer rate between the proton pools of the macromolecules and the water [28]. This is done using a saturation transfer preparation sequence in which the more rigid macromolecular proton pool is first selectively irradiated and the decrease in the steady-state magnetization of the water proton signal is observed in the subsequent imaging step. Theories relating the magnitude of the attenuation to the magnetization transfer rate, the intrinsic relaxation times of the water and macromolecules and the resonance offset frequency (i.e. the frequency difference between the selective saturation pulse and the water) have been presented [29] and can be used to generate time dependent maps of the effective transfer rate.

C.8 Diffusion Weighted Imaging

Maps of the effective water self-diffusion coefficient can be obtained using a standard pulsed gradient spin-echo (PGSE) sequence such as that of Stejskal and Tanner [31] either as a preparation sequence, or by inserting gradient pulses into the spin-echo imaging sequence which is the protocol used in 'dynamic NMR microscopy' [32]. The principle behind the PGSE method has been discussed in earlier chapters and in a number of texts [1] and reviews [33]. The observed echo amplitude, $S(q, \Delta, \tau)$, in the conventional (i.e. non-imaging) two-pulse Stejskal–Tanner sequence is a function of the three independent variables q, the wave vector corresponding to the pulsed gradient area, $q = (2\pi)^{-1} g \delta$; Δ, the diffusion time (pulsed gradient separation); and τ, the 90–180° pulse spacing. The functional form of $S(q, \Delta, \tau)$ for water in a food material depends on both the microstructure of the food and on the microdynamics of the water transport and this will be formalized in section C.13. Here it is sufficient to note that the sequence can be used to define 'effective' water self-diffusion coefficients by assuming that relaxation and diffusion are simply factorizable and that the expression for diffusive attenuation for unbounded diffusion can be applied, with an effective diffusion coefficient, in situations where diffusion may well be restricted or hindered. The theoretical form of $S(q, \Delta, \tau)$ for such unbounded diffusion is then

$$S(q, \Delta, \tau) = \exp(-T_2^{-1}t) \cdot \exp(-4\pi^2 q^2 D_{eff}\Delta) \qquad (10C.10)$$

Here D_{eff} is the effective water diffusion coefficient.

The maximum diffusion time attainable in PGSE experiments is limited by the water proton relaxation time to about 1–2 seconds. More usually relaxation and dephasing in susceptibility gradients limits practical diffusion times to only a few tens of milliseconds. Even in two seconds water molecules undergoing unrestricted diffusion can diffuse an average distance of only about 100 microns so that the PGSE method measures effective diffusivity on a microscopic distance scale. The resulting effective PGSE diffusion coefficients are not therefore necessarily the same as the macroscopic diffusion coefficient measured over distances of millimetres or more, especially if there is heterogeneity over distance scales exceeding 100 microns. Fortunately, truly macroscopic water diffusion coefficients can be conveniently measured by imaging the diffusive exchange of D_2O and H_2O between two layers consisting of the sample containing D_2O and H_2O respectively. In proton imaging the deuterated water gives no signal so the interdiffusion is readily imaged. This method has been applied to a variety of heterogeneous model food materials including gels and cellular tissue [34]. Comparison of the PGSE and macroscopic water self-diffusion coefficient can give useful information about the system microstructure [35].

C.9   Flow Imaging

Coherent flow in a PGSE sequence introduces phase shifts that can be used to derive velocity maps [1]. The (non-Newtonian) flow dynamics of the complex liquids, suspensions and pastes used in food manufacture can be studied by imaging lamellar flow down a cylindrical pipe. Figure 10C.6 shows an example of this type of experiment. Similar protocols can be used to test models for liquid flow in complex geometries [36]. In plant physiology flow imaging allows vascular flow in the plant at various stages of development to be monitored *in vivo*. The potential of diffusion and velocity-weighted microimaging has been illustrated beautifully by Jenner *et al.* [37], who studied the circulation of water within a wheat grain and by Xia [38]. The transport of water in food materials can, in principle, be determined by imaging the transport of D_2O either using the loss of water proton signal as H_2O is replaced by D_2O or directly using deuterium imaging [39]. Both methods have been used to study water flow in plant materials.

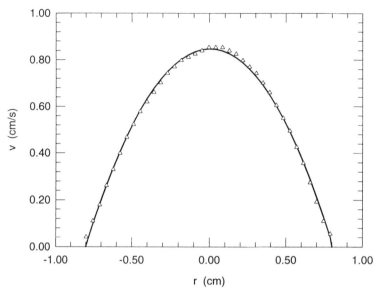

Figure 10C.6 Velocity profile measured for water flowing through a 1.2 cm tube. Experimental points are indicated by triangles. The line is the expected parabolic velocity profile for laminar flow through a straight cylindrical tube [86]

C.10   Temperature Mapping and Heat Transport

Besides mass transport and phase changes, MRI can be used to map temperature changes in food materials undergoing processes such as freezing, drying, sterilization and pasteurization. The time dependence of the temperature maps can then be modelled in terms of the intrinsic thermal properties of foods such as thermal conductivities, specific heats and latent heats. Thermal boundary conditions at the food surface can also be monitored in an analogous way to the surface moisture content.

In MRI, temperature maps are measured indirectly from the temperature dependence of other NMR parameters, usually the longitudinal relaxation time or the effective water self-diffusion coefficient discussed above. The water proton longitudinal relaxation time was found to vary roughly linearly with temperature in human tissue [40] and in model agarose gels [41] which permits temperature maps to be measured. On the other hand, since, according to the Stokes–Einstein relation, the water self-diffusion coefficient is inversely proportional to the shear viscosity, an exponential temperature dependence is expected for the water self-diffusion coefficient, $D \propto \exp(-E_a/k_B T)$. This is observed in MRI studies of a polyacrylamide gel [42] and in a model food gel [43] for which temperature maps have been reported. In these reports MRI temperature studies have been made in the absence of mass transport. When there is simultaneous mass and heat transport it will be necessary to carefully calibrate the system since relaxation times and the effective diffusion coefficient depend on both temperature and water content.

C.11   Chemical Shift Imaging

The principles of chemical shift imaging have been described in a number of reviews and books [1, 44, 50]. In clinical practice chemical shift imaging can be used to resolve fat and water signals. Similar protocols have been used to follow the ingress of oil into water-rich foods during frying [45]. With heterogeneous food materials problems can arise from the broadening due to a wide distribution of relaxation times and strong internal susceptibility gradients. Methods have been proposed to alleviate these problems [46,47] and have been used successfully to image oil and water distribution in sandstones [47] and cheeses [48]. The Dixon chemical shift resolved imaging sequence [49] has been used to map separately the spatial distribution of the liquid-like lipids and water in milk and the separation of cream from milk over a 9-hour period [48]. The changing distribution of either the water or oil components in emulsions, including meat/fat emulsions has been reported [51]. Most recently the fat and water components in a cocoa bean have been separately imaged using the Dixon imaging sequence with a pixel resolution of only 80 microns [52].

C.12   Dynamic q-space Microscopy

In Section C.8 a PGSE preparation sequence was used to weight the image by the effective water self-diffusion coefficient and in Section C.10 the temperature dependence of this effective diffusion coefficient was used to obtain temperature maps. PGSE methods can however be used in their own right to probe structure on a microscopic distance scale ranging from about 0.5–10 microns, which is smaller than the best microimaging resolutions currently reported (*ca* 5 microns). This is possible because PGSE methods use the dynamic propagator, $P_s(\mathbf{r}/\mathbf{r}', t)$, to probe the local microstructure. If the two-pulse Stejskal–Tanner sequence is used it can be shown [1] that the echo amplitude, $S(\mathbf{q}, \Delta, \tau)$ in the absence of relaxation, is given as

$$S(\mathbf{q}, \Delta, \tau) = \int d\mathbf{r} \int d\mathbf{r}' P_s(\mathbf{r}/\mathbf{r}', \Delta).\exp[i2\pi\mathbf{q}.(\mathbf{r}' - \mathbf{r})] \qquad (10C.11)$$

where the 'narrow-gradient' limit has been assumed ($\delta \ll \Delta$). Here $P_s(\mathbf{r}/\mathbf{r}', t)$ is the dynamic propagator defined as the conditional probability that a spin initially at a point $\mathbf{r}$ is located at a point $\mathbf{r}'$ after a time t. Equation (10C.11) can be rewritten by introducing the dynamic displacement $\mathbf{R} = \mathbf{r}' - \mathbf{r}$, in which case

$$S(\mathbf{q}, \Delta) = \int d\mathbf{R} P_s(\mathbf{R}, \Delta)\exp(i2\pi\mathbf{q}.\mathbf{R}) \qquad (10C.12)$$

where the average propagator $P_s(\mathbf{R}, \Delta)$ has been defined as

$$P_s(\mathbf{R}, t) = \int d\mathbf{r}\rho(\mathbf{r}).P_s(\mathbf{r}/\mathbf{r} + \mathbf{R}, t) \tag{10C.13}$$

Equation (10C.12) can be compared with equation (10C.1) relating signal intensity to image intensity in a standard imaging experiment and shows that whereas the conventional k-space imaging experiment images the spin density $\rho(\mathbf{r})$ (neglecting relaxation and diffusive attenuation), the PGSE experiment in q-space is imaging the dynamic displacements of the spins. This makes the PGSE experiment a useful probe of microstructure whenever the average dynamic propagator, $P_s(\mathbf{R}, t)$, obtained by Fourier inversion of equation (10C.12) is restricted by the presence of barriers, such as cell walls, lipid membranes, oil-water interfaces, or solid surfaces. In the limit of long diffusion times (i.e. $\Delta \gg a^2/D$) the dynamic propagator, $\mathbf{P}(\mathbf{r}/\mathbf{r}'\Delta)$ becomes equal to the starting fluid density $\rho(\mathbf{r}')$ so equation (10C.12) becomes

$$S(q, \text{long } \Delta) = \left| \int d\mathbf{r}\rho(\mathbf{r})\exp(-i2\pi.\mathbf{q}.\mathbf{r}) \right|^2 = |S(q)|^2 \tag{10C.14}$$

where $|S(q)|^2$ is the static structure factor. This equation embodies the principle of 'q-space imaging' which has been pioneered by Callaghan and coworkers [53] and discussed in a recent monograph [1]. Equation (10C.14) shows that, at long diffusion times, the PGSE signal of a fluid confined in a droplet or pore should show interference effects when plotted against increasing wave vector $\mathbf{q}$, in an analogous way to light diffraction from a single slit. Such 'diffraction' effects have been observed in packed assemblies of polystyrene and Latex spheres [54].

Despite the formal elegance of dynamic q-space imaging there are a number of problems that must be overcome before it can be routinely used to determine food microstructure. First there is the complicating effect of (multiple exponential) relaxation which has been neglected in the above equations. Many food materials have transverse relaxation times of only a few milliseconds so that it may not be possible to attain the long diffusion time limit required for the validity of equation (10C.14). The use of a stimulated echo sequence may help in these cases since longitudinal relaxation times can be much longer. Even when the long diffusion time regime can be attained, one cannot assume, in heterogeneous systems, that relaxation and diffusion are simply factorizable because different regions and surfaces in the sample may be associated with different relaxation times. This makes theoretical analysis extremely complicated. Indeed, analytic solutions for the propagator $P_s(\mathbf{r}/\mathbf{r}', t)$ in structured systems only exist for a few simple geometries such as a rectangular box and a sphere [56]. Numerical methods must be used to calculate the propagators in more complex geometries and this is a far from trivial exercise, especially when combined surface relaxation and restricted diffusion have to be taken into account. Nevertheless progress in this direction is being made. A numerical model of a plant cell has been developed to interpret relaxation and q-space data in plant tissue and has been used to deduce the tonoplast membrane permeability in apple tissue [57].

C.13 Water Proton Relaxometry of Foods

As we have seen, a proper understanding of water proton relaxation mechanisms in biological and food materials is essential if relaxation contrast (and magnetization transfer contrast) in clinical and food imaging is to be interpreted and used diagnostically. Relaxation also has an important influence in dynamic q-space microscopy as we have just discussed. In this section we therefore look in greater detail at the molecular mechanisms determining water relaxation in food materials. We begin by considering water proton relaxation in the simple case of spatially homogeneous protein and polysaccharide solutions and gels.

C.13.1 PROTEIN AND POLYSACCHARIDE SOLUTIONS AND GELS

Despite their lack of spatial heterogeneity, the molecular mechanisms of water proton relaxation in dilute protein and polysaccharide solutions has been a source of controversy for many years. The difficulty arises because there are at least two possible relaxation pathways. The first involves the lengthening of the rotational and translational correlation times of water molecules interacting with the surface of dissolved protein or polysaccharide molecules. For many years this was thought to be the only relaxation mechanism and unreasonably long correlation times and 'bound' water lifetimes were required to explain the observed decrease in water proton relaxation times with increasing macromolecule concentration. The second relaxation mechanism, which is now thought to be the dominant relaxation pathway, at least in water-rich protein and polysaccharide systems, is fast proton exchange between water and exchangeable macromolecule protons such as those in hydroxyl and amino groups. At room temperature and neutral pH the mean lifetime of a macromolecule amino or hydroxyl proton is of the order of a millisecond, which is fast on the relaxation timescale so that only a single, averaged signal from the exchangeable proton pool will be observed in a standard CPMG or inversion-recovery experiment. This exchangeable proton pool will be characterized by effective transverse and longitudinal relaxation times, T_{2obs} and T_{1obs}, that depend on factors such as the exchange rate, k_b, the fraction of exchangeable macromolecule protons, P_b, and their associated intrinsic relaxation times, T_{2b} and T_{1b}. To analyse the relaxation times quantitatively it is necessary to consider transverse, longitudinal and rotating frame relaxation separately, beginning with transverse relaxation.

C.13.1.1 Transverse relaxation

The quantities determining transverse relaxation time, T_{2obs}, of the exchangeable proton pool are depicted in Figure 10C.7. Because there is a resonance frequency difference, $\delta\omega$ Hz between macromolecule amino and water protons, T_{2obs} is a function of the spectrometer frequency, ω_0, and τ, the 90–180° pulse spacing, in the CPMG sequence used to measure transverse relaxation times. The rather complicated analytical expressions for this dependence have been presented in a number of papers [58, 59] and predict a dispersive relationship between T_{2obs} and pulsing rate, $a = 1/\tau$ (see Chapter 3, Section 3.8), that has been observed in a variety of protein and carbohydrate systems [60, 61] and agrees quantitatively with the analytical theory (see Figure 10C.8). The theory accounts for the reduction in transverse relaxation times observed when a dilute protein or polysaccharide solution is gelled and seen in Figure 10C.8. This is caused primarily by the reduction in T_{2b}, the intrinsic transverse relaxation rate of the exchangeable macromolecule protons resulting from their loss of rotational and translational mobility. It should be noted that, with transverse magnetization, there is no possibility of magnetization transfer by dipolar cross-relaxation [62]. The pool of non-exchanging macromolecule protons therefore appears as a faster relaxing component in the CPMG echo decay envelope and is uncoupled from the water relaxation. This is not, however, the case with longitudinal magnetization to which we now turn.

C.13.1.2 Longitudinal relaxation

Figure 10C.9 illustrates the situation for longitudinal relaxation. Apart from the absence of effects arising from resonance frequency differences, the major

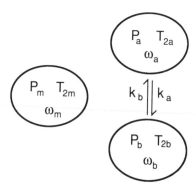

Figure 10C.7 A schematic diagram showing the three proton pools in an aqueous biopolymer system. Pool a is the water protons of fraction P_a transverse relaxation time T_{2a} and resonance frequency ω_a. Pool b are the exchangeable biopolymer protons and pool m the non-exchanging bipolymer protons. k_b and k_a are the proton exchange rates [64]

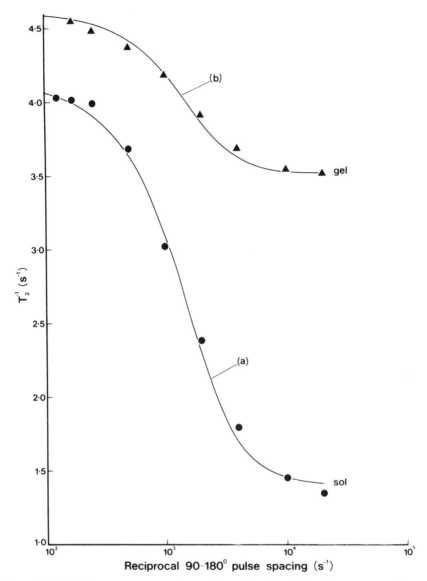

Figure 10C.8 The transverse water proton relaxation dispersions measured as a function of pulsing frequency with the CPMG pulse sequence for the sol and gel states of a 20% gelatine solution at pH 7 [60]

difference between Figures 10C.7 and 10C.9 is the possibility of secular dipolar cross-relaxation between the three proton pools. This is the so-called 'flip-flop' spin interaction because it involves the mutual flipping of two proton spins of opposite sign under the secular dipolar hamiltonian [62]. The dipolar

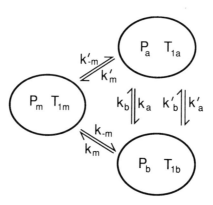

Figure 10C.9   The schematic diagram for the proton exchange cross-relaxation model for longitudinal water proton relaxation in aqueous biopolymer systems. This differs from Figure 10C.7 in the inclusion of the secular dipolar cross-relaxation rates k'_b, k_m and k'_m, etc. and the absence of resonance frequency offsets [64]

cross-relaxation pathways from water are indicated by the primed rate constants. Before T_{1obs} can be calculated it is necessary to develop the theory for the secular dipolar cross-relaxation rate by analogy with the theory for the NOE effect which has been discussed in previous chapters. Figure 10C.10, taken from references [63] and [64], relates the dipolar cross-relaxation rate to the correlation time, τ, of the motion responsible for modulating the interaction. This figure shows that direct dipolar cross-relaxation from water at the macromolecule surface is negligibly small because it is known to have a residence lifetime of only a few picoseconds. On the other hand, the magnitude of the secular dipolar cross-relaxation rate between exchangeable and non-exchangeable macromolecule protons will not necessarily be negligible, but will depend on the correlation times characterizing the macromolecule chain motion. Rotational correlation times of globular proteins in dilute solution are typically only a few tens of nanoseconds. This implies that dipolar cross-relaxation rates between exchangeable and non-exchangeable macromolecular protons will be very slow, of the order of $1\ \mathrm{s}^{-1}$. This leaves fast proton chemical exchange between water and exchangeable macromolecule protons as the major longitudinal cross-relaxation pathway in dilute globular protein solutions. Because the longitudinal relaxation time of the exchangeable protein protons, T_{1b}, depends on the spectrometer frequency, ω_0, this model predicts that the observed 'water' longitudinal relaxation rate, T_{1obs}, should also show a frequency dispersion as the spectrometer frequency is varied. Such dispersions have been observed for globular protein solutions in classic experiments by Koenig and coworkers [66]. Gelling the protein or polysaccharide should cause a dramatic increase in the correlation time modulating the cross-relaxation between exchangeable and non-exchangeable macromolecule protons. The

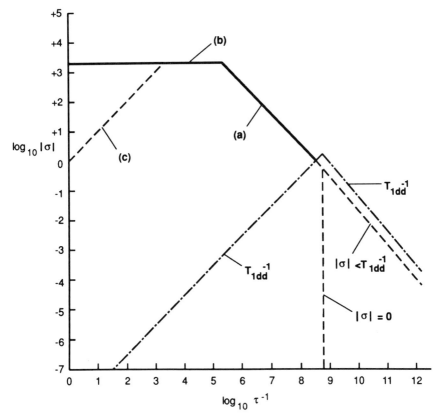

Figure 10C.10 The dependence of the pairwise secular dipolar cross-relaxation rate, σ, on the correlation time, τ of the molecular motion modulating the interaction. Curve (b) is the maximum cross-relaxation rate in the rigid lattice; curve (c) is the cross-relaxation caused by direct proton exchange. Curve (a) is calculated by analogy with the theory of the NOE [64]

dipolar cross-relaxation rate should therefore increase by several orders of magnitude, an effect that is now well documented [60, 63]. The effect of gelation on T_{1obs}, is, however, less dramatic than with transverse relaxation because gelation causes much smaller changes in the macromolecules intrinsic longitudinal relaxation time. The proton exchange cross-relaxation model has also been formulated for the rotating frame relaxation time, $T_{1\rho}(\omega_0, \omega_1, \omega_z)$ which is especially rich in dynamic information because it depends, in general, on the spectrometer frequency, ω_0, the radio-frequency field strength ω_1, and on the resonance frequency offset, ω_z. Despite this dynamic richness, the potential of rotating frame relaxation weighting in food imaging has yet to be fully realized.

C.13.2   COMPARTMENTALIZED SYSTEMS

Unlike simple protein and polysaccharide solutions or gels, biological tissue and most processed foods have a compartmentalized, heterogeneous structure over a wide range of distance scales ranging from a few microns to several millimetres. It is therefore necessary to generalize the analysis of water proton relaxation mechanisms to compartmentalized systems. On a theoretical level the relaxation behaviour in such heterogeneous systems can, at least in principle, be analysed by solving the Bloch–Torrey equations [67] describing the combined effects of diffusion and relaxation on the components of the water proton magnetization density $M_n(r, t)$ in each compartment n,

$$\partial M_n/\partial t = -i\omega_n M_n - \gamma_n M_n + D_n \nabla^2 M_n \qquad (10C.15)$$

Here $M_n(r, t)$ is the complex transverse magnetization density defined as $M_x(r, t) + iM_y(r, t)$, and each compartment is associated with an intrinsic resonance frequency, ω_n, transverse relaxation rate, γ_n, and water self-diffusion coefficient, D_n (see Figure 10C.11). If surfaces or interfaces behave as relaxation sinks then these too must be assigned a surface relaxation strength (Figure 10C.11). The coupled Bloch–Torrey equations for each compartment

DIFFUSIVE EXCHANGE IN HETEROGENEOUS SYSTEMS

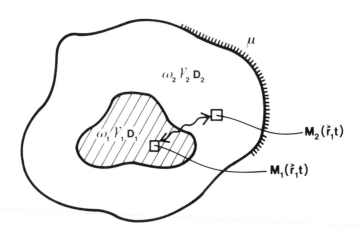

TORREY H.C. PHYS. REV. 104 (1956) 563

$$\frac{\delta M_n}{\delta t} = -i\omega_n M_n - \gamma_n M_n + D_n \nabla^2 M_n$$

$$n = 1, 2 \dots$$

Figure 10C.11 Schematic diagram illustrating compartmentalized and surface relaxation in heterogeneous systems. The two compartments are characterized by intrinsic relaxation rates γ_1 and γ_2, resonance frequencies ω_1 and ω_2 and water self-diffusion coefficients D_1 and D_2. The surface is characterized by a relaxation strength μ

are then solved with spatial boundary conditions appropriate to the food microstructure.

In general the longitudinal and transverse relaxation in compartmentalized, heterogeneous systems will be multiple exponential, and can be analysed either using a finite number of exponentials or as a continuous distribution of exponentials. The continuous distribution is, in many ways, the most useful in complex food materials since it can be considered analogously to an ordinary one-dimensional NMR spectrum. This is illustrated in Figure 10C.12 which shows the continuous distribution water proton transverse relaxation time 'spectrum' for potato tissue, including tentative peak assignments for several temperatures [68]. The relaxation time spectrum at 258 K shows how water inside the starch and cell wall compartments remains unfrozen even at this low temperature. It was this observation that suggested the possibility of imaging the freeze-drying kinetics of potato referred to previously [27].

If diffusive exchange of water between compartments is slow on the observational timescale then each relaxation time peak can be assigned to a particular compartment. However, as the rate of diffusive exchange between compartments increases the separate relaxation time peaks broaden and eventually merge as a single peak at the average relaxation time. This exchange averaging is entirely analogous to chemical exchange well known in high resolution spectroscopy when there are molecular conformational rearrangements on the NMR timescale. This was discussed in a previous chapter. In relaxometry, however, the diffusive exchange rate is sensitive to the size and morphology of the compartments and to the presence of permeability barriers such as lipid bilayer membranes [69]. In the following this connection between microstructure and relaxation will be illustrated for a variety of representative systems.

C.13.2.1   Saturated, high permeability materials

Frozen-thawed gels are representative of this class of system and consist of water pools formed by the melting of ice crystals, surrounded by unfrozen gel. Since, according to the previous discussion, the water and gel phases have different intrinsic water proton relaxation times, the signal decay is, in general, multiple exponential. However, the observed relaxation time distribution (or 'spectrum') bears no apparent connection with the two intrinsic relaxation times of the separated water and gel phases. This is because diffusion of water molecules between the water and gel phases averages the intrinsic relaxation of the two phases to an extent that depends on the complex microstructure and intrinsic diffusion coefficients in the two phases. This makes quantitative analysis extremely difficult. A related model heterogeneous gel system where the morphology is more controlled so that it can be quantitatively analysed consists of a randomly packed, water-saturated bed of spherical Sephadex beads of controlled bead radius and gel cross-linking density. In this system the

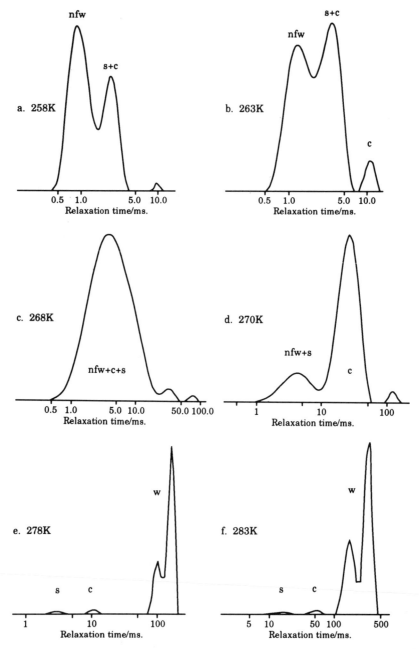

Figure 10C.12 Distributions of water proton transverse relaxation times for potato tissue at various temperatures. Tentative peak assignments are as follows: nfw, non-freezing water associated with dissolved biopolymers and metabolites; s', unfrozen water in membrane-bound starch granules; c, unfrozen water associated with cell walls; w, water in the cytoplasm and non-starch vacuoles [68]

observed multiple exponential relaxation agrees tolerably well with numerical solutions to the Bloch–Torrey equations with boundary conditions appropriate to spherical beads [70, 71].

C.13.2.2 Unsaturated, high permeability materials

If an initially water-saturated bed of packed Sephadex beads is dried, osmotic and capillary forces dictate that air should first replace the water outside the water-swollen Sephadex beads and that this should occur first in the largest inter-bead pores. Only after all extra-bead water has been removed will water begin to be removed from within the gelled beads and cause bead shrinkage. These expectations accord reasonably well with the observed changes in water proton transverse and longitudinal time distributions calculated using the Bloch–Torrey equations and simple models of the changing air–water distribution [72]. This suggests that air–water redistribution on a microscopic distance scale during food processing can be studied using water proton relaxometry. This complements the air–water redistribution observed on a macroscopic distance scale by the imaging method discussed previously. At sub-zero temperatures the microscopic air–ice–unfrozen water distribution can similarly be deduced by relaxation methods as has been demonstrated recently with the model unsaturated Sephadex bead beds at low temperatures [73].

C.13.2.3 Saturated, low permeability materials

Another type of relaxation behaviour arises when one or more phase is solid and has a surface which acts as a relaxation sink for water proton magnetization. Water-saturated granular beds, porous glasses and sandstones are representative of this class of system. Because water relaxation in these materials proceeds mainly by diffusion to the solid surface, larger water-filled subregions or pores are associated with longer relaxation times than smaller water-filled pores so that the observed multiple exponential relaxation behaviour reflects the pore size distribution and the ease with which water in one pore can diffuse through connecting throats or channels into a neighbouring pore (the 'pore connectivity'). The relaxation behaviour in these pore-connected systems can be modelled by numerically solving the Bloch–Torrey equations with appropriate surface boundary conditions. Fitting the observed and calculated multiple exponential relaxation can then be used to deduce the pore size distribution and pore connectivity [74, 75].

C.13.2.4 Unsaturated beds of low permeability materials

If an initially water-saturated porous material is dried the air-water distribution will depend on the hydrophilic or hydrophobic nature of the solid surfaces and on the pore size distribution. With hydrophilic materials, capillary forces dictate that air will progressively replace water in the large pores before small pores until every pore contains an air bubble but there is still a continuous phase of water throughout the whole material. This is known as the 'funicular' state. Further replacement of water by air progressively breaks the water connectivity until the pendular state is reached where only the throats and channels are water filled. If the water-surface interaction has a strongly hydrophobic character, as with fat or lipid surfaces, then capillary theory suggests that air will first displace water in the throats and small pores before the large pores. Clearly these changes will be reflected in the distribution of water proton relaxation times which can be used to follow the microscopic dynamics of the air– water redistribution. This has been successfully applied to the drying of model beds of silica [76] and glass beads but remains to be applied to real food materials such as the drying of protein or polysaccharide powders and milk products.

C.13.2.5 Plant tissue

In many ways water relaxation in plant tissue combines features from each of the above mentioned classes of material. The cytoplasmic and vacuolar fluid consists, for the most part, of a solution of carbohydrates, amino acids and proteins and can give frequency dispersion behaviour in its transverse and longitudinal relaxation times. The cell wall acts as a surface relaxation sink while the tonoplast and plasmalemma membranes serve as water permeability barriers controlling the water transport dynamics between the (sub-)cellular compartments [77]. Fortunately plant tissue morphology differs from the previous types of material in having a more regularly repeating cellular structure which is easy to characterize with light or electron microscopy. This means that NMR relaxometry (and PGSE diffusometry) can be used to study the sub- and inter-cellular water transport. For example, by fitting the observed relaxation and PGSE data with solutions of the Bloch–Torrey equations appropriate to the observed cell morphology, it has been possible to deduce values for the tonoplast membrane permeability in apple tissue [57]. The plasmalemma membrane permeability was also derived by fitting the relaxation time changes observed when the intrinsic relaxation time of the extracellular water was reduced by addition of a low concentration of a paramagnetic ion [78], a method originally proposed by Conlon and Othred [79]. This type of experiment has enormous potential for investigating the effect of processes such as ripening, dehydration and freeze-thawing on the microscopic distribution and transport of water in cellular tissue.

C13.3 LOW WATER CONTENT MATERIALS

C.13.3.1 Dynamics of adsorbed water

Up to this point we have considered water relaxation (and diffusion) in the 'water-rich' regime of food hydration. Even in the pendular state of granular beds there is still a substantial amount of bulk water. For example, the model 400 microns glass bead bed referred to previously is characterized by a degree of saturation S of about 0.2 in the pendular state. Drying materials beyond the pendular state removes all bulk water and leaves only a surface layer of adsorbed water a few molecules thick. This residual surface adsorbed water has an important role in food science since it can lower the activation energy to biopolymer chain motion (a process known as plasticization), and this can change the texture of foods and facilitate crystallization and retrogradation processes responsible for food (including bread) staling.

High power NMR is a powerful tool for investigating the dynamics of surface adsorbed water and biopolymer chain motion in such low water-content materials. Figure 10C.13 shows the proton spectrum of waxy maize starch containing 10% w/w adsorbed water at various temperatures. The narrow width

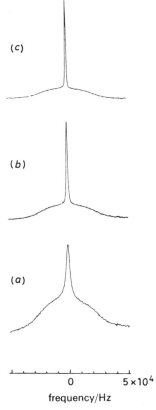

Figure 10C.13 Proton spectra of native waxy maize starch containing 10% adsorbed water at (a) 200 K (b) 303 K and (c) 333 K. There is an arbitrary vertical expansion for all spectra

of the central peak, which arises from adsorbed water visually demonstrates its high mobility and the correlation times characterizing this motion can be deduced by a careful analysis of the longitudinal and rotating frame water proton (cross-)relaxation rates [80]. The plasticization of the amylose and amylopectin chains with increasing water content can also be seen in Figure 10C.13 by the change in shape of the broader peak arising from the starch protons.

C.13.3.2   Dynamics of water in glassy states

NMR studies of the dynamics of glassy and rubbery states in food materials are still in their infancy and the physical chemistry of glassy states is beyond the scope of this chapter, but has been extensively reviewed [81]. However it is worth noting that, like the surface adsorbed water just discussed, water in glasses appears to be remarkably mobile. For example, a maltose glass at 273 K containing 10% by weight water gives rise to a narrow peak in the proton spectrum, superimposed on the broader rigid-lattice spectrum of the maltose protons. As with the adsorbed water, a careful analysis of the relaxation times in this system, together with the cross-relaxation rates between water and maltose protons can, in principle, yield the correlation times of the water molecules in the glass. The preliminary picture that emerges for the glassy state from such studies involves the 'trapping' of water molecules in rigid cages of maltose such that the probability of water molecules diffusing between cages is very low, thus explaining the very low macroscopic self-diffusion coefficient of water in glasses. However, inside the cages water molecules retain high rotational and translational mobility.

C.14   High Resolution NMR Spectroscopy in Heterogeneous Foods

In many applications it would be desirable to be able to monitor the amounts of low molecular weight metabolites such as sugars, amino acids and lipids in fruit, vegetables and meats in the intact tissue without the need for lengthy extraction and separation. This is especially important for monitoring food quality during ripening, storage and distribution and can be done either for the whole sample or for selected sub-volumes within the sample. Both methods will be considered.

C.14.1   WHOLE VOLUME SPECTROSCOPY

The spectra obtained by a naive application of high resolution NMR techniques to foods are usually of very low quality, showing broad, poorly resolved peaks of low signal-to-noise ratio. There are three main reasons for this which will be discussed in turn.

C.14.1.1 Water peak suppression

The first cause of difficulties in proton spectra of most foods is the overwhelming abundance of water. This gives rise to a massive water peak which obscures the much smaller metabolite peaks of interest. Fortunately there are several techniques available for suppressing this water signal. Perhaps the simplest is direct presaturation of the water magnetization by selective irradiation using a long radio-frequency preparation pulse before spectrum acquisition. Inversion recovery nulling can be used if the water T_1 is significantly different from that of the metabolites of interest. An elaboration of this idea involves selective inversion of the water magnetization with a soft 180° pulse, followed by its T_1 nulling and selective excitation of the spectral region of interest. A binomial selective non-excitation pulse sequence has been used by Hore [52] to selectively excite the spectral region on either side of the water peak. In some cases it is possible to add a low concentration of paramagnetic ions such as manganese. This selectively shortens the water proton relaxation time so that a spin-echo spectrum acquired with an echo time much longer than the water relaxation time will show little or no water signal. This protocol has been successfully applied to fruit juices and milk products [83].

C.14.1.2 High molecular weight components

High molecular weight food components such as proteins and polysaccharides give rise to broad proton peaks because of their short transverse relaxation times. Fortunately it is relatively straightforward to suppress these peaks by using spin-echo spectroscopy with an echo delay longer than the transverse relaxation time of the macromolecules, but short compared with that of the low molecular weight metabolites of interest.

C.14.1.3 Susceptibility broadening

Inhomogeneous peak broadening by local internal magnetic field gradients created by susceptibility discontinuities is another major difficulty with intact foods. This 'susceptibility broadening' effect can often be eliminated by slow spinning at the magic angle which has been used successfully to obtain high resolution proton and ^{13}C spectra in plant tissue [84, 85].

C.14.2 VOLUME-SELECTIVE SPECTROSCOPY

In concluding this chapter we come full circle by considering the challenge of obtaining high resolution spectra from subregions of heterogeneous food samples using volume selective spectroscopy which combines high resolution and imaging techniques. This is an exciting field of research and a number of pulse sequences have been proposed differing in their mode of selectivity. The first type is based on selective excitation of a particular subregion with 90 degree pulses and orthogonal field gradients, either directly or via the stimulated echo pulse sequence (VOSY [1]). The stimulated echo pulse sequence is very flexible in this respect since it permits either volume or chemical shift selective excitation (sometimes called the CHESS sequence [1]). Another approach is to destroy longitudinal magnetization outside the subvolume of interest and this is the idea behind the SPACE [1] and DIGGER [2] pulse sequences. Yet another class of sequences (ISIS [1] and OSIRIS [1]) uses selective inversion by three orthogonal soft 180° pulses. Volume selective spectroscopy is widely used in clinical imaging but there have been few applications to foods. Nevertheless the potential of volume selective spectroscopy in food science is enormous, especially if techniques such as volume selective spectroscopy, water suppression and removal of susceptibility broadening are combined.

References

1. P. T. Callaghan, *Principles of Nuclear Magnetic Resonance Microscopy*, Oxford Science Publications, Clarendon Press, 1991.
2. P. Mansfield and I. L. Pykett, *J. Magn. Reson.*, **29**, 355 (1978).
3. A. Haase, *Magn. Reson. Med.*, **13**, 771 (1990).
4. P. Majors and A. Caprihan, *J. Magn. Reson.*, **94**, 225–233 (1991).
5. B. P. Hills and F. Babonneau, *Magn. Reson. Imaging*, **12**, 1065–1074 (1994).
6. G. W. Schrader and J. B. Litchfield, *American Society of Agricultural Engineers*, paper no. 91–6055,1991.
7. E. Perez, R. Kavten and M. J. McCarthy, in A. S. Mujumbo (ed.), *Drying '89*, Hemisphere, New York, 1989.
8. R. Ruan, S. J. Schmidt, A. R. Schmidt and J. B. Litchfield, *J. Food Process Eng.*, **14**, 297–313 (1991).
9. H. Song and J. B. Litchfield, *Cereal Chem.*, **67**, 580 (1989).
10. S. R. DeGroot and P. Mazur, Chapter XI, in *Nonequilibrium Thermodynamics*, North-Holland, Amsterdam, 1969.
11. J. E. Maneval, M. J. McCarthy and S. Whitaker, *Matter. Res. Soc. Symp. Proc.*, **195**, 531 (1990).
12. B. P. Hills, K. M. Wright, J. J. Wright, T. A. Carpenter and L. D. Hall, *Magn. Reson. Imaging*, **12**, 1053–1063 (1994).
13. S. L. Duce, T. A. Carpenter, L. D. Hall and B. P. Hills, *Magn. Reson. Imaging*, **10**, 289 (1992).
14. M. J. McCarthy and R. J. Kauten, *Trends Food Sci. Technol.*, 134 (1990).
15. G. Gassner, in P. E. Pfeffer and W. V. Gerasimowicz (eds), Nuclear Magnetic Resonance in Agriculture, p. 405. CRC Press, Boca Raton, USA, (1989).
16. P. A. Bottomley, H. H. Rogers and T. H.Foster, *Proc. Natl. Acad. Sci. USA*, **84**, 2752 (1987).
17. A. Connelly, J. A. B. Lohman, B. C. Loughman, H. Quiquampoix and R. G. Ratcliffe, *J. Exp. Bot.*, **38**, 1713 (1987).
18. F. Fuller, M. A. Foster and J. M. S. Hutchinson, *Magn. Reson. Imaging*, **2**, 187 (1984).

19. E. Groeneveld, E. Kallweit, M. Henning and A. Pfau, in D. Lister (ed.), *In-vivo Measurement of Body Composition in Meat Animals*, p. 84. Elsevier, London, 1984.
20. J. R. Heil, W. E. Perkins and M. J. McCarthy, *J. Food Sci.*, **55**, 763 (1990).
21. R. J. Kauten, J. E. Maneval and M. J. McCarthy, *J. Food Sci.*, **56**, 355 (1991).
22. P. Chen, M. J. McCarthy and R. Kausen, *Trans. American Soc. of Agric. Engineers*, **32**, 1747 (1989).
23. S. L. Duce, T. A. Carpenter and L. D. Hall, in E. O. Stejskal (ed.), *31st Experimental NMR Spectroscopy Conference Proceedings*, p. 193, Asilonar California, 1990.
24. C. Y. Wang and P. C. Wang, *Hort. Sci.*, **24**, 106 (1989).
25. S. Y. Wang, P. C. Wang and M. Faust, *Sci. Hortic.*, **35**, 227 (1988).
26. M. J. McCarthy, S. Charoenrein, J. R. Heil and D. S. Reid, in D. S. Reid (ed.), *Proceedings of International Conference on Technical Innovations in Freezing and Refrigeration of Fruits and Vegetables* pp. 136–140, Univ. of California, Davis CA, USA.
27. D. N. Rutledge, F. Rene, B. P. Hills and L. Foucat, *J. Food Process Engineering* (in press).
28. S. D. Wolff and R. S. Balaban, *Magnetic Reson. in Medicine*, **10**, 135–144 (1989).
29. J. Grad and R. G. Bryant, *J. Magn. Reson.*, **90**, 1–8 (1990).
30. R. A. Jones and T. E. Southon, *J. Magn. Reson.*, **97**, 171–176 (1992).
31. E. O. Stejskal and J. E. Tanner, *J. Chem. Phys.*, **42**, 288 (1965).
32. P. T. Callaghan and Y. Xia, *J. Magn. Reson.*, **91**, 326–352 (1991).
33. J. Karger, H. Pfeifer and W. Heik, *Advances in Magn. Reson.*, **12**, 1–89 (1988).
34. B. P. Hills, S. C. Smart, V. M. Quantin and P. S. Belton, *Magn. Reson. Imaging*, **11**, 1175–1184 (1993).
35. B. P. Hills, in E. Dickinson and P. Walstra (eds), *Food Colloids and Polymers*, Royal Soc. Chem., Cambridge, 1993.
36. B. E. Hammer, C. A. Heath, S. D. Mirer and G. Belfort, *Biotechnology*, **8**, 327 (1990).
37. C. F. Jenner, Y. Xia, C. D. Eccles and P. T. Callaghan, *Nature*, **336**, 399 (1988).
38. Y. Xia, (1991) unpublished Ph.D. thesis, Massey University, Palmerston North, New Zealand.
39. J. Link and J. Seelig, *J. Magn. Reson.*, **89**, 310–330 (1990).
40. R. J. Dickinson, A. S. Hall, A. J. Hind and I. R. Young, *Computer Assisted Tomography*, **10**, 468–472 (1986).
41. F. A. Howe, *Magn. Reson. Imaging*, **6**, ,263–270 (1988).
42. D. Le Bihan, J. Delannoy and R. L. Levin, *Radiology*, **171**, 853–857 (1989).
43. X. Sun, J. B. Litchfield and S. J. Schmidt, *J. Food Sci.*, **58**, 168–172 (1993).
44. Z. H. Cho and H. W. Park, *Advances in Magn. Reson. Imaging*, **1**, 1–48 (1989).
45. S. L. Duce, T. A. Carpenter and L. D. Hall, DTI-MAFF Link scheme, unpublished results.
46. P. Bornert and W. Dreher, *J. Magn. Reson.*, **87**, 220 (1990).
47. M. A. Horsfield, C. Hall and L. D. Hall, *J. Magn. Reson.*, **87**, 319 (1990).
48. S. L. Duce, A. Amin, M. A. Horsfield, M. Tyszka and L. D. Hall, *International Dairy Journal* (in press)
49. W. T. Dixon, *Radiology*, **153**, 189 (1984).
50. J. M. Pope, in B. Blumich and W. Kuhn (eds), *Magnetic Resonance Microscopy* VCH Publ., Weinheim, 1992.
51. S. L. Duce, S. Ablett, T. M. Guiheneuf, M. A. Horsfield and L. D. Hall, *J. Food Science* (in press).
52. J. J. Wright and L. D. Hall, DTI-MAFF Link scheme, unpublished results.

53. P. T. Callaghan, D. MacGowan, K. J. Packer and F. O. Zelaya, *J. Magn. Reson.*, **90**, 177–182 (1990).
54. G. A. Barrall, L. Frydman and G. C. Chingas, *Science*, **255**, 714–717 (1992).
55. J. E. Tanner and E. O. Stejskal, *J. Chem. Phys.*, **49**, 1768 (1968).
56. J. S. Murday and R. M. Cotts, *J. Chem. Phys.*, **48**, 288 (1965).
57. B. P. Hills and J. E. M. Snaar, *Molec. Phys.*, **76**, 979 (1992).
58. J. P. Carver and R. E. Richards, *J. Magn. Reson.*, **6**, 89 (1972).
59. B. P. Hills, S. F. Tacaks and P. S. Belton, *Molec. Phys.*, **67**, 903–918 (1989).
60. B. P. Hills, *Molec. Phys.*, **76**, 509 (1992).
61. B. P. Hills, C. Cano and P. S. Belton, *Macromolecules*, **24**, 2944–2950 (1991).
62. A. Abragam, *The Principles of Nuclear Magnetism*, Oxford, Clarendon Press, 1961.
63. H. T. Edzes and E. T. Samulski, *J. Magn. Reson.*, **31**, 207 (1978).
64. B. P. Hills, *Molec. Phys.*, **76**, 489 (1992).
65. G. Otting, E. Liepinsh and K. Wuthrich, *Science*, **254**, 974 (1991).
66. S. H. Koenig, R. D. Brown,D. Adams, D. Emerson and C. G. Harrison, *Investigative Radiology*, **19**, 76 (1984).
67. H. C. Torrey, *Phys. Rev.*, **104**, 563 (1956).
68. B. P. Hills and G. LeFloch, *Food Chemistry*, **51**, 331–336 (1994).
69. B. P. Hills and P. S. Belton, *Ann. Rep. NMR Spectr.*, **21**, 99–159 (1989).
70. B. P. Hills, K. M. Wright and P. S. Belton, *Molec. Phys.*, **67**, 193–208 (1989).
71. B. P. Hills, K. M. Wright and P. S. Belton, *Molec. Phys.*, **67**, 1309–1326 (1989).
72. B. P. Hills and F. Babonneau, *Magn. Reson. Imaging*, **12**, 909–922 (1994).
73. B. P. Hills and G. LeFloch, *Molec. Phys.*, **82**, 751–763 (1994).
74. S. Davies and K. J. Packer, *J. Appl. Phys.* (1990).
75. B. P. Hills, P. S. Belton and V. M. Quantin, *Molec. Phys.*, **78**, 893–908 (1993).
76. B. P. Hills and V. M. Quantin, *Molec. Phys.*, **79**, 77–93 (1993).
77. B. P. Hills and S. L. Duce, *Magn. Reson. Imaging*, **8**, 321–331 (1990).
78. J. E. M. Snaar and H. Van As, *Biophys. J.*, **63**, 1654–1658 (1992).
79 T. Conlon and R. Othred, *Biochim. Biophys. Acta*, **288**, 354 (1972).
80. S. F. Tanner, B. P. Hills and R. Parker, *J. Chem. Soc. Faraday Trans.*, **87**, 2613–2621 (1991).
81. L. Slade and H. Levine, in J. M. V. Blanchard and P. J. Lillford (eds), *Glassy State in Food*, Nottingham University Press, Loughborough.
82. P. J. Hore, *J. Magn. Reson.*, **54**, 539 (1983).
83. T. M. Eads and R. G. Bryant, *J. Agric. and Food Chem.*, **34**, 834 (1986).
84. Q. W. Ni and T. M. Eads, *J. Agric. and Food Chem.*, **40**, 1507 (1992).
85. T. M. Eads, in R. Chandrasekaran (ed.), *Frontiers in Carbohydrate Research*, Elsevier, New York, pp. 128–140, .
86. S. J. Gibbs, J. A. Derbyshire, T. A. Carpenter and L. D. Hall, DTI-MAFF LINK scheme, unpublished results.

INDEX